Mathematics

Phase 6

NOTION PRESS

NOTION PRESS

India. Singapore. Malaysia.

ISBN xxx-x-xxxxx-xx-x

Dedicated to our Beloved Prime Minister Shree Narendra Modi ji, whose life gives the Inspiration to every Individual

*If a Man **Decides** to achieve something in his life nothing is Impossible*

CHAPTERS

1. INDEFINITE INTEGRATION

1. INTRODUCTION

Integration is a reverse process of differentiation. The integral or primitive of a function f(x) with respect to x is a differential function φ(x) such that the derivative of φ(x) with respect to x is the given function f(x). It is expressed symbolically as $\int f(x)dx = \phi(x)$

Thus. $\int f(x)dx = \phi(x) \Leftrightarrow \dfrac{d}{dx}\left[\phi(x)\right] = f(x)$.

The process of finding the integral of a function is called Integration and the given function is called Integrand. Now, it is obvious that the operation of integration is the inverse operation of differentiation. Hence the integral of a function is also named as the anti-derivative of that function.

Further we observe that

$$\left.\begin{array}{rcl} \dfrac{d}{dx}\left(x^2\right) &=& 2x \\[2mm] \dfrac{d}{dx}\left(x^2+2\right) &=& 2x \\[2mm] \dfrac{d}{dx}\left(x^2+k\right) &=& 2x \end{array}\right\} \Rightarrow \int 2x\,dx = x^2 + \text{constant}$$

So we always add a constant to the integral of function, which is called the constant of Integration. It is generally denoted by c. Due to the presence of this arbitrary constant such an integral is called an Indefinite Integral.

2. ELEMENTERY INTEGRATION

The following integrals are directly obtained from the derivatives of standard functions.

(a) $\int 0.dx = c$

(b) $\int 1.dx = x + c$

(c) $\int k.dx = kx + c (k \in R)$

(d) $\int x^n dx = \dfrac{x^{n-1}}{n+1} + c(n \neq -1)$

(e) $\int \dfrac{1}{x}dx = \log_e x + c$

(f) $\int e^x dx = e^x + c$

(g) $\int a^x dx = \dfrac{a^x}{\log_e a} + c = a^x \log_a e + c$

(h) $\int \sin x\, dx = -\cos x + c$

(i) $\int \cos x\, dx = \sin x + c$

3. BASIC THEOREMS OF INTEGRATION

If $f(x)$, $g(x)$ are two functions of a variable x and k is a constant, then

(a) $\int k\, f(x) dx = k \int f(x) dx$

(b) $\int \left[f(x) \pm g(x) \right] dx = \int f(x) dx \pm \int g(x) dx$

(c) $\dfrac{d}{dx} \left(\int f(x) dx \right) = f(x)$

(d) $\int \left(\dfrac{d}{dx} f(x) \right) dx = f(x) + c$

> **NOMORECLASS CONCEPTS**
>
> The results of integration are very different from differentiation. There is no standard formula for integration.
>
> Always make sure to write the constant of integration. NEVER assume it as zero from your side.

Illustration 1: Evaluate: $\int \dfrac{1 - \sin x}{\cos^2 x} dx$ (JEE MAIN)

Sol: As we know, $\int \left[f(x) \pm g(x) \right] dx = \int f(x) dx \pm \int g(x) dx$ therefore we can split

$\int \dfrac{1 - \sin x}{\cos^2 x} dx$ as $\int \dfrac{1}{\cos^2 x} dx - \int \dfrac{\sin x}{\cos^2 x} dx$ and then by solving we can get result.

$\int \dfrac{1 - \sin x}{\cos^2 x} dx = \int \dfrac{1}{\cos^2 x} dx - \int \dfrac{\sin x}{\cos^2 x} dx = \int \sec^2 x\, dx \ - \int \tan x \sec x\, dx \ = \tan x - \sec x + c$

Illustration 2: Evaluate: $\int \sqrt{1 + \sin 2x}\ dx$ (JEE MAIN)

Sol: Here $\sin^2 x + \cos^2 x = 1$ and $\sin 2x = 2 \sin x \cos x$, therefore by using these formulae and solving we will get the result.

$\int \sqrt{1 + \sin 2x}\ dx = \int \sqrt{\sin^2 x + \cos^2 x + 2 \sin x \cos x}\ dx = \int \sqrt{(\sin x + \cos x)^2} dx$

$= \int (\sin x + \cos x) dx = \int \sin x\, dx + \int \cos x\, dx$

$= -\cos x + \sin x + c$

Illustration 3: Evaluate: $\int \sin^4 x \ dx$

Sol: Here as we know, $\sin^2 x = \dfrac{1-\cos 2x}{2}$, Now by putting this in the above integration and solving we will get the

term $\dfrac{1}{4}\int (1 - 2\cos 2x + \cos^2 2x)dx$, After that by using the formula

$\cos^2 2x = \dfrac{1+\cos 4x}{2}$ we can solve the problem given above.

$$\int \sin^4 x \ dx = \int \left(\dfrac{1-\cos 2x}{2}\right)^2 dx = \dfrac{1}{4}\int(1 - 2\cos 2x + \cos^2 2x)dx$$

$$= \dfrac{1}{4}\int \left(1 - 2\cos 2x + \dfrac{1+\cos 4x}{2}\right)dx = \dfrac{1}{8}\int(3 - 4\cos 2x + \cos 4x)dx = \dfrac{1}{8}\left[3x - 2\sin 2x + \dfrac{\sin 4x}{4}\right] + C$$

Illustration 4: If $f'(x) = 4x^3 - \dfrac{3}{x^4}$ such that $f(2)=0$, then, find $f(x)$

Sol: Here $f'(x) = 4x^3 - \dfrac{3}{x^4}$ therefore $f(x) = \int\left(4x^3 - \dfrac{3}{x^4}\right)dx$ hence by splitting this integration and solving we will get the result.

We have, $\dfrac{d}{dx}f(x) = 4x^3 - \dfrac{3}{x^4} \Rightarrow f(x) = \int\left(4x^3 - \dfrac{3}{x^4}\right)dx = \int 4x^3 dx - \int \dfrac{3}{x^4}dx = 4\int x^3 dx - 3\int x^{-4}dx$

$$= 4\dfrac{x^{3+1}}{3+1} - 3\dfrac{x^{-4+1}}{-4+1} + C = x^4 + \dfrac{1}{x^3} + C \qquad \qquad \text{...(i)}$$

Given $f(2) = 2^4 + \dfrac{1}{2^3} + C = 0 \Rightarrow 0 = 16 + \dfrac{1}{8} + C \Rightarrow C = -\dfrac{129}{8}$

Putting the value of C in (i), we get $f(x) = x^4 + \dfrac{1}{x^3} - \dfrac{129}{8}$

4. METHODS OF INTEGRATION

When the integrand can't be reduced into some standard form then integration is performed using following methods

4.1 Integration by Substitution

4.1.1 Integrand is a Function of Another Function

If the integral is of the form $\int f\big[\phi(x)\big]\phi'(x)dx$, then we put $\phi(x) = t$ so that $\phi'(x)\,dx=dt$. Now integral is reduced $\int f(t)\,dt$.

NOMORECLASS CONCEPTS

In this method the function is broken into two factors so that one factor can be expressed in terms of the function whose differential coefficient is the second factor.

In case of objective questions in which direct indefinite integration is asked, function being very complicated to integrate, then try differentiating the options.

If $I = \int \dfrac{dx}{\sin(x-a)\cos(x-b)}$, then I is Equal to

(a) $\dfrac{1}{\sin(a-b)}\log\left|\dfrac{\sin(x-a)}{\cos(x-b)}\right| + C$

(b) $\dfrac{1}{\cos(a-b)}\log\left|\dfrac{\sin(x-a)}{\cos(x-b)}\right| + C$

(c) $\dfrac{1}{\sin(a+b)}\log\left|\dfrac{\sin(x-a)}{\cos(x-b)}\right| + C$

(d) $\dfrac{1}{\cos(a+b)}\log\left|\dfrac{\sin(x-a)}{\cos(x-b)}\right| + C$

Illustration 5: Evaluate: $\int x\tan x^2 \sec x^2 dx$ (JEE MAIN)

Sol: This problem is based on integration using substitution method. In this we can put $x^2 = t$ and therefore $2x$ dx=dt and then solving we will get the result.

Let $x^2 = t$

$\Rightarrow 2x\,dx=dt \Rightarrow x\,dx = \dfrac{1}{2}dt$ $\therefore \int x\tan x^2 \sec x^2 dx = \dfrac{1}{2}\int \tan t \sec t\, dt = \dfrac{1}{2}\sec t + c = \dfrac{1}{2}\sec x^2 + c$

4.1.2 Integrand is the Product of Function and its Derivative

If the integral is of the form $I = \int f'(x)\ f(x)\ dx$ we put $f(x) = t$ and convert it into a standard integral.

Illustration 6: Evaluate: $\int \tan x \sec^2 x\, dx$ (JEE MAIN)

Sol: Here $\sec^2 x$ is a derivatives of $\tan x$ hence we can put $\tan x = t$ and $\sec^2 x.dx = dt$ thereafter we can solve the given problem.

Let $\tan x = t \Rightarrow \sec^2 x.dx = dt$

$\therefore I = \int \tan x \sec^2 x\, dx = \int t\, dt = \dfrac{t^2}{2} + c = \dfrac{\tan^2 x}{2} + c$

4.1.3 Integrand is a Function of the Form f(ax+b)

Here we put $ax+b = t$ and convert it into a standard integral. Now if,

$\int f(x)dx = \phi(x),$ then $\int f(ax+b)dx = \dfrac{1}{a}\phi(ax+b)$

Illustration 7: Evaluate: $\int \cos 3x \cos 5x\, dx$ (JEE MAIN)

Sol: By multiplying and dividing by 2 in the given integration and using the formula

$2\cos A.\cos B = \cos(A+B) + \cos(A-B)$ we can solve it.

$I = \int \cos 3x \cos 5x\, dx = \dfrac{1}{2}\int(\cos 8x + \cos 2x)dx = \dfrac{1}{2}\left[\dfrac{1}{8}\sin 8x + \dfrac{1}{2}\sin 2x\right] + c$

Illustration 8: Evaluate: $I = \int \dfrac{x\,dx}{x^4 + x^2 + 1}$ $\hfill$ (JEE ADVANCED)

Sol: Here by putting $x^2 = t \Rightarrow dt = 2x\,dx$ we will get the term $\dfrac{1}{2}\int\dfrac{dt}{t^2+t+1} = \dfrac{1}{2}\int\dfrac{dt}{\left(t+(1/2)\right)^2+\left(\sqrt{3}/2\right)^2}$ and then by

putting $t + \dfrac{1}{2} = \dfrac{\sqrt{3}}{2}\tan\theta$, we can solve it.

Let $x^2 = t \Rightarrow dt = 2x\,dx$ $\quad \therefore I = \dfrac{1}{2}\int\dfrac{dt}{t^2+t+1} = \dfrac{1}{2}\int\dfrac{dt}{\left(t+(1/2)\right)^2+\left(\sqrt{3}/2\right)^2}$

$\displaystyle\int\dfrac{1}{\left[f(x)\right]^2+a^2}dx = \dfrac{1}{a}\tan^{-1}\left[\dfrac{f(x)}{a}\right]\times\dfrac{1}{f'(x)}+c$ $\quad \therefore I = \dfrac{1}{2}\left[\dfrac{1}{\frac{\sqrt{3}}{2}}\tan^{-1}\left(\dfrac{t+\frac{1}{2}}{\frac{\sqrt{3}}{2}}\right)\right] - c$

$= \dfrac{1}{\sqrt{3}}\left[\tan^{-1}\left(\dfrac{2t+1}{\sqrt{3}}\right)\right]+c = \dfrac{1}{\sqrt{3}}\left[\tan^{-1}\left(\dfrac{2x^2+1}{\sqrt{3}}\right)\right]-c$

Now put $t + \dfrac{1}{2} = \dfrac{\sqrt{3}}{2}\tan\theta \Rightarrow dt = \dfrac{\sqrt{3}}{2}\sec^2\theta\,d\theta$

$\therefore \dfrac{1}{2}\int\dfrac{(\sqrt{3}/2)\sec^2\theta\,d\theta}{(3/4)\left(\tan^2\theta+1\right)} = \dfrac{1}{\sqrt{3}}\int d\theta = \dfrac{1}{\sqrt{3}}\theta + c = \dfrac{1}{\sqrt{3}}\tan^{-1}\left(\dfrac{2t+1}{\sqrt{3}}\right)+c = \dfrac{1}{\sqrt{3}}\tan^{-1}\left(\dfrac{2x^2+1}{\sqrt{3}}\right)+c$

Standard integration results

(a) $\quad \displaystyle\int\dfrac{f'(x)}{f(x)}dx = \log_e\left[f(x)\right]+c$

(b) $\quad \displaystyle\int\left[f(x)\right]^n f'(x)dx = \dfrac{\left[f(x)\right]^{n+1}}{n+1}+c$ (provided $n\neq-1$)

(c) $\quad \displaystyle\int\dfrac{f'(x)}{\sqrt{f(x)}}dx = 2\sqrt{f(x)}+c$

Illustration 9: Evaluate: $\displaystyle\int\dfrac{\sec^2 x}{\sqrt{\tan x}}dx$ $\hfill$ (JEE MAIN)

Sol: Here simply substituting $t = \tan x \Rightarrow dt = \sec^2 x\,dx$ we can solve it.

Let $t = \tan x \Rightarrow dt = \sec^2 x\,dx$

$\therefore I = \displaystyle\int\dfrac{dt}{\sqrt{t}} = 2t^{\frac{1}{2}} + c = 2\sqrt{\tan x}+c$

4.1.4 Integral of the Form

$\displaystyle\int\dfrac{dx}{a\sin x + b\cos x}$ then substitute $a = r\cos\theta$ and $b = r\sin\theta$, $\tan\theta = \dfrac{b}{a} \Rightarrow \theta = \tan^{-1}\left(\dfrac{b}{a}\right)$, we get

$I = \displaystyle\int\dfrac{dx}{r\sin(x+\theta)} = \dfrac{1}{r}\int\text{cosec}(x+\theta)dx = \dfrac{1}{r}\log\tan\left(\dfrac{x+\theta}{2}\right)+c = \dfrac{\log\tan\left((x/2)+(1/2)\tan^{-1}(b/a)\right)}{\sqrt{a^2+b^2}}+c$

$\int \sin^m x \cos^n x \ dx$, where m, n $\in$ N

$\Rightarrow$ If m is odd put cos x=t

If n is odd put sin x = t

If both m and n are odd, put sin x=t if m $\geq$ n and cos x=t otherwise.

If both m and n are even, use power reducing formulae

$\sin^2 x = \dfrac{1 - \cos 2x}{2}$ or $\cos^2 x = \dfrac{1 + \cos 2x}{2}$

If m+n is a negative even integer, put tan x=t

Illustration 10: Evaluate: $\int \dfrac{1}{\sin x + \cos x} dx$ **(JEE ADVANCED)**

Sol: As we know, if integration is in the form of $\int \dfrac{dx}{a\sin x + b\cos x}$ then we can put

a=r cos θ and b=r sin θ hence the integration will be $\dfrac{1}{r}\log \tan\left(\dfrac{x+\theta}{2}\right) + c$.

Here a=1 & b=1

So $\int \dfrac{1}{\sin x + \cos x} dx = \dfrac{1}{\sqrt{1+1}}\log\tan\left(\dfrac{x}{2} + \dfrac{1}{2}\tan^{-1}1\right) + c = \dfrac{1}{\sqrt{2}}\log\tan\left(\dfrac{x}{2} + \dfrac{\pi}{8}\right) + c$

4.1.5 Standard Substitutions

The following standard substitutions will be useful

Integrand form	Substitutions
$\sqrt{a^2 - x^2}$ or $\dfrac{1}{\sqrt{a^2 - x^2}}$	x = a sin θ or x = a cos θ
$\sqrt{x^2 + a^2}$ or $\dfrac{1}{\sqrt{x^2 + a^2}}$	x = a tan θ or x = a cot θ or x=a sinh θ
$\sqrt{x^2 - a^2}$ or $\dfrac{1}{\sqrt{x^2 - a^2}}$	x = a sec θ or x = acosec θ
$\sqrt{\dfrac{x}{a+x}}$ or $\sqrt{\dfrac{a+x}{x}}$ or $\sqrt{x(a+x)}$ or $\dfrac{1}{\sqrt{x(a+x)}}$	x=a tan^2 θ
$\sqrt{\dfrac{x}{a-x}}$ or $\sqrt{\dfrac{a-x}{x}}$ or $\sqrt{x(a-x)}$ or $\dfrac{1}{\sqrt{x(a-x)}}$	x = a sin^2 θ or x= a cos^2 θ

$\sqrt{\dfrac{x}{x-a}}$ or $\sqrt{\dfrac{x-a}{x}}$ or $\sqrt{x(x-a)}$ or $\dfrac{1}{\sqrt{x(x-a)}}$	$x=a\sec^2\theta$ or $x=a\csc^2\theta$
$\sqrt{\dfrac{a-x}{a+x}}$ or $\sqrt{\dfrac{a+x}{a-x}}$	$x=a\cos2\theta$
$\sqrt{\dfrac{x-\alpha}{\beta-x}}$ or $\sqrt{(x-\alpha)(\beta-x)}$ $\qquad(\beta>\alpha)$	$x=\alpha\cos^2\theta+\beta\sin^2\theta$

Some Standard Integrals

(a) $\displaystyle\int\tan x\,dx=\log\sec x+c=-\log\cos x+c$

(b) $\displaystyle\int\cot x\,dx=\log\sin x+c=-\log\csc x+c$

(c) $\displaystyle\int\sec x\,dx=\log(\sec x+\tan x)+c=-\log(\sec x-\tan x)+c=\log\tan\left(\dfrac{\pi}{4}+\dfrac{x}{2}\right)+c$

(d) $\displaystyle\int\csc x\,dx=-\log(\csc x+\cot x)+c=\log(\csc x-\cot x)+c=\log\tan\left(\dfrac{x}{2}\right)+c$

(e) $\displaystyle\int\sec x\tan x\,dx=\sec x+c$

(f) $\displaystyle\int\csc x\cot x\,dx=-\csc x+c$

(g) $\displaystyle\int\sec^2 x\,dx=\tan x+c$

(h) $\displaystyle\int\csc^2 x\,dx=-\cot x+c$

(i) $\displaystyle\int\log x\,dx=x\log\left(\dfrac{x}{e}\right)+c=x\left(\log x+1\right)+c$

NOMORECLASS CONCEPTS

If the integral is of the form $\displaystyle\int R\left(x^{\frac{1}{p}},x^{\frac{1}{q}},x^{\frac{1}{r}}......\right)dx$, where R is a rational function then,

Let $a = $ lcm of $(p,q,r,.......)$ and put $x = t^a$

Illustration 11: Prove that: $\displaystyle\int\dfrac{dx}{\sqrt{x^2-a^2}}=\log(x+\sqrt{x^2-a^2})+C$ $\qquad$ **(JEE ADVANCED)**

Sol: By putting $x=a\sec\theta\Rightarrow dx=a\sec\theta\tan\theta d\theta$, we can solve the problem given above.

Let $x=a\sec\theta\Rightarrow dx=a\sec\theta\tan\theta d\theta\Rightarrow\displaystyle\int\dfrac{dx}{\sqrt{x^2-a^2}}=\int\dfrac{a\sec\theta\tan\theta d\theta}{a\tan\theta}=\int\sec\theta d\theta=\log(\sec\theta+\tan\theta)+C$

$=\log\left(\dfrac{\sqrt{x^2+a^2}}{a}+\dfrac{x}{a}\right)+C=\log(x+\sqrt{x^2+a^2})+C'$

Illustration 12: Evaluate: $\int \cos 2x.\cos 4x.\cos 6x dx$ (JEE MAIN)

Sol: By multiplying and dividing by 2 in the given integration and then by using $2\cos A.\cos B = \cos(A+B)+\cos(A-B)$ we can solve it.

Let $I = \int \cos 2x.\cos 4x.\cos 6x dx$ $= \dfrac{1}{2}\int (2\cos 2x.\cos 4x)\cos 6x dx$ $= \dfrac{1}{2}\int (\cos 6x + \cos 2x)\cos 6x dx$

$\left[\because 2\cos A\cos B = \cos(A+B)+\cos(A-B)\right]$

$= \dfrac{1}{2}\int \left(\cos^2 6x + \cos 6x - \cos 2x\right)dx$ $= \dfrac{1}{4}\int \left(2\cos^2 6x + 2\cos 6x - \cos 2x\right)dx$

$= \dfrac{1}{4}\left[\int (1+\cos 12x)+(\cos 8x + \cos 4a)dx\right]$ $= \dfrac{1}{4}\int dx + \dfrac{1}{4}\int \cos 12x dx + \dfrac{1}{4}\int \cos 8x dx + \dfrac{1}{4}\int \cos 4x dx$

$\because \int \cos\left[f(x)dx\right] = \dfrac{\sin\left[f(x)\right]}{f'(x)} + C$

$I = \dfrac{1}{4}\left[x + \dfrac{\sin 12x}{12} + \dfrac{\sin 8x}{8} + \dfrac{\sin 4x}{4}\right] + C$

Illustration 13: Evaluate: $\int \dfrac{\cos 2x - \cos 2\alpha}{\cos x - \cos \alpha}dx$ (JEE ADVANCED)

Sol: Here in this problem by using the formulae

$\cos C - \cos D = -2\sin\dfrac{C+D}{2}.\sin\dfrac{C-D}{2}$, $\sin 2A = 2\sin A \cos A$ and $2\cos C \cos D = \cos(C+D)+\cos(C-D)$

We can solve the problem above step by step.

We have,

$\int \dfrac{\cos 2x - \cos 2\alpha}{\cos x - \cos \alpha}dx = \int \dfrac{-2\sin(x+\alpha)\sin(x-\alpha)}{-2\sin((x+\alpha)/2)\sin((x-\alpha)/2)}dx$ $= \int \dfrac{\sin(x+\alpha)\sin(x-\alpha)}{\sin((x+\alpha)/2)\sin((x-\alpha)/2)}dx$

$= \int \dfrac{2\sin((x+\alpha)/2)\cos((x+\alpha)/2).2\sin((x-\alpha)/2)\cos((x-\alpha)/2)}{\sin((x+\alpha)/2)\sin((x-\alpha)/2)}dx$ (sin 2A = 2 sin A cos A)

$= 4\int \cos\left(\dfrac{x+\alpha}{2}\right)\cos\left(\dfrac{x-\alpha}{2}\right)dx$ $\because$ [2 cos C cos D = cos (C+D)+cos (C-D)]

$= 2\int (\cos x + \cos \alpha)dx = 2\int \cos x dx + 2\cos \alpha \int dx = 2\sin x + 2x\cos \alpha + C$

Illustration 14: Evaluate: $\int \dfrac{\sin^8 x - \cos^8 x}{1 - 2\sin^2 x.\cos^2 x}dx$ (JEE ADVANCED)

Sol: Here by using the formula $a^2 - b^2 = (a+b)(a-b)$ and putting $(\sin^2 x + \cos^2 x)^2$ in place of 1 in the denominator, we can reduce the above integration and then using $\cos 2x = \cos^2 x - \sin^2 x$ we can solve it.

We have, $\int \dfrac{\sin^8 x - \cos^8 x}{1 - 2\sin^2 x.\cos^2 x}dx = \int \dfrac{(\sin^4 x + \cos^4 x)(\sin^4 x - \cos^4 x)}{(\sin^2 x + \cos^2 x)^2 - 2\sin^2 x.\cos^2 x}dx$

$$= \int \frac{(\sin^4 x + \cos^4 x)(\sin^2 x + \cos^2 x)(\sin^2 x - \cos^2 x)}{(\sin^4 x + \cos^4 x)} dx = \int 1.(\sin^2 x - \cos^2 x)dx = -\int \cos 2x dx \left[\because \cos 2x = \cos^2 x - \sin^2 x\right]$$

$$= -\frac{\sin 2x}{2} + C$$

4.2 Integration by Parts

If u and v are two functions of x, then

$$\int (u.v)dx = u\left(\int v \, dx\right) - \int\left(\frac{du}{dx}\right)\left(\int v \, dx\right)dx$$

This is also known as uv rule of integration. This method of integrating is called integration by parts.

NOMORECLASS CONCEPTS

- From the first letter of the words inverse circular, logarithmic, Algebraic, Trigonometric, Exponential functions, we get a word ILATE. Therefore the preference of selecting the u function will be according to the order ILATE.

- In some problems we have to give preference to logarithmic function over inverse trigonometric functions. Hence sometimes the word LIATE is used for reference.

- For the integration of Logarithmic or Inverse trigonometric functions alone, take unity (1) as the v function.

Illustration 15: Evaluate: $\int (1 + x)\log x dx$ **(JEE MAIN)**

Sol: Here we can integrate the given problem by using Integration by parts i.e.

$$\int (u.v)dx = u\left(\int v \, dx\right) - \int\left(\frac{du}{dx}\right)\left(\int v \, dx\right)dx$$

Here u = $\log x$ and v = $(1 + x)$.

Let I= $\int (1 + x)\log x dx$

Integrating by parts, taking log x as 1st function, (by LIATE rule) we get

$$I= \log x \int (1 + x)dx - \int\left[\frac{d}{dx}(\log x).\int (1+x)dx\right]dx = \log x \left(x + \frac{x^2}{2}\right) - \int \frac{1}{x}.\left(x + \frac{x^2}{2}\right)dx = \left(x + \frac{x^2}{2}\right)\log x - \int\left(1 + \frac{x}{2}\right)dx$$

$$= \left(x + \frac{x^2}{2}\right)\log x - \left(x + \frac{x^2}{4}\right) + C$$

Illustration 16: Evaluate: $\int \sec^3 x dx$ **(JEE ADVANCED)**

Sol: Here we can solve by integrating by parts, taking sec x as the first function.

$$I = \int \sec^3 x dx = \int \sec x.\sec^2 x dx \qquad \text{Let } u = \sec x \ \& \ v = \sec^2 x$$

$$I = \sec x \tan x - \int (\sec x \tan x).\tan x \, dx = \sec x \tan x - \int \sec x \tan^2 x dx = \sec x \tan x - \int \sec x(\sec^2 x - 1)dx$$

$$I = \sec x \tan x - \int \sec^3 x \, dx + \int \sec x \, dx \quad \Rightarrow \quad I = \sec x + \tan x - I + \int \sec x \, dx$$

$$\Rightarrow 2I = \sec x \cdot \tan x + \log(\sec x + \tan x) + C \quad \Rightarrow I = \frac{1}{2}\left[\sec x \tan x + \log(\sec x \tan x)\right] + C$$

Illustration 17: Evaluate : $\int (\sin^{-1} x)^2 \, dx$ **(JEE ADVANCED)**

Sol: We can write the given integration as $\int (\sin^{-1} x)^2 \cdot 1 \, dx$ and then taking $u = \left(\sin^{-1} x\right)^2$ & $v = 1$ solving by integration by parts.

$$I = (\sin^{-1} x)^2 \cdot x - \int \left\{ \frac{d}{dx}(\sin^{-1} x)^2 \cdot x \right\} dx = (\sin^{-1} x)^2 \cdot x - \left\{ 2(\sin^{-1} x) \cdot \frac{1}{\sqrt{1-x^2}} \cdot x \right\} dx$$

Now, putting $\sin^{-1} x = t \Rightarrow x = \sin t$ so that $\dfrac{dx}{\sqrt{1-x^2}} = dt$

$$\Rightarrow I = x(\sin^{-1} x)^2 - 2\int t \cdot \sin t \, dt = x(\sin^{-1} x)^2 - 2\left\{ -t \cos t + \int \cos t \, dt \right\} \text{ (again Integrating by parts)}$$

$$= x(\sin^{-1} x)^2 - 2\left\{ -t \cos t + \sin t \right\} + C = x(\sin^{-1} x)^2 + 2t \cos t - 2\sin t + C = x(\sin^{-1} x)^2 + 2\sin^{-1} x \cdot \sqrt{1-x^2} - 2x + C$$

Illustration 18: Evaluate: $\int \dfrac{\sin^{-1}\sqrt{x} - \cos^{-1}\sqrt{x}}{\sin^{-1}\sqrt{x} + \cos^{-1}\sqrt{x}} dx$ **(JEE ADVANCED)**

Sol: By using the formula $\sin^{-1}\sqrt{x} + \cos^{-1}\sqrt{x} = \dfrac{\pi}{2}$, we can solve the above problem.

$$\text{Let } I = \int \frac{\sin^{-1}\sqrt{x} - \cos^{-1}\sqrt{x}}{\sin^{-1}\sqrt{x} + \cos^{-1}\sqrt{x}} dx = \int \frac{\sin^{-1}\sqrt{x} - \left((\pi/2) - \sin^{-1}\sqrt{x}\right)}{(\pi/2)} dx \quad \left[\because \sin^{-1}\sqrt{x} + \cos^{-1}\sqrt{x} = \frac{\pi}{2} \right]$$

$$= \frac{2}{\pi} \int \left(2\sin^{-1}\sqrt{x} - \frac{\pi}{2} \right) dx = \frac{4}{\pi} \int \sin^{-1}\sqrt{x} \, dx - \int 1 \, dx$$

$$= \frac{4}{\pi} \int \sin^{-1}\sqrt{x} \, dx - x \qquad \qquad \qquad \qquad \qquad \qquad \qquad \qquad \qquad \qquad \qquad \text{... (i)}$$

Putting $\sin^{-1}\sqrt{x} = \theta \Rightarrow x = \sin^2\theta$ so that $dx = 2\sin\theta \cdot \cos\theta \, d\theta = \sin 2\theta \, d\theta$.

$$\therefore \int \sin^{-1}\sqrt{x} \, dx = \int \theta \cdot \sin 2\theta \, d\theta \text{ Let } u = \theta \text{ & } V = \sin 2\theta, \text{ then integing by parts we get}$$

$$= \frac{-\theta}{2}\left(1 - 2\sin^2\theta\right) + \frac{1}{4} 2\sin\theta \cdot \cos\theta$$

$$= -\theta \cdot \frac{\cos 2\theta}{2} + \frac{1}{2}\int \cos 2\theta \, d\theta = -\frac{\theta}{2}\cos 2\theta + \frac{1}{4}\sin 2\theta = -\frac{1}{2}\theta(1 - 2\sin^2\theta) + \frac{1}{2}\sin\theta\sqrt{1 - \sin^2\theta}$$

$$= -\frac{1}{2}(\sin^{-1}\sqrt{x})(1 - 2x) + \frac{1}{2}\sqrt{x} \cdot \sqrt{1-x} + C \qquad \qquad \qquad \qquad \qquad \qquad \text{... (ii)}$$

From (i) and (ii), we get

$$I = \frac{4}{\pi}\left\{ -\frac{1}{2}(1-2x)\sin^{-1}\sqrt{x} + \frac{1}{2}\sqrt{x-x^2} \right\} - x + C = \frac{2}{\pi}\left\{ \sqrt{x-x^2} - (1-2x)\sin^{-1}\sqrt{x} \right\} - x + C$$

4.2.1 Integration by Cancellation

Illustration 19: Evaluate : $\int \left(3x^2 \tan\frac{1}{x} - x\sec^2\frac{1}{x} \right) dx$ **(JEE MAIN)**

Sol: Let $\int \left(3x^2 \tan\frac{1}{x} - x\sec^2\frac{1}{x} \right)dx = \int 3x^2 \tan\frac{1}{x}dx - \int x\sec^2\frac{1}{x}dx$ and then by using the integration by parts formula

i.e. $\int (u.v)dx = u\left(\int v\ dx\right) - \int \left(\frac{du}{dx}\right)\left(\int v\ dx\right)dx$ we can solve the problem above.

$\int \left(3x^2 \tan\frac{1}{x} - x\sec^2\frac{1}{x} \right)dx = \int 3x^2 \tan\frac{1}{x}dx - \int x\sec^2\frac{1}{x}dx = \tan\frac{1}{x}x^3 - \int \left(\sec^2\frac{1}{x}\right)\left(-\frac{1}{x^2}\right)x^3 dx - \int x\sec^2\frac{1}{x}dx = x^3 \tan\frac{1}{x} + c$

4.2.2 Integration of the Form:

If the integral is of the form $\int e^x \left[f(x) + f'(x) \right]dx$, then use the formula;

$\int e^x \left[f(x) + f'(x) \right]dx = e^x f(x) + c$

Illustration 20: Evaluate: $\int e^x (\log x + 1/x)dx$ (JEE MAIN)

Sol: Solution of this problem is based on the method mentioned above, here f(x) = log x and f'(x)

$= 1/x$. $= \int e^x \left[\log x + \frac{1}{x} \right]dx$

$I = \int \left(e^x \log x + \frac{e^x}{x} \right)dx = e^x \log x + c$ $\quad \left[\begin{array}{l} \text{Here, } f(x) = \log x \\ \& f'(x) = 1/x \end{array} \right]$

If the integral is of the form $\int \left[xf'(x) + f(x) \right]dx$ then use the formula; $\int \left[xf'(x) + f(x) \right]dx = xf(x) + c$

Illustration 21: Evaluate : $\int (x\sec^2 x + \tan x)dx$ (JEE MAIN)

Sol: Similar to the problem above.

Here $I = \int (x\sec^2 x + \tan x)dx = \int \left[xf'(x) + f(x) \right]dx$ [where $f(x) = \tan x$] $= x.\tan x + c$

4.2.3 Special Integrals

$\int e^{ax} \sin bx\,dx = \frac{e^{ax}}{a^2 + b^2}(a\sin bx - b\cos bx) + c$

$\int e^{ax} \cos bx\,dx = \frac{e^{ax}}{a^2 + b^2}(b\sin bx + a\cos bx) + c$

$\int e^{ax} \cos bx\,dx = \frac{e^{ax}}{a^2 + b^2}(b\sin bx + a\cos bx) + c$

Illustration 22: Evaluate : $\int e^{\sin^{-1} x}dx$ (JEE MAIN)

Sol: By putting $\sin^{-1} x = t \Rightarrow x = \sin t \Rightarrow dx = \cos t\ dt$ and then integrating by parts we can solve the given problem.

$I = \int e^{\sin^{-1} x}dx$

Let $\sin^{-1} x = t \Rightarrow x = \sin t \Rightarrow dx = \cos t\ dt$

$\Rightarrow I = \int e^t \cos t\,dt = \frac{e^t}{2}(\sin t + \cos t) + c = \frac{e^{\sin^{-1} x}}{2}(x + \sqrt{1 - x^2}) + c$

4.3 Integration of Rational Functions

4.3.1 When the Denominator can be Factorized (Using Partial Fraction)

Let the integrand be of the form $\dfrac{f(x)}{g(x)}$, where both $f(x)$ and $g(x)$ are polynomials. If degree of $f(x)$ is greater than degree of $g(x)$, then first divide $f(x)$ by $g(x)$ till the degree of the remainder becomes less than the degree of $g(x)$. Let $Q(x)$ be the quotient and $R(x)$, the remainder then

$$\frac{f(x)}{g(x)} = Q(x) + \frac{R(x)}{g(x)}$$

Now in $R(x)/g(x)$, factorize $g(x)$ and then write partial fractions in the following manner:

(a) For every non-repeated linear factor in the denominator. Write

$$\frac{1}{(x-a)(x-b)} = \frac{A}{x-a} + \frac{B}{x-b}$$

(b) For every repeated linear factor in the denominator. Write

$$\frac{1}{(x-a)^3(x-b)} = \frac{A}{(x-a)} + \frac{B}{(x-a)^2} + \frac{C}{(x-a)^3} + \frac{D}{(x-b)}$$

(c) For every non-repeated quadratic factor in the denominator. Write

$$\frac{1}{(ax^2+bx+c)(x-d)} = \frac{Ax+B}{ax^2+bx+c} + \frac{C}{x-d}$$

(d) For every repeated quadratic factor in the denominator. Write

$$\frac{1}{(ax^2+bx+c)^2(x-d)} = \frac{Ax+B}{(ax^2+bx+c)^2} + \frac{Cx+D}{ax^2+bx+c} + \frac{E}{x-d}$$

NOMORECLASS CONCEPTS

Consider $f(x)$ as the function we need to factorize

1. For non- repeated linear factor in the denominator.

Let $f(x) = \dfrac{1}{(x-a)(x-b)} = \dfrac{A}{(x-a)} + \dfrac{B}{(x-b)}$

To obtain the value of A remove $(x-a)$ from $f(x)$ and find $f(a)$.

Similarly, to obtain value of B, remove $(x-b)$ from $f(x)$ and find $f(b)$.

2. For repeated linear factor in the denominator.

Let $f(x) = \dfrac{1}{(x-a)^3(x-b)} = \dfrac{A}{(x-a)} + \dfrac{B}{(x-a)^2} + \dfrac{C}{(x-a)^3} + \dfrac{D}{(x-b)}$

To obtain value of D remove $(x-b)$ from $f(x)$ and find $f(b)$.

To obtain value of c remove $(x-a)^3$ from $f(x)$ and find $f(a)$.

Now that we have reduced the number of unknowns from 4 to 2, we can find A and B easily by equating.

Now let's try this method for $\dfrac{x^4+x^3+2x^2-x+4}{x(x^2+2)(x^2-1)^3}$

Partial fraction will be of the form

$$\frac{x^4+x^3+2x^2-x+4}{x(x^2+2)(x^2+1)^3}=\frac{A}{x}+\frac{Bx+C}{(x^2+2)}+\frac{Dx+E}{(x^2+1)}+\frac{Fx+G}{(x^2+1)^2}+\frac{Hx+I}{(x^2+1)^3}$$

Now remove x and put x=0, we get A=2

Now remove $(x^2+1)^3$ and put $x^2=-1$ i.e. x = i (you can also substitute x = −i).

We get Hi+I = −3i −2. Hence H = −3 and I = −2.

Now remove (x^2+2) and put x= $\sqrt{2}i$. We get B ($\sqrt{2}i$) +C=2$\sqrt{2}i$ +3. Hence B = 2 and C = 3

Now the number of unknowns have reduced from 9 to 4 and the remaining unknowns can be solved easily.

This method very useful instead of solving for all the unknowns at the same time.

Also remember that substituting an imaginary number for x is not discussed anywhere in NCERT. So, use this method only for competitive exams.

Illustration 23: Evaluate : $\displaystyle\int\frac{x}{x^2-x-2}dx$ (JEE MAIN)

Sol: Here the given integration is in the form of $\dfrac{1}{(x-a)(x-b)}$, hence by using partial fractions we can split it as

$\dfrac{A}{(x-a)}+\dfrac{B}{(x-b)}$ and then by solving we will get the required result.

Here I = $\displaystyle\int\frac{x}{(x-2)(x+1)}dx=\int\frac{1}{3}\left(\frac{2}{x-2}+\frac{1}{x+1}\right)dx=\frac{1}{3}\Big[2\log(x-2)+\log(x+1)\Big]+c=\frac{1}{3}\log\Big[(x-2)^2(x+1)\Big]+c$

Illustration 24: Evaluate : $\displaystyle\int\frac{xdx}{3x^4-18x^2+11}$ (JEE ADVANCED)

Sol: Here simply by putting t= $x^2 \Rightarrow$ dt = 2x dx and then by using partial fractions we can solve the given problem.

$$I=\int\frac{xdx}{3x^4-18x^2+11}dx=\int\frac{\frac{1}{2}dt}{3t^2-18t+11}\quad\text{(Put } t=x^2\Rightarrow dt=2x\,dx)$$

$$=\frac{1}{6}\int\frac{dt}{t^2-6t+(11/3)}=\frac{1}{6}\int\frac{dt}{(t-3)^2-(16/3)}=\frac{1}{6}\int\frac{dt}{(t-3)^2-\left(4/\sqrt{3}\right)^2}$$

$$=\frac{1}{6}\frac{1}{2\times(4/\sqrt{3})}\log\frac{(t-3)-(4/\sqrt{3})}{(t-3)+(4/\sqrt{3})}+C=\frac{\sqrt{3}}{48}\log\frac{\sqrt{3}t-3\sqrt{3}-4}{\sqrt{3}t-3\sqrt{3}+4}+C=\frac{\sqrt{3}}{48}\log\frac{\sqrt{3}x^2-3\sqrt{3}-4}{\sqrt{3}x^2-3\sqrt{3}+4}+C$$

4.3.2 When the Denominator cannot be Factorized

In this case the integral may be in the form

(i) $\int \dfrac{dx}{ax^2 + bx + c}$ (ii) $\int \dfrac{(px+q)}{ax^2 + bx + c} dx$

Method:

(i) Here taking the coefficient of x^2 common from the denominator, write

$$x^2 + (b/a)x + c/a = (x + b/2a)^2 - \dfrac{b^2 - 4ac}{4a^2}$$

Now the integrand obtained can be evaluated easily by using standard formulae.

(ii) Here suppose that $px + q = A\left[\dfrac{d}{dx}(ax^2 + bx + c)\right] + B = A(2ax+b) + B$ .....(i)

Now comparing coefficient of x and constant terms.

We get A=p/2a, B=q-(pb/2a)

$$\therefore I = \dfrac{p}{2a}\int \dfrac{2ax+b}{ax^2bx+c} dx + \left(q - \dfrac{pb}{2a}\right)\int \dfrac{dx}{ax^2 + bx + c}$$

Now we can integrate it easily.

4.3.3 Integrand Containing Only Even Powers of x

To find integral of such functions, first we divide numerator and denominator by x^2, then express the numerator as $d(x \pm 1/x)$ and the denominator as a function of $(x \pm 1/x)$. The following examples illustrate it.

NOMORECLASS CONCEPTS

$\int R(\sin x, \cos x)dx$ where R is a rational function (universal substitution tan(x/2)=t)

Special cases:

(a) If $R(-\sin x, \cos x) = -R(\sin x, \cos x)$

Put cos x=t

(b) If $R(\sin x, \cos x) = -R(\sin x, \cos x)$

Put sin x=t

(c) If $R(-\sin x, -\cos x) = R(\sin x, \cos x)$

Put tan x=t

Illustration 25: Evaluate: $\int \dfrac{x^2 + 1}{x^4 + 1} dx$ **(JEE ADVANCED)**

Sol: Here dividing the numerator and denominator by x^2, we get $\int \dfrac{1 + (1/x^2)}{x^2 + (1/x^2)} dx = \int \dfrac{1 + (1/x^2)}{\left[\{x - (1/x)\}^2 + 2\right]} dx$ and then by putting $x - 1/x = t \Rightarrow [1 + 1/x^2]dx = dt$, we can solve it.

$$I = \int \frac{1+(1/x^2)}{x^2+(1/x^2)}dx = \int \frac{1+(1/x^2)}{\left[\{x-(1/x)\}^2+2\right]}dx$$

Now taking $x-1/x=t \Rightarrow [1+1/x^2]dx = dt$, we get

$$I = \int \frac{dt}{t^2+2} = \frac{1}{\sqrt{2}}\tan^{-1}\left(\frac{t}{\sqrt{2}}\right)+c = \frac{1}{\sqrt{2}}\tan^{-1}\left(\frac{x^2-1}{\sqrt{2}x}\right)+c$$

4.4 Integration of Irrational Functions

If any one term in numerator or denominator is irrational then it is made rational by a suitable substitution. Also if the integral is of the form

$$\int \frac{dx}{\sqrt{ax^2+bx+c}} \quad \text{or} \quad \int \sqrt{ax^2+bx+c}\, dx$$

Then we integrate it by expressing $ax^2+bx+c = (x+\alpha)^2+\beta^2$

Also for integrals of the form $\int \frac{px+q}{\sqrt{ax^2+bx+c}}dx$ or $\int (px+q)\sqrt{ax^2bx+c}\, dx$

First we express px+q in the form

$px+q = A\left[\frac{d}{dx}(ax^2bx+c)\right]+B$ and then proceed as usual with standard form.

Illustration 26: Evaluate : $\int \frac{e^x}{\sqrt{5-4e^x-e^{2x}}}dx$ (JEE MAIN)

Sol: Simply by putting $e^x = t$, then $e^x dx = dt$, we can solve the given problem.

Put $e^x = t$, then $e^x dx = dt$

$$\therefore \int \frac{e^x}{\sqrt{5-4e^x-e^{2x}}}dx = \int \frac{dt}{\sqrt{5-4t-t^2}} = \int \frac{dt}{\sqrt{5-(t^2+4t)}} = \int \frac{dt}{\sqrt{5-(t^2+4t+4)+4}} = \int \frac{dt}{\sqrt{9-(t+2)^2}}$$

$$= \int \frac{dt}{\sqrt{(3)^2-(t+2)^2}} = \sin^{-1}\left(\frac{t+2}{3}\right)+C = \sin^{-1}\left(\frac{e^x-2}{3}\right)+C$$

Illustration 27: Evaluate : $\int \frac{1}{\sqrt{(x-a)(x-b)}}dx$ (JEE ADVANCED)

Sol: Here first expand $(x-a)(x-b)$ and then adding and subtracting by $\left(\frac{a+b}{2}\right)^2$, we can reduce the above integration. After that by putting $x-\left(\frac{a+b}{2}\right)=u$, we can solve the given problem.

$$\text{Let, } I = \int \frac{1}{\sqrt{(x-a)(x-b)}}dx = \int \frac{1}{\sqrt{x^2-(a+b)x+ab}}dx = \int \frac{dx}{\sqrt{x^2-(a+b)x+((a+b)/2)^2-((a+b)/2)^2+ab}}$$

$$= \int \frac{dx}{\sqrt{(x-((a+b)/2))^2-\left[((a^2+b^2+2ab)/4)-ab\right]}} = \int \frac{dx}{\sqrt{(x-((a+b)/2))^2-\left[(a-b)^2/4\right]}}$$

$$= \int \frac{dx}{\sqrt{(x-((a+b)/2))^2-((a-b)/2)^2}} \qquad \dots \text{(i)}$$

On putting $x - \left(\dfrac{a+b}{2}\right) = u$ so that $dx = du$ in (i), we get

$$I = \int \dfrac{du}{\sqrt{u^2 - \left((a-b)/2\right)^2}} \quad \left[\int \dfrac{1}{\sqrt{x^2 - a^2}} dx = \log\left|x + \sqrt{x^2 - a^2}\right|\right] = \log\left|u + \sqrt{u^2 - \left(\dfrac{a-b}{2}\right)^2}\right| + c$$

Putting $u = \left(x - \left(\dfrac{a+b}{2}\right)\right)$, we get

$$I = \log\left|\left(x - \left(\dfrac{a+b}{2}\right)\right) + \sqrt{\left(x - \left(\dfrac{a+b}{2}\right)\right)^2 - \left(\dfrac{a-b}{2}\right)^2}\right| + c = \log\left|x - \dfrac{a+b}{2} + \sqrt{(x-a)(x-b)}\right| + c$$

5. STANDARD INTEGRALS

(a) $\displaystyle\int \dfrac{1}{x^2 + a^2} dx = \dfrac{1}{a}\tan^{-1}\left(\dfrac{x}{a}\right) + c$

(b) $\displaystyle\int \dfrac{1}{x^2 - a^2} dx = \dfrac{1}{2a}\log\left|\dfrac{x-a}{x+a}\right| + c$

(c) $\displaystyle\int \dfrac{1}{a^2 - x^2} dx = \dfrac{1}{2a}\log\left|\dfrac{a+x}{a-x}\right| + c$

(d) $\displaystyle\int \dfrac{1}{\sqrt{a^2 - x^2}} dx = \sin^{-1}\left(\dfrac{x}{a}\right) + c = -\cos^{-1}\left(\dfrac{x}{a}\right) + c$

(e) $\displaystyle\int \dfrac{1}{\sqrt{x^2 + a^2}} dx = \sinh^{-1}\left(\dfrac{x}{a}\right) + c = \log\left(x + \sqrt{x^2 + a^2}\right) + c$

(f) $\displaystyle\int \dfrac{1}{\sqrt{x^2 - a^2}} dx = \cosh^{-1}\left(\dfrac{x}{a}\right) + c = \log\left(x + \sqrt{x^2 - a^2}\right) + c$

(g) $\displaystyle\int \sqrt{a^2 - x^2}\, dx = \dfrac{x}{2}\sqrt{a^2 - x^2} + \dfrac{a^2}{2}\sin^{-1}\dfrac{x}{a} + c$ (Substitute $x = a\cos\theta$ or $x = a\sin\theta$ and proceed)

(h) $\displaystyle\int \sqrt{x^2 + a^2}\, dx = \dfrac{x}{2}\sqrt{x^2 + a^2} + \dfrac{a^2}{2}\log\left|x + \sqrt{x^2 + a^2}\right| + c$ (Substitute $x = a\tan\theta$ or $x = a\cot\theta$ and proceed)

(i) $\displaystyle\int \sqrt{x^2 - a^2}\, dx = \dfrac{x}{2}\sqrt{x^2 - a^2} - \dfrac{a^2}{2}\log\left|x + \sqrt{x^2 - a^2}\right| + c$ (Substitute $x = a\sec\theta$ or $x = a\,\mathrm{cosec}\,\theta$ and proceed)

(j) $\displaystyle\int \dfrac{1}{x\sqrt{x^2 - a^2}} dx = \dfrac{1}{a}\sec^{-1}\dfrac{x}{a} + c$ \qquad (Valid for $x > a > 0$)

(k) $\displaystyle\int e^{ax}\sin bx\, dx = \dfrac{e^{ax}}{a^2 + b^2}(a\sin bx - b\cos bx) + c = \dfrac{e^{ax}}{\sqrt{a^2 + b^2}}\sin\left\{bx - \tan^{-1}\left(\dfrac{b}{a}\right)\right\} + c$

(I) $\int e^{ax} \cos bx\, dx = \dfrac{e^{ax}}{a^2 + b^2}(a\cos bx + b\sin bx) + c = \dfrac{e^{ax}}{\sqrt{a^2 + b^2}}\cos\left\{bx - \tan^{-1}\dfrac{b}{a}\right\} + c$

Integration of irrational algebraic functions:

Type 1: (a) $\int \dfrac{dx}{(x - \alpha)\sqrt{(x - \alpha)(\beta - x)}}$ (Put : $x = a\cos^2 q + b\sin^2 q$)

(b) $\int \dfrac{dx}{(x - \alpha)\sqrt{(x - \beta)}}$ (Put : $x = a\sec^2 q - b\tan^2 q$)

Type 2: $\int \dfrac{dx}{(ax + b)\sqrt{px + q}}$ (Put: $px + q = t^2$)

Type 3: $\int \dfrac{dx}{(ax + b)\sqrt{px^2 + qx + r}}$ (Put: $ax + b = \dfrac{1}{t}$)

Type 4: $\int \dfrac{dx}{(ax^2 bx + c)\sqrt{px + q}}$ (Put: $px + q = t^2$)

Type 5: $\int \dfrac{dx}{(ax^2 + bx + c)\sqrt{px^2 + qx + r}}$

Case I: When $(ax^2 bx + c)$ breaks up into two linear factors, e.g.

$I = \int \dfrac{dx}{\left(x^2 - x - 2\right)\sqrt{x^2 + x + 1}}\,dx$ then $= \int\left(\dfrac{A}{x - 2} + \dfrac{B}{x + 1}\right)\dfrac{1}{\sqrt{x^2 + x + 1}}\,dx = A\int\dfrac{dx}{\left(x - 2\right)\sqrt{x^2 + x + 1}} + B\int\dfrac{cx}{(x + 1)\sqrt{x^2 + x + 1}}$

Put $x - 2 = \dfrac{1}{t}$ Put $x + 1 = \dfrac{1}{t}$

Case II: If $ax^2 + bx + c$ is a perfect square say $(lx - m)^2$, then put $lx + m = \dfrac{1}{t}$

Case III: If b = 2, q = 0

e.g. $\int \dfrac{dx}{(ax^2 + b)\sqrt{px^2 + r}}$ then, put $x = \dfrac{1}{t}$ or trigonometric substitutions are also helpful.

Integral of the form $\int \dfrac{dx}{P\sqrt{Q}}$, where P, Q are linear or quadratic functions of x.

Integral	Substitutions
$\int \dfrac{1}{(ax + b)\sqrt{cx + d}}\,dx$	$cx - d = z^2$
$\int \dfrac{dx}{\left(ax^2 + bx + c\right)\sqrt{px + q}}$	$px + q = z^2$
$\int \dfrac{dx}{(px + q)\sqrt{ax^2 + bx + c}}$	$px + q = \dfrac{1}{z}$
$\int \dfrac{dx}{\left(ax^2 + b\right)\sqrt{cx^2 + d}}$	$x = \dfrac{1}{z}$

$$\int \frac{dx}{(ax+b)^m \sqrt{ax^2+bx+c}} \qquad\qquad ax+b = 1/t$$

Illustration 28: Evaluate : $\displaystyle\int \frac{dx}{(x+1)\sqrt{(x-2)}}$ (JEE MAIN)

Sol: Simply by putting $x-2 = t^2$, $\therefore dx = 2t\,dt$ we can solve the given problem by using the appropriate formula.

$$\int \frac{dx}{(x+1)\sqrt{(x-2)}}$$

Put $x-2 = t^2$

$\therefore dx = 2t\,dt$

$$\therefore I = \int \frac{2t\,dt}{(t^2+3)t} = 2\int \frac{dt}{t^2+(\sqrt{3})^2} = \frac{2}{\sqrt{3}}\tan^{-1}\left(\frac{t}{\sqrt{3}}\right)+c = \frac{2}{\sqrt{3}}\tan^{-1}\left(\sqrt{\left(\frac{x-2}{3}\right)}\right)+c \;(\because t = \sqrt{(x-2)})$$

Illustration 29: Evaluate : $\displaystyle\int \frac{dx}{(x^2-4)\sqrt{x}}$ (JEE MAIN)

Sol: Here first put $x = t^2$ therefore $dx = 2t\,dt$ and then using partial fractions we reduce the given integration in standard form. After that by solving we will get the result.

Let $I = \displaystyle\int \frac{dx}{(x^2-4)\sqrt{x}}$

Put $x = t^2$ $\therefore dx = 2t\,dt$ then $I = \displaystyle\int \frac{2t}{(t^4-4)t}dt = 2\int \frac{dt}{(t^2+2)(t^2-2)}$

Put $t^2 = z$ $\therefore$ $\dfrac{1}{(t^2+2)(t^2-2)} = \dfrac{1}{(z+2)(z-2)} = \dfrac{A}{z+2}+\dfrac{B}{z-2}$

$A = -\dfrac{1}{4}$ and $B = \dfrac{1}{4}$ $\Rightarrow$ $\dfrac{1}{(t^2+2)(t^2-2)} = \dfrac{1}{4(t^2+2)}+\dfrac{1}{4(t^2-2)}$

$$\therefore I = 2\int \frac{1}{(t^2+2)(t^2-2)} = -\frac{1}{2}\int \frac{dt}{t^2+2}+\frac{1}{2}\int \frac{dt}{t^2-2} = -\frac{1}{2\sqrt{2}}\tan^{-1}\left(\frac{t}{\sqrt{2}}\right)+\frac{1}{4\sqrt{2}}\log\left|\frac{t-\sqrt{2}}{t+\sqrt{2}}\right|+c$$

$$= -\frac{1}{2\sqrt{2}}\tan^{-1}\left(\frac{x}{2}\right)+\frac{1}{4\sqrt{2}}\log\left|\frac{\sqrt{x}-\sqrt{2}}{\sqrt{x}+\sqrt{2}}\right|+c \;(\because t = \sqrt{x})$$

6. SPECIAL TRIGONOMETRIC FUNCTIONS

Here we shall study the methods for evaluation of the following types of integrals.

Type 1

(i) $\displaystyle\int \frac{dx}{a+b\sin^2 x}$ (ii) $\displaystyle\int \frac{dx}{a+b\cos^2 x}$ (iii) $\displaystyle\int \frac{dx}{a\cos^2 x+b\sin x\cos x+c\sin^2 x}$ (iv) $\dfrac{dx}{\int(a\sin x+b\cos x)^2}$

Method: Divide the numerator and denominator by $\cos^2 x$ in all such types of integrals and then put $\tan x = t$

Illustration 30: Evaluate : $\int \dfrac{dx}{1+3\sin^2 x}$ (JEE MAIN)

Sol: Here dividing the numerator and denominator by $\cos^2 x$ we can solve it.

$$I = \int \frac{\sec^2 x\, dx}{\sec^2 x + 3\tan^2 x} = \int \frac{\sec^2 x\, dx}{1+4\tan^2 x} = \frac{1}{2}\tan^{-1}(2\tan x) + c$$

Type 2

(i) $\int \dfrac{dx}{a+b\cos x}$ (ii) $\int \dfrac{dx}{a+b\sin x}$ (iii) $\int \dfrac{dx}{a\cos x + b\sin x}$ (iv) $\int \dfrac{dx}{a\sin x + b\cos x + c}$

Method: In such types of integrals we use the following substitutions

$$\sin x = \frac{2\tan(x/2)}{1+\tan^2(x/2)} = \frac{2t}{1+t^2}, \quad \cos x = \frac{1-\tan^2(x/2)}{1+\tan^2(x/2)} = \frac{1-t^2}{1+t^2}; \quad dx = \frac{2dt}{1+t^2}$$

and integrate another method for the evaluation of the integral.

Illustration 31: Evaluate: $\int \dfrac{dx}{5+4\cos x}dx$ (JEE MAIN)

Sol: Here by putting $\cos x = \dfrac{1-\tan^2(x/2)}{1+\tan^2(x/2)}$ and then by taking tan (x/2) = t we can solve the given problem

$$I = \int \frac{dx}{5+4\left[(1-\tan^2(x/2))/(1+\tan^2(x/2))\right]} = \int \frac{\sec^2(x/2)}{9+\tan^2(x/2)}dx = 2\int \frac{ct}{3^2+t^2} \qquad \text{where } \tan(x/2) = t$$

$$= 2\left(\frac{1}{3}\tan^{-1}\left(\frac{t}{3}\right)\right) + C = 2\left[\frac{1}{3}\tan^{-1}\left(\frac{\tan x/2}{3}\right)\right] + C$$

Type 3

(i) $\int \dfrac{p\sin x + q\cos x}{a\sin x + b\cos x}dx$ (ii) $\int \dfrac{p\sin x}{a\sin x + b\cos x}dx$ (iii) $\int \dfrac{q\cos x}{a\sin x + b\cos x}dx$

For their integration, we first express numerator as follows-

Numerator = A (denominator) + B (derivative of denominator)

Then integral = Ax + B log (denominator) + C

Illustration 32: Evaluate : $\int \dfrac{6+3\sin x + 14\cos x}{3+4\sin x + 5\cos x}dx$ (JEE ADVANCED)

Sol: By using partial fractions, we can reduce the given integration to the standard form.

$$\int \frac{6+3\sin x + 14\cos x}{3+4\sin x + 5\cos x}dx$$

$\Rightarrow 6+3\sin x + 14\cos x = A (3+4\sin x + 5\cos x) + B (4\cos x - 5\sin x) + c$

Solving R.H.S. & comparing both sides, we get $4A - 5B = 3\ 5A + 4B = 14$

Also, $3A+C=6$ $\therefore \int \dfrac{A(3+4\sin x + 5\cos x) + B(4\cos x - 5\sin x) + c}{3+4\sin x + 5\cos x}$

$\Rightarrow Ax + \log(3 + 4\sin x + 5\cos x) + \int \dfrac{C \; dx}{\underbrace{3 + 4\sin x + 5\cos x}_{\text{this is of type 2}}}$

Illustration 33: Evaluate : $\int \dfrac{2\sin 2\phi - \cos\phi}{6 - \cos^2\phi - 4\sin\phi} d\phi$ **(JEE ADVANCED)**

Sol: Here we can write the given integration as $\int \dfrac{2(\sin 2\phi - 4\cos\phi) + 7\cos\phi}{6 - \cos^2\phi - 4\sin\phi} d\phi$ and as we know $2(\sin 2\phi - 4\cos\phi)$ is

the derivative of $6 - \cos^2\phi - 4\sin\phi$ hence by putting $6 - \cos^2\phi - 4\sin\phi = t$, we can solve the given problem.

$I = \int \dfrac{2\sin 2\phi - \cos\phi}{6 - \cos^2\phi - 4\sin\phi} d\phi = \dfrac{d}{d\phi}(6 - \cos^2\phi - 4\sin\phi)$

$= 2\cos\phi\sin\phi - 4\cos\phi = \sin 2\phi - 4\cos\phi = 2\sin 2\phi - \cos\phi = 2(\sin 2\phi - 4\cos\phi) + 7\cos\phi$

$I = \int \dfrac{2(\sin 2\phi - 4\cos\phi) + 7\cos\phi}{6 - \cos^2\phi - 4\sin\phi} d\phi = \int \dfrac{2(\sin 2\phi - 4\cos\phi)d\phi}{6 - \cos^2\phi - 4\sin\phi} + \int \dfrac{7\cos\phi d\phi}{6 - \cos^2\phi - 4\sin\phi} = 2\int \dfrac{dt}{t} + \int \dfrac{7\cos\phi d\phi}{6 - (1 - \sin^2\phi) - 4\sin\phi}$

$= 2\log t + C_1 + \int \dfrac{7\cos\phi d\phi}{5 + \sin^2\phi - 4\sin\phi} = 2\log(6 - \cos^2\phi - 4\sin\phi) + C_1 + \int \dfrac{7dx}{x^2 - 4x + 5}$ $(\sin\phi = x)$

$= 2\log(6 - \cos^2\phi - 4\sin\phi) + C_1 + \int \dfrac{7dx}{(x-2)^2 + 1} = 2\log(6 - \cos^2\phi - 4\sin\phi) + C_1 + 7\tan^{-1}\dfrac{x-2}{1} + C_2$

$= 2\log(6 - \cos^2\phi - 4\sin\phi) + 7\tan^{-1}(\sin\phi - 2) + C$

7. SPECIAL EXPONENTIAL FUNCTIONS

(a) $\int \dfrac{ae^x}{b + ce^x} dx$ [put $e^x = t$]

(b) $\int \dfrac{1}{1 + e^x} dx$ [Multiply and divide by e^{-x} and $e^{-x} = t$]

(c) $\int \dfrac{1}{1 - e^x} dx$ [Multiply and divide by e^{-x} and $e^{-x} = t$]

(d) $\int \dfrac{1}{e^x - e^{-x}} dx$ [Multiply and divide by e^x]

(e) $\int \dfrac{e^x - e^{-x}}{e^x + e^{-x}} dx$ $\left[\dfrac{f'(x)}{f(x)} \text{form}\right]$

(f) $\int \dfrac{e^x + 1}{e^x - 1} dx$ [Multiply and divide by $e^{-x/2}$]

(g) $\int \dfrac{1}{(1 + e^x)(1 - e^{-x})} dx$ [Multiply and divide by e^x and put $e^x = t$]

(h) $\int \dfrac{1}{\sqrt{1 - e^x}} dx$ [Multiply and divide by $e^{-x/2}$]

(i) $\int \dfrac{1}{\sqrt{1+e^x}}dx$ [Multiply and divide by $e^{-x/2}$]

(j) $\int \dfrac{1}{\sqrt{e^x-1}}dx$ [Multiply and divide by $e^{-x/2}$]

(k) $\int \dfrac{1}{\sqrt{2e^x-1}}dx$ [Multiply and divide by $\sqrt{2}e^{-x/2}$]

(l) $\int \sqrt{1-e^x}dx$ [Integrand $= (1-e^x)/\sqrt{1-e^x}$]

(m) $\int \sqrt{1+e^x}dx$ [Integrand $= (1+e^x)/\sqrt{1+e^x}$]

(n) $\int \sqrt{e^x-1}dx$ [Integrand $= (e^x-1)/\sqrt{e^x-1}$]

(o) $\int \sqrt{\dfrac{e^x+a}{e^x-a}}dx$ [Integrand $= (e^x+a)/\sqrt{e^{2x}-a^2}$]

Illustration 34: Evaluate : $\int \sqrt{e^x-1}dx$ (JEE MAIN)

Sol: Here by multiplying and dividing by $\sqrt{e^x-1}$ in the given integration and then by putting $e^x-1=t^2$ we can evaluate the given integration.

Here $I = \int \sqrt{e^x-1}dx = \int \dfrac{e^x-1}{\sqrt{e^x-1}}dx = \int \dfrac{e^x}{\sqrt{e^x-1}}dx - \int \dfrac{1}{\sqrt{e^x-1}}dx$

Let $e^x-1 = t^2$, then $e^x = dx = 2t$ dt

$\therefore I = 2\int dt - \int \dfrac{2}{t^2+1}dt = 2t - 2\tan^{-1}(t) + c = 2\left[\sqrt{e^x-1} - \tan^{-1}\sqrt{e^x-1}\right] + c$

Illustration 35: Evaluate : $\int \dfrac{e^x}{\sqrt{5-4e^x-e^{2x}}}dx$ (JEE MAIN)

Sol: We have, $\int \dfrac{e^x}{\sqrt{5-4e^x-e^{2x}}}dx$

Put $e^x = t$, then $e^x\, dx = dt$

$\therefore \int \dfrac{e^x}{\sqrt{5-4e^x-e^{2x}}}dx = \int \dfrac{dt}{\sqrt{5-4t-t^2}}dx$

$= \int \dfrac{dt}{\sqrt{5-(t^2+4t)}} = \int \dfrac{dt}{\sqrt{5-(t^2+4t+4)+4}}$

$= \int \dfrac{dt}{\sqrt{9-(t+2)^2}} = \int \dfrac{dt}{\sqrt{(3)^2-(t+2)^2}} = \sin^{-1}\left(\dfrac{t+2}{3}\right) + C = \sin^{-1}\left(\dfrac{e^x+2}{3}\right) + C$

PROBLEM-SOLVING TACTICS

Integration by Parts

(a) Integration by parts is useful for dealing with integrals of the products of the following functions

$$u \ll \tan^{-1}x, \ \sin^{-1}x, \cos^{-1}x \ \ (\log x)^k \ \ \sin x, \ \cos x \quad e^x \gg dv$$

Priority for choosing u and dv: ILATE

(b) Integration by parts is sometimes useful for finding integrals of functions involving inverse functions such as n x and $\sin^{-1}x$.

(c) Sometimes when dealing with integrals, the integrand involves inverse functions (like $\sin^{-1}x$), it is useful to substitute x = the inverse of that inverse function (like x = sin u), then do integration by parts.

(d) Sometimes you will have to do integration by parts more than once (for example, $\int x^2 e^x dx$ and $\int x^3 \sin x dx$.

Sometimes you need to do it twice by parts, then manipulate the equation (for example, $\int e^x \sin x dx$).

(e) Try u – substitution first before integration by parts.

Trigonometric Integral

(a) Integral Type : $\int \sin^m x \cos^n x dx$

Case 1: One of m or n are even, and the other odd

Use u – substitution by setting u = sin or cos that with an even power. Use the identity $\sin^2 x + \cos^2 x = 1$.

Case 2: Both m and n are odd

Use u – substitution by setting u = sin or cos that with a higher power. Use the identity $\sin^2 x + \cos^2 x = 1$.

Case 3: Both m and n are even (hard case)

Do not use u – substitution. Use the half double angle formula to reduce the integrand into case 1 o r2:

$$\sin x \cos x = \frac{1}{2}\sin 2x \ ; \ \sin^2 x = \frac{1}{2}(1 - \cos 2x) \ ; \ \cos^2 x = \frac{1}{2}(1 - \cos 2x)$$

(Note: 0 is also an even number. For example, $\sin^3 x = \sin^3 x \cos^0 x$, so it is in case 1)

Just remember that when both are even, you can't use u-substitution, but you can use the half – double angle formula. When it is not that case, let u = sin x or cos x, and one will work (at the end there is no square root term after substitution).

(b) Integral type : $\int \tan^m x \sec^n x dx$

Case 1: sec is odd power, tan is even power.

Hard to do, we omit (most likely won't pop out in the exam).

Case 2: Else

Set u = sec x or tan x, and use $1 + \tan^2 x = \sec^2 x$. One will work at the end (there is no square root term after substitution).

(c) Integral type : $\int \sin(Ax)\cos(Bx)dx$, $\int \cos(Ax)\cos(Bx)dx$, $\int \sin(Ax)\sin(Bx)dx$

Use the product to sum formula:

$$\cos\theta\cos\phi = \frac{1}{2}(\cos\theta - \phi) + (\cos\theta + \phi)) \ ; \sin\theta\cos\phi = \frac{1}{2}(\cos\theta - \phi) - (\cos\theta + \phi))$$

$$\sin\theta\cos\phi = \frac{1}{2}(\sin\theta - \phi) + (\sin\theta + \phi))$$

Reduce product into sum and then integrate.

Trigonometric Substitution

(a) Trigonometric substitution is useful for quadratic form with square root:

$\sqrt{a^2 - x^2}$: Let $x = a\sin\theta$

$\sqrt{x^2 + a^2}$: Let $x = a\tan\theta$

$\sqrt{x^2 - a^2}$: Let $x = a\sec\theta$

(b) General procedure for doing trig sub:

Step 1: Draw the right triangle, and decide what trigonometric function to substitute for x.

Step 2: Find dx, then substitute the integrand using triangle, convert integral into trigonometric integral.

Step 3: Solve the trigonometric integral.

Step 4: Substitute back using triangle.

  (i) If the quadratic form is not in the Pythagoras form (for example, $\sqrt{2 + 2x + x^2}$, then use the perfecting the square method to transform it into Pythagoras form).

  (ii) Try u – substitution before trigonometric substitution.

  (iii) Integrals involving $(1 - x^2)$ and $(x^2 - 1)$ without square roots can be solved easily with partial fractions. So don't use trigonometric substitution.

Rational Integral and Partial Fraction

(a) General step for solving rational integral:

Step 1: Do long division for the rational function if the degree of the numerator is higher than the denominator.

Step 2: Do partial fraction decomposition.

Step 3: Evaluate the integral of each simple fraction.

(b) General step for partial fraction:

Step 1: Factorize the denominator.

Step 2: Set the partial fraction according to "rule".

Step 3: Solve the unknown of the numerator of the partial fraction.

Improper Integral

(a) General steps for evaluating improper integral:

Step 1: Change the improper integral into the appropriate limit. [Change $\pm\infty$ or singular point (where) to appropriate limit.]

Step 2: Evaluate the integral.

Step 3: Find the limit.

(b) The very first step to test improper integral involving ∞ is to check its limit. If its limit is not zero, then the integral diverges.

(c) Whenever you see improper integrals involving the quotient of a rational or irrational function, such as

$$\int_a^\infty \frac{x^3 + \sqrt{3x}}{(8x^3 + 7x)^{3/2}}\,dx$$

Use **limit comparison test**. The appropriate comparing function can be found by looking at the Integrand (quotient of rational irrational). "Discard" the lower degree terms.

(d) Sometimes, using u – substitution before using any test will be easier.

(e) Sometimes, to determine if an improper integral converges or diverges, directly evaluating the improper integral is easier.

(f) When doing a **comparison test**, beware of the comparing function that you choose. It might not give an appropriate conclusion if the comparing function is not correct.

(g) Try the **limit comparison test** before the **comparison test**.

(h) Useful comparing function, which is good to know their convergence or divergence

$$\int_a^\infty x^k e^{-\beta x} dx < \infty \quad \text{For } k \geq 0, \beta > 0$$

$$\int_a^\infty \frac{1}{x^p} dx \begin{cases} < \infty & \text{if } p > 1; \\ = \infty & \text{if } p \leq 1; \end{cases}$$

$$\int_a^\infty \frac{1}{x^p} dx \begin{cases} < \infty & \text{if } p < 1; \\ = \infty & \text{if } p \geq 1; \end{cases}$$

FORMULAE SHEET

Basic theorems of Integration:

1. $\int k f(x) dx = k \int f(x) dx$	**2.** $\int \left[f(x) \pm g(x) \right] dx = \int f(x) dx \pm \int g(x) dx$
3. $\frac{d}{dx} \left(\int f(x) dx \right) = f(x)$	**4.** $\int \left(\frac{d}{dx} f(x) \right) dx = f(x)$

Elementary Integration:

1. $\int 0.dx = c$	**2.** $\int 1.dx = x + c$
3. $\int k.dx = kx + c (k \in R)$	**4.** $\int x^n dx = \frac{x^{n+1}}{n+1} + c (n \neq -1)$
5. $\int \frac{1}{x} dx = \log_e x + c$	**6.** $\int e^x dx = e^x + c$
7. $\int a^x dx = \frac{a^x}{\log_e a} + c = a^x \log_a e + c$	**8.** $\int \sin x\, dx = -\cos x + c$
9. $\int \cos x\, dx = \sin x + c$	**10.** $\int (ax + b)^n dx = \frac{(ax + b)^{n+1}}{a(n+1)} + c$
11. $\int \frac{c}{ax + b} dx = \frac{c}{a} \log |ax + b| + c$	**12.** $\int f'(x) e^{f(x)} dx = e^{f(x)} + c$
13. $\int \log x\, dx = x \log x - x + c$	**14.** $\int \log_a x\, dx = x \log_a x - \frac{x}{\log a} + c$

Standard substitution:

1. $\sqrt{a^2 - x^2}$ or $\dfrac{1}{\sqrt{a^2 - x^2}}$	$x = a\sin\theta$ or $x = a\cos\theta$
2. $\sqrt{x^2 + a^2}$ or $\dfrac{1}{\sqrt{x^2 + a^2}}$	$x = a\tan\theta$ or $x = a\cot\theta$
3. $\sqrt{x^2 - a^2}$ or $\dfrac{1}{\sqrt{x^2 - a^2}}$	$x = a\sec\theta$ or $x = a\operatorname{cosec}\theta$
4. $\sqrt{\dfrac{x}{a+x}}$, $\sqrt{\dfrac{a+x}{x}}$, $\sqrt{x(a+x)}$ and $\dfrac{1}{\sqrt{x(a+x)}}$	$x = a\tan^2\theta$
5. $\sqrt{\dfrac{x}{a-x}}$ or $\sqrt{\dfrac{a-x}{x}}$ $\sqrt{x(a-x)}$ and $\dfrac{1}{\sqrt{x(a-x)}}$	$x = a\sin^2\theta$ or $x = a\cos2$
6. $\sqrt{\dfrac{x}{x-a}}$ or $\sqrt{\dfrac{x-a}{x}}$ or $\sqrt{x(x-a)}$ or $\dfrac{1}{\sqrt{x(x-a)}}$	$x = a\sec^2\theta$ or $x = a\operatorname{cosec}^2\theta$
7. $\sqrt{\dfrac{a-x}{a+x}}$ and $\sqrt{\dfrac{a+x}{a-x}}$	$x = a\cos2\theta$
8. $\sqrt{\dfrac{x-\alpha}{\beta-x}}$ or $\sqrt{(x-\alpha)(\beta-x)}$ $(\beta > \alpha)$	$x = \alpha\cos^2\theta + \beta\sin^2\theta$

Some standard Integrals:

1. $\int \tan x\,dx = \log\sec x + c = -\log\cos x + c$	**2.** $\int \cot x\,dx = \log\sin x + c$
3. $\int \sec x\,dx = \log(\sec x + \tan x) + c$ $= -\log(\sec x - \tan x) + c = \log\tan\left(\dfrac{\pi}{4} + \dfrac{x}{2}\right) + c$	**4.** $\int \operatorname{cosec} x\,dx = -\log(\operatorname{cosec} x + \cot x) + c$ $= \log(\operatorname{cosec} x - \cot x) + c = \log\tan\left(\dfrac{x}{2}\right) + c$
5. $\int \sec x\tan x\,dx = \sec x + c$	**6.** $\int \operatorname{cosec} x\cot x\,dx = -\operatorname{cosec} x + c$
7. $\int \sec^2 x\,dx = \tan x + c$	**8.** $\int \operatorname{cosec}^2 x\,dx = -\cot x + c$
9. $\int \log x\,dx = x\log\left(\dfrac{x}{e}\right) + c = x(\log x + 1) + c$	**10.** $\int \sin^2 x\,dx = \dfrac{1}{2}\left(x - \dfrac{\sin 2x}{2}\right) + C$ $= \dfrac{1}{2}(x - \sin x\cos x) + C$

11. $\int \cos^2 x dx = \dfrac{1}{2}\left(x + \dfrac{\sin 2x}{2}\right) + C =$ $\dfrac{1}{2}(x + \sin x \cos x) + C$	**12.** $\int \sec^3 x dx = \dfrac{1}{2}\sec x \tan x$ $+ \dfrac{1}{2}\log\|\sec x + \tan x\| + c$
13. $\int \sin^n x dx = -\dfrac{\sin^{n-1} x \cos x}{n} +$ $\dfrac{n-1}{n}\int \sin^{n-2} x dx$	**14.** $\int \cos^n x dx = -\dfrac{\cos^{n-1} x \sin x}{n} +$ $\dfrac{n-1}{n}\int \cos^{n-2} x dx$

Integration by Parts:

1. $\int (u.v)dx = u\left(\int v\ dx\right) - \int \left(\dfrac{du}{dx}\right)\left(\int v\ dx\right)dx$	**2.** $\int e^x\left[f(x) + f'(x)\right]dx = e^x f(x) + c$

Standard Integrals:

1. $\int \dfrac{1}{x^2 + a^2}dx = \dfrac{1}{a}\tan^{-1}\left(\dfrac{x}{a}\right) + c$
2. $\int \dfrac{1}{x^2 - a^2}dx = \dfrac{1}{2a}\log\left\|\dfrac{x-a}{x+a}\right\| + c$
3. $\int \dfrac{1}{a^2 - x^2}dx = \dfrac{1}{2a}\log\left\|\dfrac{a+x}{a-x}\right\| + c$
4. $\int \dfrac{1}{\sqrt{a^2 - x^2}}dx = \sin^{-1}\left(\dfrac{x}{a}\right) + c = -\cos^{-1}\left(\dfrac{x}{a}\right) + c$
5. $\int \dfrac{1}{\sqrt{x^2 + a^2}}dx = \sinh^{-1}\left(\dfrac{x}{a}\right) + c = \log\left(x + \sqrt{x^2 + a^2}\right) + c$
6. $\int \dfrac{1}{\sqrt{x^2 - a^2}}dx = \cosh^{-1}\left(\dfrac{x}{a}\right) + c = \log\left(x + \sqrt{x^2 - a^2}\right) + c$
7. $\int \sqrt{a^2 - x^2}dx = \dfrac{x}{2}\sqrt{a^2 - x^2} + \dfrac{a^2}{2}\sin^{-1}\dfrac{x}{a} + c$
8. $\int \sqrt{x^2 + a^2}dx = \dfrac{x}{2}\sqrt{x^2 + a^2} + \dfrac{a^2}{2}\log\left\|x + \sqrt{x^2 + a^2}\right\| + c$
9. $\int \sqrt{x^2 - a^2}dx = \dfrac{x}{2}\sqrt{x^2 - a^2} - \dfrac{a^2}{2}\log\left\|x + \sqrt{x^2 - a^2}\right\| + c$

10. $\int \dfrac{1}{x\sqrt{x^2 - a^2}}\,dx = \dfrac{1}{a}\sec^{-1}\dfrac{x}{a} + c$ (Valid for $x > a > 0$)

11. $\int e^{ax}\sin bx\ \ dx = \dfrac{e^{ax}}{a^2 + b^2}(a\sin bx - b\cos bx) + c = \dfrac{e^{ax}}{\sqrt{a^2 + b^2}}\sin\left\{bx - \tan^{-1}\left(\dfrac{b}{a}\right)\right\} + c$

12. $\int e^{ax}\cos bx\ \ dx = \dfrac{e^{ax}}{a^2 + b^2}(a\cos bx + b\sin bx) + c = \dfrac{e^{ax}}{\sqrt{a^2 + b^2}}\cos\left\{bx - \tan^{-1}\dfrac{b}{a}\right\} + c$

Solved Examples

JEE Main/Boards

Example 1: Evaluate : $\int\dfrac{x + \sin x}{1 + \cos x}dx$

Sol: Here by using the formula

$\sin x = 2\sin\dfrac{x}{2}\cos\dfrac{x}{2}$ and $1 + \cos x = 2\cos^2\dfrac{x}{2}$

we can solve the given problem.

$\int\dfrac{x + \sin x}{1 + \cos x}dx = \int\dfrac{x + 2\sin x/2\cos x/2}{2\cos^2 x/2}dx$

$= \int\dfrac{x}{2}\sec^2 x/2 + \tan\dfrac{x}{2}dx = x\tan x/2 + c$

Example 2: Evaluate : $\int\sqrt{x^2 + a^2}\,dx$

Sol: By applying integration by parts and taking $\sqrt{x^2 + a^2}$ as the first function we can solve the given problem.

$\int\sqrt{x^2 + a^2}\,dx =$

$= \sqrt{x^2 + a^2}\,x - \int\dfrac{(x^2 + a^2) - a^2}{\sqrt{x^2 + a^2}}dx\ \ \sqrt{x^2 + a^2}\,x - \int\dfrac{x^2}{\sqrt{x^2 + a^2}}dx$

$= \dfrac{x\sqrt{x^2 + a^2} + a^2\int\dfrac{dx}{\sqrt{x^2 + a^2}}}{2}$

Put $x = a\tan\theta$

$\int\dfrac{dx}{\sqrt{x^2 + a^2}} = \int\sec\theta\,d\theta = \log\left(\sec\theta + \tan\theta\right)$

$= \log\left(\dfrac{\sqrt{x^2 + a^2}}{a} + \dfrac{x}{a}\right)$

$\therefore I = \dfrac{1}{2}\left(x\sqrt{x^2 + a^2} + \log\left(\dfrac{\sqrt{x^2 + a^2}}{a} + \dfrac{x}{a}\right)\right) + c$

Example 3: Evaluate : $\int\tan^{-1}\sqrt{\dfrac{1 - \sin x}{1 + \sin x}}dx$

Sol: Here first write $\cos\left((\pi/2) - x\right)$ at the place of $\sin x$ then by using the formula $1 - \cos x = 2\sin^2\dfrac{x}{2}$

And $1 + \cos x = 2\cos^2\dfrac{x}{2}$ we can solve it.

$I = \int\tan^{-1}\sqrt{\dfrac{1 - \sin x}{1 + \sin x}}dx = \int\tan^{-1}\sqrt{\dfrac{1 - \cos\left((\pi/2) - x\right)}{1 + \cos\left((\pi/2) - x\right)}}dx$

$= \int\tan^{-1}\sqrt{\dfrac{2\sin^2\left((\pi/4) - (x/2)\right)}{2\cos^2\left((\pi/4) - (x/2)\right)}}dx$

$= \int\tan^{-1}\tan\left(\dfrac{\pi}{4} - \dfrac{x}{2}\right)dx = \int\left(\dfrac{\pi}{4} - \dfrac{x}{2}\right)dx = \dfrac{\pi}{4}x - \dfrac{x^2}{4} + C$

Example 4: Evaluate : $\int\log(2 + x^2)dx$

Sol: Here integrating by parts by taking $\log(2 + x^2)$ as the first function we can solve the given problem.

$$I = \int \log(2+x^2)dx = \int \log(2+x^2).1dx$$

Taking $\log(2+x^2)$ as first function and integrating by parts, we get

$$I = \left[\log(2+x^2)\right]x - \int x.\frac{2x}{2+x^2}dx$$

$$= x\log(2+x^2) - 2\int \frac{(x^2+2)-2}{2+x^2}dx$$

$$= x\log(2+x^2) - 2\int \left[1 - \frac{2}{x^2+2}\right]dx$$

$$= x\log(x^2+2) - 2\left[x - \sqrt{2}\tan^{-1}\left(\frac{x}{\sqrt{2}}\right)\right] + c$$

Example 5: Evaluate : $\int \dfrac{e^{2x} - e^{-2x}}{e^{2x} + e^{-2x}}dx$

Sol: Simply put $e^{2x} + e^{-2x} = t \Rightarrow (e^{2x} - e^{-2x})dx = \dfrac{dt}{2}$ and then solving we will get the result.

$$I = \int \frac{e^{2x} - e^{-2x}}{e^{2x} + e^{-2x}}dx$$

Put $e^{2x} + e^{-2x} = t \Rightarrow (e^{2x} - e^{-2x})dx = \dfrac{dt}{2}$

$$\therefore I = \frac{1}{2}\int \frac{dt}{t} = \frac{1}{2}\log|t| + C = \frac{1}{2}\log\left|e^{2x} + e^{-2x}\right| + c$$

Example 6: Evaluate : $\int \dfrac{x^3 - 1}{x^3 + x}dx$

Sol: By splitting the given integration as

$$\int \frac{x^3}{x(x^2+1)}dx - \int \frac{1}{x(x^2+1)}dx$$

We can solve the given problem.

$$\int \frac{x^3 - 1}{x^3 + x}dx = \int \frac{x^3}{x(x^2+1)}dx - \int \frac{1}{x(x^2+1)}dx$$

$$= \int \frac{x^2}{x^2+1}dx - \int \frac{1}{x} - \frac{x}{(x^2+1)}dx$$

$$= \int \left(1 - \frac{1}{x^2+1}\right)dx - \int \frac{1}{x}dx + \int \frac{x}{x^2+1}dx$$

$$= x - \tan^{-1}x - \log x + \log\sqrt{x^2+1} + c$$

Example 7: Evaluate: $\int \dfrac{dx}{(1+x^2)\sqrt{1-x^2}}$

Sol: By putting $x=\sin\theta \Rightarrow dx = \cos\theta\,d\theta$, we will reduce the given integration as

$$\int \frac{\sec^2\theta}{1+2\tan^2\theta}d\theta$$ and then put $t = \tan\theta \Rightarrow dt = \sec^2\theta\,d\theta$

and solve it.

Put $x=\sin\theta \Rightarrow dx = \cos\theta\,d\theta$

$$\Rightarrow I = \int \frac{dx}{(1+x^2)\sqrt{1-x^2}} = \int \frac{1}{1+\sin^2\theta}d\theta$$

$$= \int \frac{\sec^2\theta}{1+2\tan^2\theta}d\theta$$

Again put $t = \tan\theta \Rightarrow dt = \sec^2\theta\,d\theta$

$$I = \int \frac{1}{1+2t^2}dt = \frac{1}{\sqrt{2}}\int \frac{1}{t^2 + \left(1/\sqrt{2}\right)^2}dt$$

$$= \frac{1}{2}\left(\frac{1}{1/\sqrt{2}}\right)\tan^{-1}\left(\frac{t}{1/\sqrt{2}}\right) + c$$

$$= \frac{1}{2}\tan^{-1}(\sqrt{2}\tan\theta) + c = \frac{1}{2}\tan^{-1}\left(\frac{x\sqrt{2}}{\sqrt{1-x^2}}\right) + c$$

Example 8: Evaluate : $\int \dfrac{xdx}{(x-1)(x^2+4)}$

Sol: By partial fractions, we can reduce the given fraction as a sum of two fractions which will be easier to integrate.

$$\frac{x}{(x-1)(x^2+4)} = \frac{A}{x-1} + \frac{Bx+C}{x^2+4}$$

$x = 1 \Rightarrow A = 1/5$

$x = 2i \Rightarrow B = -1/5, C = 4/5$

$$\therefore I = \int \left(\frac{1}{5(x-1)} + \frac{4-2x}{5(x^2+4)}\right)dx$$

$$= \frac{1}{5}\log(x-1) - \frac{1}{10}\left[\frac{2x}{x^2+4} - \frac{8}{x^2+4}\right]$$

$$= \frac{1}{5}\log(x-1) - \frac{\log(x^2+4)}{10} + \frac{2}{5}\tan^{-1}\frac{x}{2}$$

Example 9: Evaluate: $\int \dfrac{\sin x}{\sin 4x} dx$

Sol: By using the formula $\sin 2x = 2\sin x.\cos x$, we can reduce the given fraction and then by putting $\sin x = t$ we can solve it.

$$\int \frac{\sin x}{\sin 4x} dx = \int \frac{\sin x\, dx}{2\cos 2x \sin 2x} = \int \frac{dx}{4\cos x \cos 2x}$$

$$= \int \frac{\cos x\, dx}{4(1-\sin^2 x)(1-2\sin^2 x)}$$

Put $\sin x = t$

$$\Rightarrow I = \frac{1}{4}\int \frac{dt}{(1-t^2)(1-2t^2)}$$

$$= \frac{1}{4}\int \left(\frac{1}{(t^2-1)} - \frac{2}{(2t^2-1)} \right) dt$$

$$= \frac{1}{4}\left(\frac{1}{2}\log\left|\frac{t-1}{t+1}\right| - \frac{1}{\sqrt{2}}\log\left|\frac{\sqrt{2}t-1}{\sqrt{2}t+1}\right| \right) + c$$

$$= \frac{1}{8}\log\left|\frac{\sin x - 1}{\sin x + 1}\right| + \frac{1}{4\sqrt{2}}\log\left|\frac{\sqrt{2}\sin x + 1}{\sqrt{2}\sin x - 1}\right| + c$$

Example 10: Evaluate : $\int \dfrac{dx}{(x)(x^4-1)}$

Sol: Here we can write the given integration as

$\int \dfrac{x^{-5}}{(1-x^{-4})} dx$ and then by putting $1-x^{-4} = t$

$\Rightarrow 4x^{-5} = dt$ we can solve it.

$$\int \frac{dx}{(x)(x^4-1)} = \int \frac{x^{-5}}{(1-x^{-4})} dx$$

Put $1-x^{-4} = t \Rightarrow 4x^{-5} = dt$

$$\Rightarrow = \int \frac{dt}{4t} = \frac{1}{4}\log|t| + c = \frac{1}{4}\log|1-x^{-4}| + c$$

JEE Advanced/Boards

Example 1: Evaluate : $\int x\sin^{-1} x\, dx$

Sol: Integrating by parts taking $\sin^{-1} x$ as the first term we can solve the given integration.

$\int x\sin^{-1} x\, dx$

Let $u = \sin^{-1} x, v = x$

$$= \frac{x^2}{2}\sin^{-1} x - \int \frac{1}{\sqrt{1-x^2}}.\frac{x^2}{2} dx = \frac{x^2}{2}\sin^{-1} x + \frac{1}{2}\int \frac{(1-x^2)-1}{\sqrt{1-x^2}} dx$$

$$= \frac{x^2}{2}\sin^{-1} x + \frac{1}{2}.\int \sqrt{1-x^2}\, dx - \frac{1}{2}\int \frac{1}{\sqrt{1-x^2}} dx$$

$$= \frac{x^2}{2}\sin^{-1} x + \frac{x\sqrt{1-x^2}}{4} - \frac{1}{4}\sin^{-1} x + c$$

$$= \frac{2x^2-1}{4}\sin^{-1} x + \frac{x}{4}\sqrt{1-x^2} + c$$

Example 2: Evaluate : $\int \dfrac{e^x(2-x^2)dx}{(1-x)\sqrt{1-x^2}}$

Sol: We can split the given fraction as

$\int e^x \left\{ \dfrac{1+x}{\sqrt{1-x^2}} + \dfrac{1}{(1-x)\sqrt{1-x^2}} \right\} dx$ and this integration is

in the form of $e^x(f(x) + f'(x))$

$$I = \int \frac{e^x(2-x^2)dx}{(1-x)\sqrt{1-x^2}} = \int e^x \frac{(1-x^2)+1}{(1-x)\sqrt{1-x^2}} dx$$

$$= \int e^x \left\{ \frac{1+x}{\sqrt{1-x^2}} + \frac{1}{(1-x)\sqrt{1-x^2}} \right\} dx$$

But $\dfrac{d}{dx}\left(\dfrac{1+x}{\sqrt{1-x^2}} \right) = \dfrac{1}{\sqrt{1-x^2}} + (1+x)\dfrac{x}{(1-x^2)^{3/2}}$

$$= \frac{1}{\sqrt{1-x^2}} + \frac{(1+x)x}{(1-x)(1+x)\sqrt{1-x^2}}$$

$$= \frac{1}{\sqrt{1-x^2}} + \frac{x}{(1-x)\sqrt{1-x^2}} = \frac{1-x+x}{(1-x)\sqrt{1-x^2}}$$

$$= \frac{1}{(1-x)\sqrt{1-x^2}}$$

Hence, the integrand is of type $e^x(f(x) + f'(x))$

$$\therefore I = e^x \frac{1+x}{\sqrt{1-x^2}} + C$$

Example 3: Evaluate : $\int \dfrac{\cos^3 x + \cos^5 x}{\sin^2 x + \sin^4 x} dx$

Sol: Here by taking $\cos^3 x$ and $\sin^2 x$ common from the numerator and denominator respectively and then by putting $\sin x = t$ we can solve the given problem.

$$I = \int \frac{\cos^3 x(1+\cos^2 x)}{\sin^2 x(1+\sin^2 x)} dx$$

Since power of cos x is odd, put sin x = t;

Then, cos x dx = dt

$$I = \int \frac{(1-t^2)(1+1-t^2)}{t^2(1+t^2)} dt = \int \frac{(1-t^2)(2-t^2)}{t^2(1+t^2)} dt$$

$$= \int \left(1 + \frac{2}{t^2} - \frac{6}{1+t^2}\right) dt = t - \frac{2}{t} - 6\tan^{-1} t + c$$

$$= \sin x - 2\csc x - 6\tan^{-1}(\sin x) + c$$

Example 4: Evaluate : $\int \frac{4e^x + 6e^{-x}}{9e^x - 4e^{-x}} dx$

Sol: By partial fractions we can reduce the given fraction as a sum of two fractions and then by integrating them we will get the result.

$$I = \int \frac{4e^x + 6e^{-x}}{9e^x - 4e^{-x}} dx$$

Let $4e^x + 6e^{-x} = A\left(9e^x - 4e^{-x}\right) + B\left(\frac{d}{dx}\left(9e^x - 4e^{-x}\right)\right)$

By comparing the coefficients of e^x and e^{-x}, we get

$$A = \frac{-19}{36}, B = \frac{35}{36}$$

$$\therefore I = \int \frac{A(9e^x - 4e^{-x}) + B(9e^x + 4e^{-x})}{9e^x - 4e^{-x}} dx$$

$$= A\int dx + B\int \frac{9e^x + 4e^{-x}}{9e^x - 4e^{-x}} dx = Ax + B\log\left|9e^x - 4e^{-x}\right| + c$$

$$= -\frac{19}{36}x + \frac{35}{36}\log\left|9e^x - 4e^{-x}\right| + c$$

Example 5: Evaluate : $\int \frac{1+x^2}{1-x^2} \frac{dx}{\sqrt{1+3x^2+x^4}}$ for x > 0

Sol: Dividing by x^2 in the numerator and denominator and then putting $x - 1/x = t$ we can solve the given problem.

$$I = \int \frac{1+x^2}{1-x^2} \frac{dx}{\sqrt{1+3x^2+x^4}} = \int \frac{\left((1/x^2)+1\right)dx}{\left((1/x)-x\right)\sqrt{\left(x-(1/x)\right)^2+5}}$$

Put $x - 1/x = t$; $\left(1 + \frac{1}{x^2}\right) dx = dt$

$$I = -\int \frac{dt}{t\sqrt{t^2+5}}$$

Put $t^2 + 5 = z^2 \Rightarrow 2t\, dt = 2z\, dz$

$$\Rightarrow I = -\int \frac{dt}{z^2-5} = -\frac{1}{2\sqrt{5}}\log\left|\frac{z-\sqrt{5}}{z+\sqrt{5}}\right| + c$$

$$= -\frac{1}{2\sqrt{5}}\log\left|\frac{\sqrt{(x^2+(1/x^2)+3)} - \sqrt{5}}{\sqrt{(x^2+(1/x^2)+3)} + \sqrt{5}}\right| + c$$

Example 6: Evaluate :

$$\int \csc^2 x.\log\left(\cos x + \sqrt{\cos 2x}\right)dx \text{ for } \sin x > 0$$

Sol: By substituting $\cos^2 x - \sin^2 x$ in place of $.\cos 2x$. we can reduce the given integration as the sum of two integrations and then by integrating them separately we will obtain the result.

$$\int \csc^2 x.\log\left(\cos x + \sqrt{\cos 2x}\right)dx$$

$$= \int \csc^2 x\log\left(\cos x + \sqrt{\cos^2 x - \sin^2 x}\right)dx$$

$$= \int \csc^2 x\log\left(\sin x(\cot x + \sqrt{\cot^2 x - 1})\right)dx$$

$$= \int \csc^2 x.\log\sin x\, dx + \int \csc^2 x.\log[\cot x + \sqrt{\cot^2 x - 1}]dx$$

$$= I_1 + I_2$$

$$I_1 = \int \csc^2 x.\log\sin x\, dx$$

$$= (-\cot x).\log\sin x - \int(-\cot x).\cot x\, dx$$

$$= -\cot x.\log\sin x + \int(\csc^2 x - 1)dx$$

$$= -\cot x.\log\sin x - \cot x - x + C_1$$

$$I_2 = \int \csc^2 x.\log[\cot x + \sqrt{\cot^2 x - 1}]dx$$

Put $\cot x = t$; $-\csc^2 x\, dx = dt$

$$I_2 = -\int \log[t + \sqrt{t^2 - 1}]dt$$

(Integrate by parts)

$$= -t.\log\left(t + \sqrt{t^2 - 1}\right) + \int t.\frac{1 + (t/\sqrt{t^2 - 1})}{t + \sqrt{t^2 - 1}}dt$$

$$= -t.\log\left(t + \sqrt{t^2 - 1}\right) + \int \frac{t}{\sqrt{t^2 - 1}}dt$$

$$= -t.\log\left(t + \sqrt{t^2 - 1}\right) + \sqrt{t^2 - 1} + C_2$$

$$= -\cot x\log\left(\cot x + \sqrt{\cot^2 x - 1}\right) + \sqrt{\cot^2 x - 1} + C_2$$

Example 7: Evaluate : $\int \frac{\sin x}{\sin^3 x + \cos^3 x}dx$

Sol: By taking $\cos^3 x$ common from the denominator and then by putting $\tan x = t$ we can solve it.

Integrand contains odd powers in sin x and cos x. So, put $\tan x = t$.

$$\Rightarrow I = \int \frac{1}{\cos^3 x}\frac{\sin x}{1+\tan^3 x}dx$$

$$= \int \frac{\tan x.\sec^2 x}{1+\tan^3 x}dx \text{ (put } \tan x = t)$$

$$= \int \frac{t}{1+t^3}dt \quad = -\frac{1}{3}\int \frac{dt}{1+t} + \frac{1}{3}\int \frac{t+1}{t^2-t+1}dt$$

$$= -\frac{1}{3}\log|t+1| + \frac{1}{6}\int \frac{(2t-1)+3}{t^2-t+1}dt$$

$$= -\frac{1}{3}\log|t+1| + \frac{1}{6}\log|t^2-t+1| +$$

$$\frac{1}{2}\frac{dt}{(t-(1/2))^2+(3/4)}$$

$$= -\frac{1}{3}\log|t+1| + \frac{1}{6}\log|t^2-t+1| +$$

$$\frac{1}{2}\frac{2}{\sqrt{3}}\tan^{-1}\left(\frac{t-(1/2)}{\sqrt{3}/2}\right) + C$$

$$= \frac{1}{6}\log\left|\frac{1-\tan x+\tan^2 x}{(1+\tan x)^2}\right| + \frac{1}{\sqrt{3}}\tan^{-1}\left(\frac{2\tan x-1}{\sqrt{3}}\right) + C$$

Example 8: If $I_{m,n} = \int \cos^m x.\cos nx\, dx$, show that $(m+n)I_{m,n} = \cos^m x\sin nx + mI_{m-1,n-1}$

Sol: By using integration by parts and by taking $\cos^m x$ as the first term we can prove the given equation.

Integrating by parts,

$$I_{m,n} = \cos^m x\frac{\sin nx}{n} +$$

$$\frac{m}{n}\int \cos^{m-1}x\sin x\sin nx\, dx \qquad \text{... (i)}$$

But $\cos(n-1)x=\cos(nx-x)$

$= \cos nx \cos nx + \sin nx \sin x$

$\Rightarrow \sin x \sin nx = \cos(n-1)x - \cos nx \cos x \qquad \text{... (ii)}$

From (i) and (ii):

$$I_{m,n} = \frac{1}{n}\cos^m x\sin nx +$$

$$\frac{m}{n}\int \cos^{m-1}x[\cos(n-1)x - \cos nx\cos x]dx$$

$$= \frac{1}{n}\cos^m x\sin nx + \frac{m}{n}I_{m-1,n-1} - \frac{m}{n}I_{m,n}$$

$$\Rightarrow (m+n)I_{m,n} = \cos^m x\sin nx + mI_{m-1,n-1}$$

Example 9: Evaluate : $I = \int\left(x+\sqrt{1+x^2}\right)^n dx$

Sol: Simply putting $x+\sqrt{1+x^2} = t$ and integrating we can solve the given problem.

Let $x+\sqrt{1+x^2} = t$, then

$$\left(1+\frac{x}{\sqrt{1+x^2}}\right)dx = dt \Rightarrow \frac{t}{\sqrt{1+x^2}}dx = dt$$

As $\sqrt{1+x^2}+x = t$

$$\frac{1}{t} = \frac{1}{\sqrt{1+x^2}+x} = \frac{\sqrt{1+x^2}-x}{1}$$

$$\Rightarrow 2\sqrt{1+x^2} = t+\frac{1}{t} = \frac{t^2+1}{t}$$

Thus $I = \int t^n\left(\frac{t^2+1}{2t}\right)\frac{dt}{t}$

$$= \frac{1}{2}\int t^{n-2}(t^2+1)dt = \frac{1}{2}\int(t^n+t^{n-2})dt = \frac{1}{2}\left(\frac{t^{n+1}}{n+1}+\frac{t^{n-1}}{n-1}\right)+C$$

Where $t = x+\sqrt{1+x^2}$

Example 10: Evaluate:

$$I = \int \frac{2\sin^3(x/2)dx}{(\cos(x/2))\sqrt{\cos^3 x+3\cos^2 x+\cos x}} \text{ for } \cos x > 0$$

Sol: Here we can reduce the given fraction by using the formula $\sin x = 2\sin\frac{x}{2}\cos\frac{x}{2}$ and then by putting $\cos x = t$ we can solve it.

$$I = \int \frac{(2\sin(x/2)\cos(x/2))(2\sin^2(x/2))dx}{(2\cos^2(x/2))\sqrt{\cos^3 x+3\cos^2 x+\cos x}}$$

$$= \int \frac{(1-\cos x)\sin x\ dx}{(1+\cos x)\sqrt{\cos^3 x+3\cos^2 x+\cos x}}$$

[Put $\cos x = t$]

$$\Rightarrow I = \int \frac{(t-1)dt}{(1+t)\sqrt{t^3+3t^2+t}}$$

$$= \int \frac{\left(t^2-1\right)dt}{(1+t)^2 t\sqrt{t+3+(1/t)}}$$

$$= \int \frac{t^2\left(1-(1/t^2)\right)dt}{t(t^2+2t+1)\sqrt{t+(1/t)+3}}$$

$$= \int \frac{\left(1-(1/t^2)\right)dt}{(t+(1/t)+2)\sqrt{(t+(1/t))+3}}$$

Put $t+\dfrac{1}{t}+3 = z^2 : z > 0$;

Then $\left(1-\dfrac{1}{t^2}\right)dt = 2zdz$

$$\Rightarrow I = \int \frac{2zdz}{(z^2-1)z} = 2\int \frac{dz}{z^2-1} = \log\left|\frac{z-1}{z+1}\right| + c$$

$$\Rightarrow I = \log\left|\frac{\sqrt{\cos x + \sec x + 3}-1}{\sqrt{\cos x + \sec x + 3}+1}\right| + c$$

Example 11: Evaluate:

$$I = \int \frac{(\sin x - \cos x)\ dx}{(\sin x + \cos x)\sqrt{\sin x \cos x + \sin^2 x \cos^2 x}}$$

Sol: Here by putting sin x + cos x = t we can integrate the given fraction using the appropriate formula.

Let $\sin x + \cos x = t$

$\Rightarrow (\cos x - \sin x)\ dx = dt$

Also, $t^2 = (\sin x + \cos x)^2 = 1 + 2\sin x \cos x$

$\therefore \sin x \cos x = \dfrac{t^2-1}{2}$

$$\Rightarrow I = -\int \frac{dt}{t\sqrt{\left((t^2-1)/2\right)\left(1+\left((t^2-1)/2\right)\right)}}$$

$$= -\int \frac{dt}{t\sqrt{((t^2-1)(t^2+1))/4}} = -2\int \frac{t^3 dt}{t^4\sqrt{t^4-1}}$$

Put $t^4 - 1 = z^2 : z > 0$

$$\Rightarrow I = -2\int \frac{1}{4}\frac{2z\,dz}{(z^2+1)z} = -\int \frac{dz}{1+z^2}$$

$$= -\tan^{-1} z + c = -\tan^{-1}\sqrt{t^4-1} + c$$

$$= -\tan^{-1}\sqrt{(1+\sin 2x)^2 - 1} + c$$

$$= -\tan^{-1}\sqrt{\sin^2 2x + 2\sin 2x} + c$$

Example 12: Evaluate : $I = \int \dfrac{dx}{1+\sqrt{x^2+x+2}}$

Sol: We can reduce the given fraction as

$$\int \frac{dx}{1+\sqrt{\left(x+(1/2)\right)^2+(7/4)}} \text{ and then by putting}$$

$x+\dfrac{1}{2} = \dfrac{\sqrt{7}}{2}\tan\theta : -\dfrac{\pi}{2} < \theta < \dfrac{\pi}{2}$; and using appropriate integration formula we can integrate the given fraction.

$$I = \int \frac{dx}{1+\sqrt{\left(x+(1/2)\right)^2+(7/4)}}$$

Put $x+\dfrac{1}{2} = \dfrac{\sqrt{7}}{2}\tan\theta : -\dfrac{\pi}{2} < \theta < \dfrac{\pi}{2}$; then

$$dx = \frac{\sqrt{7}}{2}\sec^2\theta\, d\theta$$

$$\Rightarrow I = \int \frac{\sqrt{7}}{2}\frac{\sec^2\theta d\theta}{1+(\sqrt{7}/2)\sec\theta}$$

$$= \frac{\sqrt{7}}{2}\int \frac{d\theta}{\cos\theta\left(\cos\theta+(\sqrt{7}/2)\right)}$$

$$= \int\left(\frac{1}{\cos\theta} - \frac{1}{\cos\theta+(\sqrt{7}/2)}\right)d\theta$$

$$= \log\left|\sec\theta + \tan\theta\right| - \int \frac{d\theta}{a+\cos\theta}\ ;\ a = \frac{\sqrt{7}}{2}$$

$$I = \log\left|\sec\theta + \tan\theta\right| - I_1 \qquad \text{....(i)}$$

Where, $I_1 = \int \dfrac{d\theta}{a+\cos\theta}$

Put $\tan\dfrac{\theta}{2} = t; \cos\theta = \dfrac{1-t^2}{1+t^2}$

$$I_1 = \int \frac{2dt}{1+t^2}\frac{1}{a+\left((1-t^2)/(1+t^2)\right)}$$

$$= 2\int \frac{dt}{a(1+t^2)+1-t^2}$$

$$= \frac{2}{a-1}\int \frac{dt}{\left((a+1)/(a-1)\right)+t^2}$$

$$= \frac{2}{a-1}\sqrt{\frac{a-1}{a+1}}\tan^{-1}\left(\sqrt{\frac{a-1}{a+1}}t\right) + c$$

$$= \frac{2}{\sqrt{a^2-1}}\tan^{-1}\left(\frac{\sqrt{a-1}}{a+1}\tan\frac{\theta}{2}\right) + c \qquad \text{....(ii)}$$

From (i) and (ii), we get I.

Exercise 1

Q.1 $\int \dfrac{\sec x}{\sec x + \tan x} dx$

Q.2 $\int \left(1 + \dfrac{1}{1+x^2} - \dfrac{2}{\sqrt{1-x^2}} + 5\dfrac{1}{|x|\sqrt{x^2-1}} + a^x\right) dx$

Q.3 $\int \tan^{-1}\left(\dfrac{\sin 2x}{1+\cos 2x}\right) dx : x \in \left(-\dfrac{\pi}{2}, \dfrac{\pi}{2}\right)$

Q.4 $\int \dfrac{1+\tan x}{x + \log \sec x} dx$

Q.5 $\int \dfrac{2\cos x - 3\sin x}{3\cos x + 2\sin x} dx$

Q.6 $\int \dfrac{2x-1}{\sqrt{x^2-x-1}} dx$

Q.7 $\int \dfrac{dx}{\sqrt{1-3x} - \sqrt{5-3x}}$

Q.8 $\int x^2 e^{x^3} \cos(e^{x^3}) dx$

Q.9 $\int \dfrac{\sec^2(2\tan^{-1} x)}{1+x^2} dx$

Q.10 $\int \dfrac{dx}{(2\sin x + 3\cos x)^2}$

Q.11 $\int \cos^{3/5} x \sin^3 x \, dx$

Q.12 $\int \dfrac{\log x}{x^2} dx$

Q.13 $\int \sin^{-1}\sqrt{\dfrac{x}{a+x}} dx : a > 0$

Q.14 $\int e^x \dfrac{2+\sin 2x}{1+\cos 2x} dx$

Q.15 $\int \dfrac{dx}{x\left[6(\log x)^2 + 7\log x + 2\right]}$

Q.16 $\int \dfrac{x^2+1}{(x+3)(x-1)^2} dx$

Q.17 $\int \dfrac{1}{1-\tan x} dx$

Q.18 If $f'(x) = x - \dfrac{1}{x^2}$ and $f(1) = \dfrac{1}{2}$, find $f(x)$.

Q.19 For any natural number evaluate m

$\int \left(x^{3m} + x^{2m} + x^m\right)\left(2x^{2m} + 3x^m + 6\right)^{1/m} dx, x > 0$

Q.20 $\int \dfrac{x^3+3x+2}{(x^2+1)^2(x+1)} dx$

Q.21 $\int \dfrac{dx}{\sin x + \sec x}$

Q.22 $\int \dfrac{\sqrt{\cos 2x}}{\sin x} cx : \cos x > 0$

Q.23 $\int \dfrac{\sqrt{x^2+1}(\log(x^2+1) - 2\log x)}{x^4} dx$

Q.24 $\int \dfrac{\sin x}{\sin x - \cos x} dx$

Q.25 $\int x\sin^{-1}\left(\dfrac{1}{2}\dfrac{\sqrt{2a-x}}{a}\right) dx$

Q.26 $\int \sec^4 x \csc^2 x \, dx$

Q.27 $\int \dfrac{d\theta}{(a+b\cos\theta)^2} : a > b > 0$

Q.28 Evaluate $\int \dfrac{dx}{x\sqrt{x^4-1}}$

Q.29 $\int \dfrac{dx}{\sqrt{\sin^3 x \sin(x+\alpha)}}$

Q.30 $\int \dfrac{dx}{(1+\sqrt{x})\sqrt{x-x^2}}$

Q.31 $\int \dfrac{\cos 8x - \cos 7x}{1+2\cos 5x} dx$

Q.32 $\int \dfrac{x^3+1}{x(x-1)^3} dx$

Q.33 $\int \dfrac{dx}{\sqrt[3]{\sin^{11}x \cdot \cos x}}$

Q.34 Evaluate $\int \dfrac{e^x}{\sqrt{e^{2x}-4}}dx$

Q.35 Evaluate $\int \dfrac{\log x}{(1+x)^3}dx$

Q.36 Evaluate $\int \dfrac{f(x)}{x^3-1}dx$, where f(x) is polynomial of the second degree in x such that f(0) = f(1) = 3f(2) = −3

Exercise 2

Single Correct Choice Type

Q.1 $\int \left(\dfrac{x}{1+x^5}\right)^{3/2} dx$ equals-

(A) $\dfrac{2}{5}\sqrt{\dfrac{x^5}{1+x^5}}+c$ (B) $\dfrac{2}{5}\sqrt{\dfrac{x}{1+x^5}}+c$

(C) $\dfrac{2}{5}\dfrac{1}{\sqrt{1+x^5}}+c$ (D) None of these

Q.2 $\int \dfrac{\cos^8 x - \sin^8 x}{1-2\sin^2 x\cos^2 x}dx$ equals-

(A) $-\dfrac{\sin 2x}{2}+c$ (B) $\dfrac{\sin 2x}{2}+c$

(C) $\dfrac{\cos 2x}{3}+c$ (D) $-\dfrac{\cos 2x}{1}+c$

Q.3 Identify the correct expression

(A) $x\int \log x\, dx = x^2\log -x^2 + c$

(B) $x\int \log|x|dx = xe^x + c$

(C) $x\int e^x dx = xe^x + cx$

(D) $\int \dfrac{dx}{\sqrt{a^2+x^2}} = \dfrac{1}{a}\tan^{-1}\left[\dfrac{x}{a}\right]+c$

Q.4 Primitive of $\sqrt{1+2\tan x(\sec x + \tan x)}$ w.r.t. x is

(A) $\log(\sec x + \tan x) + \log\cos x + c$

(B) $\log(\sec x + \tan x) + \log\sec x + c$

(C) $2\log\left(\sec\dfrac{x}{2}+\tan\dfrac{x}{2}\right)+c$

(D) $\log\sqrt{1+\sin x} + c$

Q.5 $\int \dfrac{\sin 2x}{\sin^4 x + \cos^4 x}dx$ is equal to

(A) $\cot^{-1}(\cot^2 x) + c$ (B) $-\cot^{-1}(\tan^2 x) + c$

(C) $\tan^{-1}(\tan^2 x) + c$ (D) $-\tan^{-1}(\cos 2x) + c$

Q.6 The value of integral $\int \dfrac{d\theta}{\cos^3\theta\sqrt{\sin 2\theta}}$ can be expressed as irrational function of $\tan\theta$ as

(A) $\dfrac{\sqrt{2}}{5}\left(\sqrt{\tan^2\theta + 5}\right)\tan\theta + c$

(B) $\dfrac{2}{5}\left(\tan^2\theta + 5\right)\sqrt{\tan\theta} + c$

(C) $\dfrac{\sqrt{2}}{5}\left(\tan^2\theta + 5\right)\sqrt{\tan\theta} + c$

(D) $\dfrac{\sqrt{2}}{5}\left(\tan^2\theta + 5\right)^{\sqrt{\tan\theta}} + c$

Q.7 $\int \dfrac{dx}{a+bx^2}$ a,b≠0 and a/b > 0

(A) $\dfrac{1}{\sqrt{ab}}\tan^{-1}x\sqrt{\dfrac{b}{a}}+c$ (B) $\sqrt{\dfrac{b}{a}}\tan^{-1}x\sqrt{\dfrac{b}{a}}+c$

(C) $\sqrt{\dfrac{a}{b}}\tan^{-1}x\sqrt{\dfrac{a}{b}}+c$ (D) $\sqrt{ab}\tan^{-1}x\sqrt{\dfrac{b}{a}}+c$

Q.8 $\int \dfrac{1-x^7}{x(1+x^7)}dx$ equals

(A) $\log x + \dfrac{2}{7}\log(1+x^7)+c$ (B) $\log x - \dfrac{2}{4}\log(1-x^7)+c$

(C) $\log x - \dfrac{2}{7}\log(1+x^7)+c$ (D) $\log x + \dfrac{2}{4}\log(1-x^7)+c$

Q.9 $\int \dfrac{\log|x|}{x\sqrt{1+\log|x|}}dx$ equal

(A) $\dfrac{2}{3}\sqrt{1+\log|x|}(\log|x|-2)+c$

(B) $\dfrac{2}{3}\sqrt{1+\log|x|}(\log|x|+2)+c$

(C) $\frac{1}{3}\sqrt{1+\log|x|}(\log|x|-2)+c$

(D) $2\sqrt{1+\log|x|}(3\log|x|-2)+c$

Q.10 If $\int\dfrac{x^4+1}{x(x^2+1)^2}dx = A\log|x|+\dfrac{B}{1+x^2}+c$,

Where c is the constant of integration then:

(A) A=1; B=-1   (B) A=-1; B=1

(C) A=1; B=1   (D) A=-1; B=-1

Q.11 $\int 4\sin x\cos\dfrac{x}{2}\cos\dfrac{3x}{2}dx$ equals

(A) $\cos x - \dfrac{1}{2}\cos 2x + \dfrac{1}{3}\cos 3x + c$

(B) $\cos x - \dfrac{1}{2}\cos 2x - \dfrac{1}{3}\cos 3x + c$

(C) $\cos x + \dfrac{1}{2}\cos 2x + \dfrac{1}{3}\cos 3x + c$

(D) $\cos x + \dfrac{1}{2}\cos 2x - \dfrac{1}{3}\cos 3x + c$

Q.12 $\int \sin x.\cos x.\cos 2x.\cos 4x.\cos 8x.\cos 16x\,dx$ equals

(A) $\dfrac{\sin 16x}{1024}+c$ (B) $-\dfrac{\cos 32x}{1024}+c$

(C) $\dfrac{\cos 32x}{1096}+c$ (D) $-\dfrac{\cos 32x}{1096}+c$

Q.13 If $\int f(x)\,dx = F(x)$, then $\int x^3 f(x^2)dx$ is equal to

(A) $\dfrac{1}{2}\left((F(x))^2 - \int(f(x))^2\,dx\right)$ (B) $\dfrac{1}{2}\left(x^2F(x^2)- \int(f(x^2)\,d(x^2)\right)$

(C) $\dfrac{1}{2}\left(F(x)-\dfrac{1}{2}\int(F(x^2)\,dx\right)$ (D) None of these

Q.14 If $\int\dfrac{4e^x+6e^{-x}}{9e^x-4e^{-x}}dx = Ax + B\log(9e^{2x}-4)+C$ then

A, B and C are

(A) $A=\dfrac{3}{2}, B=\dfrac{36}{35}, C=\dfrac{3}{2}\log 3$ +constant

(B) $A=\dfrac{3}{2}, B=\dfrac{35}{36}, C=\dfrac{-3}{2}\log 3$ +constant

(C) $A=-\dfrac{3}{2}, B=\dfrac{35}{36}, C=\dfrac{3}{2}1\log 3$ +constant

(D) None of these

Q.15 $\int\dfrac{\sqrt{\cot x}-\sqrt{\tan x}}{\sqrt{2}(\cos x+\sin x)}dx$ equals

(A) $\sec^{-1}(\sin x + \cos x)+c$

(B) $\sec^{-1}(\sin x - \cos x)+ c$

(C) $\log[(\sin x+\cos x)+ \sqrt{\sin 2x}]+c$

(D) $\log[(\sin x-\cos x)+ \sqrt{\sin 2x}]+c$

Previous Years' Questions

Q.1 The value of $\int\dfrac{(x^2-1)dx}{x^3\sqrt{2x^4-2x^2+1}}$ *(2006)*

(A) $2\sqrt{2-\dfrac{2}{x^2}+\dfrac{1}{x^4}}+c$ (B) $2\sqrt{2+\dfrac{2}{x^2}+\dfrac{1}{x^4}}+c$

(C) $\dfrac{1}{2}\sqrt{2-\dfrac{2}{x^2}+\dfrac{1}{x^4}}+c$ (D) None of these

Q.2 If $\int\dfrac{4e^x+6e^{-x}}{9e^x-4e^{-x}}dx = Ax + B\log(9e^{2x}-4)-c$, then

A=........., B=...... .. and C=........   *(1989)*

Q.3 Integrate $\dfrac{1}{1-\cot x}$ or $\dfrac{\sin x}{\sin x-\cos x}$ *(1978)*

Q.4 Integrate the curve $\dfrac{x}{1+x^4}$ *(1978)*

Q.5 Integrate

$\sin x\cdot\sin 2x\cdot\sin 3x+\sec^2 x\cdot\cos^2 2x+\sin^4 x\cdot\cos^4 x.$ *(1979)*

Q.6 Integrate $\dfrac{x^2}{(a+bx)^2}$ *(1979)*

Q.7 Evaluate $\int(\sqrt{\tan x} + \sqrt{\cot x})dx$.   *(1988)*

Q.8 Evaluate $\int\dfrac{(x+1)}{x(1+xe^x)^2}dx$.   *(1996)*

Q.9 Integrate the following   *(1997)*

$\int\left(\dfrac{1-\sqrt{x}}{1+\sqrt{x}}\right)^{1/2}\dfrac{dx}{x}$

Q.10 The value of $\sqrt{2}\,\dfrac{\sin x\,dx}{\sin\left(x-\dfrac{\pi}{4}\right)}$ is **(2008)**

(A) $x+\log\left|\cos\left(x-\dfrac{\pi}{4}\right)\right|+c$ (B) $x-\log\left|\sin\left(x-\dfrac{\pi}{4}\right)\right|+c$

(C) $x+\log\left|\sin\left(x-\dfrac{\pi}{4}\right)\right|+c$ (D) $x-\log\left|\sin\cos\left(x-\dfrac{\pi}{4}\right)\right|+c$

Q.11 If $\dfrac{dy}{dx}=y+3>0$ and $y(0)=2$, then $y(\log 2)$ is equal to **(2011)**

(A) 5   (B) 13   (C) -2   (D) 7

Q.12 If the integral

$$\int\frac{5\tan x}{\tan x-2}\,dx=x+a\log|\sin x-2\cos x|+k$$

then a is equal to **(2012)**

(A) -1   (B) -2   (C) 1   (D) 2

Q.13 If $\int f(x)\,dx=\psi(x)$, then $\int x^5 f(x^3)\,dx$ is equal to **(2013)**

(A) $\dfrac{1}{3}\left[x^3\psi(x^3)-\int x^2\psi(x^3)\,dx\right]+c$

(B) $\dfrac{1}{3}x^3\psi(x^3)-\int x^2\psi(x^3)\,dx+c$

(C) $\dfrac{1}{3}x^3\psi(x^3)-\int x^2\psi(x^3)\,dx+c$

(D) $\dfrac{1}{3}\left[x^3\psi(x^3)-\int x^2\psi(x^3)\,dx\right]+c$

Q.14 The integral $\left(1+x-\dfrac{1}{x}\right)e^{x+\frac{1}{x}}\,dx$ is equal to **(2014)**

(A) $(x+1)^{x+\frac{1}{x}}+c$ (B) $-xe^{x+\frac{1}{x}}+c$

(C) $(x-1)^{x+\frac{1}{x}}+c$ (D) $xe^{x+\frac{1}{x}}+c$

Q.15 The integral $\displaystyle\int\frac{dx}{x^2\left(x^4+1\right)^{3/4}}$ equals **(2015)**

(A) $\left(\dfrac{x^4+1}{x^4}\right)^{1/4}+c$ (B) $\left(x^4+1\right)^{1/4}+c$

(C) $-\left(x^4+1\right)^{1/4}+c$ (D) $-\left(\dfrac{x^4+1}{x^4}\right)^{1/4}+c$

Q.16 The integral $\displaystyle\int\frac{2x^{12}+5x^9}{\left(x^5+x^3+1\right)^3}\,dx$ is equal to **(2016)**

(A) $\dfrac{x^{10}}{2\left(x^5+x^3+1\right)^2}+c$ (B) $\dfrac{x^5}{2\left(x^5+x^3+1\right)^2}+c$

(C) $\dfrac{-x^{10}}{2\left(x^5+x^3+1\right)^2}+c$ (D) $\dfrac{-x^5}{2\left(x^5+x^3+1\right)^2}+c$

JEE Advanced/Boards

Exercise 1

Q.1 (i) $\displaystyle\int\frac{dx}{\cot(x/2)\cdot\cot(9x/3)\cdot\cot(x/6)}$;

(ii) $\displaystyle\int\frac{\tan(\log x)\,\tan(\log(x/2))\,\tan(\log 2)}{x}\,dx$

Q.2 $\displaystyle\int\frac{dx}{(x-\alpha)\sqrt{(x-\alpha)(x-\beta)}}$

Q.3 $\displaystyle\int\frac{\log\left(\log((1+x)/(1-x))\right)}{1-x^2}\,dx$

Q.4 $\displaystyle\int\left[\left(\frac{x}{e}\right)^x+\left(\frac{e}{x}\right)^x\right]\log x\,dx$

Q.5 $\displaystyle\int\sqrt{\frac{\cos(x-a)}{\sin(x+a)}}\,dx$

Q.6 $\displaystyle\int\frac{x^5+3x^4-x^3+8x^2-x+8}{x^2+1}\,dx$

Q.7 $\displaystyle\int\frac{(\sqrt{x}+1)\,dx}{\sqrt{x}\,(\sqrt[3]{x}+1)}$

Q.8 $\int \sin^{-1} \sqrt{\dfrac{x}{a+x}}\, dx$

Q.9 $\int \dfrac{x \ln x}{(x^2-1)^{3/2}}\, dx$

Q.10 $\int \dfrac{\log 6 \sqrt[6]{((\sin x)6^{\cos x})}\cos x}{\sin x}\, dx$

Q.11 $\int \left[\dfrac{\sqrt{x^2+1}\left[\log(x^2+1)-2\log x\right]}{x^4} \right] dx$

Q.12 If $f(x)=\dfrac{\sin x}{\sin^2 x+4\cos^2 x}$ the antiderivative of

$\left(\dfrac{1}{\sqrt{3}}\right)\tan^{-1}\left(\left(\dfrac{1}{\sqrt{3}}\right)g(x)\right)+c$ is then g (x) is equal to

Q.13 $\int \dfrac{\cot x\, dx}{(1-\sin x)(\sec x+1)}$

Q.14 $\int \dfrac{3x^2+1}{(x^2-1)^3}\, dx$

Q.15 $\int \dfrac{(ax^2-b)\,dx}{x\sqrt{c^2x^2-(ax^2+b)^2}}$

Q.16 Evaluate $\int \dfrac{e^{m\,\tan^{-1}x}}{(1+x^2)^{3/2}}\, dx$

Q.17 Evaluate $\int \sin^{-1}\sqrt{\dfrac{x}{a+x}}\, dx$

Q.18

$\int \dfrac{(e^{\sqrt{x}}-e^{-\sqrt{x}})\cos\left(e^{\sqrt{x}}+e^{-\sqrt{x}}+\dfrac{\pi}{4}\right)+\left(e^{\sqrt{x}}+e^{-\sqrt{x}}\right)\cos\left(e^{\sqrt{x}}-e^{-\sqrt{x}}-\dfrac{\pi}{4}\right)}{\sqrt{x}}\, dx$

Q.19 $\int \dfrac{x^2+x}{(e^x+x+1)^2}\, dx$

Q.20 $\int \dfrac{e^{\cos x}(x\sin^3 x+\cos x)}{\sin^2 x}\, dx$

Q.21 $\int \dfrac{5x^4+4x^5}{(x^5+x+1)^2}\, dx$

Q.22 $\int \dfrac{a^2\sin^2 x+b^2\cos^2 x}{a^4\sin^2 x+b^4\cos^2 x}\, dx$

Q.23 $\int \dfrac{\cos^2 x}{1+\tan x}\, cx$

Q.24 $\int \log x.\sin^{-1}x\, dx$

Q.25 Evaluate $\int \dfrac{(x^2+1)e^x}{(x+1)^2}\, dx$

Q.26 Evaluate $\int \dfrac{e^{\sin z}}{\cos^2 x}(x\cos^3 x-\sin x)dx$

Q.27 Evaluate $\int \dfrac{dx}{\sqrt{1-2x-x^2}}$

Q.28 $\int \dfrac{dx}{\sec x+\csc x}$

Q.29 Evaluate $\int \sqrt{2x^2+3x+4}\,dx$

Q.30 Evaluate $\int \dfrac{dx}{x(x^n+1)}$

Q.31 $\int \dfrac{\cos x-\sin x}{7-9\sin 2x}\, dx$

Q.32 $\int \dfrac{\sqrt{\cot x}-\sqrt{\tan x}}{1+3\sin 2x}\, dx$

Q.33 $\int \dfrac{4x^5-7x^4+8x^3-2x^2+4x-7}{x^2(x^2+1)^2}\, dx$

Q.34 $\int \dfrac{dx}{\cos^3 x-\sin^3 x}$

Q.35 $\int \dfrac{x^2}{(x\cos x-\sin x)(x\sin x+\cos x)}\, dx$

Q.36 $\int \cos 2\theta.\ln\dfrac{\cos\theta+\sin\theta}{\cos\theta-\sin\theta}\, d\theta$

Q.37 Match the columns

Column I	Column II
(A) $\int \dfrac{x^4-1}{x^2\sqrt{x^4+x^2+1}}\, dx$	(p) $\log\left(\dfrac{(x^2+1)+\sqrt{x^4+1}}{x}\right)+c$
(B) $\int \dfrac{x^2-1}{x\sqrt{1+x^4}}\, dx$	(q) $c-\dfrac{1}{2}\log\left(\dfrac{\sqrt{x^4+1}-2\sqrt{x}}{(x^2-1)}\right)$
(C) $\int \dfrac{1+x^2}{(1-x^2)\sqrt{1+x^4}}\, dx$	(r) $c-\tan^{-1}\left(\sqrt{\sqrt{1+\dfrac{1}{x^4}}-1}\right)$

Column I	Column II
(D) $\int \dfrac{1}{(1+x^4)\sqrt{\sqrt{1+x^4-x^2}}}\,dx$	(s) $\dfrac{\sqrt{x^4+x^2+1}}{x}+c$

Exercise 2

Single Correct Choice Type

Q.1 If $\int \dfrac{\tan^{-1}x - \cot^{-1}x}{\tan^{-1}x + \cot^{-1}x}\,dx$ is equal to :

(A) $\dfrac{4}{\pi}x\tan^{-1}x + \dfrac{2}{\pi}\log(1+x^2) - x + c$

(B) $\dfrac{4}{\pi}x\tan^{-1}x - \dfrac{2}{\pi}\log(1+x^2) + x + c$

(C) $\dfrac{4}{\pi}x\tan^{-1}x + \dfrac{2}{\pi}\log(1+x^2) + x + c$

(D) $\dfrac{4}{\pi}x\tan^{-1}x - \dfrac{2}{\pi}\log(1+x^2) - x + c$

Q.2 $\int \dfrac{x^2-4}{x^4+24x^2+16}\,dx$ equals

(A) $\dfrac{1}{4}\tan^{-1}\left(\dfrac{x^2+4}{4x}\right)+c$

(B) $-\dfrac{1}{4}\cot^{-1}\left(\dfrac{x^2+4}{x}\right)+c$

(C) $-\dfrac{1}{4}\cot^{-1}\left(\dfrac{4x^2+4}{x}\right)+c$

(D) $\dfrac{1}{4}\cot^{-1}\left(\dfrac{x^2+4}{x}\right)+c$

Q.3 $\int \dfrac{(x-1)^2}{x^4+2x^2+1}\,dx$ equals

(A) $\dfrac{x^3}{3}+x+\dfrac{x}{x^2+1}+c$

(B) $\dfrac{x^5+x^3+x+3}{3(x^2+1)}+c$

(C) $\dfrac{x^5+4x^3+3x+3}{3(x^2+1)}+c$

(D) None of these

Q.4 $\int \dfrac{x^4-4}{x^2\sqrt{4+x^2+x^4}}\,dx$ equals

(A) $\dfrac{\sqrt{4+x^2+x^4}}{x}+c$

(B) $\sqrt{4+x^2+x^4}+c$

(C) $\dfrac{\sqrt{4+x^2+x^4}}{2}+c$

(D) $\dfrac{\sqrt{4+x^2+x^4}}{2x}+c$

Q.5 $\int \dfrac{\sec x + \tan x - 1}{\tan x - \sec x + 1}\,dx$ equals

(A) $\log[(\sec x + \tan x)] + \log \sec x + c$

(B) $\log[\sec x - \tan x] - \log \cos x + c$

(C) $\log(\sec x + \tan x) - \log \sec x + c$

(D) $-\log(\sec x + \tan x) + \log \cos x + c$

Q.6 $\int e^x \dfrac{(x^2-3)}{(x+3)^2}\,dx$

(A) $e^x \cdot \dfrac{x}{x+3} + c$

(B) $e^x\left(2 - \dfrac{6}{x+3}\right) + c$

(C) $e^x\left(1 - \dfrac{6}{x+3}\right) + c$

(D) $e^x \dfrac{3}{x+3} + c$

Q.7 $\int \sqrt{\dfrac{1-\cos x}{\cos \alpha - \cos x}}\,dx$ where $0 < \alpha < x < \pi$, equals

(A) $2\log\left(\cos\dfrac{\alpha}{2} - \cos\dfrac{x}{2}\right) + c$

(B) $2\cos^{-1}\left(\dfrac{\cos(x/2)}{\cos(\alpha/2)}\right) + c$

(C) $2\sqrt{2}\log\left(\cos\dfrac{\alpha}{2} - \cos\dfrac{x}{2}\right) + c$

(D) $-2\sin^{-1}\left(\dfrac{\cos(x/2)}{\cos(\alpha/2)}\right) + c$

Q.8 Primitive of $\dfrac{3x^4-1}{(x^4+x+1)^2}$ w.r.t. x is :

(A) $\dfrac{x}{x^4+x+1} + c$

(B) $-\dfrac{x}{x^4+x+1} + c$

(C) $\dfrac{x+1}{x^4+x+1} + c$

(D) $-\dfrac{x+1}{x^4+x+1} + c$

Q.9 If $\int e^{3x} \cos 4x\,dx = e^{3x}(A\sin 4x + B\cos 4x) + c$ Then

(A) $4A=3b$

(B) $2A=3B$

(C) $3A=4B$

(D) $4B+3A=1$

Q.10 The evaluation of $\int \dfrac{px^{p+2q-1} - qx^{q-1}}{x^{2p+2q} + 2x^{p+q} + 1}\,dx$ is

(A) $-\dfrac{x^p}{x^{p+q}+1} + c$

(B) $\dfrac{x^q}{x^{p+q}+1} + c$

(C) $-\dfrac{x^q}{x^{p+q}+1} + c$

(D) $\dfrac{x^p}{x^{p+q}+1} + c$

Q.11 If $\int f(x)dx = g(x)$, then $\int f^{-1}(x)dx$ is equal to –

(A) $g^{-1}(x)$ (B) $xf^{-1}(x) - g(f^{-1}(x))$

(C) $xf^{-1}(x)g^{-1}(x)$ (D) $f^{-1}(x)$

Q.12 Primitive of $\sqrt[3]{\dfrac{x}{(x^4-1)^4}}$ w.r.t x is –

(A) $\dfrac{3}{4}\left(1 + \dfrac{1}{x^4-1}\right)^{\frac{1}{3}} + c$ (B) $-\dfrac{3}{4}\left(1 + \dfrac{1}{x^4-1}\right)^{\frac{1}{3}} + c$

(C) $\dfrac{4}{3}\left(1 + \dfrac{1}{x^4-1}\right)^{\frac{1}{3}} + c$ (D) $-\dfrac{4}{3}\left(1 + \dfrac{1}{x^4-1}\right)^{\frac{1}{3}} + c$

Q.13 If $\int e^u . \sin 2x\, dx$ can be found in terms of known functions of x then u can be :

(A) x   (B) sin x   (C) cos x   (D) cos 2x

Previous Years' Questions

Q.1 The value of the integral $\int \dfrac{\cos^3 x + \cos^5 x}{\sin^2 x + \sin^4 x}dx$ is   *(1995)*

(A) $\sin x - 6\tan^{-1}(\sin x) + c$

(B) $\sin x - 2(\sin x)^{-1} + c$

(C) $\sin x - 2(\sin x)^{-1} - 6\tan^{-1}(\sin x) + c$

(D) $\sin x - 2(\sin x)^{-1} + 5\tan^{-1}(\sin x) + c$

Q.2 Let $f(x) = \dfrac{x}{(1+x^n)^{1/n}}$ for $n \geq 2$ and

$g(x) = \underbrace{f o f \ldots. o f}_{f \text{ occurs n times}}(x)$. Then, $\int x^{n-2}g(x)dx$ equals *(2007)*

(A) $\dfrac{1}{n(n-1)}(1+nx^n)^{1-\frac{1}{n}} + c$

(B) $\dfrac{1}{n-1}(1+nx^n)^{1-\frac{1}{n}} + c$

(C) $\dfrac{1}{n(n+1)}(1+nx^n)^{1+\frac{1}{n}} + c$

(D) $\dfrac{1}{n+1}(1+nx^n)^{1-\frac{1}{n}} + c$

Q.3 Let $I = \int \dfrac{e^x}{e^{4x}+e^{2x}+1}dx$, $J = \int \dfrac{e^{-x}}{e^{-4x}+e^{-2x}-1}dx$

Then, for an arbitrary constant c, the value of J-I equals *(2008)*

(A) $\dfrac{1}{2}\log\left|\dfrac{e^{4x}-e^{2x}+1}{e^{4x}+e^{2x}+1}\right| + c$ (B) $\dfrac{1}{2}\log\left|\dfrac{e^{2x}+e^x+1}{e^{2x}-e^x+1}\right| + c$

(C) $\dfrac{1}{2}\log\left|\dfrac{e^{2x}-e^x+1}{e^{2x}+e^x+1}\right| + c$ (D) $\dfrac{1}{2}\log\left|\dfrac{e^{4x}+e^{2x}+1}{e^{4x}-e2^x+1}\right| + c$

Q.4 $f(x)$ is the integral of

$\dfrac{2\sin x - \sin 2x}{x^3}, x \neq 0$ find $\lim\limits_{x\to 0} f'(x)$.   *(1979)*

Q.5 Evaluate the following integrals   *(1980)*

(i) $\int \sqrt{1 + \sin\left(\dfrac{1}{2}x\right)}dx$ (ii) $\int \dfrac{x^2}{\sqrt{1-x}}dx$

Q.6 Evaluate $\int (e^{\log x} + \sin x)\cos x\, dx$.   *(1981)*

Q.7 Evaluate $\int \dfrac{(x-1)e^x}{(x+1)^3}dx$.   *(1983)*

Q.8 Evaluate $\int \dfrac{dx}{x^2(x^4+1)^{3/4}}$.   *(1984)*

Q.9 $\int \sqrt{\dfrac{1-\sqrt{x}}{1+\sqrt{x}}}dx$ *(1985)*

Q.10 Evaluate $\int \dfrac{\sin^{-1}\sqrt{x} - \cos^{-1}\sqrt{x}}{\sin^{-1}\sqrt{x} + \cos^{-1}\sqrt{x}}dx$.   *(1986)*

Q.11 Evaluate $\int \dfrac{(\cos 2x)^{1/2}}{\sin x}dx$.   *(1987)*

Q.12 Find the indefinite integral

$\int \left(\dfrac{1}{\sqrt[3]{x} + \sqrt[4]{x}} + \dfrac{\log(1 + \sqrt[6]{x})}{\sqrt[3]{x} + \sqrt{x}}\right)dx$ *(1992)*

Q.13 Integrate $\int \dfrac{x^3 + 3x + 2}{(x^2+1)^2(x+1)}dx$ *(1999)*

Q.14 Evaluate $\int \sin^{-1}\left(\dfrac{2x+2}{\sqrt{4x^2+8x+13}}\right)dx$.   *(2000)*

Q.15 For any natural number m, evaluate   *(2002)*

$\int(x^{3m}+x^{2m}+x^m)(2x^{2m}+3x^m+6x^m)^{1/m}dx, x>0$.

Q.16 Let $I = \int \dfrac{e^x}{e^{4x}+e^{2x}+1}dx$, $J = \int \dfrac{e^{-x}}{e^{-4x}+e^{-2x}+1}dx$.

Then, for an arbitrary constant C, the value of $J-I$ equals   *(2008)*

(A) $\dfrac{1}{2}\log\left(\dfrac{e^{4x}-e^{2x}+1}{e^{4x}+e^{2x}+1}\right)+c$ (B) $\dfrac{1}{2}\log\left(\dfrac{e^{2x}+e^x+1}{e^{2x}-e^x+1}\right)+c$

(C) $\dfrac{1}{2}\log\left(\dfrac{e^x-e^{-x}-1}{e^x+e^{-x}+1}\right)+c$ (D) $\dfrac{1}{2}\log\left(\dfrac{e^{4x}+e^{2x}+1}{e^{4x}-e^x+1}\right)+c$

Q.17 The integral $\int \dfrac{\sec^2 x}{(\sec x + \tan x)^{9/2}}dx$ equals

(for some arbitrary constant K)   *(2012)*

(A) $-\dfrac{1}{(\sec x + \tan x)^{11/2}}\left\{\dfrac{1}{11}-\dfrac{1}{7}(\sec x+\tan x)^2\right\}+K$

(B) $\dfrac{1}{(\sec x + \tan x)^{11/2}}\left\{\dfrac{1}{11}-\dfrac{1}{7}(\sec x+\tan x)^2\right\}+K$

(C) $-\dfrac{1}{(\sec x + \tan x)^{11/2}}\left\{\dfrac{1}{11}+\dfrac{1}{7}(\sec x+\tan x)^2\right\}+K$

(D) $\dfrac{1}{(\sec x + \tan x)^{11/2}}\left\{\dfrac{1}{11}+\dfrac{1}{7}(\sec x+\tan x)^2\right\}+K$

Important Questions

JEE Main/Boards

Exercise 1

Q.14	Q.19	Q.23	Q.25
Q.29	Q.32	Q.35	Q.38

Exercise 2

Q.4	Q.10	Q.13	Q.15

Previous Years' Questions

Q.5	Q.7	Q.9

JEE Advanced/Boards

Exercise 1

Q.3	Q.11	Q.18	Q.20
Q.30	Q.37	Q.38	

Exercise 2

Q.1	Q.5	Q.7	Q.10
Q.12			

Previous Years' Questions

Q.1	Q.2	Q.4	Q.10
Q.12			

Answer Key

JEE Main/Boards

Exercise 1

Q.1 $\tan x - \sec x + A$

Q.2 $x + \tan^{-1} x - 2\sin^{-1} x + 5\sec^{-1} x + \dfrac{a^x}{\log a} + A$

Q.3 $\dfrac{x^2}{2} + A$

Q.4 $\log\left|x + \log(\sec x)\right| + A$

Q.5 $\log\left|3\cos x + 2\sin x\right| + A$

Q.6 $2\sqrt{x^2 - x - 1} + A$

Q.7 $\dfrac{1}{18}\left[(1 - 3x)^{3/2} + (5 - 3x)^{3/2}\right] + A$

Q.8 $\dfrac{1}{3}\sin(e^{x^3}) + A$

Q.9 $\dfrac{x}{1 + x^2} + c$

Q.10 $-\dfrac{1}{2(2\tan x + 3)} + A$

Q.11 $-\dfrac{5}{8}\cos^{8/5} x + \dfrac{5}{18}\cos^{18/5} x + A$

Q.12 $\dfrac{-\log x}{x} - \dfrac{1}{x} + A$

Q.13 $\left(\tan^{-1}\sqrt{\dfrac{x}{a}}\right)(a + x) - a\sqrt{\dfrac{x}{a}} + c$

Q.14 $e^x \tan x + A$

Q.15 $\log\left|2\log x + 1\right| - \log\left|3\log x + 2\right| + A$

Q.16 $\dfrac{5}{8}\log|x + 3| + \dfrac{3}{8}\log|x - 1| - \dfrac{1}{2(x - 1)} + A$

Q.17 $\dfrac{1}{2}x - \dfrac{1}{2}\log\left|\cos x - \sin x\right| + A$

Q.18 $f(x) = \dfrac{x^2}{2} + \dfrac{1}{x} - 1$

Q.19 $\dfrac{1}{6(m + 1)}(2x^{3m} + 3x^{2m} + 6x^m)^{m+1/m}$

Q.20 $\dfrac{1}{4}\log(x^2 + 1) - \dfrac{1}{2}\log(x + 1) + \dfrac{3}{2}\tan^{-1} x + \dfrac{x}{x^2 + 1}$

Q.21 $\dfrac{1}{2\sqrt{3}}\log\left|\dfrac{\sqrt{3} + p}{\sqrt{3} - p}\right| + \tan^{-1}(q) + c$

Q.22 $-\log\left|\cot x + \sqrt{\cot^2 x - 1}\right| + \dfrac{1}{\sqrt{2}}\log\left|\dfrac{\sqrt{2} + \sqrt{1 - \tan^2 x}}{\sqrt{2} - \sqrt{1 - \tan^2 x}}\right| + c$

Q.23 $-\dfrac{1}{3}\left(1 + \dfrac{1}{x^2}\right)^{3/2}\left[\log\left(1 + \dfrac{1}{x^2}\right) - \dfrac{2}{3}\right] + c$

Q.24 $\dfrac{1}{2}x + \dfrac{1}{2}\log\left|\sin x - \cos x\right| + c$

Q.25 $\dfrac{x^2}{2}\sin^{-1}\left(\dfrac{1}{2}\sqrt{\dfrac{2a - x}{a}}\right) + \dfrac{a^2}{2}\left(\sin^{-1}\dfrac{x}{2a} - \dfrac{x}{2a}\sqrt{1 - \dfrac{x^2}{4a^2}}\right) + c$

Q.26 $\dfrac{1}{3}\tan^3 x + 2\tan x - \cot x + c$

Q.27 $\dfrac{2}{\sqrt{a^2 - b^2}}\tan^{-1}\left[\sqrt{\dfrac{a - b}{a + b}}\tan(\theta/2)\right] + c_1$

Q.28 $\int \dfrac{1}{2}d\theta = \dfrac{1}{2}\theta + c = \dfrac{1}{2}\sec^{-1}(x^2) + c$

Q.29 $\dfrac{-2}{\sin \alpha}\sqrt{\cos \alpha + \cot x \ \sin \alpha} + c$

Q.30 $2\left(\sqrt{\dfrac{x}{1 - x}} - \dfrac{1}{\sqrt{1 - x}}\right) + c$

Q.31 $\dfrac{1}{2}\int(\cos 3x - \cos 2x)dx = \dfrac{1}{3}\sin 3x - \dfrac{1}{2}\sin 2x + c$

Q.32 $-\log|x| + 2\log|x - 1| + \dfrac{1}{x - 1} - \dfrac{1}{(x - 1)^2} + c$

Q.33 $2\log t - \dfrac{3}{2}\log(2t-1) - \dfrac{3}{2}\dfrac{1}{(2t-1)} + c$, where $t = x + \sqrt{x^2 - x + 1}$

Q.34 $\log(e^x + \sqrt{e^{2x} - 4}) - \log 2 + c = \log(e^x + \sqrt{e^{2x} - 4}) + c'$

Q.35 $I = -\dfrac{1}{2}\dfrac{\log x}{(1+x)^2} + \dfrac{1}{2}\log\dfrac{x}{1+x} + \dfrac{1}{2(1+x)} + c$

Q.36 $\log\dfrac{x^2 + x + 1}{|x-1|} + \dfrac{2}{\sqrt{3}}\tan^{-1}\dfrac{2x+1}{\sqrt{3}} + c$

Exercise 2

Single Correct Choice Type

Q.1 A	**Q.2** B	**Q.3** C	**Q.4** B	**Q.5** C	**Q.6** C
Q.7 A	**Q.8** C	**Q.9** A	**Q.10** C	**Q.11** B	**Q.12** B
Q.13 B	**Q.14** D	**Q.15** A			

Previous Years' Questions

Q.1 C   **Q.2** $A = -\dfrac{3}{2}$, $B = \dfrac{35}{36}$ and $C \in R$ **Q.3** $\dfrac{1}{2}\log(\sin x - \cos x) + \dfrac{x}{2} + c$

Q.4 $\dfrac{1}{2}\tan^{-1}(x^2) + c$ **Q.5** $-\dfrac{\cos 4x}{16} - \dfrac{\cos 2x}{8} + \dfrac{\cos 6x}{24} + \sin 2x + \tan x - 2x + \dfrac{3x}{128} - \dfrac{\sin 4x}{128} + \dfrac{\sin 8x}{1024}$

Q.6 $\dfrac{1}{b^3}\left(a + bx - 2a\log(a+bx) - \dfrac{a^2}{a+bx} + c\right)$ **Q.7** $\sqrt{2}\tan^{-1}\left(\dfrac{\sqrt{\tan x} - \sqrt{\cot x}}{\sqrt{2}}\right) + c$

Q.8 $\displaystyle\int\dfrac{dt}{t^2(t-1)} = \log\left|\dfrac{xe^x}{1+xe^x}\right| + \dfrac{1}{1+xe^x} + c$ **Q.9** $2[\cos^{-1}\sqrt{x} - \log|1 + \sqrt{1-x}| - \dfrac{1}{2}\log|x|] + c$

Q.10 C	**Q.11** D	**Q.12** D	**Q.13** C	**Q.14** D	**Q.15** D

Q.16 A

JEE Advanced/Boards

Exercise 1

Q.1 (i) $2\log\left(\sec\dfrac{x}{2}\right) - 3\log\left(\sec\dfrac{x}{3}\right) - 6\log\left(\sec\dfrac{x}{6}\right) + + c$ (ii) $\log\left(\dfrac{\sec(\log x)}{\sec\left(\log\dfrac{x}{2}\right)}\right) - x\tan\log|2| + c$

Q.2 $\dfrac{-2}{\alpha-\beta}\cdot\dfrac{\sqrt{x-\beta}}{x-\alpha}+c$

Q.3 $\log\left(\dfrac{1+x}{1-x}\right)\log\left(\dfrac{1+x}{1-x}\right)\log\left(\dfrac{1+x}{1-x}\right)+c$

Q.4 $\left(\dfrac{x}{e}\right)^x-\left(\dfrac{e}{x}\right)^x+c$

Q.5 $\cos a.\arccos\left(\dfrac{\cos x}{\cos a}\right)-\sin a.\log\left(\sin x+\sqrt{\sin^2 x-\sin^2 a}\right)+c$

Q.6 $\dfrac{x^4}{4}+x^3-x^2+5x+\dfrac{1}{2}\log(x^2+1)+3\tan^{-1}x+c$
$+\,c$ where $t=x^{1/6}$

Q.7 $6\left[\dfrac{t^4}{4}-\dfrac{t^2}{2}+t+\dfrac{1}{2}\log(1+t^2)-\tan^{-1}t\right]+C$ where $t=x^{1/6}$

Q.8 $(a+x)\tan^{-1}\left[\dfrac{\sqrt{x}}{a}-\sqrt{ax}\right]+c$

Q.9 $\sec^{-1}x-\dfrac{\log x}{\sqrt{x^2-1}}+c$

Q.10 $\dfrac{1}{6}\left[\dfrac{\log^2(\sin x)}{\log 36}+\log\left[\tan\dfrac{x}{2}+\cos x\right]\right]+c$

Q.11 $-\dfrac{1}{3}\left[1+\dfrac{1}{x^2}\right]^{3/2}\left[\log\left(1+\dfrac{1}{x^2}\right)-\dfrac{2}{3}\right]+c$

Q.12 $\int f(x)\,dx=\int\dfrac{dt}{t^2+3}=\dfrac{1}{\sqrt{3}}\tan^{-1}\dfrac{\sec x}{\sqrt{3}}+c$

Q.13 $\dfrac{1}{2}\log\left|\tan\dfrac{x}{2}\right|+\dfrac{1}{4}\tan^2\dfrac{x}{2}+\tan\dfrac{x}{2}+c$

Q.14 $-\dfrac{x}{(x^2-1)^2}+C$

Q.15 $\sin^{-1}\left(\dfrac{ax^2+b}{cx}\right)+k$

Q.16 $1=\dfrac{e^{m\tan^{-1}x}}{m^2+1}\left(m\cdot\dfrac{1}{\sqrt{x^2+1}}+\dfrac{x}{\sqrt{x^2+1}}\right)-c=\dfrac{e^{m\tan^{-1}x}(m+x)}{(m^2+1)\sqrt{x^2+1}}+c$

Q.17 $a\left[\tan^{-1}\sqrt{\dfrac{x}{a}}\left(\dfrac{a+x}{a}\right)-\sqrt{\dfrac{x}{a}}\right]+c$

Q.18 $\sqrt{2}\left(\cos\left(e^{-\sqrt{x}}\right)\right)\left(\sin(e\sqrt{x})+\cos\left(e^{\sqrt{x}}\right)\right)+c$

Q.19 $-\log\left(\dfrac{x+1}{e^x}+1\right)-\dfrac{1}{\left(\dfrac{x+1}{e^x}+1\right)}+C$

Q.20 $-e^{\cos x}(x+\cosec x)+c$

Q.21 $C-\dfrac{x+1}{x^5+x+1}$ or $c+\dfrac{x^5}{x^x+x+1}$

Q.22 $\dfrac{1}{a^2+b^2}\left(x+\tan^{-1}\left(\dfrac{a^2\tan x}{b^2}\right)\right)+c$

Q.23 $\dfrac{1}{4}\log(\cos x+\sin x)+\dfrac{x}{2}+\dfrac{1}{8}(\sin 2x+\cos 2x)+c$

Q.24 $(x\log x-x)\sin^{-1}x+\sqrt{1-x^2}\log x-\log\left(\dfrac{1-\sqrt{1-x^2}}{x}\right)+c$

Q.25 $e^x f(x)+c=e^x\left(\dfrac{x-1}{x+1}\right)+c$

Q.26 $e^z f(z)+c=e^z\left(\sin^{-1}z-\dfrac{1}{\sqrt{1-z^2}}\right)+c=e^{\sin x}(x-\sec x)+c$

Q.27 $\sin^{-1}\left(\dfrac{1+x}{\sqrt{2}}\right)+c$

Q.28 $\dfrac{1}{2}\left[\sin x-\cos x-\dfrac{1}{\sqrt{2}}\log\tan\left(\dfrac{x}{2}+\dfrac{\pi}{8}\right)\right]+Cc$

Q.29 $I=\dfrac{4x+3}{8}\sqrt{2x^2+3x+4}+\dfrac{23}{16\sqrt{2}}\log\left[\dfrac{4x+3}{4}+\dfrac{\sqrt{2x^2+3x+4}}{\sqrt{2}}\right]+c$

Q.30 $\frac{1}{n}\log\left|\frac{z-1}{z}\right|+c = \frac{1}{n}\log\left|\frac{x^n}{x^n+1}\right|+c$

Q.31 $\frac{1}{24}\log\frac{(4+3\sin x+3\cos x)}{4-3\sin x-3\cos x)}+c$

Q.32 $\tan^{-1}\left(\frac{\sqrt{2}\sin 2x}{\sin x+\cos x}\right)+c$

Q.33 $4\log x+\frac{7}{x}+6\tan^{-1}(x)+\frac{6x}{1+x^2}+c$

Q.34 $\frac{2}{3}\tan^{-1}(\sin x+\cos x)+\frac{1}{3\sqrt{2}}\log\left|\frac{\sqrt{2}+\sin x+\cos x}{\sqrt{2}-\sin x-\cos x}\right|+c$

Q.35 $\log\left|\frac{x\sin x+\cos x}{x\cos x-\sin x}\right|+c$

Q.36 $\frac{1}{2}(\sin 2\theta)\log\left(\frac{\cos\theta+\sin\theta}{\cos\theta-\sin\theta}\right)-\frac{1}{2}\log(\sec 2\theta)+c = \frac{2}{\cos 2\theta}+c$

Q.37 $A \to s;\ B \to p;\ C \to q;\ D \to r$

Exercise 2

Q.1 D   **Q.2** A   **Q.3** D   **Q.4** A   **Q.5** A   **Q.6** C   **Q.7** B

Q.8 B   **Q.9** C   **Q.10** C   **Q.11** B   **Q.12** B   **Q.13** A

Previous Years' Questions

Q.1 C   **Q.2** A   **Q.3** C   **Q.4** 1

Q.5 (i) $4\sin\frac{x}{4}-4\cos\frac{x}{4}+c$ (ii) $-2\left\{\sqrt{1-x}-\frac{2}{2}(1-x)^{3/2}+\frac{1}{5}(1-x)^{5/2}\right\}+c$

Q.6 $x\sin x+\cos x-\frac{\cos 2x}{4}+c$

Q.7 $\frac{e^x}{(x+1)^2}+c$

Q.8 $-\frac{(x^4+1)^{1/4}}{x}+c$

Q.9 $-2\sqrt{1-x}+\cos^{-1}\sqrt{x}+\sqrt{x(1-x)}+c$

Q.10 $\frac{2}{\pi}[\sqrt{x-x^2}-(1-2x)\sin^{-1}\sqrt{x}]-x+c$

Q.11 $-\log|\cot x+\sqrt{\cot^2 x-1}|+\frac{1}{\sqrt{2}}\log\left|\frac{\sqrt{2}+\sqrt{1-\tan^2 x}}{\sqrt{2}-\sqrt{1-\tan^{2x}}}\right|+c$

Q.12 $\frac{3}{2}x^{2/3}-\frac{12}{7}x^{7/12}-\frac{4}{3}x^{1/2}-\frac{12}{5}x^{5/12}+\frac{1}{2}x^{1/3}-4x^{1/4}-7x^{1/6}-12x^{1/12}+(2x^{1/2}-3x^{1/3}$
$-6x^{1/6}+11)\log(1+x^{1/6})+12\log(1+x^{1/2})-3[\log(1+^{1/6})]^2+c$

Q.13 $-\frac{1}{2}\log|x+1|+\frac{1}{4}\log|x^2+1|+\frac{3}{2}\tan^{-1}x+\frac{x}{x^2+1}+c$

Q.14 $(x+1)\tan^{-1}\left(\frac{2x+2}{3}\right)-\frac{3}{4}\log(4x^2+8x+13)+c$

Q.15 $\frac{1}{6(m+1)}\cdot(2x^{3m}+3x^{2m}+6x^m)^{(m+1)/m}+c$

Q.16 C   **Q.17** C

JEE Main/Boards

Exercise 1

Sol 1: $\int \dfrac{\sec x}{\sec x + \tan x}dx = \int \dfrac{\sec x(\sec x - \tan x)}{\sec^2 x - \tan^2 x}dx$

$= \int (\sec^2 x - \sec x \tan x)dx = \tan x - \sec x + C$

Sol 2: $\int \left(1 + \dfrac{1}{1+x^2} - \dfrac{2}{\sqrt{1-x^2}} + 5\dfrac{1}{|x|\sqrt{x^2-1}} + a^x\right)dx$

$= \int dx + \int \dfrac{dx}{1+x^2} - \int \dfrac{2dx}{\sqrt{1-x^2}} + 5\int \dfrac{dx}{|x|\sqrt{x^2-1}} + \int a^x dx$

$= x + \tan^{-1}x - 2\sin^{-1}x + 5\sec^{-1}x + \dfrac{a^x}{\log a} + c$

Sol 3: $\dfrac{\sin 2x}{1 + \cos 2x} = \tan x$

$\int \tan^{-1}\left(\dfrac{\sin 2x}{1+\cos 2x}\right)dx = \int \tan^{-1}\tan x dx$

$= \int x dx = \dfrac{x^2}{2} + c$

Sol 4: $\int \dfrac{1 + \tan x}{x + \log \sec x}dx$

Put $x + \log \sec x = t$

$\Rightarrow \left(1 + \dfrac{1}{\sec x}\sec x \tan x\right)dx = dt$

$\therefore \int \dfrac{dt}{t} = \log t + c = \log(x + \log \sec x) + c$

Sol 5: Put $3\cos x + 2\sin x = t$

$-3\sin x + 2\cos x = \dfrac{dt}{dx}$

$\therefore I = \int \dfrac{dt}{t} = \log t = \log(3\cos x + 2\sin x)$

Sol 6: $\int \dfrac{2x-1}{\sqrt{x^2 - x - 1}}dx$

Put $x^2 - x - 1 = t$

$\Rightarrow (2x - 1)dx = dt$

$\int \dfrac{dt}{\sqrt{t}} = \int t^{-\frac{1}{2}}dt = 2\sqrt{t} + c = 2\sqrt{x^2 - x - 1} + c$

Sol 7: $\int \dfrac{dx}{\sqrt{1-3x} - \sqrt{5-3x}}$

$= \int \dfrac{\sqrt{1-3x} + \sqrt{5-3x}}{(-4)}dx$

$= -\dfrac{1}{4}\int \sqrt{1-3x}dx - \dfrac{1}{4}\int \sqrt{5-3x}dx$

$= \dfrac{1}{12} \times \dfrac{2}{3}\left(\sqrt{1-3x}\right)^{3/2} + \dfrac{1}{12} \times \dfrac{2}{3}\left(\sqrt{5-3x}\right)^{3/2}$

$= \dfrac{1}{18}\left[(1-3x)^{3/2} + (5-3x)^{3/2}\right] + c$

Sol 8: $\int x^2 e^{x^3}\cos(e^{x^3})dx$

$e^{x^3} = t \Rightarrow 3x^2 e^{x^3}dx = dt$

$\therefore \dfrac{1}{3}\int \cos t dt = \dfrac{1}{3}\sin t + c = \dfrac{1}{3}\sin e^{x^3} + c$

Sol 9: $\int \dfrac{\sec^2(2\tan^{-1}x)}{1+x^2}dx$

$2\tan^{-1}x = t \Rightarrow \dfrac{2}{1+x^2}dx = dt$

$\dfrac{1}{2}\int \sec^2 t dt = \dfrac{1}{2}\tan t + c = \dfrac{1}{2}\tan(2\tan^{-1}x) + c$

$= \dfrac{1}{2}\dfrac{2\tan(\tan^{-1}x)}{1+\tan^2(\tan^{-1}x)} + c = \dfrac{x}{1+x^2} + c$

Sol 10: $\int \dfrac{dx}{(2\sin x + 3\cos x)^2} = \int \dfrac{\sec^2 x}{(2\tan x + 3)}dx$

Put $2\tan x + 3 = t \Rightarrow 2\sec^2 x dx = dt$

$\therefore \dfrac{1}{2}\int \dfrac{dt}{t^2} = -\dfrac{1}{2t} + c = -\dfrac{1}{2} \times \dfrac{1}{(2\tan x + 3)} + c$

Sol 11: $\int \cos^{3/5}x \sin^3 x dx$

$= \int \cos^{3/5}x(1 - \cos^2 x)\sin x dx$

$$= \int \left(\cos^{3/5} x - \cos^{13/5} x \right) \sin x\, dx$$

Put $\cos x = t; -\sin x\, dx = dt$

$$= \int \left(t^{13/5} - t^{3/5} \right) dt \ = \ \frac{5}{18} t^{18/5} - \frac{5}{8} t^{8/5} + c$$

$$= \frac{5}{18} (\cos x)^{18/5} - \frac{5}{8} (\cos x)^{8/5} + c$$

Sol 12: $\int \dfrac{\log x}{x^2} dx$

$$\log x \int \frac{1}{x^2} dx - \int \left(\frac{d\log x}{dx} \int \frac{1}{x^2} dx \right) dx$$

$$= (\log x) \left(-\frac{1}{x} \right) - \int \frac{1}{x} \times \left(-\frac{1}{x} \right) dx + c$$

$$= -\frac{1}{x} \log x + \int \frac{1}{x^2} dx + c \ = \ -\frac{1}{x} (\log x + 1) + c$$

Sol 13: $\int \sin^{-1} \sqrt{\dfrac{x}{a+x}} dx$, $a > 0$

$x = a\tan^2\theta \Rightarrow dx = 2a\tan\theta\sec^2\theta\, d\theta$

$$\int \sin^{-1} \sqrt{\frac{a\tan^2\theta}{a\sec^2\theta}} \ 2a\tan\theta\sec^2\theta\, d\theta$$

$$= 2a \int (\sin^{-1}\sin\theta)\tan\theta\sec^2\theta\ d\theta$$

$\tan\theta = t \Rightarrow \sec^2\theta\, d\theta = dt$

$$= 2a \int t\tan^{-1} t\, dt$$

$$= 2a \left[\tan^{-1} t \int t\, dt - \frac{1}{2} \int \frac{1}{1+t^2} \times t^2 dt \right]$$

$$= 2a \left[(\tan^{-1} t)\frac{t^2}{2} - \frac{1}{2} \left[t - \tan^{-1} t \right] \right] + c$$

$$= at^2 \tan^{-1} t + \left[\frac{1}{2} \tan^{-1} t - \frac{1}{2} t \right] 2a + c$$

$\therefore t = \tan\theta = \dfrac{\sqrt{x}}{\sqrt{a}}$

$$\therefore I = a\frac{x}{a}\tan^{-1}\sqrt{\frac{x}{a}} + \left[\frac{1}{2}\tan^{-1}\sqrt{\frac{x}{a}} - \frac{1}{2}\sqrt{\frac{x}{a}} \right] 2a + c$$

$$= x\tan^{-1}\sqrt{\frac{x}{a}} + \left[\frac{1}{2}\tan^{-1}\sqrt{\frac{x}{a}} - \frac{1}{2}\sqrt{\frac{x}{a}} \right] 2a + c$$

$$= x\tan^{-1}\sqrt{\frac{x}{a}} + a\tan^{-1}\sqrt{\frac{x}{a}} - a\sqrt{\frac{x}{a}} + c$$

$$= \left(\tan^{-1}\sqrt{\frac{x}{a}} \right)(a+x) - a\sqrt{\frac{x}{a}} + c$$

Sol 14: $\int e^x \left(\dfrac{2 + \sin 2x}{1 + \cos 2x} \right) dx$

$1 + \cos 2x = 2\cos^2 x$

$2 + \sin 2x = 2 + 2\sin x \cos x$

$$\Rightarrow \int e^x \left(\frac{2 + 2\sin x \cos x}{2\cos^2 x} \right) = \int e^x (\sec^2 x + \tan x) dx$$

$\tan x = t \Rightarrow \sec^2 x\, dx = dt$

$\therefore$ This is of form $\int e^x (f(x) + f'(x)) dx \ = \ e^x f(x)$

$\therefore I = e^x \tan x + c$

Sol 15: $\int \dfrac{dx}{x[6(\log x)^2 + 7\log x + 2]}$

$\log x = t; \dfrac{1}{x} dx = dt$

$$\int \frac{dt}{(6t^2 + 7t + 2)} \ = \ \int \frac{dt}{(6t^2 + 3t + 4t + 2)}$$

$$= \int \frac{dt}{(3t+2)(2t+1)} \ = -\int \left[\frac{3}{(3t+2)} - \frac{2}{(2t+1)} \right] dt$$

$$= -3 \int \frac{1}{3t+2} dt + 2 \int \frac{1}{2t+1} dt$$

$$= -3 \frac{1}{3} \log(3t+2) + \frac{2}{2} \log(2t+1) + c$$

$$= \frac{2}{2} \left(\frac{2t+1}{3t+2} \right) + c \ = \ \frac{2}{2} \log \left(\frac{2\log x + 1}{3\log x + 2} \right) + c$$

$$= \log |\, 2\log x + 1\,| - \log |\, 3\log x + 2\,| + c$$

Sol 16: $\int \dfrac{x^2 + 1}{(x+3)(x-1)^2} dx$

$$I = \int \frac{5}{8} \times \frac{1}{(x+3)} dx + \int \frac{3}{8(x-1)} dx + \frac{1}{2} \int \frac{1}{(x-1)^2} dx$$

$$= \frac{5}{8} \log(x+3) + \frac{3}{8}\log(x-1) - \frac{1}{2(x-1)} + c$$

Sol 17: $\int \dfrac{1}{(1 - \tan x)} dx$

Put $\tan x = t \Rightarrow \sec^2 x\, dx = dt$

or $dx = \dfrac{dt}{1 + \tan^2 x} = \dfrac{dt}{1 + t^2}$

$$\therefore I = \int \frac{1}{(1-t)(1+t^2)}dt$$

$$= \frac{1}{2}\int \frac{1}{(1-t)}dt + \frac{1}{2}\int \frac{t+1}{t^2+1}dt$$

$$= -\frac{1}{2}\log(1-t) + \frac{1}{4}\log(1+t^2) + \frac{1}{2}\tan^{-1}t + c$$

$$= \frac{1}{2}\log\left(\frac{1}{1-\tan x}\right) + \frac{1}{2}\log\sqrt{1+\tan^2 x} + \frac{x}{2} + c$$

$$= \frac{1}{2}\log\left(\frac{\sec x}{1-\tan x}\right) + \frac{x}{2} + c$$

$$= \frac{x}{2} - \frac{1}{2}\log|\cos x - \sin x| + c$$

Sol 18: $f'(x) = x - \dfrac{1}{x^2}$

$$f(x) = \int f'(x)dx = \int\left(x - \frac{1}{x^2}\right)dx = \frac{x^2}{2} + \frac{1}{x} + c$$

$$f(1) = \frac{1}{2} + 1 + c = \frac{1}{2} \Rightarrow c = -1$$

$$\therefore f(x) = \frac{x^2}{2} + \frac{1}{x} - 1$$

Sol 19: $\int (x^{3m} + x^{2m} + x^m)(2x^{2m} + 3x^m + 6)^{1/m}dx$

Put $x^m = t$ and integrate.

Sol 20: $\int \dfrac{x^3 + 3x + 2}{(x^2+1)^2(x+1)}dx$

$x^3 + 3x + 2 = x^3 + x + 2x + 2$

$= x(x^2+1) + 2(x+1)$

$$I = \int \frac{x(x^2+1) + 2(x+1)}{(x^2+1)^2(x+1)}dx$$

$$= \int \frac{x}{(x^2+1)(x+1)} + 2\int \frac{1}{(1+x^2)^2}dx$$

$$= \int \frac{(x+1)-1}{(x^2+1)(x+1)}dx + 2\int \left(\frac{1}{1+x^2}\right)^2 dx$$

$$= 2\int \frac{1}{(1+x^2)^2}dx - \int \frac{dx}{(x^2+1)(x+1)} + \int \frac{1}{(1+x^2)}dx$$

Put $x = \tan\theta$

$dx = \sec^2\theta d\theta$

$$= 2\int \frac{\sec^2\theta}{\sec^4\theta}d\theta + \tan^{-1}x + \frac{1}{2}\int\left(\frac{(x-1)}{(1+x^2)} - \frac{1}{x+1}\right)dx$$

$$= \frac{2}{2}\int(\cos 2\theta + 1)d\theta + \tan^{-1}x$$

$$= \frac{1}{2}\left[\frac{1}{2}\tan^{-1}(1-x^2) - \tan^{-1}x - \log(x+1)\right] + c$$

$$= \frac{1}{4}\tan^{-1}(1+x^2) - \frac{1}{2}\tan^{-1}x - \frac{1}{2}\log(x+1) + c$$

Sol 21: $I = \int \dfrac{1}{\sin x + \sec x}dx$

$$\Rightarrow \int \frac{2\cos x}{2 + 2\sin x\cos x}dx = \int \frac{(\cos x + \sin x) + (\cos x - \sin x)}{2 + \sin 2x}dx$$

$$\Rightarrow \int \frac{\cos x + \sin x}{2 + 2\sin x}dx + \int \frac{\cos x - \sin x}{2 + 2\sin 2x}dx$$

$$\Rightarrow \int \frac{1}{2 + [1 - (\sin x + \cos x)^2]} \times d(\sin x - \cos x)$$

$$+ \int \frac{1}{2 + [(\sin x + \cos x)^2 - 1]} \times d(\sin x + \cos x)$$

$$\Rightarrow \int \frac{1}{(\sqrt{3})^2 - (\sin x - \cos x)^2} \times d(\sin x - \cos x)$$

$$+ \int \frac{1}{(1)^2 + (\sin x + \cos x)^2} \times d(\sin x + \cos x)$$

$$\Rightarrow \int \frac{1}{(\sqrt{3})^2 - p^2}dp + \int \frac{1}{1^2 + q^2}dq$$

$$\Rightarrow \frac{1}{2\sqrt{3}}\log\left|\frac{\sqrt{3}+p}{\sqrt{3}-p}\right| + \tan^{-1}(q) + c$$

Where $p = \sin x - \cos x$ & $q = \sin x + \cos x$

Sol 22:

$$I = \int \frac{\sqrt{\cos 2x}}{\sin x}dx = \int \sqrt{\frac{\cos^2 x - \sin^2 x}{\sin^2 x}}dx = \int \sqrt{\cot^2 x - 1}dx$$

On putting

$\cot x = \sec\theta$ & $-\csc^2 dx = \sec\theta\tan\theta d\theta$

We get, $I = \int \sqrt{\sec^2\theta - 1} \times \dfrac{\sec\theta\tan\theta}{-\csc^2 x}d\theta$

$$= -\int \frac{\sec\theta\tan^2\theta}{1 + \sec^2\theta}d\theta$$

$$= -\int \frac{\sin^2\theta}{\cos\theta + \cos^3\theta}d\theta = -\int \frac{1-\cos^2\theta}{\cos\theta(1+\cos^2\theta)}d\theta$$

$$= \int \frac{(1+\cos^2\theta) - 2\cos^2\theta}{\cos\theta(1+\cos^2\theta)} = -\int \sec\theta\, d\theta + 2\int \frac{\cos\theta}{1+\cos^2\theta}d\theta$$

$$\Rightarrow \quad -\int \sec\theta\, d\theta + 2\int \frac{d(\sin\theta)}{1+\cos^2\theta}$$

$$\Rightarrow \quad -\int \sec\theta\, d\theta + 2\int \frac{d(\sin\theta)}{2-\sin^2\theta}$$

$$\Rightarrow \quad -\log|\sec\theta + \tan\theta| + 2 \times \frac{1}{\sqrt{2}}\log\left|\frac{\sqrt{2}+\sin\theta}{\sqrt{2}-\sin\theta}\right| + c$$

$$\Rightarrow \quad -\log|\sec\theta + \sqrt{\sec^2\theta - 1}| + \frac{1}{\sqrt{2}}\log\left|\frac{\sqrt{2}+\sqrt{1-\cos^2\theta}}{\sqrt{2}-\sqrt{1-\cos^2\theta}}\right| + c$$

$$\Rightarrow \quad -\log|\cot x + \sqrt{\cot^2 x - 1}| + \frac{1}{\sqrt{2}}\log\left|\frac{\sqrt{2}+\sqrt{1-\tan^2 x}}{\sqrt{2}-\sqrt{1-\tan^2 x}}\right| + c$$

Sol 23: $\int \dfrac{\sqrt{x^2+1}\left(\log(x^2+1) - \log x^2\right)}{x^4}dx$

$$= \int \frac{\left(\sqrt{1+\frac{1}{x^2}}\log\left(1+\frac{1}{x^2}\right)\right)}{x^3}dx$$

$1 + \dfrac{1}{x^2} = t \Rightarrow -\dfrac{2}{x^3}dx = dt$

$$-\frac{1}{2}\int \sqrt{t}\log t\, dt = -\frac{1}{2}\left[(\log t)\frac{t^{3/2}}{3/2} - \int \frac{1}{t} \times \frac{t^{3/2}}{3/2}dt\right]$$

$$= \frac{1}{2}\left[-\frac{2}{3}(\log)t^{3/2} + \frac{2}{3}\int t^{1/2}dt\right]$$

$$= \frac{1}{2}\left[-\frac{2}{3}\left[\log\left(1+\frac{1}{x^2}\right)\left[1+\frac{1}{x^2}\right]^{3/2} + \frac{4}{9}\left(1+\frac{1}{x^2}\right)^{3/2} + c\right]\right]$$

$$= -\frac{1}{3}\left[1+\frac{1}{x^2}\right]^{3/2}\left[\log\left(1+\frac{1}{x^2}\right) - \frac{2}{3}\right] + c$$

Sol 24: $\int \dfrac{\sin x}{\sin x - \cos x}dx$

$$= \frac{1}{2}\int \frac{\sin x - \cos x + \sin x + \cos x}{(\sin x - \cos x)}dx = \frac{1}{2}\int 1\, dx + \frac{1}{2}\int \frac{dt}{t}$$

Put $\sin x - \cos x = t$

$(\cos x + \sin x)dx = dt$

$$\Rightarrow \frac{1}{2}x + \frac{1}{2}\log(\sin x - \cos x) + c$$

$$\because \int \frac{1}{t}dt = \log t = \log(\sin x - \cos x)$$

Sol 25: $\int x\sin^{-1}\left(\dfrac{1}{2}\cdot\dfrac{\sqrt{2a-x}}{a}\right)dx$

$$\sin^{-1}\left(\frac{1}{2}\cdot\frac{\sqrt{2a-x}}{a}\right)\int x\, dx - \int\left[\left(\frac{d\sin^{-1}\frac{1}{2}\frac{\sqrt{2a-x}}{a}}{dx}\right)\int x\, dx\right]dx$$

$$\Rightarrow \frac{x^2}{2}\sin^{-1}\left(\frac{1}{2}\frac{\sqrt{2a-x}}{a}\right)$$

$$-\int\left(\frac{x^2}{2} \times \frac{1}{\sqrt{1-\left(\frac{2a-x}{4a^2}\right)}} \times \frac{1}{2a} \times \frac{-1}{2\sqrt{2a-x}}\right)dx$$

$$\Rightarrow \frac{x^2}{2}\sin^{-1}\left(\frac{1}{2}\frac{\sqrt{2a-x}}{a}\right)$$

$$+\frac{2a}{8a}\int \frac{x^2}{\sqrt{4a^2-2a+x}} \times \frac{1}{\sqrt{2a-x}}dx$$

Sol 26: $\int \dfrac{1}{\cos^{4x}\sin^2 x}dx$

$$\int \frac{(\sin^2 x + \cos^2 x)^2}{\sin^2 x\cos^4 x}dx$$

$$= \int \left(\frac{\sin^4 x + \cos^4 x + 2\sin^2 x\cos^2 x}{\sin^2 x\cos^4 x}\right)dx$$

$$= \int \left(\frac{\sin^2 x}{\cos^4 x} + \frac{1}{\sin^2 x} + \frac{2}{\cos^2 x}\right)dx$$

$$= \int \tan^2 x\sec^2 x\, dx + \int \cosec^2 x\, dx + 2\int \sec^2 x\, dx$$

$$= \frac{\tan^3 x}{3} - \cot x + 2\tan x + c$$

Sol 27: $I = \int \dfrac{1}{(a+b\cos\theta)^2}d\theta$

Let $P = \dfrac{\sin\theta}{(a+b\cos\theta)}$

1.48

$$\frac{dP}{d\theta} = \frac{(a+b\cos\theta)\cos\theta - \sin\theta(0 - b\sin\theta)}{(a+b\cos\theta)^2}$$

$$= \frac{a\cos\theta + b(\cos^2\theta + \sin^2\theta)}{(a+b\cos\theta)^2} = \frac{a\cos\theta + b}{(a+b\cos\theta)^2}$$

Let $a + b\cos\theta = Y \Rightarrow \cos\theta = \dfrac{Y-a}{b}$

$$\frac{dP}{d\theta} = \frac{a\left(\dfrac{Y-a}{b}\right) + b}{\left[a + b\left(\dfrac{Y-a}{b}\right)\right]^2} = \frac{aY + b^2 - a^2}{bY^2}$$

$$\Rightarrow \quad \frac{dP}{d\theta} = \frac{a}{b}\left(\frac{1}{Y}\right) + \left(\frac{b^2 - a^2}{b}\right)\frac{1}{Y^2}$$

$$\Rightarrow \quad \frac{dP}{d\theta} = \frac{a}{b}\left(\frac{1}{a+b\cos\theta}\right) + \left(\frac{b^2 - a^2}{b}\right)\left(\frac{1}{a+b\cos\theta}\right)^2$$

Integrating,

$$P = \frac{a}{b}\int\frac{1}{a+b\cos\theta}d\theta + \frac{b^2-a^2}{b}\int\frac{1}{(a+b\cos\theta)}d\theta$$

$$\Rightarrow \quad \frac{-(b^2-a^2)}{b}\int\frac{1}{(a+b\cos\theta)^2}d\theta = \frac{a}{b}\int\frac{1}{a+b\cos\theta}d\theta - P$$

$$\Rightarrow \quad \int\frac{1}{(a+b\cos\theta)^2}d\theta = \frac{b}{(a^2-b^2)}\left[\frac{a}{b}\int\frac{1}{a+b\cos\theta}d\theta - P\right]$$

$$= \frac{b}{a^2-b^2}\left[\frac{a}{b}I_1 - \frac{\sin\theta}{(a+b\cos\theta)}\right] + c$$

Where $I_1 = \displaystyle\int\frac{1}{a+b\cos\theta}d\theta$

$$= \frac{2}{\sqrt{a^2-b^2}}\tan^{-1}\left[\sqrt{\frac{a-b}{a+b}}\tan(\theta/2)\right] + c_1$$

Sol 28: [Here, $\sqrt{x^4-1} = \sqrt{(x^2)^2 - 1}$, which is of the form, $\sqrt{x^2-a^2}$ hence substitution $x^2 = \sec\theta$ may be tried]

Now, $\displaystyle\int\frac{dx}{x\sqrt{x^4-1}} = \int\frac{dx}{x\sqrt{(x^2)^2 - 1}}$ ...(i)

Let $x^2 = \sec\theta$, then $2x\,dx = \sec\theta\,\tan\theta\,d\theta$

$$dx = \frac{\sec\theta\tan\theta}{2x}d\theta = \frac{\sec\theta\tan\theta}{2\sqrt{\sec\theta}}d\theta$$

Now, from (i)

$$\int\frac{dx}{x\sqrt{x^4-1}} = \int\left(\frac{1}{\sqrt{\sec\theta}\sqrt{\sec^2\theta - 1}}\right)\frac{\sec\theta\cdot\tan\theta}{2\sqrt{\sec\theta}}d\theta$$

$$= \int\left(\frac{1}{\sqrt{\sec\theta}\cdot\tan\theta}\right)\frac{\sec\theta\tan\theta}{2\sqrt{\sec\theta}}d\theta$$

$$= \int\frac{1}{2}d\theta = \frac{1}{2}\theta + c = \frac{1}{2}\sec^{-1}(x^2) + c$$

Sol 29: $I = \displaystyle\int\frac{dx}{\sqrt{\sin^3 x\sin(x+\alpha)}}$

$$= \int\frac{1}{\sqrt{\sin^3 x[\sin x\cos\alpha + \sin\alpha\cos x]}}dx$$

$$= \int\frac{1}{\sqrt{\sin^4 x(\cos\alpha + \cot x\sin\alpha)}}dx$$

$$= -\frac{1}{\sin\alpha}\int\frac{1}{\sqrt{\cos\alpha + \cot x\sin\alpha}}d(\cos\alpha + \cot x\sin\alpha)$$

$$= -\frac{1}{\sin\alpha}\int\frac{1}{\sqrt{t}}dt \quad ; \quad \text{where } t = \cos\alpha + \cot x\sin\alpha$$

$$= \frac{-2}{\sin\alpha}\sqrt{\cos\alpha + \cot x\sin\alpha} + c$$

Sol 30: $I = \displaystyle\int\frac{1}{(1+\sqrt{x})\sqrt{x - x^2}}dx$

put $x = \sin^2\theta$ & $dx = 2\sin\theta\cos\theta\,d\theta$

$$\Rightarrow I = \int\frac{2\sin\theta\cos\theta\,d\theta}{(1+\sin\theta)\sqrt{\sin^2\theta - \sin^4\theta}} = 2\int\frac{1-\sin\theta}{\cos^2\theta}d\theta$$

$$\Rightarrow = 2(\tan\theta - \sec\theta) + c = 2\left(\sqrt{\frac{x}{1-x}} - \frac{1}{\sqrt{1-x}}\right) + c$$

Sol 31: $I = \displaystyle\int\frac{\cos 8x - \cos 7x}{1 + 2\cos 5x}dx$

$$= \frac{1}{2}\int\frac{2\sin\dfrac{5x}{2}\cos 8x - 2\sin\dfrac{5x}{2}\cos 7x}{\sin\dfrac{5x}{2} + 2\cos 5x\sin\dfrac{5x}{2}}dx$$

$$= \frac{1}{2}\int \frac{\left(\sin\frac{21x}{2} - \sin\frac{11x}{2}\right) - \left(\sin\frac{19x}{2} - \sin\frac{9x}{2}\right)}{\sin\frac{15x}{2}}$$

$$= \frac{1}{2}\int \frac{2\sin\frac{15x}{2}\cos 3x - 2\sin\frac{15x}{2}\cos 2x}{\sin\frac{15x}{2}}dx$$

$$= \frac{1}{2}\int (\cos 3x - \cos 2x)dx = \frac{1}{3}\sin 3x - \frac{1}{2}\sin 2x + c$$

Sol 32: $I = \int \frac{x^3 + 1}{x(x-1)^3}dx$

$$\frac{x^3 + 1}{x(x-1)^3} = \frac{A}{x} + \frac{B}{x-1} + \frac{C}{(x-1)^2} + \frac{D}{(x-1)^3}$$

put $x = 1 \Rightarrow D = 2$

put $x = 0 \Rightarrow A = -1$

put $x = -1$ & $x = 2 \Rightarrow B = 2$ & $c = 1$

$$\therefore \quad I = \int \frac{-1}{x}dx + \int \frac{2}{x-1}dx + \int \frac{1}{(x-1)^2}dx + \int \frac{2}{(x-1)^3}dx$$

$$= -\log|x| + 2\log|x-1| + \frac{1}{x-1} - \frac{1}{(x-1)^2} + c$$

Sol 33: $I = \int \frac{1}{x + \sqrt{x^2 - x + 1}}dx$

put $t = x + \sqrt{x^2 - x + 1}$

$$\Rightarrow x = \frac{t^2 - 1}{2t - 1} \quad \text{and} \quad dx = \frac{2t^2 - 2t + 2}{(2t-1)^2}dt$$

$$\Rightarrow I = 2\int \frac{t^2 - t + 1}{t(2t-1)^2}dt$$

let $\frac{t^2 - t + 1}{t(2t-1)^2} = \frac{A}{t} + \frac{B}{2t-1} + \frac{C}{(2t-1)^2}$

Solving by partial fraction method, we get

$A = 1, \quad C = \frac{3}{2} \quad \text{and} \quad B = -\frac{3}{2}$

$$I = 2\log t - \frac{3}{2}\log(2t-1) - \frac{3}{2}\frac{1}{(2t-1)} + c$$

where $t = x + \sqrt{x^2 - x + 1}$

Sol 34: [Here $\sqrt{e^{2x} - 4} = \sqrt{(e^x)^2 - 2^2}$, which is of the form $\sqrt{x^2 - a^2}$, hence substitution $e^x = 2\sec\theta$ may be tried]

Now, $\int \frac{e^x}{\sqrt{e^{2x} - 4}}dx = \int \frac{e^x}{\sqrt{(e^x)^2 - 2^2}}dx$ ...(i)

Let $e^x = 2\sec\theta$, then $e^x dx = 2\sec\theta\tan\theta d\theta$

Now from (i),

$$\int \frac{e^x}{\sqrt{e^{2x} - 4}}dx = \int \frac{2\sec\theta\tan\theta}{\sqrt{4\sec^2\theta - 4}}d\theta = \int \frac{2\sec\theta\tan\theta}{2\tan\theta}d\theta$$

$$= \int \sec\theta d\theta = \log|\sec\theta + \tan\theta| + c$$ .....(ii)

$\because \quad e^x = 2\sec\theta \quad \therefore \quad \sec\theta = \frac{e^x}{2}$

$$\therefore \quad \tan\theta = \sqrt{\sec^2\theta - 1} = \sqrt{\frac{e^{2x}}{4} - 1} = \sqrt{\frac{e^{2x} - 4}{2}}$$

From (ii), $\int \frac{e^x}{\sqrt{2^{2x} - 4}}dx = \log\left|\frac{e^x}{2} + \frac{\sqrt{e^{2x} - 4}}{2}\right| + c$

$$= \log\left(\frac{e^x + \sqrt{e^{2x} - 4}}{2}\right) + c \qquad [\because e^x + \sqrt{e^{2x} - 4} > 0]$$

$$= \log(e^x + \sqrt{e^{2x} - 4}) - \log 2 + c = \log(e^x + \sqrt{e^{2x} - 4}) + c'$$

Sol 35: $I = (\log x)\cdot\left[-\frac{1}{2(1+x)^2}\right] - \int\left(-\frac{1}{2(1+x)^2 x}\right)dx$

[Taking $u = \log x$]

$$= -\frac{1}{2}\cdot\frac{\log x}{(1+x)^2} + \frac{1}{2}\int \frac{dx}{x(1+x)^2}$$ ....(i)

Now, $\frac{1}{x(1+x)^2} = \frac{1+x-x}{x(1+x)^2} = \frac{1}{x(1+x)} - \frac{1}{(1+x)^2}$

$$= \frac{1+x-x}{x(1+x)} - \frac{1}{(1+x)^2} = \frac{1}{x} - \frac{1}{1+x} - \frac{1}{(1+x)^2}$$

$$\therefore \quad \int \frac{dx}{x(1+x)^2} = \int\left[\frac{1}{x} - \frac{1}{1+x} - \frac{1}{(1+x)^2}\right]dx$$

$$= \log x - \log(1+x) + \frac{1}{1+x} = \log\frac{x}{1+x} + \frac{1}{1+x}$$

[Here $x > 0$ a $\log x$ occurs in the integrand]

$\therefore$ From (i), $I = -\frac{1}{2}\frac{\log x}{(1+x)^2} + \frac{1}{2}\log\frac{x}{1+x} + \frac{1}{2(1+x)} + c$

Sol 36: Let $f(x) = ax^2 + bx + c$

$f(0) = c = -3$

$f(1) = a + b + c = -3$

or $a + b = 0$ $\because$ $a = 1, b = -1$

$f(2) = 4a + 2b + c = -1$

or $4a + 2b = 2$

$\therefore f(x) = x^2 - x - 3$

$\int \dfrac{x^2 - x - 3}{x^3 - 1}dx = \int \dfrac{x(x-1)}{(x-1)(x^2+x+1)}dx - \int \dfrac{3}{(x^3-1)}dx$

$= \int \dfrac{x}{(x^2+x+1)}dx - \int \dfrac{3}{(x^3-1)}dx$

$\dfrac{1}{2}\left[\int \dfrac{2x+1}{(x^2+x+1)}dx - \int \dfrac{dx}{\left(x+\dfrac{1}{2}\right)^2 + \left(\dfrac{\sqrt{3}}{2}\right)^2}\right] - \int \dfrac{3}{(x^3-1)}dx$

$= \dfrac{1}{2}\log(x^2+x+1) - \dfrac{1}{2} \times \dfrac{2}{\sqrt{3}}\tan^{-1}\dfrac{x+\dfrac{1}{2}}{\dfrac{\sqrt{3}}{2}}$

$- \int \left(\dfrac{1}{x-1} - \dfrac{x+2}{(x^2+x+1)}\right)dx$

$= \dfrac{1}{2}\log(x^2+x+1) - \dfrac{1}{\sqrt{3}}\tan^{-1}\dfrac{2x+1}{\sqrt{3}} - \log(x-1)$

$+ \dfrac{1}{2}\int \left(\dfrac{2x+1}{x^2+x+1} + \dfrac{3}{(x^2+x+1)}\right)dx$

$= \log(x^2+x+1) - \dfrac{1}{\sqrt{3}}\tan^{-}\dfrac{2x+1}{\sqrt{3}} - \log(x-1)$

$+ \dfrac{3}{2 \times \dfrac{\sqrt{3}}{2}}\tan^{-}\left(\dfrac{2x+1}{\sqrt{3}}\right)$

$= \log \dfrac{(x^2+x+1)}{|x-1|} + \dfrac{2}{\sqrt{3}}\tan^{-1}\dfrac{2x+1}{\sqrt{3}} + c$

Exercise 2

Single Correct Choice Type

Sol 1: (A) $\int \left(\dfrac{x}{1+x^5}\right)^{3/2}dx$

$\int \dfrac{x^{3/2}}{x^{15/2}\left(1+\dfrac{1}{x^5}\right)^{3/2}}dx = \int \dfrac{x^{-6}}{\left(1+\dfrac{1}{x^5}\right)^{3/2}}dx$

Put $1 + \dfrac{1}{x^5} = t$

$\Rightarrow -5x^{-6}dx = dt$ or $x^{-6}dx = -\dfrac{1}{5}dt$

$\therefore I = -\dfrac{1}{5}\int \dfrac{dt}{t^{3/2}} = -\dfrac{1}{5} \times (-2)\dfrac{1}{\sqrt{t}} + c$

$\therefore I = \dfrac{2}{5}\dfrac{1}{\sqrt{1+\dfrac{1}{x^5}}} + c = \dfrac{2}{5}\sqrt{\dfrac{x^5}{1+x^5}} + c$

Sol 2: (B) $-\int \dfrac{(\sin^8 x - \cos^8 x)}{1 - 2\sin^2 x\cos^2 x}dx$

$= -\int \dfrac{(\sin^2 x - \cos^2 x)(\sin^2 x + \cos^2 x)(\sin^4 x + \cos^4 x)}{1 - 2\sin^2 x\cos^2 x}dx$

$= -\int \dfrac{(\sin^2 x - \cos^2 x)\left[(\sin^2 x + \cos^2 x)^2 - 2\sin^2 x\cos^2 x\right]}{1 - 2\sin^2 x\cos^2 x}dx$

$= -\int (\sin^2 x - \cos^2 x)dx = -\int(-\cos 2x)dx = \dfrac{1}{2}\sin 2x + c$

Sol 3: (C)

(A) $x\int \log x\, dx = x\left[(\log x)x - \int x \times \dfrac{1}{x}dx\right] = x^2 \log x - x^2 + cx$

(B) $x\int \ell|x|dx = x^2 \log|x| - x^2 + cx$

(C) $x\int e^x dx = x\left[e^x + c\right] = xe^x + cx$

(D) $\int \dfrac{dx}{\sqrt{a^2 + x^2}}$

$x = a\tan\theta,\ dx = a\sec^2\theta\, d\theta$

$\Rightarrow I = \int \dfrac{\sec^2\theta c\theta}{a\sec\theta} = \int \sec\theta\, d\theta$

$= \log|\sec\theta + \tan\theta| + c = \log\left|\dfrac{x}{a} + \dfrac{\sqrt{x^2+a^2}}{a}\right| + c$

Sol 4: (B) $\int \sqrt{1 + 2\tan x(\sec x + \tan x)}\,dx$

$= \int \sqrt{1 + 2\tan x \sec x + 2\tan^2 x}\quad dx$

$= \int \sqrt{\tan^2 x + \sec^2 x + 2\tan x \sec x}\,dx$

$= \int (\tan x + \sec x)\,dx = \int \tan x\,dx + \int \sec x\,dx$

$= -\log|\cos x| + \log|\sec x + \tan x| + c$

$= \log|\sec x| + \log|\sec x + \tan x| + c$

Sol 5: (C) $I = \int \dfrac{2\sin x \cos x}{\sin^4 x + \cos^4 x}\,dx = \int \dfrac{2\tan x \sec^2 x}{\tan^4 x + 1}\,dx$

Let $\tan^2 x = t$

$\therefore I = 2\tan x \sec^2 x\,dx = dt$

$\therefore I = \int \dfrac{1}{t^2 + 1}\,dt$

$\therefore I = \tan^{-1}(t) + c = \tan^{-1}(\tan^2 x) + c$

$\tan^{-1}(\tan^2 x) = \tan^{-1}\left(\dfrac{1}{\cot^2 x}\right) = \cot^{-1}(\cot^2 x)$

$\tan^{-1}(\tan^2 x) = \dfrac{\pi}{2} - \cot^{-1}(\tan^2 x) + c$

$= -\cot - (\tan^2 x) + c_1$

Sol 6: (C) $\int \dfrac{1}{\cos^3 \theta \sqrt{2\sin\theta\cos\theta}}\,d\theta$

$= \int \dfrac{1}{\sqrt{2}\cos^{7/2}\theta \sin^{1/2}\theta}\,d\theta$

Dividing and multiplying by $\cos^4\theta$

$I = \int \dfrac{\sec^4\theta}{\sqrt{2}\tan^{1/2}\theta}$

let $\tan^{1/2}\theta = t$

$\therefore \dfrac{1}{2\sqrt{\tan\theta}} \times \sec^2\theta\,d\theta = dt$

$\therefore I = \int \sqrt{2}\sec^2\theta\,dt$

$= \int \sqrt{2}\left(1 + t^4\right)dt = \sqrt{2} \times \left(t + \dfrac{t^5}{5}\right) + c$

$= \dfrac{\sqrt{2}}{5}\left(5\sqrt{\tan\theta} + \tan^2\theta \times \sqrt{\tan\theta}\right) + c$

$I = \dfrac{\sqrt{2}}{5}(\tan^2\theta + 5)\sqrt{\tan\theta} + c$

Sol 7: (A) $\int \dfrac{dx}{b\left(\dfrac{a}{b} + x^2\right)} = \dfrac{1}{b \times \sqrt{\dfrac{a}{b}}}\tan^{-1}\left(\dfrac{x}{\sqrt{\dfrac{a}{b}}}\right)$

$= \dfrac{1}{\sqrt{ab}}\tan^{-1}\left(x\sqrt{\dfrac{b}{a}}\right) + c$

Sol 8: (C) $I = \int \dfrac{1 + x^7 - 2x^7}{x(1 + x^7)}\,dx$

$= \int \dfrac{1}{x} - \dfrac{2x^6}{1 + x^7}\,dx = \log x - \int \dfrac{2x^6}{1 + x^7}\,dx$

let $1 + x^7 = t \Rightarrow 7x^6 dx = dt$

$I = \log x - \dfrac{2}{7}\int \dfrac{1}{t}\,dt = \log x - \dfrac{2}{7}\ell n\,t + c$

$= \log x - \dfrac{2}{7}\ell n(1 + x^7) + c$

Sol 9: (A) Let $\log|x| = t \Rightarrow \dfrac{1}{x}\,dx = dt$

$\therefore I = \int \dfrac{t}{\sqrt{1+t}}\,dt = \int \sqrt{1+t} - \dfrac{1}{\sqrt{1+t}}\,dt$

$= \dfrac{(t+1)^{3/2}}{3} \times 2 - (t+1)^{1/2} \times 2 + c$

$= 2(t+1)^{1/2}\left(\dfrac{t+1}{3} - 1\right) + c = \dfrac{2}{3}(t+1)^{1/2}(t-2) + c$

$= \dfrac{2}{3}(\log|x| + 1)^{1/2}(\log|x| - 2) + c$

Sol 10: (C)

$I = \int \dfrac{x^4 + 2x^2 + 1 - 2x^2}{x(x^2 + 1)^2}\,dx = \int \dfrac{(x^2 + 1)^2 - 2x^2}{x(x^2 + 1)^2}\,dx$

$= \int \dfrac{1}{x} - \dfrac{2x}{(x^2 + 1)^2}\,dx = \log|x| - \int \dfrac{2x}{(x^2 + 1)^2}\,dx$

let $x^2 + 1 = t \Rightarrow 2x\,dx = dt$

$\therefore I = \log|x| - \int \dfrac{1}{t^2}\,dt$

$= \log|x| + \dfrac{1}{t} + c = \log|x| + \dfrac{1}{1 + x^2} + c$

$A = 1, B = 1$

Sol 11: (B) $I = \int 2\sin x(\cos 2x + \cos x)\,dx$

$= \int 2\sin x(2\cos^2 x - 1 + \cos x)\,dx$

1.52

Let $\cos x = t \Rightarrow -\sin x\,dx = dt$

$\therefore I = \int 2(1-t-2t^2)dt = 2t - t^2 - \dfrac{4t^3}{3} + c$

$= 2\cos x - \cos^2 x - \dfrac{4\cos^3 x}{3} + c$

$= \cos x - \left(\dfrac{4\cos^3 x - 3\cos x}{3}\right) - \cos^2 x + c$

$= \cos x - \dfrac{1}{3}\cos 3x - \cos^2 x - \dfrac{1}{2} + c$

$= \cos x - \dfrac{1}{3}\cos 3x - \dfrac{\cos 2x}{2} + c$

Sol 12: (B) $I = \int \dfrac{1}{2}\sin 2x\cos 2x\cos 4x\cos 18x\cos 16x\,dx$

$= \dfrac{1}{32}\int \sin 32x\,dx = -\dfrac{\cos 32x}{1024} + c$

Sol 13: (B) $I = \int x^3 f(x^2)dx$

let $x^2 = t$

$\therefore I = \dfrac{1}{2}\int t f(t)dt$

$= \dfrac{t}{2}\int f(t)dt - \int \dfrac{1}{2}\left(\int f(t)dt\right)dt$

$= \dfrac{1}{2}\left[x^2 F(x^2) - \int f(x^2)d(x^2)\right]$

Sol 14: (D) $I = \int \dfrac{4e^{2x}+6}{9e^{2x}-4}dx$

$\Rightarrow 4e^{2x} + 6 = a(9e^{2x}-4) + b \times 18 \times e^{2x}$

$\Rightarrow 9a + 18b = 4$

$-4a = 6$

$\therefore a = -\dfrac{3}{2}$

$18b = 4 + \dfrac{27}{2} \Rightarrow b = \dfrac{35}{36}$

$\therefore I = \int -\dfrac{3}{2}dx + \dfrac{35}{36}\dfrac{18e^{2x}}{9e^{2x}-4}dx$

$= -\dfrac{3x}{2} + \dfrac{35}{36}\log(9e^{2x}-4) + c$

$A = -\dfrac{3}{2}$ and $B = \dfrac{35}{36}$

Sol 15: (A) $I = \int \dfrac{\cos x - \sin x}{\sqrt{2\sin x\cos x}\,(\cos x + \sin x)}dx$

$2\sin x\cos x = (\cos x + \sin x)^2 - 1$

And $d(\cos x + \sin x) = (-\sin x + \cos x)dx$

$\therefore$ let $\cos x + \sin x = t$

$\therefore I = \int \dfrac{1}{\sqrt{t^2-1}\,(t)}dt$

$\therefore I = \sec^{-1}(t) + c = \sec^{-1}(\cos x + \sin x) + c$

Previous Years' Questions

Sol 1: (C) Let $I = \int \dfrac{(x^2-1)dx}{x^3\sqrt{2x^4-2x^2+1}}$, dividing numerator

and denominator by x^5

$= \int \dfrac{\left(\dfrac{1}{x^3} - \dfrac{1}{x^5}\right)dx}{\sqrt{2 - \dfrac{2}{x^2} + \dfrac{1}{x^4}}}$

Put $2 - \dfrac{2}{x^2} + \dfrac{1}{x^4} = t \Rightarrow \left(\dfrac{4}{x^3} - \dfrac{4}{x^5}\right)dx = dt$

$\therefore I = \dfrac{1}{4}\int \dfrac{dt}{\sqrt{t}} = \dfrac{1}{4}\cdot\dfrac{t^{\frac{1}{2}}}{\dfrac{1}{2}} + c$

$= \dfrac{1}{2}\sqrt{2 - \dfrac{2}{x^2} + \dfrac{1}{x^4}} + c$

Sol 2: Given, $\int \dfrac{4e^x + 6e^{-x}}{9e^x - 4e^{-x}}dx = Ax + B\log(9e^{2x}-4) + c$

LHS $= \int \dfrac{4e^{2x}+6}{9e^{2x}-4}dx$

Let $4e^{2x} + 6 = A(9e^{2x}-4) + B(18e^{2x})$

$\Rightarrow 9A + 18B = 4$

and $-4A = 6$

$\Rightarrow A = -\dfrac{3}{2}$ and $3 = \dfrac{35}{36}$

$\therefore \int \dfrac{A(9e^{2x}-4) + B(18e^{2x})}{9e^{2x}-4}dx$

$= A\int 1dx + B\int \dfrac{1}{t}dt$, where $t = 9e^{2x} - 4$

$= Ax + B\log(9e^{2x}-4) + c$

$= -\dfrac{3}{2}x + \dfrac{35}{36}\log(9e^{2x}-4) + c$

$\therefore A = -\dfrac{3}{2},\ B = \dfrac{35}{36}$

and C = any real number

Sol 3: Let $I = \displaystyle\int \dfrac{\sin x}{\sin x - \cos x}\,dx$

Again, let $\sin x = A(\cos x + \sin x) + B(\sin x - \cos x)$,

then $A + B = 1$ and $A - B = 0$

$\Rightarrow A = \dfrac{1}{2},\ B = \dfrac{1}{2}$

$\therefore I = \displaystyle\int \dfrac{\dfrac{1}{2}(\cos x + \sin x) + \dfrac{1}{2}(\sin x - \cos x)}{(\sin x - \cos x)}\,dx$

$= \dfrac{1}{2}\displaystyle\int \dfrac{\cos x + \sin x}{\sin x - \cos x}\,dx + \dfrac{1}{2}\displaystyle\int 1\,dx + c$

$= \dfrac{1}{2}\log(\sin x - \cos x) + \dfrac{1}{2}x + c$

Sol 4: Let $I = \displaystyle\int \dfrac{x\,dx}{1 + x^4}$

$= \dfrac{1}{2}\displaystyle\int \dfrac{2x}{1 + (x^2)^2}\,dx$

Put $x^2 = u \Rightarrow 2x\,dx = du$

$\therefore I = \dfrac{1}{2}\displaystyle\int \dfrac{du}{1 + u^2} = \dfrac{1}{2}\tan^{-1}(u) + c = \dfrac{1}{2}\tan^{-1}(x^2) + c$

Sol 5: Let $I_1 = \displaystyle\int \sin x \sin 2x \sin 3x\,dx$

$= \dfrac{1}{4}\displaystyle\int (\sin 4x + \sin 2x - \sin 6x)\,dx$

$= -\dfrac{\cos 4x}{16} - \dfrac{\cos 2x}{8} + \dfrac{\cos 6x}{24}$

$I_2 = \displaystyle\int \sec^2 x \cdot \cos^2 2x\,dx = \displaystyle\int \sec^2 x(2\cos^2 x - 1)^2\,dx$

$= \displaystyle\int (4\cos^2 x + \sec^2 x - 4)\,dx = \displaystyle\int (2\cos 2x + \sec^2 x - 2)\,dx$

$= \sin 2x + \tan x - 2x$

$\therefore \displaystyle\int \dfrac{A(9e^{2x} - 4) + B(18e^{2x})}{9e^{2x} - 4}\,dx$

$= A\displaystyle\int 1\,dx + B\displaystyle\int \dfrac{1}{t}\,dt,\ \text{ where } t = 9e^{2x} - 4$

$= Ax + B\log(9e^{2x} - 4) + c$

$= -\dfrac{3}{2}x + \dfrac{35}{36}\log(9e^{2x} - 4) + c$

$\therefore A = -\dfrac{3}{2},\ B = \dfrac{35}{36}$

and C = any real number

and $I_3 = \displaystyle\int \sin^4 x \cos^4 x\,dx$

$= \dfrac{1}{128}\displaystyle\int (3 - 4\cos 4x + \cos 8x)\,dx = \dfrac{3x}{128} - \dfrac{\sin 4x}{128} + \dfrac{\sin 8x}{1024}$

$\therefore I = I_1 + I_2 + I_3$

$= -\dfrac{\cos 4x}{16} - \dfrac{\cos 2x}{8} + \dfrac{\cos 6x}{24} + \sin 2x + \tan x - 2x$

$+ \dfrac{3x}{128} - \dfrac{\sin 4x}{128} + \dfrac{\sin 8x}{1024}$

Sol 6: Let $I = \dfrac{x^2}{(a + bx)^2}$

Put $a + bx = t$

$\Rightarrow b\,dx = dt$

$\therefore I = \displaystyle\int \dfrac{\left(\dfrac{t - a}{b}\right)^2}{t^2} \cdot \dfrac{dt}{b} = \dfrac{1}{b^3}\displaystyle\int \left(\dfrac{t^2 - 2at + a^2}{t^2}\right)dt$

$= \dfrac{1}{b^3}\displaystyle\int \left(1 - \dfrac{2a}{t} + \dfrac{a^2}{t^2}\right)dt = \dfrac{1}{b^3}\left(t - 2a\log t - \dfrac{a^2}{t}\right) + c$

$= \dfrac{1}{b^3}\left(a + bx - 2a\log(a + bx) - \dfrac{a^2}{a + bx} + c\right)$

Sol 7: Let $I = \displaystyle\int (\sqrt{\tan x} + \sqrt{\cot x})\,dx = \displaystyle\int \dfrac{\tan x + 1}{\sqrt{\tan x}}\,dx$

Put $\tan x = t^2 \Rightarrow \sec^2 x\,dx = 2t\,dt$

$\Rightarrow dx = \dfrac{2t}{1 + t^4}\,dt$

$\therefore I = \displaystyle\int \dfrac{t^2 + 1}{\sqrt{t^2}} \cdot \dfrac{2t}{t^4 + 1}\,dt = 2\displaystyle\int \dfrac{t^2 + 1}{t^4 + 1}\,dt$

$= 2\displaystyle\int \dfrac{1 + \dfrac{1}{t^2}}{t^2 + \dfrac{1}{t^2} - 2 + 2}\,dt = 2\displaystyle\int \dfrac{1 + \dfrac{1}{t^2}}{\left(t - \dfrac{1}{t}\right)^2 + (\sqrt{2})^2}\,dt$

Put $t - \dfrac{1}{t} = u \Rightarrow 1 + \dfrac{1}{t_2}\,dt = du$

$\therefore I = 2\displaystyle\int \dfrac{du}{u^2 + (\sqrt{2})^2}$

$\Rightarrow I = \dfrac{2}{\sqrt{2}}\tan^{-1}\left(\dfrac{u}{\sqrt{2}}\right) + c$

$$= \sqrt{2}\tan^{-1}\left(\frac{\sqrt{\tan x} - \sqrt{\cot x}}{\sqrt{2}}\right) + c$$

Sol 8: $\int \frac{(x+1)}{x(1+xe^x)^2}dx$

This can be rewritten as $\int \frac{e^x(x+1)}{2e^x(1+xe^x)^2}dx$

let $1 + xe^x = t \Rightarrow e^x(1+x)dx = dt$

Now integration becomes $\int \frac{dt}{t^2(t-1)}$

$\Rightarrow \quad \frac{1}{t^2(t-1)} = \frac{A}{t-1} + \frac{Bt+C}{t^2}$ (using partial fraction)

$\Rightarrow \quad 1 = t^2(A+B) + (C-B)t - C$

Comparing, we get C = -1, B = -1 and A = 1

Now our integration becomes

$$\int \frac{dt}{t^2(t-1)} = \int \frac{1}{t-1}dt - \int \frac{t+1}{t^2}dt = \int \frac{1}{t-1}dt - \int \frac{1}{t}dt - \int t^{-2}dt$$

$$= \log(t-1) - \log(t) - \frac{t^{-2+1}}{-2+1} + C = \log\frac{t-1}{t} + \frac{1}{t} + c$$

Putting $t = 1 + xe^x$, we get

$$\int \frac{dt}{t^2(t-1)} = \log\left|\frac{xe^x}{1+xe^x}\right| + \frac{1}{1+xe^x} + c$$

Sol 9: Let $I = \int \left(\frac{1-\sqrt{x}}{1+\sqrt{x}}\right)^{\frac{1}{2}} \cdot \frac{dx}{x}$

Put $x = \cos^2\theta \Rightarrow dx = -2\cos\theta \sin\theta \, d\theta$

$\therefore I = \int \left(\frac{1-\cos\theta}{1+\cos\theta}\right)^{\frac{1}{2}} \cdot \frac{-2\cos\theta\cdot\sin\theta}{\cos^2\theta}d\theta$

$$= \int \frac{\sin\frac{\theta}{2}}{\cos\frac{\theta}{2}} \cdot \frac{-2\sin\theta}{\cos\theta}d\theta = -\int \frac{2\sin\frac{\theta}{2} \cdot 2\sin\frac{\theta}{2} \cdot \cos\frac{\theta}{2}}{\cos\frac{\theta}{2} \cdot \cos\theta}d\theta$$

$$= -2\int \frac{2\sin^2\frac{\theta}{2}}{\cos\theta}d\theta = -2\int \frac{1-\cos\theta}{\cos\theta}d\theta$$

$$= 2\int(1-\sec\theta)d\theta = 2[\theta - \log|\sec\theta + \tan\theta|] + c$$

$$\Rightarrow I = 2\left[\cos^{-1}\sqrt{x} - \log\left|\frac{1}{\sqrt{x}} + \sqrt{\frac{1}{x}-1}\right|\right] + c$$

$$I = 2\left[\cos^{-1}\sqrt{x} - \log\left|1+\sqrt{1-x}\right| - \frac{1}{2}\log|x|\right] + c$$

Sol 10: (C) $\sqrt{2}\frac{\sin x \, dx}{\sin\left(x-\frac{\pi}{4}\right)} = \sqrt{2}\int \frac{\sin\left(x - \frac{\pi}{4} + \frac{\pi}{4}\right)dx}{\sin\left(x - \frac{\pi}{4}\right)}$

$$= \sqrt{2}\int\left(\cos\frac{\pi}{4} + \cot\left(x-\frac{\pi}{4}\right)\sin\frac{\pi}{4}\right)dx$$

$$= \int dx + \int \cot\left(x - \frac{\pi}{4}\right)dx = x + \ell n\left|\sin\left(x - \frac{\pi}{4}\right)\right| + c$$

Sol 11: (D) $\frac{dy}{dx} = y + 3$

$\Rightarrow \frac{dy}{y+3} = dx$

$\log(y+3) = x - c$

$x = 0 \Rightarrow y = 2$

$\Rightarrow \log 5 = 0 + c$

$c = \log 5$

$\log(y+3) = x + \log 5$

$y + 3 = e^{x-\log 5} \Rightarrow y + 3 \, e^{\log 2 + \log 5}$

$y + 3 = 10 \Rightarrow y = 7$

Sol 12: (D)

$\int \frac{5\sin x}{\sin x - 2\cos x}dx$

$\Rightarrow \int \left[\frac{2(\cos x + 2\sin x) + (\sin x - 2\cos x)}{\sin x - 2\cos x}\right]dx$

$= \int \left(\frac{\cos x + 2\sin x}{\sin x - 2\cos}\right)dx + \int dx + k$

$= 2\log|\sin x - 2\cos x| + x + k$

$\therefore a = 2$

Sol 13: (C) $\int f(x)dx = \psi(x)$

$I = \int x^5 f(x^3)dx$

Put $x^3 = t \Rightarrow x^2 dx = \frac{dt}{3} = \frac{1}{3}\int tf(t)dt$

$= \frac{1}{3}\left[t\psi(t)dt\right.$

$$= \frac{1}{3}\left[x^3\psi\left(x^3\right) - 3\int x^2\psi\left(x^3\right)dx\right] + c = \frac{1}{3}x^3\psi\left(x^3\right)dx + c$$

Sol 14: (D) $I = \int\left\{e^{\left(x+\frac{1}{x}\right)} + x\left(1 - \frac{1}{x^2}\right)e^{x+\frac{1}{x}}\right\}dx$

$= x.e^{x+\frac{1}{x}} + c$

As $\int\left(xf'(x) + f(x)\right)dx = xf(x) + c$

Sol 15: (D) $\int\dfrac{dx}{x^2\left(x^4+1\right)^{3/4}}$

$\int\dfrac{dx}{x^3\left(1+\dfrac{1}{x^4}\right)^{3/4}} \Rightarrow 1 + \dfrac{1}{x^4} = t^4$

$-4\dfrac{1}{x^5} = dx = 4t^3dt$

$\dfrac{dx}{x^3} = t^3dt$

$\int\dfrac{-t^3dt}{t^3} = -t + c = -\left(1 + \dfrac{1}{x^4}\right)^{1/4} + c$

Sol 16: (A) $\int\dfrac{2x^{12}+5x^3}{\left(x^5+x^3+1\right)}dx$

$\int\dfrac{\left(\dfrac{2}{x^3}+\dfrac{5}{x^6}\right)}{\left(1+\dfrac{1}{x^2}+\dfrac{1}{x^5}\right)^3}dx$

Let $1 + \dfrac{1}{x^2} + \dfrac{1}{x^5} = t$

$\dfrac{dt}{dx} = \dfrac{-2}{x^3} - \dfrac{5}{x^6}$

$\int\dfrac{-dt}{t^3} = \dfrac{1}{2t^2} + c = \dfrac{1}{2\left(1+\dfrac{1}{x^2}+\dfrac{1}{x^5}\right)^2} + c = \dfrac{x^{10}}{2\left(x^5+x^3+1\right)^2} + c$

JEE Advanced/Boards

Exercise 1

Sol 1: (i) $\int\tan\dfrac{x}{2}\tan\dfrac{x}{3}\tan\dfrac{x}{6}dx$

$= \int\tan\dfrac{x}{2}\left[1 - \dfrac{\left(\tan\dfrac{x}{3}+\tan\dfrac{x}{6}\right)}{\tan\dfrac{x}{2}}\right]dx$

$= \int\left(\tan\dfrac{x}{2} - \tan\dfrac{x}{3} - \tan\dfrac{x}{6}\right)dx$

$= \int\tan\dfrac{x}{2}dx - \int\tan\dfrac{x}{3}dx - \int\tan\dfrac{x}{6}dx$

$= 2\log\sec\dfrac{x}{2} - 3\log\sec\dfrac{x}{3} - 6\log\sec\dfrac{x}{6} + c$

(ii) $\int\dfrac{\tan(\log x)\tan\left(\log\dfrac{x}{2}\right)\tan(\log 2)}{x}dx$

Put $\log x = t \Rightarrow \dfrac{1}{x}dx = dt$

$\int\tan t\tan(t - \log 2)\tan(\log 2)dt$

$= \int\left[\tan t - \tan(t - \log 2) - \tan(\log 2)\right]dt$

$= \log\sec t - \log\sec(t - \log 2) - x\tan(\log 2) + c$

$= \log\dfrac{\sec(\log x)}{\sec\left(\log\dfrac{x}{2}\right)} - x\tan\log|2| + c$

Sol 2: Put $x = \alpha\sec 2\theta - \beta\tan 2\theta$

$\int\dfrac{2(\alpha-\beta)\sec^2\theta\tan\theta d\theta}{(\alpha-\beta)\tan^2\theta\sqrt{(\alpha-\beta)\tan^2\theta(\alpha-\beta)\sec^2\theta}}$

$= \int\dfrac{2\sec^2\theta\tan\theta}{\tan^2\theta\times(\alpha+\beta)\tan\theta\sec\theta}d\theta$

$= \dfrac{2}{\alpha-\beta}\int\dfrac{\sec\theta}{\tan^2\theta}d\theta$

Put $\tan\theta = t$ $\sec^2\theta d\theta = dt$

Or $\sec\theta d\theta = \dfrac{dt}{\sqrt{1+t^2}}$

$$\therefore I = \frac{2}{(\alpha-\beta)}\int \frac{dt}{t^2\sqrt{1+t^2}} = \frac{2}{(\alpha-\beta)}\int \frac{t^{-3}dt}{\sqrt{t^{-2}+1}}$$

$$\therefore 1 + t^{-2} = u \Rightarrow -2t^{-3}dt = du$$

Or $t^{-3}dt = -\frac{1}{2}du$

$$= \frac{1}{2}\times\frac{2}{(\alpha-\beta)}\int \frac{-du}{\sqrt{u}} = \frac{-1}{(\alpha-\beta)}\times 2\sqrt{u}$$

$$= \frac{-2}{(\alpha-\beta)}\sqrt{1+\frac{1}{t^2}}$$

$$= \frac{-2}{(\alpha-\beta)}\sqrt{1+\frac{1}{\tan^2\theta}} = \frac{-2}{(\alpha-\beta)}\sqrt{\frac{\sec^2\theta}{\tan^2\theta}}$$

$$= \frac{-2}{\alpha-\beta}\sqrt{\frac{(\alpha-\beta)\sec^2\theta}{(\alpha-\beta)\tan^2\theta}} = \frac{-2}{\alpha-\beta}\sqrt{\frac{(x-\beta)}{(x-\alpha)}}$$

Sol 3: $\int \dfrac{\ell n\left(\ell n\left(\dfrac{1+x}{1-x}\right)\right)dx}{1-x^2}$

Put $\log\left(\dfrac{1+x}{1-x}\right) = t$

$$\Rightarrow \left(\left(\frac{1-x}{1+x}\right)\times\frac{1-x+1+x}{(1-x)^2}\right)dx = dt$$

Or $\left(\dfrac{2}{1-x^2}\right)dx = dt$

$$\therefore I = \int \log(t)\frac{dt}{2}$$

$$= \frac{1}{2}\int 1.\log(t)\frac{dt}{2}$$

Integration by parts,

$$I = \frac{1}{2}\left[\log(t)\int 1dt - \left[\int \frac{d}{dt}(\log(t))\int 1dt\right]dt\right]$$

$$= \frac{1}{2}\left[t(\log t - 1)\right] + c$$

$$= \frac{1}{2}\left[\log\left(\frac{1+x}{1-x}\right)\left\{\log\left(\log\left(\frac{1+x}{1-x}\right)\right)-1\right\}\right] + c$$

Sol 4: $d\left(\dfrac{x}{e}\right)^x = \left(\dfrac{x}{e}\right)^x[\log x]$ and $d\left(\dfrac{e}{x}\right)^x = \left(\dfrac{e}{x}\right)^x[-\log x]$

$$\therefore \int \left(\frac{x}{e}\right)^x \log x\,dx + \int \left(\frac{e}{x}\right)^x \log x\,dx = \left(\frac{x}{e}\right)^x - \left(\frac{e}{x}\right)^x + c$$

Sol 5: $\int \sqrt{\dfrac{\cos(x-a)}{\sin(x+a)}}dx$

$$\int \sqrt{\frac{\cos x\cos a - \sin x\sin a}{\sin x\cos a + \cos x\sin a}}dx$$

$$= \int \sqrt{\frac{1-\tan x\tan a}{\tan x + \tan a}}dx = \int \sqrt{\cot(a+x)}\ dx$$

Sol 6: $\int \dfrac{x^5 + 3x^4 - x^3 + 8x^2 - x + 8}{(x^2+1)}dx$

$$= \int \frac{(x^3 + 3x^2 - 2x + 5)(x^2+1)}{(x^2+1)}dx + \int \frac{x+3}{x^2+1}dx$$

$$= \frac{x^4}{4} + x^3 - x^2 + 5x + \frac{1}{2}\log(x^2+4) + 3\tan^{-1}x + c$$

Sol 7: $\int \dfrac{(\sqrt{x}+1)}{\sqrt{x}(^3\sqrt{x}+1)} = \int \dfrac{(x)^{1/2}+1}{x^{1/2}\left(x^{1/3}+1\right)}dx$

Put $x^{1/6} = t \Rightarrow dx = 6t^5 dt$

$$6\int \frac{(t^3+1)t^5}{t^3(t^2+1)}dt = 6\int \frac{(t^3+1)t^2}{(t^2+1)}dt$$

$$= 6\int \left[\frac{(t^3-t+1)(t^2+1)}{(t^2+1)} + \frac{t-1}{(t^2+1)}\right]dt$$

$$= 6\left[\frac{t^4}{4} - \frac{t^2}{2} + t\right] + 3\log(1+t^2) - 6\tan^{-1}t + c$$

Where $t = x^{1/6}$

Sol 8: $\int \sin^{-1}\sqrt{\dfrac{x}{a+x}}dx$, $a > 0$

$x = a\tan^2\theta$

$dx = 2a\tan\theta\sec^2\theta\,d\theta$

$$\int \sin^{-1}\sqrt{\frac{a\tan^2\theta}{a\sec^2\theta}}\ 2a\tan\theta\sec^2\theta\,d\theta$$

$$2a\int (\sin^{-1}\sin\theta)\tan\theta\sec^2\theta\ d\theta$$

$\tan\theta = t \Rightarrow \sec^2\theta\,d\theta = dt$

$$2a\int t\tan^{-1}t\,dt$$

$$2a\left[\tan^{-1}t\int t\,dt - \frac{1}{2}\int \frac{1}{1+t^2}\times t^2 dt\right]$$

$$= 2a\left[(\tan^{-1} t)\frac{t^2}{2} - \frac{1}{2}\left[t - \tan^{-1} t\right]\right] + c$$

$$= at^2\tan^{-1}t + \left[\frac{1}{2}\tan^{-1} t - \frac{1}{2}t\right]2a + c$$

$$\therefore t = \tan\theta = \frac{\sqrt{x}}{\sqrt{a}}$$

$$\therefore I = a\frac{x}{a}\tan^{-1}\sqrt{\frac{x}{a}} + \left[\frac{1}{2}\tan^{-1}\sqrt{\frac{x}{a}} - \frac{1}{2}\sqrt{\frac{x}{a}}\right]2a + c$$

$$= x\tan^{-1}\sqrt{\frac{x}{a}} + \left[\frac{1}{2}\tan^{-1}\sqrt{\frac{x}{a}} - \frac{1}{2}\sqrt{\frac{x}{a}}\right]2a + c$$

$$= x\tan^{-1}\sqrt{\frac{x}{a}} + a\tan^{-1}\sqrt{\frac{x}{a}} - a\sqrt{\frac{x}{a}} + c$$

$$= \left(\tan^{-1}\sqrt{\frac{x}{a}}\right)(a + x) - a\sqrt{\frac{x}{a}} + c$$

$$= \sqrt{x}\left[\sqrt{x}\tan^{-1}\sqrt{\frac{x}{a}} - \sqrt{a}\right] + a\tan^{-1}\sqrt{\frac{x}{a}} + c$$

Sol 9: $\displaystyle\int \frac{x\log x}{(x^2 - 1)^{3/2}}dx$

$$\therefore \int\left(\frac{x\log x}{(x^2-1)^{3/2}} - \frac{1}{x\sqrt{x^2-1}}\right)dx + \int\frac{1}{x\sqrt{x^2-1}}dx$$

$$\int\left(-d\frac{\log x}{\sqrt{x^2-1}}\right) + \sec^{-1}x + c = -\frac{\log x}{\sqrt{x^2-1}} + \sec^{-1}x + c$$

Sol 10: $\displaystyle\int\frac{\log_6[(\sin x)6^{\cos x}]^{1/6}\cos x}{\sin x}dx$

$$\int\frac{\frac{1}{6}\left[\log_6(\sin x) + \log_6 6^{\cos x}\right]\cos x}{\sin x}dx$$

$$\int\left(\left(\frac{\frac{1}{6}\log(\sin x)}{(\log 6)\sin x} + \frac{1}{6}\frac{\cos x}{\sin x}\right)\cos x\right)dx$$

$$= \int\left[\frac{1}{6}\frac{1}{\log 6}\frac{\cos x}{\sin x}\log(\sin x) + \frac{1}{6}\frac{\cos^2 x}{\sin x}\right]dx$$

$$= \frac{1}{6}\frac{\log^2(\sin x)}{\log 36} + \frac{1}{6}\int(\csc x - \sin x)dx$$

$$= \frac{1}{6}\left[\frac{\log^2(\sin x)}{\log 36} + \ell n\tan\frac{x}{2} + \cos x\right] + c$$

Sol 11: $\displaystyle\int\frac{\sqrt{x^2+1}\left(\log(x^2+1) - \log x^2\right)}{x^4}dx$

$$\int\frac{\left(\sqrt{1+\frac{1}{x^2}}\log\left(1+\frac{1}{x^2}\right)\right)}{x^3}dx$$

$$1 + \frac{1}{x^2} = t \Rightarrow -\frac{2}{x^3}dx = dt$$

$$-\frac{1}{2}\int\sqrt{t}\log t\,dt = -\frac{1}{2}\left[(\log t)\frac{t^{3/2}}{3/2} - \int\frac{1}{t}\times\frac{t^{3/2}}{3/2}dt\right]$$

$$= \frac{1}{2}\left[-\frac{2}{3}(\log t)t^{3/2} + \frac{2}{3}\int t^{1/2}dt\right]$$

$$= \frac{1}{2}\left[-\frac{2}{3}\left[\log\left(1+\frac{1}{x^2}\right)\left[1+\frac{1}{x^2}\right]^{3/2} + \frac{4}{9}\left(1+\frac{1}{x^2}\right)^{3/2} + c\right]\right]$$

$$= -\frac{1}{3}\left[1+\frac{1}{x^2}\right]^{3/2}\left[\log\left(1+\frac{1}{x^2}\right) - \frac{2}{3}\right] + c$$

Sol 12: $f(x) = \dfrac{\sin x}{\sin^2 x + 4\cos^2 x} = \dfrac{\tan x\sec x}{\tan^2 x + 4}$

$$= \frac{\tan x\sec x}{\sec^2 x + 3}$$

Putting $\sec x = t$, $dx\sec x\tan x = dt$ so

$$\int f(x)dx = \int\frac{dt}{t^2+3} = \frac{1}{\sqrt{3}}\tan^{-1}\frac{\sec x}{\sqrt{3}} + c$$

Sol 13: $\displaystyle\int\frac{\cos^2 x}{\sin x(1-\sin x)(1+\cos x)}dx = \int\frac{(1+\sin x)}{\sin x(1+\cos x)}dx$

$$= \int\frac{\left(\sin\frac{x}{2} + \cos\frac{x}{2}\right)^2}{2\sin\frac{x}{2}\cos\frac{x}{2}\times 2\cos^2\frac{x}{2}}dx = \frac{1}{4}\int\frac{\left(\tan\frac{x}{2}+1\right)^2}{\tan\frac{x}{2}\cos^2\frac{x}{2}}dx$$

Let $\tan\frac{x}{2} = t \Rightarrow \sec^2\frac{x}{2}dx = 2dt$

$$= \frac{1}{2}\int\frac{(t+1)^2}{t}dt = \frac{1}{2}\int\left(t+2+\frac{1}{t}\right)dt$$

$$= \frac{t^2}{4} + t + \frac{1}{2}\log|t| + c$$

Sol 14: $\int \dfrac{3x^2+1}{(x^2-1)^3}\,dx$

$\int \left(\dfrac{3(x^2-1)+4}{(x^2-1)^3} \right) dx$ or $\int \dfrac{(3x^2+1)(x^2-1)}{(x^2-1)^4}\,dx$

$= \int \left[\dfrac{3x^2(x^2-1)+(x^2-1)}{(x^2-1)^4} \right] dx = \int \dfrac{3x^4-2x^2-1}{(x^2-1)^4}\,dx$

$= \int -\left[\dfrac{x^4+1-2x^2-4x^4+4x^2}{(x^2-1)^4} \right] dx$

$= \int -\left[\dfrac{(x^2-1)^2-2.2x^2(x^2-1)}{(x^2-1)^4} \right] dx$

$= \int -d\left(\dfrac{x}{(x^2-1)^2} \right) = -\dfrac{x}{(x^2-1)^2}+c$

Sol 15: $\int \dfrac{(ax^2-b)dx}{x\sqrt{c^2x^2-(ax^2+b)^2}}$

$\int \dfrac{(ax^2-b)}{cx^2\sqrt{1-\dfrac{(ax^2+b)^2}{c^2x^2}}} \Rightarrow \int \dfrac{acx^2-bc}{c^2x^2\sqrt{1-\dfrac{(ax^2+b)^2}{c^2x^2}}}dx$

$= \int \dfrac{2acx^2-(acx^2+bc)}{(cx^2)\sqrt{1-\dfrac{(ax^2+b)^2}{c^2x^2}}}dx$

$= \int \dfrac{2acx^2-(ax^2+b)c}{(cx)^2} \times \dfrac{1}{\sqrt{1-\left(\dfrac{ax^2+b}{cx}\right)^2}}dx$

Put $\dfrac{ax^2+b}{cx}=t = \left(\dfrac{(2ax)cx-c(ax^2+b)}{(cx)^2} \right)dx$

$\therefore \int \dfrac{dt}{\sqrt{1-t^2}} = \sin^{-1}t+c = \sin^{-1}\left(\dfrac{ax^2+b}{cx} \right)-c$

Sol 16: Put $z=\tan^{-1}x$, then $dz = \dfrac{1}{1+x^2}dx$ and $x=\tan z$

Now, $I = \int \dfrac{e^{mz}}{\sqrt{1+\tan^2 z}}dz = \int e^{mz}\cos z\,dz$

$= \dfrac{e^{mz}}{m}\cos z - \int \dfrac{e^{mz}}{m}(-\sin z)dz$

$= \dfrac{e^{mz}}{m}\cos z + \dfrac{1}{m}\int e^{mz}\sin z\,dz$

$= \dfrac{e^{mz}}{m}\cos z + \dfrac{1}{m}\left[\dfrac{e^{mz}}{m}\sin z - \int \dfrac{e^{mz}}{m}\cos z\,dz \right]$

$= \dfrac{e^{mz}}{m}\cos z + \dfrac{e^{mz}}{m^2}\sin z - \dfrac{1}{m^2}1$

Or, $\left(1+\dfrac{1}{m^2} \right)1 = \dfrac{e^{mz}}{m^2}(m\cos z+\sin z)$

$\therefore 1 = \dfrac{e^{mz}(m\cos z+\sin z)}{m^2+1}+c$

Or,

$1 = \dfrac{e^{m\tan^{-1}x}}{m^2+1}\left(m\cdot\dfrac{1}{\sqrt{x^2+1}}+\dfrac{x}{\sqrt{x^2+1}} \right)+c = \dfrac{e^{m\tan^{-1}x}(m+x)}{(m^2+1)\sqrt{x^2+1}}+c$

Sol 17: [Here $\sqrt{\dfrac{x}{a+x}}$ occurs, $\therefore$ put $x=a\tan^2\theta$]

Put $x=a\tan^2\theta$, then $dx = 2a\tan\theta\sec^2\theta d\theta$

Now,

$I = \int \sin^{-1}(\sin\theta)2a\tan\theta\sec^2\theta d\theta = 2a\int \theta\cdot(\tan\theta\sec^2\theta)d\theta$

$= 2a\left[\theta\dfrac{\sec^2\theta}{2} - \int 1\dfrac{\sec^2\theta}{2}d\theta \right]+c$

$[\int \tan\theta\sec^2\theta d\theta = \int z\,dz,$ where $z=\sec\theta]$

$= a[\theta\sec^2\theta-\tan\theta]+c$

$= a\left[\tan^{-1}\sqrt{\dfrac{x}{a}}\left(\dfrac{a+x}{a}\right) - \sqrt{\dfrac{x}{a}} \right]+c$

Sol 18: $\dfrac{d\sin\left(e^{\sqrt{x}}+e^{\sqrt{x}}+\dfrac{\pi}{4} \right)}{dx}$

$= \cos\left(e^{\sqrt{x}}+e^{-\sqrt{x}}+\dfrac{\pi}{4} \right)\left[\dfrac{e^{\sqrt{x}}}{2\sqrt{x}} - \dfrac{e^{-\sqrt{x}}}{2\sqrt{x}} \right]$

Also $\dfrac{d}{dx}\sin\left(e^{\sqrt{x}}-e^{-\sqrt{x}}+\dfrac{\pi}{4} \right)$

$= \cos\left(e^{\sqrt{x}}-e^{-\sqrt{x}}+\dfrac{\pi}{4} \right)\left[\dfrac{e^{\sqrt{x}}}{2\sqrt{x}} + \dfrac{2e^{-\sqrt{x}}}{2\sqrt{x}} \right]$

$$\therefore I = 2\int\left[\begin{array}{l} d\sin\left(e^{\sqrt{x}} + e^{-\sqrt{x}} + \dfrac{\pi}{4}\right) \\[2mm] + d\sin\left(e^{\sqrt{x}} - e^{-\sqrt{x}} + \dfrac{\pi}{4}\right) \end{array}\right]$$

$$= 2\left[\sin\left(e^{\sqrt{x}} + e^{-\sqrt{x}} + \frac{\pi}{4}\right) + \sin\left(e^{\sqrt{x}} - e^{-\sqrt{x}} + \frac{\pi}{4}\right)\right]$$

$$= 2\sin\left(e^{\sqrt{x}} + \frac{\pi}{4}\right)\cos\left(e^{-\sqrt{x}}\right)$$

$$= \sqrt{2}\left(\sin\left(e^{\sqrt{x}}\right) + \cos\left(e^{\sqrt{x}}\right)\right)\cos\left(e^{-\sqrt{x}}\right) + c$$

Sol 19: $\int \dfrac{(x^2 + x)}{(e^x + x + 1)^2}\,dx$

$$= \int\left(\frac{x(e^x + x + 1) - xe^x}{(e^x + x + 1)^2}\right)dx$$

$$= \int\left(\frac{x}{e^x + x + 1} - \frac{xe^x}{(e^x + x + 1)^2}\right)dx$$

$$= \int\left[\left(\frac{1}{1 + \left(\frac{x+1}{e^x}\right)}\right) \times \frac{x}{e^x} - \frac{1}{\left(1 + \frac{x+1}{e^x}\right)^2}\left(\frac{x}{e^x}\right)\right]dx$$

Put $\dfrac{x+1}{e^x} + 1 = t \Rightarrow \dfrac{-x}{e^x}dx = dt$

$$\int\left(-\frac{dt}{t} + \frac{1}{t^2}\right)dt = -\log t - \frac{1}{t} + c$$

$$= -\log\left(\frac{x+1}{e^x} + 1\right) - \frac{1}{\left(\frac{x+1}{e^x} + 1\right)} + c$$

Sol 20: $\int \dfrac{e^{\cos x}(x\sin^3 x + \cos x)}{\sin^2 x}\,dx$

$$\int e^{\cos x}\left(\frac{\cos x}{\sin^2 x} + x\sin x\right)dx$$

$$\int e^{\cos x}(\cot x\, \mathrm{cosec}\, x + x\sin x)dx$$

$$\Rightarrow -\int e^{\cos x}(1 - \cot x\, \mathrm{cosec}\, x - x\sin x - 1)dx$$

$$= -\int\left[e^{\cos x}(1 - \mathrm{cosec}\, x\cot x) + (x + \mathrm{cosec}\, x)e^{\cos x}(-\sin x)\right]dx$$

$$= -\int d e^{\cos x}(x + \mathrm{cosec}\, x) = -e^{\cos x}(x + \mathrm{cosec}\, x) + c$$

Sol 21: $\int \dfrac{5x^4 + 4x^5}{(x^5 + x + 1)^2}\,dx$

$$= \int \frac{5x^5 + 5x^4 + x + 1 - x^5 - x - 1}{(x^5 + x + 1)^2}\,dx$$

$$= \int -\frac{(x^5 + x + 1) + (5x^4 + 1)(x + 1)}{(x^5 + x + 1)^2}\,dx$$

$$= \int -d\left(\frac{x+1}{x^5 + x + 1}\right) = -\frac{x+1}{x^5 + x + 1} + c \quad \text{or} \quad \frac{x^5}{x^5 + x + 1} + c$$

Sol 22: $\int\left(\dfrac{a^2\tan^2 x + b^2}{a^4\tan^2 x + b^4}\right)dx$

$$\Rightarrow \frac{1}{(a^2 + b^2)}\int \frac{b^2(a^2 + b^2) + a^2(b^2 + a^2)\tan^2 x}{a^4\tan^2 x + b^4}\,dx$$

$$\Rightarrow \frac{1}{a^2 + b^2}\int\left(\frac{a^2 b^2(1 + \tan^2 x)}{a^4\tan^2 x + b^4} + 1\right)dx$$

$$\Rightarrow \frac{1}{a^2 + b^2}\int\left(\frac{a^2 b^2 + b^4 + (a^2 b^2 + a^4)\tan^2 x}{a^4\tan^2 x + b^4}\right)dx$$

$$\Rightarrow \frac{1}{a^2 + b^2}\int\left(\left(\frac{a^2}{b^2}\right)\sec^2 x \times \frac{1}{\left(\frac{a^4}{b^4}\tan^2 x + 1\right)} + 1\right)dx$$

$$\Rightarrow \frac{1}{a^2 + b^2}x + \frac{1}{a^2 + b^2}\int \frac{\left(\frac{a}{b}\right)^2\sec^2 x}{1 + \left(\frac{a^2}{b^2}\tan x\right)^2}\,dx$$

Put $\dfrac{a^2}{b^2}\tan x = t \Rightarrow \dfrac{a^2}{b^2}\sec^2 x\,dx = dt$

$$\therefore \frac{1}{a^2 + b^2}x + \frac{1}{a^2 + b^2}\int \frac{dt}{1 + t^2}$$

Or $\dfrac{1}{a^2 + b^2}\left[x + \tan^{-1}\dfrac{a^2}{b^2}\tan x\right] + c$

Sol 23: $\int \dfrac{\cos^2 x}{1 + \tan x}\,dx = \int \dfrac{\cos^3 x}{\sin x + \cos x}\,dx$

$$= \frac{1}{4}\int \frac{3\cos x}{\sin x + \cos x}\,dx + \frac{1}{4}\int \frac{\cos 3x}{\sin x + \cos x}\,dx$$

$$= \frac{1}{4}\int \frac{\cos x\cos 2x - \sin x\sin 2x}{(\sin x + \cos x)}dx + \frac{3}{4}\int \frac{\cos x}{\sin x + \cos x}dx$$

$$= \frac{1}{4}\int\left[(\cos 2x - \sin 2x) + \left(\frac{\sin 2x\cos x - \cos 2x\sin x}{\sin x + \cos x}\right)\right]dx$$

$$+ \frac{3}{4}\int \frac{\cos x}{\sin x + \cos x}dx$$

$$= \frac{1}{4}\int(\cos 2x - \sin 2x)dx + \frac{1}{4}\int dx$$

$$- \frac{1}{4}\int \frac{\cos x\,dx}{\sin x + \cos x} + \frac{3}{4}\int \frac{\cos x\,dx}{\sin x + \cos x}$$

$$= \frac{1}{8}(\sin 2x + \cos 2x) + \frac{x}{4} + \frac{1}{2}\int \frac{\cos x\,dx}{\sin x + \cos x}$$

$$= \frac{1}{8}(\sin 2x + \cos 2x) + \frac{x}{4}$$

$$+ \frac{1}{4}\int\left(\frac{\cos x + \sin x + \cos x - \sin x}{\cos x + \sin x}\right)dx$$

$$= \frac{1}{8}(\sin 2x + \cos 2x) + \frac{x}{4} + \frac{1}{4}\int\left[1 + \left(\frac{\cos x - \sin x}{\cos x + \sin x}\right)\right]dx$$

$$= \frac{1}{8}(\sin 2x + \cos 2x) + \frac{x}{2} + \frac{1}{4}\log(\cos x + \sin x) + c$$

Sol 24:

$$I = (x\log x - x)\sin^{-1}x - \int \frac{x\log x}{\sqrt{1-x^2}}dx + \int \frac{x}{\sqrt{1-x^2}}dx \quad ...(i)$$

[Integrating by parts taking $\sin^{-1}x$ as u]

Now in order to evaluate $\int \frac{x\log x}{\sqrt{1-x^2}}dx$

Put $x = \sin\theta$, then $dx = \cos\theta\,d\theta$

$$\therefore \int \frac{x\log x}{\sqrt{1-x^2}}dx = \int \sin\theta \log\sin\theta\,d\theta$$

$$= -\cos\theta \log\sin\theta - \int -\cos\theta \cot\theta\,d\theta$$

$$= -\cos\theta\log\sin\theta + \int \frac{\cos^2\theta}{\sin\theta}d\theta$$

$$= -\cos\theta\log\sin\theta + \int \frac{1-\sin^2\theta}{\sin\theta}d\theta$$

$$= -\cos\theta\log\sin\theta + \int (\cosec\theta - \sin\theta)d\theta$$

$$= -\cos\theta\log\sin\theta + \log|\cosec\theta - \cot\theta| + \cos\theta$$

$$= -\sqrt{1-x^2}\log x + \log\left(\frac{1-\sqrt{1-x^2}}{x}\right) + \sqrt{1-x^2}$$

Again, $\int \frac{x}{\sqrt{1-x^2}}dx = -\sqrt{1-x^2}$

$\therefore$ from (i),

$$I = (x\log x - x)\sin^{-1}x + \sqrt{1-x^2}\,\log x - \log\left(\frac{1-\sqrt{1-x^2}}{x}\right) + c$$

Sol 25: $\int \frac{(x^2+1)e^x}{(x+1)^2}dx = \int e^x \frac{(x^2-1)+2}{(x+1)^2}dx$

$$= \int e^x\left[\frac{x^2-1}{(x+1)} + \frac{2}{(x+1)^2}\right]dx = \int e^x\left[\frac{x-1}{x+1} + \frac{2}{(x+1)^2}\right]dx$$

$$= \int e^x[f(x) + f'(x)]dx, \quad \text{where} \quad f(x) = \frac{x-1}{x+1}$$

$$= e^x\,f(x) + c = e^x\left(\frac{x-1}{x+1}\right) + c$$

Sol 26: [Here $e^{f(z)}$ occurs, where $f(x) = \sin x$

$\therefore$ Put $z = f(x) = \sin x$]

Put $z = \sin x$, then $dz = \cos x\,dx$

Now, $I = \int \frac{e^{\sin x}}{\cos^3 x}(x\cos^3 x - \sin x)dz$

$$= \int e^{\sin x}(x - \tan x\sec^2 x)dz$$

$$= \int e^z\left[\sin^{-1}z - \frac{z}{\sqrt{1-z^2}}\cdot\frac{1}{1-z^2}\right]dz \qquad [\because \sin x = z]$$

$$= \int e^z\left[\sin^{-1}z + \frac{1}{\sqrt{1-z^2}} - \frac{1}{\sqrt{1-z^2}} - \frac{z}{(1-z^2)^{3/2}}\right]dz$$

$$= \int e^z\left[\left(\sin^{-1}z - \frac{1}{\sqrt{1-z^2}}\right) + \left\{\frac{1}{\sqrt{1-z^2}} - \frac{z}{(1-z^2)^{3/2}}\right\}\right]dz$$

$$= \int e^z[f(z) + f'(z)]dz, \quad \text{where} \quad f(z) = \sin^{-1}z - \frac{1}{\sqrt{1-z^2}}$$

$$= e^z f(z) + c = e^z\left(\sin^{-1}z - \frac{1}{\sqrt{1-z^2}}\right) + c = e^{\sin x}(x - \sec x) + c$$

Sol 27: $I = \int \frac{dx}{\sqrt{1-(x^2+2x)}} = \int \frac{dx}{\sqrt{2-(x^2+2x+1)}}$

$$= \int \frac{dx}{\sqrt{2-(1+x)^2}} = \int \frac{dx}{\sqrt{(2)^2-(1+x)^2}} \qquad ...(i)$$

Let $z = 1 + x$, then $dz = dx$

From (i),

$$I = \int \frac{dz}{\sqrt{(\sqrt{2})^2 - z^2}} = \sin^{-1}\frac{z}{\sqrt{2}} + c = \sin^{-1}\left(\frac{1+x}{\sqrt{2}}\right) + c$$

Sol 28: $\int \frac{dx}{\sec x + \cos ecx} = \int \left(\frac{\sin x \cos x}{\sin x + \cos x}\right) dx$

$$= \frac{1}{2}\int \left(\frac{1 + 2\sin x \cos x - 1}{\sin x + \cos x}\right) dx$$

$$= \frac{1}{2}\int \frac{(\sin x + \cos x)^2 - 1}{(\sin x + \cos x)} dx$$

$$= \frac{1}{2}\int (\sin x + \cos x) dx - \frac{1}{2\sqrt{2}}\int \frac{1}{\frac{1}{\sqrt{2}}(\sin x + \cos x)} dx$$

$$= \frac{1}{2}[\sin x - \cos x] - \frac{1}{2\sqrt{2}}\int \frac{dx}{\sin\left(x + \frac{\pi}{4}\right)}$$

$$= \frac{1}{2}[\sin x - \cos x] - \frac{1}{2\sqrt{2}}\int \frac{\sec^2\left(\frac{x}{2} + \frac{\pi}{8}\right)}{2\tan\left(\frac{x}{2} + \frac{\pi}{6}\right)} dx$$

$$= \frac{1}{2}[\sin x - \cos x] - \frac{1}{2\sqrt{2}}\log\tan\left(\frac{x}{2} + \frac{\pi}{8}\right) + c$$

$$= \frac{1}{2}\left[\sin x - \cos x - \frac{1}{\sqrt{2}}\log\tan\left(\frac{x}{2} + \frac{\pi}{8}\right)\right] + c$$

Sol 29 $I = \int \sqrt{2x^2 + 3x + 4}\ dx = \int \sqrt{2\left(x^2 + \frac{3}{2}x + 2\right)} dx$

$$= \sqrt{2}\int \sqrt{x^2 + \frac{3}{2}x + 2}\ dx$$

$$= \sqrt{2}\int \sqrt{x^2 + 2\cdot x\cdot\frac{3}{4} + \left(\frac{3}{4}\right)^2 - \left(\frac{3}{4}\right)^2 + 2}\ dx$$

$$= \sqrt{2}\int \sqrt{\left(x + \frac{3}{4}\right)^2 + \frac{23}{16}}\ dx \qquad \text{...(i)}$$

Let, $z = x + \frac{3}{4}$, then $dz = dx$. Let $\sqrt{\frac{23}{16}} = \frac{\sqrt{3}}{4} = a$

Then from (i), $I = \sqrt{2}\int \sqrt{z^2 + a^2}\ dz$

Now,

$$\int \sqrt{z^2 + a^2}\,dz = \frac{z}{2}\sqrt{z^2 + a^2} + \frac{a^2}{2}\log\left(z + \sqrt{z^2 + a^2}\right) + c$$

$$= \frac{\left(-x + \frac{3}{4}\right)}{2}\sqrt{\left(x + \frac{3}{4}\right)^2 + \frac{23}{16}}$$

$$+ \frac{23}{32}\log\left[x + \frac{3}{4} + \sqrt{\left(x + \frac{3}{4}\right)^2 + \frac{23}{16}}\right] + c$$

$$I = \frac{4x + 3}{8}\sqrt{2x^2 + 3x + 4}$$

$$+ \frac{23}{16\sqrt{2}}\log\left[\frac{4x + 3}{4} + \frac{\sqrt{2x^2 + 3x + 4}}{\sqrt{2}}\right] + c$$

Sol 30: Let $z = x^n + 1$, then $dz = nx^{n-1}dx$

Now, $I = \int \frac{dx}{x\left(x^n + 1\right)} = \int \frac{dx}{nx^{n-1}\cdot x\left(x^n + 1\right)}$

$$= \frac{1}{n}\int \frac{dz}{x^n\left(x^n + 1\right)} = \frac{1}{n}\int \frac{dz}{(z-1)z} \qquad \text{...(i)}$$

Let $\frac{1}{z(z-1)} = \frac{A}{z} + \frac{B}{z-1} = \frac{A(z-1) + Bz}{z(z-1)} \qquad \text{...(ii)}$

$\therefore A(z-1) + Bz = 1 \qquad \text{...(iii)}$

Putting Z = 0 we get, $-A = 1$

$\therefore A = -1$

Putting Z = 1, we get B = 1

$\therefore$ From $I = \frac{1}{n}\int \frac{dz}{z(z-1)} = \frac{1}{n}\int \left(-\frac{1}{z} + \frac{1}{z-1}\right) dz$

$$= \frac{1}{n}[-\log|z| + \log|z-1|] + c$$

$$= \frac{1}{n}\log\left|\frac{z-1}{z}\right| + c = \frac{1}{n}\log\left|\frac{x^n}{x^n + 1}\right| + c$$

Sol 31: $\int \frac{\cos x - \sin x}{16 - 9(1 + \sin 2x)} dx$

$$\int \frac{(\cos x - \sin x)}{16 - 9(\sin x + \cos x)^2} dx$$

Let $3(\sin x + \cos x) = t$

$\Rightarrow (3\cos x - 3\sin x)dx = dt$

$\dfrac{1}{3}\displaystyle\int \dfrac{dt}{16-t^2} = \dfrac{1}{2.4}\cdot\dfrac{1}{3}\log\left|\dfrac{4+t}{4-t}\right| = \left(\dfrac{1}{24}\log\dfrac{4+3\cos x+3\sin x}{4-3\cos x-3\sin x}\right)$

Sol 32: $\displaystyle\int \dfrac{\sqrt{\cot x}-\sqrt{\tan x}}{1+3\sin 2x}dx$

$\sqrt{2}\displaystyle\int \dfrac{\cos x - \sin x}{\left(\sqrt{\sin 2x}\right)(1+3\sin 2x)}dx$

$\sqrt{2}\displaystyle\int\left[\dfrac{(\cos 2x\cos x+\sin^2 x\sin x)+(\cos 2x\sin x-\sin 2x\cos x)}{\left(\sqrt{\sin 2x}\right)(1+3\sin 2x)}\right]dx$

$2\displaystyle\int\left[\dfrac{\cos 2x(\sin x+\cos x)-\sin 2x(\cos x-\sin x)}{\sqrt{2\sin 2x}(1+3\sin 2x)}\right]dx$

$\Rightarrow \displaystyle\int \dfrac{1}{\dfrac{(\sin x+\cos x)^2+2\sin 2x}{(\sin x+\cos x)^2}}$

$\times\displaystyle\int\left[\dfrac{(\sin x+\cos x)2.\cos 2x-2\sin^2 x(\cos x-\sin x)}{(\sin x+\cos x)^2\sqrt{2\sin 2x}}\right]dx$

$\Rightarrow \displaystyle\int\left(\dfrac{1}{1+\dfrac{2\sin 2x}{(\sin x+\cos x)^2}}\right)\left(\dfrac{2\sin 2x}{\sin x+\cos x}\right)dx$

$= \tan^{-1}\left(\dfrac{\sqrt{2}\sin 2x}{\sin x+\cos x}\right)+c$

Sol 33: $\displaystyle\int \dfrac{4x^5-7x^4+8x^3-2x^2+4x-7}{x^2(x^2+1)^2}dx$

$\therefore \dfrac{4x^5-7x^4+8x^3-2x^2+4x-7}{x^2(x^2+1)^2}$

$= \dfrac{A}{x}+\dfrac{B}{x^2}+\dfrac{Cx+D}{x^2+1}+\dfrac{Ex^2+Fx+G}{(x^2+1)^2}$

$\displaystyle\int \dfrac{(4x^5-7x^2+8x^3-2x^2+4x-7)}{x^2(x^2+1)^2}dx$

$= \displaystyle\int\left(\dfrac{4}{x}+\left(-\dfrac{7}{x^2}\right)+\dfrac{6}{x^2+1}+\dfrac{6(1-x^2)}{(1+x^2)^2}\right)dx$

$= 4\log x + \dfrac{7}{x} + 6\tan^{-1}x + 6\displaystyle\int\left(\dfrac{1-x^2}{(1+x^2)^2}\right)dx$

Put $x = \tan\theta$

$\Rightarrow dx = \sec^2\theta\, d\theta$

$= 4\log x + \dfrac{7}{x} + 6\tan^{-1}x + 6\displaystyle\int\dfrac{1-\tan^2\theta}{(1+\tan^2\theta)}d\theta$

$= 4\log x + \dfrac{7}{x} + 6\tan^{-1}x + 6\displaystyle\int\cos 2\theta\, d\theta$

$= 4\log x + \dfrac{7}{x} + 6\tan^{-1}x + 6\dfrac{1}{2}\sin 2\theta$

$\because \sin 2\theta = \dfrac{2\tan\theta}{1+\tan^2\theta} = \dfrac{2x}{1+x^2}$

$\therefore I = 4\log x + \dfrac{7}{x} + 6\tan^{-1}x + \dfrac{6x}{1+x^2} + c$

Sol 34: $\displaystyle\int \dfrac{dx}{\cos^3 x - \sin^3 x}$

$\displaystyle\int \dfrac{dx}{(\cos x-\sin x)(1+\cos x\sin x)}$

$= \displaystyle\int \dfrac{(\cos x-\sin x)^2+2\sin x\cos x}{(\cos x-\sin x)(1+\cos x\sin x)}dx$

$\displaystyle\int \dfrac{(\cos x-\sin x)}{(1+\cos x\sin x)}dx + 2\displaystyle\int \dfrac{\sin x\cos x}{(\cos x-\sin x)(1+\cos x\sin x)}dx$

$= \dfrac{2}{3}\displaystyle\int \dfrac{\cos x-\sin x}{1+(\sin x+\cos x)^2}dx + \dfrac{2}{3}\displaystyle\int \dfrac{(\cos x-\sin x)}{2-(\sin x+\cos x)^2}cx$

$= \dfrac{2}{3}\tan^{-1}(\sin x-\cos x) + \dfrac{1}{3\sqrt{2}}\log\left|\dfrac{\sqrt{2}+\sin x+\cos x}{\sqrt{2}-\sin x-\cos x}\right| + c$

Sol 35: $\displaystyle\int \dfrac{x^2}{(x\cos x-\sin x)(x\sin x+\cos x)}dx$

$\displaystyle\int \dfrac{(x\cos x-\sin x)^2+(x\sin x+\cos x)^2}{(x\cos x-\sin x)(x\sin x+\cos x)}dx$

$\displaystyle\int \dfrac{x\cos x-\sin x}{x\sin x+\cos x}dx + \displaystyle\int \dfrac{x\sin x+\cos x}{x\cos x-\sin x}dx$

$+ \displaystyle\int \dfrac{-1}{(x\cos x-\sin x)(x\sin x+\cos x)}dx$

$= \displaystyle\int\left(\dfrac{x\cos x+\sin x-\sin x}{x\sin x+\cos x}\right)dx + \displaystyle\int\dfrac{x\sin x+\sin x-\cos x}{x\cos x-\sin x}dx$

$+ \displaystyle\int \dfrac{\cos^2 x+\sin^2 x-1}{(x\sin x+\cos x)(x\cos x-\sin x)}dx$

$= \log\left|\dfrac{x\sin x+\cos x}{x\cos x-\sin x}\right| + c$

Sol 36: $\int \cos 2\theta \log\left(\dfrac{\cos\theta+\sin\theta}{\cos\theta-\sin\theta}\right)d\theta$

$\Rightarrow \log\left(\dfrac{\cos\theta+\sin\theta}{\cos\theta-\sin\theta}\right)\int \cos 2\theta\, d\theta$

$$-\int\left(\dfrac{d\log\left(\dfrac{\cos\theta+\sin\theta}{\cos\theta-\sin\theta}\right)}{d\theta}\int \cos 2\theta\, d\theta\right)d\theta$$

$\Rightarrow \dfrac{1}{2}\sin 2\theta \log\left(\dfrac{\cos\theta+\sin\theta}{\cos\theta-\sin\theta}\right) - \int\left(\dfrac{2}{\cos 2\theta}\times\dfrac{\sin 2\theta}{2}\right)d\theta$

$\Rightarrow \dfrac{1}{2}\sin 2\theta \log\left(\dfrac{\cos\theta+\sin\theta}{\cos\theta-\sin\theta}\right) - \int\tan 2\theta\, d\theta$

$\Rightarrow \dfrac{1}{2}\sin 2\theta \log\left(\dfrac{\cos\theta+\sin\theta}{\cos\theta-\sin\theta}\right) - \dfrac{1}{2}\log(\sec 2\theta)d\theta$

$= \left(\dfrac{1-\tan\theta}{1+\tan\theta}\right)\left(\dfrac{(1-\tan\theta\times\sec^2\theta)}{(1-\tan\theta)^2}-(1+\tan\theta)\sec^2\theta\right)$

$= \dfrac{(1-\tan\theta)}{(1-\tan\theta)}\times\dfrac{2\sec^2\theta}{(1-\tan\theta)^2}=\dfrac{2(1+\tan^2\theta)}{(1-\tan^2\theta)}=\dfrac{2}{\cos 2\theta}$

Sol 37: A → s; B → p; C → q; D → r

(A) $\int \dfrac{x^4-1}{x^3\sqrt{x^2+\dfrac{1}{x^2}+1}}dx$

$\int \dfrac{x-x^{-3}}{\sqrt{x^2+\dfrac{1}{x^2}+1}}dx$

Put $x^2+\dfrac{1}{x^2}+1=t$

$\left(2x-\dfrac{2}{x^3}\right)dx=\dfrac{dt}{2}$ or $\left(x-\dfrac{1}{x^3}\right)dx=\dfrac{dt}{2}$

$\dfrac{1}{2}\int\dfrac{dt}{\sqrt{t}}=\sqrt{t}+c=\sqrt{x^2+\dfrac{1}{x^2}+1}+c=\sqrt{\dfrac{x^4+x^2+1}{x}}+c$

(B) $\int\dfrac{x^2-1}{x\sqrt{1+x^4}}dx$

$\int\dfrac{x^2-1}{x^2\sqrt{x^2+\dfrac{1}{x^2}}}dx \Rightarrow \int\dfrac{(1-x^{-2})}{\sqrt{\left(x+\dfrac{1}{x}\right)^2-2}}dx$

Put $x+\dfrac{1}{x}=t$

$\Rightarrow(1-x^{-2})dx=dt$

$\int\dfrac{dt}{\sqrt{t^2-2}}$

Put $t=\sqrt{2}\,\sec\theta$

$\int\dfrac{\sqrt{2}\sec\theta\tan\theta d\theta}{\sqrt{2}\tan\theta}=\log|\sec\theta+\tan\theta|$

$=\log\left|\dfrac{x^2+1}{\sqrt{2}x}+\dfrac{\sqrt{x^4+1}}{\sqrt{2}x}\right|+c$

Or $\log\left|\dfrac{x^2+1}{x}+\dfrac{\sqrt{x^4+1}}{x}\right|+c$

(C) $\int\dfrac{(1+x^2)}{x^2\left(\dfrac{1}{x}-x\right)\sqrt{x^2+\dfrac{1}{x^2}}}dx$

$\int\dfrac{(1+x^{-2})}{\left(\dfrac{1}{x}-x\right)\sqrt{\left(x-\dfrac{1}{x}\right)^2+2}}dx$

$x-\dfrac{1}{x}=t \Rightarrow (1+x^{-2})dx=dt$

$=-\int\dfrac{dt}{t\sqrt{t^2+2}}$

$t=\sqrt{2}\tan\theta$

$dt=\sqrt{2}\sec^2\theta\, d\theta$

$=-\int\dfrac{\sqrt{2}\sec^2\theta d\theta}{\sqrt{2}\tan\theta\sqrt{2}\sec\theta}=-\dfrac{1}{\sqrt{2}}\int\cosec\theta d\theta$

$=\dfrac{1}{\sqrt{2}}\log|\cosec\theta-\cot\theta|+c$

$\Rightarrow-\dfrac{1}{\sqrt{2}}\log\left|\dfrac{\sqrt{x^4+1}-\sqrt{x}}{(x^2-1)}\right|+c$

(D) $\int\dfrac{\sqrt{\sqrt{1+x^4}+x^2}}{1+x^4}dx$

$=\int\left(\dfrac{1}{\sqrt{1+\dfrac{1}{x^4}}}\right)\dfrac{1}{2\sqrt{\left(\sqrt{1+\dfrac{1}{x^4}}-1\right)}}\times\dfrac{1}{2\sqrt{1+\dfrac{1}{x^4}}}\times\left(\dfrac{-4}{x^5}\right)$

1.64

$$- \tan^{-1}\left(\sqrt{\sqrt{1+\frac{1}{x^4}}-1}\right) + c$$

$$\therefore \frac{d\tan^{-1}\left(\sqrt{\sqrt{1+\frac{1}{x^4}}-1}\right)}{dx}$$

$$= \frac{1}{1+\left(\sqrt{1+\frac{1}{x^4}}-1\right)} \times \left(\sqrt{\sqrt{1+\frac{1}{x^4}}-1}\right)$$

$$= \left(\frac{1}{\sqrt{1+\frac{x}{4}}}\right)\left(\frac{1}{2\sqrt{1+\frac{1}{x^4}}-1}\right)\left(\frac{1}{2\sqrt{1+\frac{1}{x^4}}}\right) \times \frac{-4}{x^5}$$

Exercise 2

Single Correct Choice Type

Sol 1: (D) $\int \dfrac{\tan^{-1}x - \cot^{-1}x}{\tan^{-1}x + \cot^{-1}x}dx$

$$\because \ \tan^{-1}x + \cot^{-1}x = \frac{\pi}{2}$$

$$I = \int\left(\frac{2\tan^{-1}\frac{x}{2} - \frac{\pi}{2}}{\frac{\pi}{2}}\right)dx = \frac{4}{\pi}\int \tan^{-1}x\,dx - \int dx$$

$$= \frac{4}{\pi}\int \tan^{-1}x\int dx - \int\left[\left(\left(\frac{d\tan^{-1}x}{dx}\right)\int dx\right)dx\right] - x + c$$

$$= \frac{4}{\pi}x\tan^{-1}x - \frac{4}{\pi}\int\left(\frac{x}{1+x^2}\right)dx - x + c$$

$$= \frac{4}{\pi}x\tan^{-1}x - \frac{2}{\pi}\log(1+x^2) - x + c$$

Sol 2: (A) $\int \dfrac{x^2-4}{x^4+24x^2+16}dx = \int \dfrac{x^2-4}{x^2\left(x^2-24+\frac{16}{x^2}\right)}dx$

$$= \int \frac{\left(1-\frac{4}{x^2}\right)}{\left(x^2+\frac{16}{x^2}+8\right)+16}dx = \int \frac{\left(1-\frac{4}{x^2}\right)}{\left(x+\frac{4}{x}\right)^2+(4)^2}dx$$

Put $x + \dfrac{4}{x} = t$

$$\Rightarrow \left(1-\frac{4}{x^2}\right)dx = dt$$

$$\therefore \int \frac{dt}{t^2(4)^2} = \frac{1}{4}\tan^{-1}\frac{t}{4} + c$$

$$\therefore I = \frac{1}{4}\tan^{-1}\left(\frac{x+\frac{4}{x}}{4}\right) + c$$

$$\text{Or} \quad I = \frac{1}{4}\tan^{-1}\left(\frac{x^2+4}{4x}\right) + c$$

Sol 3: (D) $\int \dfrac{(x-1)^2}{x^4-2x^2+1}dx = \int \dfrac{x^2-1-2x+2}{(x^2+1)^2}dx$

$$= \int \frac{(x^2-1)}{x^2\left(x+\frac{1}{x}\right)^2}dx + \int\frac{-2x}{(x^2+1)^2}dx + \int\frac{2dx}{(1+x^2)^2}$$

Put $x^2 + 1 = t$ and Put $x = \tan\theta$

$dx = \sec^2\theta\, d\theta$

$$= \int \frac{\left(1-\frac{1}{x^2}\right)}{\left(x+\frac{1}{x}\right)^2}dx + \int -\frac{dt}{t^2} + 2\int\frac{\sec^2\theta\, d\theta}{\sec^4\theta}$$

Put $x + \dfrac{1}{x} = u$

$$= \int\frac{du}{u^2} + \frac{1}{x^2+1} - 2\int\cos^2\theta\, d\theta$$

$$= -\frac{1}{\left(x+\frac{1}{x}\right)} + \frac{1}{(x^2+1)} + \frac{2}{2}\int(\cos2\theta+1)d\theta$$

$$= \frac{(1-x)}{x^2+1} + \frac{1}{2}\sin2\theta + \theta + c$$

$$= \frac{(1-x)}{1+x^2} + \frac{1}{2}\frac{2\times x}{1+x^2} + \tan^{-1}x + c$$

$$= \frac{(1-x)}{1+x^2} + \frac{x}{1+x^2} + \tan^{-1}x + c = \frac{1}{1+x^2} + \tan^{-1}x + c$$

Sol 4: (A) $\int \dfrac{x^4-4}{x^2\sqrt{x^4+x^2+4}}dx$

$$\int \frac{x^4-4}{x^2\times x\sqrt{x^2+1+\frac{4}{x^2}}}dx$$

$$\int \frac{\left[x - \dfrac{4}{x^3}\right]}{\sqrt{x^2 + 1 + \dfrac{4}{x^2}}} dx$$

Put $x^2 + \dfrac{4}{x^2} + 1 = t \Rightarrow \left(2x - \dfrac{8}{x^3}\right)dx = dt$

$= \dfrac{1}{2}\int \dfrac{dt}{\sqrt{t}} = \dfrac{1}{2}.2\sqrt{t} + c = \sqrt{x^2 + 1 + \dfrac{4}{x^2}} + c$

$= \dfrac{\left(\sqrt{x^4 + x^2 + 4}\right)}{x} + c$

Sol 5: (A) $\int \left(\dfrac{\sec x + \tan x - 1}{\tan x - \sec x + 1}\right)dx$

$= \int \dfrac{\sec x + \tan x - (\sec^2 x - \tan^2 x)}{(\tan x - \sec x + 1)}dx$

$= \int \dfrac{(\sec x + \tan x)[1 - \sec x + \tan x]}{[\tan x - \sec x + 1]}dx$

$= \int (\sec x + \tan x)dx$

$= \log|\sec x + \tan x| + \log \sec x + c$

$s = 1 + 2x + 3x^2 + 4x^3 + \ldots\ldots\ldots$

$s_x = x + 2x^2 + 3x^3 + \ldots\ldots\ldots$

$s(1 - x) = 1 + x + x^2 + x^3 + \ldots\ldots$

$s(1 - x) = \dfrac{1}{1 - x} \quad \therefore s = \dfrac{1}{(1 - x)^2}$

$\therefore \int \dfrac{1}{(1 - x)^2}dx = \dfrac{(1 - x)^{-1}}{-1 \times -1} + c = (1 - x)^{-1} + c$

Sol 6: (C) $\int e^x \dfrac{(x^2 - 3)}{(x + 3)^2}dx = \int e^x \dfrac{(x^2 - 9 + 6)}{(x + 3)^2}dx$

$= \int e^x \left(\dfrac{x - 3}{x + 3} + \dfrac{6}{(x + 3)^2}\right)dx$

$= \int e^x \left(f(x) + f'(x)\right)dx = e^x f(x) + c$

$= e^x \dfrac{(x - 3)}{(x + 3)} + c$

Sol 7: (B) $I = \int \sqrt{\dfrac{1 - \cos x}{\cos \alpha - \cos x}}dx$

$I = \int \dfrac{\sqrt{2}\sin\dfrac{x}{2}}{\sqrt{2\cos^2\dfrac{\alpha}{2} - 2\cos^2\dfrac{x}{2}}}$

$= \int \dfrac{\sqrt{2}\sin\dfrac{x}{2}}{\sqrt{2}\cos\dfrac{\alpha}{2}\sqrt{1 - \left(\dfrac{\cos\dfrac{x}{2}}{\cos\dfrac{\alpha}{2}}\right)^2}}dx$

$= \int \dfrac{\dfrac{1}{2} \times 2x \left(\dfrac{-\sin\dfrac{x}{2}}{\cos\dfrac{\alpha}{2}}\right)}{\sqrt{1 - \left(\dfrac{\cos\dfrac{x}{2}}{\cos\dfrac{\alpha}{2}}\right)^2}}dx$

Let $\dfrac{\cos\dfrac{x}{3}}{\cos\dfrac{\alpha}{2}} = t \quad \therefore I = \int -\dfrac{2}{\sqrt{1 - t^2}}dt$

$= 2\cos^{-1}(t) + c = 2\cos^{-1}\left(\dfrac{\cos\dfrac{x}{2}}{\cos\dfrac{\alpha}{2}}\right) + c$

Sol 8: (B) $I = \int \dfrac{3x^4 - 1}{(x^4 + x + 1)^2}dx$

$= \int \dfrac{4x^4 + x - (x^4 + x + 1)}{(x^4 + x + 1)^2}dx$

$= \int \dfrac{x(4x^2 - 1)}{(x^4 + x + 1)^2} - \dfrac{1}{(x^4 + x + 1)}dx$

Let $\int \dfrac{1}{x^4 + x + 1} = I_1$

$\therefore I = x\int \dfrac{(4x^3 + 1)}{(x^4 + x + 1)^2} - \int\int \dfrac{4x^3 + 1}{(x^4 + x + 1)^2}dx - I_1$

$= x \times \dfrac{(x^4 + x + 1)^{-1}}{-1} - \int \dfrac{(x^4 + x + 1)^{-1}}{-1}dx - I_1$

$= \dfrac{-x}{(x^4 + x + 1)} + c + I_1 - I_1$

Sol 9: (C) $\int e^{3x}\cos 4x\, dx$

Let $3x = t$

$$\therefore I = \frac{1}{3}\int e^t \cos\frac{4t}{3}dt$$

$$= \frac{1}{3}\int e^t \cos\frac{4t}{3}dt = \frac{1}{3}\int e^t\left(\cos\frac{4t}{3} - \frac{4}{3}\sin\frac{4t}{3}\right)dt$$

$$+ \frac{4}{9}\int e^t\left(\sin\frac{4\pi}{3}\right)dt$$

$$= \frac{1}{3}\int\left[e^t\cos\frac{4t}{3}\right] + \frac{4}{9}\int e^t\left(\sin\frac{4t}{3} + \frac{4}{3}\cos\frac{4t}{3}\right)dt$$

$$- \int\frac{16}{27}e^t\cos\frac{4t}{3}dt$$

$$\frac{25}{9}I = \frac{1}{3}e^t\cos\frac{4t}{3} + \frac{4}{9}e^t\sin\frac{4t}{3} + c$$

$$I = \frac{e^t}{25}\left(3\frac{4t}{3} + 4\sin\frac{4t}{3}\right) + c$$

$$\therefore 3A = 4B$$

Sol 10: (C)

$$I = \int\frac{p + x^{p+2q-1} + qx^{p+2q-1} - q(x^{q-1} + x^{p+2q-1})}{(x^{p+q} + 1)^2}dx$$

$$= \int\frac{(p+q)x^{p+q-1}x^q - qx^{q-1}(x^{p+q}) + 1}{(x^{p+q} + 1)^2}dx$$

It is of the form $\dfrac{uv' - vu'}{u^2}$

$\therefore$ Where $u = x^{p+q} + 1$ and $v = -x^q$

$$\therefore I = \frac{v}{u} + c = \frac{-x^q}{x^{p+q} + 1} + c$$

Sol 11: (B) Let $f^{-1}(x) = t$

$\therefore f(f^{-1}(x)) = x$

$\therefore f(t) = x$

$\therefore \int f^{-1}(x)dx$

$\int t\,dx = \int tf'(t)dt$

$= tf(t) - \int f(t)dt = tf(t) - g(t)$

$= f^{-1}(x)\,(x - g)(f^{-1}(x))$

Sol 12: (B) $\left(\dfrac{x}{\left(x^4 - 1\right)^4}\right)^{1/3} dx$

$$I = \int\left(\frac{x}{(x^4 - 1)^4}\right)^{1/3}dx = \int\left(\frac{x}{x^{16}\left(1 - \frac{1}{x^4}\right)^4}\right)^{1/3}dx$$

$$= \int\frac{1}{\left(1 - \frac{1}{x^4}\right)^{4/3}}\frac{1}{x^5}dx$$

Let $1 - \dfrac{1}{x^4} = t$ $\therefore \dfrac{4}{x^5}dx = dt$

$$\therefore I = \int\frac{1}{t^{4/3}} \times \frac{1}{4}dt = \frac{1}{4} \times \frac{t^{-1/3}}{-\frac{1}{3}} + c = -\frac{3}{4}t^{-1/3} + c$$

$$= -\frac{3}{4}\left(1 - \frac{1}{x^4}\right)^{-1/3} + c = -\frac{3}{4}\left(\frac{x^4}{x^4 - 1}\right)^{1/3} + c$$

Sol 13: (A) $I = \int e^u \sin 2x\,dx$

When $u = x$

$$I = \int e^x \sin 2x\,dx = \int e^x(\sin 2x + 2\cos 2x - 2\cos 2x)dx$$

$$= \int e^x(\sin 2x + 2\cos 2x) - 2\int e^x \cos 2x$$

$$= \int e^x(\sin 2x + 2\cos 2x)$$

$$- 2\int e^x(\cos 2x - 2\sin 2x + 2\sin 2x)dx$$

$$\therefore 5I = \int e^x(\sin 2x + 2\cos 2x)dx$$

$$- 2\int e^x(\cos 2x - 2\sin 2x)dx$$

$$= e^x \sin 2x - 2e^x\cos 2x + c$$

When $u = \sin x$

$$I = \int 2e^{\sin x}\cos x \sin x\,dx$$

Put $\sin x = t$ $\therefore \cos x\,dx = dt$

$\therefore I = 2\int te^t dt$ which is solvable

Previous Years' Questions

Sol 1: (C) Let $I = \int\dfrac{\cos^3 x + \cos^5 x}{\sin^2 x + \sin^4 x}dx$

$$= \int\frac{(\cos^2 x + \cos^4 x)\cdot\cos x\,dx}{(\sin^2 x + \sin^4 x)}$$

Put $\sin x = t \Rightarrow \cos x\,dx = dt$

$$\therefore I = \int \frac{[(1-t^2)+(1-t^2)^2]}{t^2+t^4} dt$$

$$\Rightarrow I = \int \frac{1-t^2+1-2t^2+t^4}{t^2+t^4} dt$$

$$\Rightarrow I = \int \frac{2-3t^2+t^4}{t^2(t^2+1)} dt . \qquad \dots(i)$$

Using partial fraction for,

$$\frac{y^2-3y+2}{y(y+1)} = 1 + \frac{A}{y} + \frac{B}{y+1} \quad (\text{ where } y = t^2)$$

$$\Rightarrow A = 2, B = -6$$

$$\therefore \frac{y^2-3y+2}{y(y+1)} = 1 + \frac{2}{y} - \frac{6}{y+1}$$

$\therefore$ Eq. (i) reduces to,

$$I = \int \left(1 + \frac{2}{t^2} - \frac{6}{1+t^2}\right) dt = t - \frac{2}{t} - 6\tan^{-1}(t) + c$$

$$= \sin x - \frac{2}{\sin x} - 6\tan^{-1}(\sin x) + c$$

Sol 2: (A) Given, $f(x) = \dfrac{x}{(1+x^n)^{1/n}}$ for $n \ge 2$

$$\therefore ff(x) = \frac{f(x)}{[1+f(x)^n]^{1/n}} = \frac{x}{(1+2x^n)^{1/n}}$$

and $fff(x) = \dfrac{x}{(1+3x^n)^{1/n}}$

$$\therefore g(x) = \underbrace{fofo...of}_{n \text{ times}}(x) = \frac{x}{(1+nx^n)^{1/n}}$$

$$\text{Let } I = \int x^{n-2} g(x) dx = \int \frac{x^{n-1}dx}{(1+nx^n)^{1/n}} = \frac{1}{n^2} \int \frac{n^2 x^{n-1}dx}{(1+nx^n)^{1/n}}$$

$$= \frac{1}{n^2} \int \frac{\frac{d}{dx}(1+nx^n)}{(1+nx^n)^{1/n}} dx$$

$$\therefore I = \frac{1}{n(n-1)}(1+nx^n)^{1-\frac{1}{n}} + c$$

Sol 3: (C) Since, $I = \int \dfrac{e^x}{e^{4x}+e^{2x}+1} dx$

$$J = \int \frac{e^{3x}}{1+e^{2x}+e^{4x}} dx$$

$$\therefore J - I = \int \frac{(e^{3x}-e^x)}{1+e^{2x}+e^{4x}} dx$$

Put $e^x = u \Rightarrow e^x dx = du$

$$\therefore J - I = \int \frac{(u^2-1)}{1+u^2+u^4} du$$

$$= \int \frac{\left(1-\frac{1}{u^2}\right)}{1+\frac{1}{u^2}+u^2} du = \int \frac{\left(1-\frac{1}{u^2}\right)}{\left(u+\frac{1}{u}\right)^2-1} du$$

Put $u + \dfrac{1}{u} = t$

$$\Rightarrow \left(1-\frac{1}{u^2}\right) du = dt = \int \frac{dt}{t^2-1}$$

$$= \frac{1}{2}\log\left|\frac{t-1}{t+1}\right| + c = \frac{1}{2}\log\left|\frac{u^2-u+1}{u^2+u+1}\right| + c$$

$$= \frac{1}{2}\log\left|\frac{e^{2x}-e^x+1}{e^{2x}+e^x+1}\right| + c$$

Sol 4: Given, $f(x) = \int\left(\dfrac{2\sin x - \sin 2x}{x^3}\right) dx$

On differentiating w.r.t. x, we get

$$f'(x) = \frac{2\sin x - \sin 2x}{x^3} = \frac{2\sin x}{x}\left(\frac{1-\cos x}{x^2}\right)$$

$$\lim_{x\to 0} f'(x) = \lim_{x\to 0} 2\left(\frac{\sin x}{x}\right)\left(\frac{2\sin^2\frac{x}{2}}{x^2}\right)$$

$$= 4 \cdot 1 \cdot \lim_{x\to 0}\left(\frac{\sin^2\frac{x}{2}}{4\times\left(\frac{x^2}{2}\right)}\right) = 1$$

Sol 5: (i) Let $I = \int\sqrt{1+\sin\dfrac{x}{2}}\, dx$

$$\Rightarrow I = \int\sqrt{\cos^2\frac{x}{4}+\sin^2\frac{x}{4}+2\sin\frac{x}{4}\cos\frac{x}{4}}\, dx$$

$$\Rightarrow I = \int\left(\cos\frac{x}{4}+\sin\frac{x}{4}\right) dx$$

$$= 4\sin\frac{x}{4} - 4\cos\frac{x}{4} + c$$

(ii) Let $I = \int \dfrac{x^2}{\sqrt{1-x}}\,dx$

Put $1 - x = t^2 \Rightarrow -dx = 2t\,dt$

$\therefore I = \int \dfrac{(1-t^2)^2 \cdot (-2t)}{t}\,dt$

$= -2\int (1 - 2t^2 + t^4)\,dt = -2\left(t - \dfrac{2t^3}{3} + \dfrac{t^5}{5}\right) + c$

$= -2\left\{\sqrt{1-x} - \dfrac{2}{3}(1-x)^{3/2} + \dfrac{1}{5}(1-x)^{5/2}\right\}$

Sol 6: Let $I = \int (e^{\log x} + \sin x)\cos x\,dx$

$= \int (x + \sin x)\cos x\,dx$

$\therefore I = \int x\cos x\,dx + \dfrac{1}{2}\int (\sin 2x)\,dx$

$= (x\cdot\sin x - \int 1\cdot \sin x\,dx) - \dfrac{\cos 2x}{4} + c$

$= x\sin x + \cos x - \dfrac{\cos 2x}{4} + c$

Sol 7: Let $I = \int \dfrac{(x-1)e^x}{(x+1)^3}\,dx$

$\Rightarrow I = \int \left\{\dfrac{x+1-2}{(x+1)^3}\right\}e^x\,dx$

$= \int \left\{\dfrac{1}{(x+1)^2} - \dfrac{2}{(x+1)^3}\right\}e^x\,dx$

$= \int e^x \cdot \dfrac{1}{(x+1)^2}\,dx - 2\int e^x \cdot \dfrac{1}{(x+1)^3}\,dx$

Applying integration by parts,

$= \left\{\dfrac{1}{(x+1)^2}\cdot e^x - \int e^x \cdot \dfrac{-2}{(x+1)^3}\,dx\right\}$

$-2\int e^x \cdot \dfrac{1}{(x+1)^3}\,dx = \dfrac{e^x}{(x+1)^2} + c$

Sol 8: Let $I = \int \dfrac{dx}{x^2(x^4+1)^{3/4}} = \int \dfrac{dx}{x^2 \cdot x^3\left(1 + \dfrac{1}{x^4}\right)^{3/4}}$

Put $1 + x^{-4} = t \Rightarrow -\dfrac{4}{x^5}\,dx = dt$

$\therefore I = -\dfrac{1}{4}\int \dfrac{dt}{t^{3/4}} = -\dfrac{1}{4}\cdot \dfrac{t^{1/4}}{1/4} + c$

$= -\left(1 + \dfrac{1}{x^4}\right)^{1/4} - c = -\dfrac{(x^4+1)^{1/4}}{x} + c$

Sol 9: Let $I = \int \sqrt{\dfrac{1-\sqrt{x}}{1+\sqrt{x}}}\,dx$

Put $x = \cos^2\theta \Rightarrow dx = -2\sin\theta\cos\theta\,d\theta$

$\therefore I = \int \sqrt{\dfrac{1-\cos\theta}{1+\cos\theta}} \cdot (-2\sin\theta\cos\theta)d\theta$

$= -\int 2\tan\dfrac{\theta}{2} \cdot \sin\theta\cos\theta\,d\theta = -2\int 2\sin^2\dfrac{\theta}{2}\cdot \cos\theta\,d\theta$

$= -2\int (1 - \cos\theta)\cos\theta\,d\theta = -2\int (\cos\theta - \cos^2\theta)\,d\theta$

$= -2\int \cos\theta\,d\theta + \int (1 + \cos 2\theta)\,d\theta$

$= -2\sin\theta + \theta + \dfrac{s\,n2\theta}{2} + c$

$= -2\sqrt{1-x} + \cos^{-1}\sqrt{x} + \sqrt{x(1-x)} + c$

Sol 10: Let $I = \int \dfrac{\sin^{-1}\sqrt{x} - \cos^{-1}\sqrt{x}}{\sin^{-1}\sqrt{x} + \cos^{-1}\sqrt{x}}\,dx$

$= \int \dfrac{\sin^{-1}\sqrt{x} - \left(\dfrac{\pi}{2} - \sin^{-1}\sqrt{x}\right)}{\dfrac{\pi}{2}}\,dx$

$= \dfrac{2}{\pi}\int \left(2\sin^{-1}\sqrt{x} - \dfrac{\pi}{2}\right)dx$

$= \dfrac{4}{\pi}\int \sin^{-1}\sqrt{x}\,dx - x + c$ ...(i)

Now, $\int \sin^{-1}\sqrt{x}\,dx$

Put $x = \sin^2\theta \Rightarrow dx = \sin 2\theta$

$= \int \theta\cdot \sin 2\theta\,d\theta$

$= -\dfrac{\theta\cos 2\theta}{2} + \int \dfrac{1}{2}\cos 2\theta\,d\theta$

$= -\dfrac{\theta}{2}\cos 2\theta + \dfrac{1}{4}\sin 2\theta$

$= -\dfrac{1}{2}\theta(1 - 2\sin^2\theta) + \dfrac{1}{2}\sin\theta\sqrt{1 - \sin^2\theta}$

$= -\dfrac{1}{2}\sin^{-1}\sqrt{x}\,(1 - 2x) + \dfrac{1}{2}\sqrt{x}\,\sqrt{1-x}$ ...(ii)

From Eqs. (i) and (ii), we get

$$I = \frac{4}{\pi}\left[-\frac{1}{2}(1-2x)\sin^{-1}\sqrt{x} + \frac{1}{2}\sqrt{x-x^2}\right] - x + c$$

$$= \frac{2}{\pi}\left[\sqrt{x-x^2} - (1-2x)\sin^{-1}\sqrt{x}\right] - x + c$$

Sol 11: Let $I = \int \dfrac{\sqrt{\cos 2x}}{\sin x}\,dx$

$$= \int \frac{\sqrt{\cos^2 x - \sin^2 x}}{\sin^2 x}\,dx = \int\sqrt{\cot^2 x - 1}\,dx$$

Put $\cot x = \sec\theta \Rightarrow -\text{cosec}^2 x\,dx = \sec\theta\tan\theta\,d\theta$

$$\therefore I = \int\sqrt{\sec^2\theta - 1}\cdot\frac{\sec\theta\cdot\tan\theta}{-(1+\sec^2\theta)}\,d\theta$$

$$= -\int\frac{\sec\theta\cdot\tan^2\theta}{1+\sec^2\theta}\,d\theta$$

$$= -\int\frac{\sin^2\theta}{\cos\theta + \cos^3\theta}\,d\theta$$

$$= -\int\frac{1-\cos^2\theta}{\cos\theta(1+\cos^2\theta)}\,d\theta$$

$$= -\int\frac{(1+\cos^2\theta) - 2\cos^2\theta}{\cos\theta(1+\cos^2\theta)}\,d\theta$$

$$= -\int\sec\theta\,d\theta + 2\int\frac{\cos\theta}{1+\cos^2\theta}\,d\theta$$

$$= -\log|\sec\theta + \tan\theta| + 2\int\frac{\cos\theta}{2-\sin^2\theta}\,d\theta$$

$$= -\log|\sec\theta + \tan\theta| + \int\frac{dt}{2-t^2}, \quad \text{where } \sin\theta = t$$

$$= -\log|\sec\theta + \tan\theta| + 2\cdot\frac{1}{2\sqrt{2}}\log\left|\frac{\sqrt{2}+\sin\theta}{\sqrt{2}-\sin\theta}\right| + c$$

$$= -\log\left|\cot x + \sqrt{\cot^2 x - 1}\right|$$

$$\qquad + \frac{1}{\sqrt{2}}\log\left|\frac{\sqrt{2}+\sqrt{1-\tan^2 x}}{\sqrt{2}-\sqrt{1-\tan^2 x}}\right| + c$$

Sol 12: Let $I = \int\left(\dfrac{1}{\sqrt[3]{x}+\sqrt[4]{x}} + \dfrac{\ln(1+\sqrt[6]{x})}{\sqrt[3]{x}+\sqrt{x}}\right)dx$

$\therefore I = I_1 + I_2,$

where $I_1 = \int\left(\dfrac{1}{\sqrt[3]{x}+\sqrt[4]{x}}\right)dx$, $I_2 = \int\dfrac{\ln(1+\sqrt[6]{x})}{\sqrt[3]{x}+\sqrt{x}}\,dx$

Now, $I_1 = \int\left(\dfrac{1}{\sqrt[3]{x}+\sqrt[4]{x}}\right)dx$

Put $x = t^{12} \Rightarrow dx = 12t^{11}dt$

$$\therefore I_1 = 12\int\frac{t^{11}}{t^4+t^3}\,dt = 12\int\frac{t^8\,dt}{t+1}$$

$$= 12\int(t^7 - t^6 + t^5 - t^4 + t^3 - t^2 + t - 1)dt + 12\int\frac{dt}{t+1}$$

$$= 12\left(\frac{t^8}{8} - \frac{t^7}{7} + \frac{t^6}{6} - \frac{t^5}{5} + \frac{t^4}{4} - \frac{t^3}{3} + \frac{t^2}{2} - 1\right) + 12\log(t+1)$$

And $I_2 = \int\left\{\dfrac{\ln(1+\sqrt[6]{x})}{\sqrt[3]{x}+\sqrt{x}}\right\}dx$

Put $x = u^6 \Rightarrow dx = 6u^5\,du$

$$\therefore I_2 = \int\frac{\log(1+u)}{u^2+u^3}6u^5\,du = \int\frac{\log(1+u)}{u^2(1+u)}\cdot 6u^5\,du$$

$$6\int\frac{u^3}{(u+1)}\log(1+u)du = 6\int\left(\frac{u^3-1+1}{u+1}\right)\log(1+u)du$$

$$= 6\int\left(u^2 - u + 1 - \frac{1}{u+1}\right)\log(1+u)du$$

$$= 6\int(u^2 - u + 1)\log(1+u)du - 6\int\frac{\log(1+u)}{(u+1)}du$$
$$\qquad\quad \text{II} \qquad\qquad\qquad\text{I}$$

$$= 6\left(\frac{u^3}{3} - \frac{u^2}{2} + u\right)\log(1+u)$$

$$\qquad -\int\frac{2u^3 - 3u^2 + 6u}{u+1}du - 6\frac{1}{2}[\log(1+u)]^2$$

$$= (2u^3 - 3u^2 + 6u)\log(1+u)$$

$$-\int\left(2u^2 - 5u + \frac{11u}{u+1}\right)du - 3[\log(1+u)]^2$$

$$= (2u^3 - 3u^2 + 6u)\log(1+u)$$

$$= -\left(\frac{2u^3}{3} - \frac{5}{2}u^2 + 11u - 11\log(u+1)\right) - 3[\log(1+u)]^2$$

$$\therefore I = \frac{3}{2}x^{2/3} - \frac{12}{7}x^{7/12} + 2x^{1/2} - \frac{12}{5}x^{5/12} + 3x^{1/3} - 4x^{1/4}$$

$$- 6x^{1/6} - 12x^{1/12} + 12\log(x^{1/12}+1)$$

$$+ (2x^{1/2} - 3x^{1/3} + 6x^{1/6}1^{111/1})\log(1+x^{1/6})$$

$$-\left[\frac{2}{3}x^{1/2} - \frac{5}{2}x^{1/3}11x^{1/6} - 11\log(1+x^{1/6})\right]$$

$-3 [\log (1 + x^{1/6})]^2 + c$

$= \dfrac{3}{2}x^{2/3} - \dfrac{12}{7}x^{7/12} + \dfrac{4}{3}x^{1/2} - \dfrac{12}{5}x^{5/12}$

$+ \dfrac{1}{2}x^{1/3} - 4x^{1/4} - 7x^{1/6} - 12x^{1/12}$

$+ (2x^{1/2} - 3x^{1/3} - 6x^{1/6} + 11)\log(1 + x^{1/6})$

$+ 12\log(1 + x^{1/2}) - 3[\log(1 + x^{1/6})]^2 + c$

Sol 13: $\dfrac{x^3 + 3x + 2}{(x^2 + 1)^2(x + 1)} = \dfrac{x^3 + 2x + x + 2}{(x^2 + 1)^2(x + 1)}$

$= \dfrac{x(x^2 + 1) + 2(x + 1)}{(x^2 + 1)^2(x + 1)} = \dfrac{x}{(x^2 + 1)(x + 1)} + \dfrac{2}{(x^2 + 1)^2}$

Again, $\dfrac{x}{(x^2 + 1)(x + 1)} = \dfrac{Ax + B}{(x^2 + 1)} + \dfrac{C}{(x + 1)}$

$\Rightarrow x = (Ax + B)(x + 1) + C(x^2 + 1)$

Putting $x = -1$, we get $-1 = 2C \Rightarrow C = -1/2$

Equation coefficient of x^2, we get

$0 = A + C \Rightarrow A = -C = 1/2$

Putting $x = 0$, we obtain

$0 = B + C \Rightarrow B = -C = 1/2$

$\dfrac{x^3 + 3x + 2}{(x^2 + 1)^2(x + 1)} = \dfrac{x + 1}{2(x^2 + 1)} - \dfrac{1}{2(x + 1)} + \dfrac{2}{(x^2 - 1)^2}$

$\therefore I = \displaystyle\int \dfrac{x^3 + 3x + 2}{(x^2 + 1)^2(x + 1)}\,dx$

$= -\dfrac{1}{2}\displaystyle\int \dfrac{dx}{x + 1} + \dfrac{1}{2}\displaystyle\int \dfrac{x + 1}{x^2 + 1}\,dx + 2\displaystyle\int \dfrac{dx}{(x^2 + 1)^2}$

$\Rightarrow I = -\dfrac{1}{2}\log|x + 1| + \dfrac{1}{4}\log|x^2 + 1| + \dfrac{1}{2}\tan^{-1}x + 2I_1 \ \ ...(i)$

where $I_1 = \displaystyle\int \dfrac{dx}{(x^2 + 1)^2}$

Put $x = \tan\theta \Rightarrow dx = \sec^2\theta\,d\theta$

$\therefore I_1 = \displaystyle\int \dfrac{\sec^2\theta\,d\theta}{(\tan^2\theta + 1)^2} = \displaystyle\int \cos^2\theta\,d\theta = \dfrac{1}{2}\displaystyle\int(1 + \cos 2\theta)\,d\theta$

$= \dfrac{1}{2}\left[\theta + \dfrac{1}{2}\sin 2\theta\right] = \dfrac{1}{2}\theta + \dfrac{1}{2}\cdot\dfrac{\tan\theta}{(1 + \tan^2\theta)}$

$= \dfrac{1}{2}\tan^{-1}x + \dfrac{1}{2}\cdot\dfrac{x}{(1 + x^2)}$

$\therefore$ From Eq. (i)

$I = -\dfrac{1}{2}\log|x + 1| + \dfrac{1}{4}\log|x^2 + 1| + \dfrac{3}{2}\tan^{-1}x + \dfrac{x}{x^2 + 1} + c$

Sol 14: Let $I = \displaystyle\int \sin^{-1}\left(\dfrac{2x + 2}{\sqrt{4x^2 + 8x + 13}}\right)dx$

$= \displaystyle\int \sin^{-1}\left(\dfrac{2x + 2}{\sqrt{(2x + 2)^2 + 9}}\right)dx$

Put $2x + 2 = 3\tan\theta \Rightarrow 2dx = 3\sec^2\theta\,d\theta$

$\therefore I = \displaystyle\int \sin^{-1}\left(\dfrac{3\tan\theta}{\sqrt{9\tan^2\theta + 9}}\right)\cdot\dfrac{3}{2}\sec^2\theta\,d\theta$

$= \displaystyle\int \sin^{-1}\left(\dfrac{3\tan\theta}{3\sec\theta}\right)\cdot\dfrac{3}{2}\sec^2\theta\,d\theta$

$= \displaystyle\int \sin^{-1}\left(\dfrac{\sin\theta}{\cos\theta\ \sec\theta}\right)\cdot\dfrac{3}{2}\sec^2\theta\,d\theta$

$= \dfrac{3}{2}\displaystyle\int \sin^{-1}(\sin\theta)\cdot\sec^2\theta\,d\theta$

$= \dfrac{3}{2}\displaystyle\int \theta\cdot\sec^2\theta\,d\theta = \dfrac{3}{2}[\theta\cdot\tan\theta - \displaystyle\int 1\cdot\tan\theta\,d\theta]$

$= \dfrac{3}{2}[\theta\tan\theta - \log\sec\theta] + c$

$= \dfrac{3}{2}\left[\tan^{-1}\left(\dfrac{2x + 2}{3}\right)\cdot\left(\dfrac{2x + 2}{3}\right) - \log\sqrt{1 + \left(\dfrac{2x + 2}{3}\right)^2}\right] + c_1$

$= (x + 1)\tan^{-1}\left(\dfrac{2x + 2}{3}\right) - \dfrac{3}{4}\log\left(1 + \left(\dfrac{2x + 2}{3}\right)^2\right) + c_1$

$= (x + 1)\tan^{-1}\left(\dfrac{2x + 2}{3}\right) - \dfrac{3}{4}\log(4x^2 + 8x + 13) + c$

$\left(\text{let }\dfrac{3}{2}\log 3 + c_1 = c\right)$

Sol 15: For any natural number m, the given integral can be written as,

$I = \displaystyle\int (x^{3m} + x^{2m} + x^m)(2x^{3m} + 3x^{2m} + 6x^m)^{1/m}\,dx$

$\Rightarrow I = \displaystyle\int (2x^{3m} + 3x^{2m} + 6x^m)^{1/m}(x^{3m-1} + x^{2m-1} + x^{m-1})\,dx$

Put $2x^{3m} + 3x^{2m} + 6x^m = t$

$\Rightarrow (6mx^{3m-1} + 6mx^{2m-1} + 6mx^{m-1})dx = dt$

$\therefore I = \int t^{1/m} \dfrac{dt}{6m} = \dfrac{1}{6m} \cdot \dfrac{t^{\frac{1}{m}+1}}{\left(\dfrac{1}{m}+1\right)}$

$= \dfrac{1}{6(m+1)} \cdot (2x^{3m} + 3x^{2m} + 6x^m)^{\frac{(m+1)}{m}} + c$

$= -\dfrac{1}{r^{11/2}}\left(\dfrac{1}{11} + \dfrac{t^2}{7}\right)$

$= -\dfrac{1}{(\sec x + \tan x)^{11/2}}\left\{\dfrac{1}{11} + \dfrac{1}{7}(\sec x + \tan x)^2\right\} + k.$

Sol 16: (C)

$J - I = \int \dfrac{e^x(e^{2x}-1)}{e^{4x} + e^{2x} + 1}dx = \int \dfrac{(z^2-1)}{z^4 + z + 1}dz$ where $z = e^x$

$J - I = \int \dfrac{\left(1 - \dfrac{1}{z^2}\right)dx}{\left(z + \dfrac{1}{z}\right) - 1} = \dfrac{1}{2}\log\left(\dfrac{e^x + e^{-x} - 1}{e^x + e^{-x} + 1}\right)$

$\therefore J - I = \dfrac{1}{2}\log\left(\dfrac{e^x + e^{-x} - 1}{e^x + e^{-x} + 1}\right) + c.$

Sol 17: (C) $I = \int \dfrac{\sec^2 x}{(\sec x + \tan x)^{9/2}}dx$

Let $\sec x + \tan x = t$

$\Rightarrow \sec x - \tan x = 1/t$

$\sec x(\sec x + \tan x)dx = dt$ S

$\sec x\, dx = \dfrac{dt}{t}, \dfrac{1}{2}\left(t + \dfrac{1}{t}\right) = \sec x$

$I = \dfrac{1}{2}\int \dfrac{\left(t + \dfrac{1}{t}\right)}{t^{9/2}}\dfrac{dt}{t} = \dfrac{1}{2}\int\left(t^{-9/2} + t^{-12/2}\right)dt$

$= \dfrac{1}{2}\left[\dfrac{t^{-9/2+1}}{-\dfrac{9}{2}+1} - \dfrac{t^{-13/2+1}}{\dfrac{13}{2}+1}\right]$

$= \dfrac{1}{2}\left[\dfrac{t^{-7/2}}{-\dfrac{7}{2}} + \dfrac{t^{-11/2}}{-\dfrac{11}{2}}\right]$

$= -\dfrac{1}{7}t^{-7/2} - \dfrac{1}{11}t^{-11/2}$

$= -\dfrac{1}{7}\dfrac{1}{t^{7/2}} - \dfrac{1}{11}\dfrac{1}{t^{11/2}}$

2. DEFINITE INTEGRATION

1. INTRODUCTION

Let f(x) be a continuous function defined on a closed interval [a, b] and $\int f(x)dx = F(x) - c$ then $\int_a^b f(x)dx = [F(x)]_a^b$ or $\int_a^b f(x)dx = F(b) - F(a)$ is called the definite integral of f(x) within limits a and b. The interval [a, b] is called the range of integration. Every definite integral has a unique solution.

Note: $\int_a^b f(x)dx = F(b) - F(a)$ also represents the net area of the curve f(x) with x-axis. $\int_0^{\pi/2} \sin^2 x\,dx$

Sol: $\int_0^{\pi/2} \sin^2 x\,dx = \int_0^{\pi/2} \left(\frac{1 - \cos 2x}{2}\right)dx = \frac{1}{2}\left[x - \frac{\sin 2x}{2}\right]_0^{\pi/2} = \frac{1}{2}\left[\frac{\pi}{2} - 0\right] = \frac{\pi}{4}$

Illustration 1: If $\int_0^1 (3x^2 + 2x + k)dx = 0$, find the value of k. **(JEE MAIN)**

Sol: Here the answer of the definite integral $\int_0^1 \left[3x^2 + 2x + k\right]dx$ is already given i.e. 0 hence by using simple integral formulas we can solve it and by comparing it to 0, we will obtain the value of k.

Here, we have, $\int_0^1 (3x^2 + 2x + k)dx = 0$

$\left[3\frac{x^3}{3} + 2\frac{x^2}{2} + kx\right]_0^1 = 0$; $\qquad \left[x^3 + x^2 + kx\right]_0^1 = 0$

$(1 + 1 + k) - (0 + 0 + 0) = 0$; $\quad 2 + k = 0 \Rightarrow k = -2$

Illustration 2: Evaluate: $\int_0^{\frac{\pi}{4}} (2\sec^2 x + x^3 + 2)dx$. **(JEE MAIN)**

Sol: As we know $\int_a^b \{f(x) \pm g(x)\}dx = \int_a^b f(x)dx \pm \int_a^b g(x)dx$. Hence by using this method we can solve the given definite integral.

We have, $\int_0^{\frac{\pi}{4}} (2\sec^2 x + x^3 + 2)dx = 2\int_0^{\frac{\pi}{4}} \sec^2 x\,dx + \int_0^{\frac{\pi}{4}} x^3 dx + 2\int_0^{\frac{\pi}{4}} dx$

$= 2\left[\tan x\right]_0^{\pi/4} + \left[\frac{x^4}{4}\right]_0^{\pi/4} + 2[x]_0^{\pi/4} = 2\left(\tan\frac{\pi}{4} - \tan 0\right) + \left[\frac{(\pi/4)^4}{4} - 0\right] + 2\left[\frac{\pi}{4} - 0\right]$

$$= 2(1-0) + \left(\frac{\pi^4}{4^5} - 0 \right) + \frac{\pi}{2} = 2 + \frac{\pi^4}{1024} + \frac{\pi}{2}$$

2. PROPERTIES OF DEFINITE INTEGRALS

Property 1

$$\int_a^b f(x)\,dx = \int_a^b f(t)\,dt = \int_a^b f(u)\,du$$

Here x is a dummy variable; it can be replaced by any other variable t, u,........

$$\int_0^{\pi/2} \sin(x)\,dx = \int_0^{\pi/2} \sin t\,dt = \int_0^{\pi/2} \sin u\,du =$$

This is similar to the summation property $\displaystyle\sum_{T=1}^{10} r^2 = \sum_{T=1}^{10} t^2 = \sum_{U=1}^{10} u^2 = \ldots\ldots$

Property 2

$$\int_a^b f(x)\,dx = -\int_b^a f(x)\,dx$$

i.e. the interchange of limits of a definite integral changes only its sign.

Property 3

$$\int_a^b f(x)\,dx = \int_a^c f(x)\,dx + \int_c^b f(x)\,dx \quad (a < c < b)$$

Generally, this property is used when the integrand has two or more rules in the integration interval

$$\Rightarrow \quad \int_a^b f(x)\,dx = \int_a^{c_1} f(x)\,dx + \int_{c_1}^{c_2} f(x)\,dx + \ldots\ldots + \int_{c_n}^b f(x)\,dx \quad \text{where } a < c_1 < c_2 < \ldots\ldots c_n < b.$$

Illustration 3: Evaluate: $\int_1^4 f(x)\,dx$, where $f(x) = \begin{cases} 2x + 8, & 1 \le x \le 2 \\ 6x, & 2 \le x \le 4 \end{cases}$ **(JEE MAIN)**

Sol: Here as we know, $\int_a^b f(x)\,dx = \int_a^c f(x)\,dx + \int_c^b f(x)\,dx$ where $(a < c < b)$. Hence by using this property and solving by using the integral formula we can solve it.

We have, $I = \int_1^4 f(x)\,dx$

$$= \int_1^2 f(x)\,dx + \int_2^4 f(x)\,dx = \int_1^2 (2x + 8)\,dx + \int_2^4 6x\,dx$$

$$= \left[x^2 + 8x \right]_1^2 + \left[3x^2 \right]_2^4 = \left[(2)^2 + 8(2) - (1)^2 - 8(1) \right] + \left[3(4)^2 - 3(2)^2 \right]$$

$$= 11 + 36 = 47.$$

Illustration 4: Evaluate : $\int_0^2 |1 - x|\,dx$ **(JEE MAIN)**

Sol: Here $|1 - x| = \begin{cases} 1 - x, & \text{when } 0 \le x \le 1 \\ x - 1, & \text{when } 1 \le x \le 2 \end{cases}$ therefore, similar to the problem above, we can solve it.

$|1 - x| = \begin{cases} 1 - x, & \text{when } 0 \le x \le 1 \\ x - 1, & \text{when } 1 \le x \le 2 \end{cases}$

$\therefore \; I = \int_0^1 (1 - x)\,dx + \int_1^2 (x - 1)\,dx = \left[x - \frac{x^2}{2} \right]_0^1 + \left[\frac{x^2}{2} - x \right]_1^2 = (1/2 - 0) + (0 + 1/2) = 1$

Property 4

$$\int_0^a f(x)\,dx = \int_0^a f(a - x)\,dx$$

Property 5

$$\int_a^b f(x)\,dx = \int_a^b f(a + b - x)\,dx$$

Application: $\displaystyle \int_a^b \frac{f(x)}{f(x) + f(a + b - x)}\,dx = \frac{b - a}{2}$

NOMORECLASS CONCEPTS

With the help of the above property, the following integrals can be obtained.

$$\int_0^{\pi/2} f(\sin x)\,dx = \int_0^{\pi/2} f(\cos x)\,dx \; ; \quad \int_0^{\pi/2} f(\tan x)\,dx = \int_0^{\pi/2} f(\cot x)\,dx$$

$$\int_0^{\pi/2} f(\sin 2x)\sin x\,dx = \int_0^{\pi/2} f(\sin 2x)\cos x\,dx \; ; \quad \int_0^1 f(\log x)\,dx = \int_0^1 f[\log(1 - x)]\,dx$$

$$\int_0^{\pi/2} \frac{\sin^n x}{\sin^n x + \cos^n x}\,dx = \int_0^{\pi/2} \frac{\cos^n x}{\cos^n x + \sin^n x}\,dx = \frac{\pi}{4}$$

$$\int_0^{\pi/2} \frac{\tan^n x}{1 + \tan^n x}\,dx = \int_0^{\pi/2} \frac{\cot^n x}{1 + \cot^n x}\,dx = \frac{\pi}{4} \; ; \quad \int_0^{\pi/2} \frac{1}{1 + \tan^n x}\,dx = \int_0^{\pi/2} \frac{1}{1 + \cot^n x}\,dx = \frac{\pi}{4}$$

$$\int_0^{\pi/2} \frac{\sec^n x}{\sec^n x + \text{cosec}^n x}\,dx = \int_0^{\pi/2} \frac{\text{cosec}^n x}{\text{cosec}^n x + \sec^n x}\,dx = \frac{\pi}{4} \; ; \quad \int_0^{\pi/4} \log(1 + \tan x)\,dx = \frac{\pi}{8}\log 2$$

$$\int_0^{\pi/2} \log \cot x\,dx = \int_0^{\pi/2} \log \tan x\,dx = 0$$

Illustration 5: Prove that $\int_0^1 \cot^{-1}(1 - x + x^2)\,dx = 2\int_0^1 \tan^{-1} x\,dx$ **(JEE MAIN)**

Sol: As we know $\cot^{-1}\left(\dfrac{a}{b}\right) = \tan^{-1}\left(\dfrac{b}{a}\right)$ and $\int_0^a f(x)\,dx = \int_0^a f(a - x)\,dx$ by using these two formulae we can solve the given problem.

$$\int_0^1 \cot^{-1}(1-x+x^2)dx = \int_0^1 \tan^{-1}\left[\frac{1}{1-x+x^2}\right] = \int_0^1 \tan^{-1}\left[\frac{1+x-x}{1-x(1-x)}\right] = \int_0^1 \tan^{-1}\left(\frac{x+(1-x)}{1-x(1-x)}\right)dx$$

$$= \int_0^1 \tan^{-1}x\,dx + \int_0^1 \tan^{-1}(1-x)dx = 2\int_0^1 \tan^{-1}x.dx$$

$$\left(\because \tan^{-1}\left(\frac{a+b}{1-ab}\right) = \tan^{-1}a + \tan^{-1}b\right)$$

Illustration 6: Find the value of $\int_0^1 \log\left(\frac{1}{x}-1\right)dx$ (JEE MAIN)

Sol: Here $\log\left(\frac{1-x}{x}\right) = \log(1-x) - \log(x)$ and $\int_0^a f(x)dx = \int_0^a f(a-x)dx$ by using these two formulae we can solve it.

$$\int_0^1 \log\left(\frac{1-x}{x}\right)dx = \int_0^1 \log(1-x)dx - \int_0^1 \log(x)dx = \int_0^1 \log\left[1-(1-x)\right]dx - \int_0^1 \log x\,dx = \int_0^1 \log x\,dx - \int_0^1 \log x\,dx$$

$$= \int_0^1 \log(x)dx - \int_0^1 \log(x)dx = 0$$

Illustration 7: Evaluate: $\int_0^{\pi/2} \frac{a\sin x + b\cos x}{\sin x + \cos x}dx$ (JEE MAIN)

Sol: As $\int_0^a f(x)dx = \int_0^a f(a-x)dx$ therefore we can write $\int_0^{\pi/2} \frac{a\sin x + b\cos x}{\sin x + \cos x}dx$ in the form of

$\int_0^{\pi/2} \frac{a\sin(\pi/2-x) + b\cos(\pi/2-x)}{\sin(\pi/2-x) + \cos(\pi/2-x)}dx$ and then adding these two equations we can solve the given problem.

$$I = \int_0^{\pi/2} \frac{a\sin x + b\cos x}{\sin x + \cos x}dx \qquad \text{... (i)}$$

$$I = \int_0^{\pi/2} \frac{a\sin(\pi/2-x) + b\cos(\pi/2-x)}{\sin(\pi/2-x) + \cos(\pi/2-x)}dx = \int_0^{\pi/2} \frac{a\cos x + b\sin x}{\sin x + \cos x}dx \qquad \text{... (ii)}$$

Adding (i) and (ii),

$$\therefore 2I = \int_0^{\pi/2} \frac{(a+b)(\sin x + \cos x)}{\sin x + \cos x}dx = \int_0^{\pi/2}(a+b)dx = (a+b)\pi/2 \Rightarrow I = (a+b)\,\pi/4$$

Illustration 8: Show that $\int_0^{\pi/2} \frac{\sin^2 x}{\sin x + \cos x}dx = \frac{1}{\sqrt{2}}\log(\sqrt{2}+1)$ (JEE ADVANCED)

Sol: This problem is similar to the problem above.

Let $I = \int_0^{\pi/2} \frac{\sin^2 x}{\sin x + \cos x}dx$ \qquad ... (i)

By property 4, we have

$$I = \int_0^{\pi/2} \frac{\sin^2\left((\pi/2)-x\right)}{\sin\left((\pi/2)-x\right) + \cos\left((\pi/2)-x\right)}dx = \int_0^{\pi/2} \frac{\cos^2 x}{\sin x + \cos x}dx \qquad \text{... (ii)}$$

Adding (i) and (ii), we get

$$2I = \int_0^{\pi/2} \frac{\sin^2 x + \cos^2 x}{\sin x + \cos x} dx \Rightarrow I = \frac{1}{2}\int_0^{\pi/2} \frac{dx}{\sin x + \cos x} = \frac{1}{2\sqrt{2}}\int_0^{\pi/2} \frac{1}{(1/\sqrt{2})\sin x + (1/\sqrt{2})\cos x} dx$$

$$= \frac{1}{2\sqrt{2}}\int_0^{\pi/2} \frac{1}{\cos(\pi/4)\sin x + \sin(\pi/4)\cos x} dx = \frac{1}{2\sqrt{2}}\int_0^{\pi/2} \frac{1}{\sin(x + (\pi/4))} dx$$

$$= \frac{1}{2\sqrt{2}}\int_0^{\pi/2} \csc\left[x + \frac{\pi}{4}\right] dx = \frac{1}{2\sqrt{2}}\left\{ \log\tan\left[\frac{x}{2} + \frac{\pi}{8}\right] \right\}_0^{\pi/2}$$

$$= \frac{1}{2\sqrt{2}}\left\{ \log\tan\left[\frac{\pi}{4} + \frac{\pi}{8}\right] - \log\tan\frac{\pi}{8} \right\} = \frac{1}{2\sqrt{2}}\log\left(\frac{\tan(3\pi/8)}{\tan(\pi/8)}\right) = \frac{1}{2\sqrt{2}}\log\left(\frac{\cot(\pi/8)}{\tan(\pi/8)}\right)$$

$$= \frac{2}{2\sqrt{2}}\log\cot\frac{\pi}{8} = \frac{1}{\sqrt{2}}\log(\sqrt{2} + 1)$$

Illustration 9: Evaluate : $\int_{-\pi/4}^{3\pi/4} \frac{\sqrt{\tan x}}{1 + \sqrt{\tan x}} dx$ (JEE ADVANCED)

Sol: By putting $\tan x = \frac{\sin x}{\cos x}$ and using the property $\int_a^b f(x)dx = \int_a^b f(a+b-x)dx$, we can solve the given problem.

Let $I = \int_{-\pi/4}^{3\pi/4} \frac{\sqrt{\tan x}}{1 + \sqrt{\tan x}} dx \Rightarrow I = \int_{-\pi/4}^{3\pi/4} \frac{\sqrt{\sin x}}{\sqrt{\cos x} + \sqrt{\sin x}} dx$ ... (i)

On applying $\int_a^b f(x)dx = \int_a^b f(a+b-x)dx$ we get

$$I = \int_{-\pi/4}^{3\pi/4} \frac{\sqrt{\sin\left((3\pi/4) - (\pi/4) - x\right)}}{\sqrt{\cos\left((3\pi/4) - (\pi/4) - x\right)} + \sqrt{\sin\left((3\pi/4) - (\pi/4) - x\right)}} dx$$

$$= \int_{-\pi/4}^{3\pi/4} \frac{\sqrt{\sin\left((\pi/2) - x\right)}}{\sqrt{\cos\left((\pi/2) - x\right)} + \sqrt{\sin\left((\pi/2) - x\right)}} dx$$

$$= \int_{-\pi/4}^{3\pi/4} \frac{\sqrt{\cos x}}{\sqrt{\sin x} + \sqrt{\cos x}} dx$$ ... (ii)

Adding (i) and (ii), we get

$$2I = \int_{-\pi/4}^{3\pi/4} \frac{\sqrt{\sin x}}{\sqrt{\sin x} + \sqrt{\cos x}} dx + \int_{-\pi/4}^{3\pi/4} \frac{\sqrt{\cos x}}{\sqrt{\sin x} + \sqrt{\cos x}} dx = \int_{-\pi/4}^{3\pi/4} \frac{\sqrt{\sin x} + \sqrt{\cos x}}{\sqrt{\sin x} + \sqrt{\cos x}} dx$$

$$= \int_{-\pi/4}^{3\pi/4} dx = [x]_{-\pi/4}^{3\pi/4} = \left[\frac{3\pi}{4} - \left(-\frac{\pi}{4}\right)\right] = \left[\frac{3\pi}{4} + \frac{\pi}{4}\right] = \pi \Rightarrow I = \frac{\pi}{2}$$

Illustration 10: The value of $\int_0^{\pi/2} \log\left(\frac{4 + 3\sin x}{4 + 3\cos x}\right) dx$ is (JEE ADVANCED)

Sol: Similar to the problems above, we can write $\int_0^{\pi/2} \log\left(\frac{4 + 3\sin x}{4 + 3\cos x}\right) dx$ as

$\int_0^{\pi/2} \log\left(\frac{4 + 3\sin\left((\pi/2) - x\right)}{4 + 3\cos\left((\pi/2) - x\right)}\right) dx$ and then by adding these two equations we can solve the given problem.

Let $I = \int_0^{\pi/2} \log\left(\dfrac{4 + 3\sin x}{4 + 3\cos x}\right) dx$

On applying property 5, we get

$I = \int_0^{\pi/2} \log\left(\dfrac{4 + 3\sin((\pi/2) - x)}{4 + 3\cos((\pi/2) - x)}\right) dx$

$= \int_0^{\pi/2} \log\left(\dfrac{4 + 3\cos x}{4 + 3\sin x}\right) dx = -\int_0^{\pi/2} \log\left(\dfrac{4 + 3\sin x}{4 + 3\cos x}\right) dx = -I \Rightarrow I = 0$

Thus, $\int_0^{\pi/2} \log\left(\dfrac{4 + 3\sin x}{4 + 3\cos x}\right) dx = 0$

Illustration 11: $I = \int_0^{\pi/2} \dfrac{dx}{4 + 5\sin x}$ (JEE ADVANCED)

Sol: Let $\sin x = \dfrac{2\tan\dfrac{x}{2}}{1 + \tan^2\dfrac{x}{2}}$ and then by putting $\tan\dfrac{x}{2} = t$, we can solve the given problem.

$I = \int_0^{\pi/2} \dfrac{dx}{4 + 5\ (2\tan(x/2)\,/\,1 + \tan^2(x/2))} = \int_0^{\pi/2} \dfrac{\sec^2(\pi/2)dx}{4 + 4\tan^2(\pi/2) + 10\tan(\pi/2)}$

Let $\tan\dfrac{x}{2} = t \Rightarrow \dfrac{1}{2}\sec^2\dfrac{x}{2} = dt$

$\Rightarrow \int_0^1 \dfrac{2dt}{4 + 4t^2 + 10t} = \dfrac{1}{2}\int_0^1 \dfrac{dt}{(t + (1/2))(t + 2)} = \dfrac{1}{3}\int_0^1 \dfrac{1}{(t + (1/2))} - \dfrac{1}{(t + 2)}dt = \dfrac{1}{3}\left[\ln\dfrac{t + (1/2)}{t + 2}\right]_0^1 = \dfrac{1}{3}\log 2$

Illustration 12: Evaluate : $\int_{\pi/6}^{\pi/3} \dfrac{dx}{1 + \sqrt{\tan x}}$ (JEE ADVANCED)

Sol: Let $\tan x = \dfrac{\sin x}{\cos x}$ and then using property $\int_a^b f(x)dx = \int_a^b f(a + b - x)dx$, we can solve the given problem.

$\int_{\pi/6}^{\pi/3} \dfrac{dx}{1 + \sqrt{\tan x}} = \int_{\pi/6}^{\pi/3} \dfrac{\sqrt{\cos x}}{\sqrt{\sin x} + \sqrt{\cos x}} dx$ ... (i)

$= \int_{\pi/6}^{\pi/3} \dfrac{\sqrt{\cos(\pi/2 - x)}}{\sqrt{\sin(\pi/2 - x)} + \sqrt{\cos(\pi/2 - x)}} dx$ [∵ here a + b = π/2]

$= \int_{\pi/6}^{\pi/3} \dfrac{\sqrt{\sin x}}{\sqrt{\cos x} + \sqrt{\sin x}} dx$ ... (ii)

$\therefore 2I = \int_{\pi/6}^{\pi/3} 1\,dx = [x]_{\pi/6}^{\pi/3} = \dfrac{\pi}{3} - \dfrac{\pi}{6} = \dfrac{\pi}{6} \Rightarrow I = \dfrac{\pi}{12}$

Property 6

$$\int_{-a}^{a} f(x)\,dx = \begin{cases} 2\int_{0}^{a} f(x)\,dx & \text{if } f(-x) = f(x) \text{ (even function)} \\ 0 & \text{if } f(-x) = -f(x) \text{ (odd function)} \end{cases}$$

Note: This property is to be used if the integrand is either an even or odd function of x

Illustration 13: $\int_{-\pi/2}^{\pi/2} \cos^2 x\, dx$ is equal to   (JEE MAIN)

Sol: As $\int_{-\pi/2}^{\pi/2} \cos^2 x\, dx = 2\int_{0}^{\pi/2} \cos^2 x\, dx$, therefore using property 7 we can solve it.

Here $I = 2 \int_{0}^{\pi/2} \cos^2 x\, dx$ $\{\because f(-x) = f(x)\}$; $\int_{0}^{\pi/2} (1 + \cos 2x) dx = \left\{ x + \dfrac{\sin 2x}{2} \right\}_{0}^{\pi/2} = \dfrac{\pi}{2}$

Illustration 14: $\int_{-1}^{1} \dfrac{x^3 \sin(1 + x^2)}{1 + x^2}\, dx$ is equal to   (JEE ADVANCED)

Sol: Here by using the property $\int_{-a}^{a} f(x)\,dx = \begin{cases} 2\int_{0}^{a} f(x)\,dx & \text{if } f(-x) = f(x) \text{ (even function)} \\ 0 & \text{if } f(-x) = -f(x) \text{ (odd function)} \end{cases}$

Here $f(x) = \dfrac{x^3 \sin(1 + x^2)}{1 + x^2}$ & $f(-x) = -\dfrac{x^3 \sin(1 - x^2)}{1 + x^2}$

$\because f(x) = -f(x)$

$\therefore I = 0$

Property 7: $\int_{0}^{2a} f(x)dx = \begin{cases} 2\int_{0}^{a} f(x)dx, & \text{if } f(2a - x) = f(x) \\ 0, & \text{if } f(2a - x) = -f(x) \end{cases}$

Note: The above property is used to halve the limits

Illustration 15: Evaluate : $\int_{0}^{2\pi} \dfrac{\sin 2\theta}{a - b\cos\theta}\, d\theta$ (JEE MAIN)

Sol: Let $\int_{0}^{2a} f(x)dx = \begin{cases} 2\int_{0}^{a} f(x)dx, & \text{if } f(2a - x) = f(x) \\ 0, & \text{if } f(2a - x) = -f(x) \end{cases}$. Hence by using this property we can solve the given problem.

Let $I = \int_{0}^{2\pi} \dfrac{\sin 2\theta}{a - b\cos\theta}\, d\theta \rightarrow$ Let $f(\theta) = \dfrac{\sin 2\theta}{a - b\cos\theta}$

$f(2\pi - \theta) = \dfrac{\sin 2(2\pi - \theta)}{a - b\cos(2\pi - \theta)} = \dfrac{-\sin 2\theta}{a - b\cos\theta} = -f(\theta)$

By property 7, we have

$\therefore \quad \int_{0}^{2\pi} \dfrac{\sin 2\theta}{a - b\cos\theta}\, d\theta = 0$

Illustration 16: Evaluate $\int_0^{2\pi} x\sin^4 x \cos^6 x\, dx$ **(JEE ADVANCED)**

Sol: Similar to the problem above.

$$I = \int_0^{2\pi} x\sin^4 x \cos^6 x\, dx = \int_0^{2\pi}(2\pi - x)\sin^4 x \cos^6 x\, dx$$

$$2I = 2\pi\int_0^{2\pi}\sin^4 x\,\cos^6 x\, dx\;;\quad I = 2\pi\int_0^{\pi}\sin^4 x\,\cos^6 x\, dx\;;$$

$$I = 4\pi\int_0^{\pi/2}\sin^4 x\,\cos^6 x\, dx\;;\quad I = 4\pi\int_0^{\pi/2}\cos^4 x\sin^6 x\;;$$

$$\Rightarrow I = \frac{2\pi}{16}\int_0^{\pi/2}(\sin 2x)^4 dx \Rightarrow 2x = t \Rightarrow dx = \frac{dt}{2}$$

$$\Rightarrow I = \frac{\pi}{16}\int_0^{\pi}\sin^4 t\, dt = \frac{\pi}{8}\int_0^{\pi/2}\sin^4 t\, dt \quad \Rightarrow I = \frac{\pi}{8}\left[\frac{1}{2}\int_0^{\pi/2}\left(\sin^4 t + \sin^4 t\right)dt\right] = \frac{\pi}{8}\cdot\frac{1}{2}\cdot\frac{3\pi}{8} = \frac{3\pi^2}{128}$$

Property 8: If $f(x) = f(x + a)$ (i.e. $f(x)$ is a function with period a), then $\int_0^{na} f(x)dx = n\int_0^{a} f(x)dx$

Illustration 17: Evaluate: $\int_0^{4\pi}\sin^8 x\, dx$ **(JEE MAIN)**

Sol: Here $\sin^8 (\pi - x) = \sin^8 x$, therefore by using this property, we can solve the given problem.

$$I = 4\int_0^{\pi}\sin^8 x\, dx = 8\int_0^{\pi/2}\sin^8 x\, dx = 8\frac{7.5.3.1}{8.6.4.2}\cdot\frac{\pi}{2} = \frac{35\pi}{32}$$

Illustration 18: Evaluate: $\int_0^{2\pi}\cos^5 x\, dx$ **(JEE ADVANCED)**

Sol: Let $I = \int_0^{2\pi}\cos^5 x\, dx$

Let $f(x) = \cos^5 x$

$f(2\pi - x) = \cos^5 (2\pi - x) = \cos^5 x = f(x)$

Then $\int_0^{2\pi}\cos^5 x\, x\, dx = 2\int_0^{\pi}\cos^5 x\, dx$

Now, $f(\pi - x) = \cos^5 (\pi - x) = (-\cos x)^3 = -\cos^5 x$

$= -f(x)$; $\int_0^{\pi}\cos^5 x\, dx = 0$

Hence $\int_0^{2\pi}\cos^5 x\, dx = 0$

Property 9

$$\int_{a}^{a+nT} f(x)dx = n\int_{0}^{T} f(x)dx \quad \text{(if } f(x + T) = f(x)\text{, and } n\hat{\imath}N \text{ i.e. } f(x) \text{ is a function with period T)}$$

$$\int_{a+mT}^{b+nT} f(x)dx = (n-m)\int_{0}^{T} f(x)dx + \int_{a}^{b} f(x)dx \qquad m,n \in I$$

Illustration 19: $I = \int_0^{200\pi} \sqrt{1 + \cos x}\, dx$ **(JEE MAIN)**

Sol: $I = \sqrt{2} \int\limits_0^{200\pi} \left| \cos\dfrac{x}{2} \right| dx \qquad \dfrac{x}{2} = t$

$$\Rightarrow I = 2\sqrt{2} \int\limits_0^{100\pi} |\cos t| \, dt = 200\sqrt{2} \int\limits_0^{\pi} |\cos t| \, dt = 400\sqrt{2}$$

Property 10: $\dfrac{d}{dx} \int\limits_{g(x)}^{h(x)} f(t) \, dt = h'(x) \, f(h(x)) - g'(x) \, f(g(x))$

Corollary (1): $\dfrac{d}{dx} \int\limits_a^{h(x)} f(t) \, dt = h'(x) \, f(h(x))$ [a is any constant independent of x]

Corollary (2): $\dfrac{d}{dx} \int\limits_a^{x} f(t) \, dt = f(x)$

Property 11: $\left| \int\limits_a^b f(x) \, dx \right| \leq \int\limits_a^b | f(x) | \, dx$

Property 12: If $f(x) \geq 0$ on [a, b], then $\int\limits_a^b f(x) dx \geq 0$

This property is also called the domination law.

There are a few more properties which might be helpful in solving problems

1. Shift property: $\int\limits_a^b f(x) dx = \int\limits_{a\pm c}^{b\pm c} f(x) \, dx$

2. Reflection property: $\int\limits_a^b f(x) dx = -\int\limits_{-a}^{-b} f(-x) \, dx$

3. Expansion/Contraction property: $\int\limits_a^b f(x) dx = k \int\limits_{a/k}^{b/k} f(x) \, dx \quad \forall \ k > 0$

NOMORECLASS CONCEPTS

$\int_\alpha^\beta \dfrac{dx}{\sqrt{(x-\alpha)(x-\beta)}} = \pi$ if ($\beta > \alpha$)

$\int_\alpha^\beta \sqrt{(x-\alpha)(x-\beta)} dx = \dfrac{\pi}{8}(\beta - \alpha)^2$

$\int_a^b \sqrt{\dfrac{x-a}{b-x}} dx = \dfrac{\pi}{2}(b-a)$

If $f(t)$ is an odd function, then $\phi(x) = \int_a^x f(t) dt$ is an even function.

If $f(x)$ is an even function, then $\phi(x) = \int_a^x f(t) dt$ is an odd function.

Every continuous function defined on [a, b] is integrable over [a, b]

Every monotonic function defined on [a, b] is integrable over [a, b]

Change of variables: If the function f(x) is continuous on [a, b] and the function x = ϕ(t) is continuously differentiable on the interval [t_1, t_2] and a = $\phi(t_1)$, b = $\phi(t_2)$, then

$$\int_a^b f(x)\,dx = \int_{t_1}^{t_2} f(\phi(t))\phi'(t)\,dt.$$

3. SOME SPECIAL INTEGRALS

3.1 Walli's Formula

$$\int_0^{\pi/2} \sin^n x \, dx = \int_0^{\pi/2} \cos^n x\, dx = \frac{(n-1)\,(n-3)\ldots2}{n(n-2)\ldots1} \qquad \text{(if n is odd positive integer)}$$

$$= \frac{(n-1)\,(n-3)\ldots1}{n(n-2)\ldots2}\left(\frac{\pi}{2}\right) \quad \text{(if n is even positive integer)}$$

Illustration 20: Evaluate $\int_0^{\pi/2} \cos^7 x \, dx$ (JEE MAIN)

Sol: By using Walli's formula we can solve the given problem.

$$I = \frac{6.4.2}{7.5.3} = \frac{16}{35}$$

3.2 Gamma Function

$$\int_0^{\pi/2} \sin^m x\cos^n x \, dx = \frac{\Gamma\big((m+1)/2\big)\Gamma\big((n+1)/2\big)}{2\Gamma\big((m+n+2)/2\big)}$$

where $\Gamma(n)$ is called the gamma function

OR

$$\int_0^{\pi/2} \sin^m \cos^n x \, dx = \frac{((m-1)\,(m-3)\ldots(2 \text{ or } 1))\,(n-1)\,((n-3)\ldots(2 \text{ or } 1))}{(m+n)\,(m+n-2)\ldots\ldots(2 \text{ or } 1)}$$

(if m and n both are not simultaneously even positive integers)

$$\frac{((m-1)\,(m-3)\ldots(1))((n-1)(n-3)\ldots(1))}{(m+n)(m+n-2)\ldots(2)}\left(\frac{\pi}{2}\right) \qquad \text{(if m and n are both even positive integers)}$$

Illustration 21: Evaluate I = $\int_0^{\pi/2} \sin^4 x \cos^5 x \, dx$. (JEE MAIN)

Sol: Using the gamma function formula i.e.

$$\int_0^{\pi/2} \sin^m x \cos^n x \, dx = \frac{\Gamma((m+1)/2)\Gamma((n+1)/2)}{2\Gamma((m+n+2)/2)}$$

We can solve it.

$$I = \frac{r((4+1)/2)\Gamma((5+1)/2)}{2\Gamma((4+5+2)/2)} = \frac{\Gamma(5/2)\,\Gamma(3)}{2\Gamma(11/2)} = \frac{((3/2).(1/2))(2.1)}{2((9/2).(7/2).(5/2).(3/2).(1/2))} = \frac{8}{315}$$

4. NEWTON LEIBNITZ FORMULA

In calculus, **Leibnitz's rule** for differentiation under the integral sign named after Gottfried Leibnitz tells us that if we have an integral $\int_{y_0}^{y_1} f(x,y)dy$ then for x in (x_0, x_1) the derivative of this integral is thus expressible as

$$\frac{d}{dx}\left(\int_{y_0}^{y_1} f(x,y)dy\right) = \int_{y_0}^{y_1} f_x(x,y)dy$$

provided that f and its partial derivative f_x are both continuous over a region in the form $[x_0, x_1] \times [y_0, y_1]$.

5. SUMMATION OF SERIES BY INTEGRATION (LIMIT AS A SUM)

To find the sum of an infinite series with the help of definite integration, the following formula is used

$$\lim_{n\to\infty} \sum_{r=0}^{n-1} f\left(\frac{r}{n}\right)\frac{1}{n} = \int_0^1 f(x)dx$$

The following method is used to solve the questions on summation of series.

(i) After writing (r – 1)th or rth term of the series, express it in the form $\frac{1}{n}f\left(\frac{r}{n}\right)$.

Therefore the given series will take the form as $\lim_{n\to\infty} \sum_{r=0}^{n-1} \frac{1}{n}f\left(\frac{r}{n}\right)$

(ii) Now write $\int$ in place of $\lim_{n\to\infty}\Sigma$ and x in place of $\frac{r}{n}$ and dx in place of n. We get summation in the form of integral $\int_0^1 f(x)dx$.

Also we can write $\int_a^b f(x)dx = \lim_{n\to\infty}\frac{b-a}{n}[f(a)+f(a+h)+....+f(a+(n-1)h)]$ $\left[\text{where } h = \frac{b-a}{n}\right]$

Illustration 22: Evaluate $\lim_{n\to\infty}\left[\frac{1}{n+1}+\frac{1}{n+2}+.....+\frac{1}{2n}\right]$ (JEE MAIN)

Sol: By using the summation of series by integration formula i.e $\lim_{n\to\infty}\sum_{r=0}^{n-1} f\left(\frac{r}{n}\right)\frac{1}{n} = \int_0^1 f(x)dx$ we can solve it.

$$\text{Limit} = \lim_{n\to\infty}\sum_{r=1}^{n}\frac{1}{n+r} = \lim_{n\to\infty}\sum\left(\frac{1}{1+(r/n)}.\frac{1}{n}\right) = \int_0^1 \frac{1}{1+x}dx = [\log(1+x)]_0^1 = \log 2$$

Illustration 23: $\lim_{n\to\infty}\frac{1^{100}+2^{100}+3^{100}........n^{100}}{n^{101}}$ (JEE MAIN)

Sol: By observing the given problem, we can say that it's a sum of an infinite series so by using the summation of series by integration formula we can solve it.

$$T_r = \frac{r^{100}}{n^{101}} = \frac{1}{n} \times \left(\frac{r}{n}\right)^{100} \quad ; S = \lim_{n \to \infty} \frac{1}{n} \sum_{r=1}^{n} \left(\frac{r}{n}\right)^{100} ; \quad = \int_0^1 x^{100} dx = \frac{1}{101}$$

Illustration 24: Find the value of $\lim_{n \to \infty} \left[\dfrac{n}{(n+1)^2} + \dfrac{n}{(n+2)^2} + \dots + \dfrac{1}{4n} \right]$ (JEE ADVANCED)

Sol: Here $t_r = \dfrac{n}{(n+r)^2} = \dfrac{1}{n} \dfrac{1}{[1+(r/n)]^2}$, therefore similar to the problem above, we can solve it.

Therefore the given series $= \lim_{n \to \infty} \sum_{r=1}^{n} \dfrac{1}{[1+(r/n)]^2} \cdot \dfrac{1}{n} = \int_0^1 \dfrac{1}{(1+x)^2} dx$

Given series $= \int_0^1 \dfrac{1}{(1+x)^2} dx = \left[-\dfrac{1}{1+x} \right]_0^1 = \dfrac{-1}{2} + 1 = \dfrac{1}{2}$

Evaluate the following definite integrals as the limit of sums.

Illustration 25: $\int_a^b \cos x \, dx$ (JEE ADVANCED)

Sol: Here $\int_a^b f(x) dx = \lim_{n \to \infty} \dfrac{b-a}{n} [f(a) + f(a+h) + \dots + f(a+(n-1)h)]$ where $f(x) = \cos x$ and $h = \dfrac{b-a}{n}$

$$\therefore \int_a^b \cos x \, dx = \lim_{n \to \infty} \frac{b-a}{n} [\cos a + \cos(a+h) + \dots + \cos(a+(n-1)h)]$$

$$= \lim_{n \to \infty} \frac{b-a}{n} \cdot \left[\frac{\cos(a+((n-1)/2) \cdot h) \cdot \sin(nh/2)}{\sin(h/2)} \right]$$

$$= \lim_{n \to \infty} \left(\frac{b-a}{n}\right) \cdot \frac{\cos\left(a + \dfrac{n-1}{2} \cdot \dfrac{(b-a)}{n}\right) \cdot \sin\left(\dfrac{n \cdot (b-a)}{2n}\right)}{\sin\left(\dfrac{b-a}{2n}\right)}$$

$$= \lim_{n \to \infty} 2 \cdot \frac{b-a}{2n} \cdot \frac{\cos(a+(1-(1/n))/2)(b-a)) \cdot \sin((b-a)/2)}{\sin((b-a)/2n)}$$

$$= \lim_{n \to \infty} 2 \cdot \frac{\cos\left(a + \left(1-(1/n)\right)\left((b-a)/2\right)\right) \cdot \sin\left((b-a)/2\right)}{\sin\left((b-a)/2n\right) / \left((b-a)/2n\right)}$$

$$= 2\cos\left(\frac{b+a}{2}\right) \sin\left(\frac{b-a}{2}\right) = \sin b - \sin a$$

Illustration 26: $\int_1^2 (x^2 + x) dx$ (JEE ADVANCED)

Sol: Similar to the problem above.

$$h = \frac{b-a}{n} = \frac{2-1}{n} = \frac{1}{n}$$

$$\int_a^b f(x) dx = \lim_{n \to \infty} \frac{b-a}{n} [f(a) + f(a+h) + \dots + f(a+(n-1)h)]$$

$$\int_1^2 (x^2 + x)\,dx = \lim_{n \to \infty} \frac{1}{n}[f(1) + f(1+h) + \dots + f(1 + (n-1)h)]$$

$$= \lim_{n \to \infty} \frac{1}{n}[(1^2 + 1) + \{(1+h)^2 + (1+h)\} + \dots + \{(1 + (n-1)h)^2 + (1 + (n-1)h)\}]$$

$$= \lim_{n \to \infty} \frac{1}{n}[1^2 \cdot n + h(1 + 2 + \dots + (n-1)) + 1 \cdot n + 2h(1 + 2 + \dots + (n-1)) + h^2(1^2 + 2^2 + \dots (n-1)^2)]$$

Here $h = \dfrac{1}{n}$

$$= \lim_{n \to \infty} \frac{1}{n}\left[n + \frac{1}{n}\frac{(n-1)(n)}{2} + n + \frac{2}{n} \cdot \frac{n(n-1)}{2} + \frac{1}{n^2}\frac{(n-1)n(2n-1)}{6}\right]$$

$$= \lim_{n \to \infty}\left[1 + \frac{(1 - (1/n))(1)}{2} + 1 + \frac{2(1 - (1/n))}{2} + \frac{(1 - (1/n))(1)(2 - (1/n))}{6}\right]$$

$$= 1 + \frac{1}{2} + 1 + 1 + \frac{1}{3} = \frac{23}{6}$$

6. INTEGRAL WITH INFINITE LIMITS

If a function f(x) is continuous for $a \le x < \infty$, then by definition,

$$\int_a^\infty f(x)\,dx = \lim_{b \to \infty} \int_a^b f(x)\,dx \qquad \dots \text{(i)}$$

If there exists a finite limit on the right-hand side of (i), then the improper integral is said to be convergent; otherwise it is divergent.

Geometrically, the improper integral (i) for f(x) > 0. is the area of the figure bounded by the graph of the function y = f(x), the straight line x = a, and the x-axis. Similarly, we can define

$$\int_{-\infty}^b f(x)dx = \lim_{a \to -\infty} \int_a^b f(x)\,dx \quad \text{and} \quad \int_{-\infty}^\infty f(x)dx = \int_{-\infty}^a f(x)dx + \int_a^\infty f(x)dx$$

7. IMPORTANT RESULTS

If $f(x) \ge 0$ and a < b, then $\displaystyle\int_a^b f(x)dx \ge 0$, e.g. $\displaystyle\int_0^{\pi/2} \sin x\,dx = 1$

If $f(x) \ge 0$ and a < b, then $\displaystyle\int_b^a f(x)dx \le 0$, e.g. $\displaystyle\int_{\pi/2}^0 \cos x\,dx = -1$

If $f(x) \le 0$ and a < b, then $\displaystyle\int_b^a f(x)dx \ge 0$, e.g. $\displaystyle\int_{\pi/2}^0 \sin x\,dx = 1$

$\displaystyle\int_0^x [x]dx = \int_0^1 (0)dx + \int_1^2 (1)dx + \int_2^3 2dx + \dots + \int_{[x]}^x [x]dx,$ where [ ] denotes the greatest integer of x.

$\displaystyle\int_0^{\pi/2} \log(\sin x)dx = \int_0^{\pi/2} \log(\cos x)dx = -\frac{\pi}{2}\log 2$

$\displaystyle\int_0^{\pi/2} \log(\tan x)dx = \int_0^{\pi/2} (\cot x)dx = 0$

$\displaystyle\int_0^{2a} f(x)dx = \int_0^a f(x)dx + \int_0^a f(2a - x)dx = \int_0^a f(x)dx + \int_0^a f(a + x)dx$

$$\int_a^b [x]dx = (b-a)\int_0^1 x\,dx, \text{where [ ] denotes the fractional part of x.}$$

e.g., $\int_0^5 [x]dx = 5\int_0^1 x\,dx = \dfrac{5}{2}$

Integral of an inverse function is given by $\int_{f(a)}^{f(b)} f^{-1}(y)dy = bf(b) - af(a) - \int_a^b f(x)dx$

Derivation of the given formula is given in the solved examples

8. GEOMETRICAL APPLICATION

The area of the figure bounded by the graphs of two continuous functions $y = f_1(x)$ and $y = f_2(x)$, $f_1(x) \le f_2(x)$, and two straight lines x= a and x = b is determined by the formula $S = \int_a^b (f_2(x) - f_1(x))dx$. It is sometimes convenient to use formulae analogous to x.with respect to y, i.e., regarding x as a function of y. In particular, the area bounded by the curve x =f(y), the y-axis and the two abscissae y = c and y = d is given by $\int_c^d f(y)dy$. The area of the figure bounded by the graphs of two continuous functions $x = f_1(y)$ and $f_2(y)$ (with $f_1(y) \le f_2(y)$), and the two straight lines y = c, y = d is given by $\int_c^d (f_2(y) - f_1(y))dy$

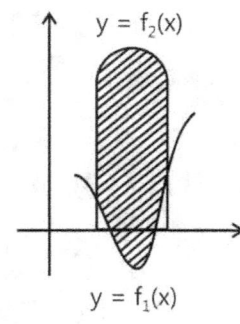

y = f₂(x)

y = f₁(x)

Figure 23.1

From the view of geometry we get an important inequality as if m ≤ f(x) ≤ M for a ≤ x ≤ b, then $m(b-a) \le \int_a^b f(x)dx \le M(b-a)$

FORMULAE SHEET

Important results

1. $\int_a^b \{f(x) \pm g(x) \pm h(x)\}dx = \int_a^b f(x)dx \pm \int_a^b g(x)dx + \int_a^b h(x)dx$	2. $\int_a^b f(x)dx = -\int_b^a f(x)dx$
3. $\int_a^b f(x)dx = \int_a^c f(x)dx + \int_c^b f(x)dx$ (a < c < b)	4. $\int_0^a f(x)dx = \int_0^a f(a-x)dx$
5. $\int_{-a}^a f(x)dx = \begin{cases} 2\int_0^a f(x)\,dx & \text{if } f(-x) = f(x) \text{ (even function)} \\ 0 & \text{if } (-x) = -f(x) \text{ (odd function)} \end{cases}$	6. $\int_a^b f(x)dx = \int_a^b f(a+b-x)dx$
7. $\int_0^{2a} f(x)dx = \begin{cases} 2\int_0^a f(x)dx, & \text{if } f(2a-x) = f(x) \\ 0, & \text{if } f(2a-x) = -f(x) \end{cases}$	8. $\dfrac{d}{dx}\int_{g(x)}^{h(x)} f(t)\,dt = h'(x)\,f(h(x)) - g'(x)\,f(g(x))$
9. $\int_a^{a+nT} f(x)dx = n\int_0^T f(x)dx$ (if f(x + T) = f(x), and n∈N i.e. f(x) is a function with period T)	10. If f(x) = f(x + a) then $\int_0^{na} f(x)dx = n\int_0^a f(x)dx$

11. $\left\|\int_a^b f(x)\,dx\right\| \leq \int_a^b \|f(x)\|\,dx$	12. $\int_a^b f(x)dx = k\int_{a/k}^{b/k} f(x)dx \quad \forall\, k > 0$
13. $\dfrac{d}{dx}\left(\int_{y_0}^{y_1} f(x,y)dy\right) = \int_{y_0}^{y_1} f_x(x,y)dy$ (Leibnitz formula)	

Definite integral of rational functions

1. $\int_0^{\infty} \dfrac{dx}{x^2 + a^2} = \dfrac{\pi}{2a}$	2. $\int_0^{\infty} \dfrac{x^{p-1}dx}{1+x} = \dfrac{\pi}{\sin(p\pi)},\ 0 < p < 1$
3. $\int_0^{\pi/2} \sin^2 x\,dx = \int_0^{\pi/2}\cos^2 x\,dx = \dfrac{\pi}{4}$	4. $\int_0^{\infty}\dfrac{\sin(px)}{x}dx = \begin{cases} \pi/2 & p>0 \\ 0 & p=0 \\ -\pi/2 & p<0 \end{cases}$
5. $\int_0^{\infty}\dfrac{\sin^2 px}{x^2} = \dfrac{\pi p}{2}$	6. $\int_0^{2x}\dfrac{dx}{a+b\sin x} = \dfrac{2\pi}{\sqrt{a^2-b^2}}$
7. $\int_0^{\infty}\sin ax^2 dx = \int_0^{\infty}\cos(ax^2)dx = \dfrac{1}{2}\sqrt{\dfrac{\pi}{2a}}$	8. $\int_0^{\infty}\dfrac{\sin x}{\sqrt{x}}dx = \int_0^{\infty}\dfrac{\cos x}{\sqrt{x}}dx = \sqrt{\dfrac{\pi}{2}}$
9. $\int_0^{\infty}\dfrac{\tan x}{x}dx = \dfrac{\pi}{2}$	

Advanced formulas

1. $\int_0^{\pi}\sin(mx)\cdot\sin(nx)dx = \begin{cases} 0 & m,n\ \text{integers and}\ m \neq n \\ \pi/2 & m,n\ \text{integers and}\ m = n \end{cases}$
2. $\int_0^{\pi}\cos(mx)\cdot\cos(nx)dx = \begin{cases} 0 & m,n\ \text{integers and}\ m \neq n \\ \pi/2 & m,n\ \text{integers and}\ m = n \end{cases}$
3. $\int_0^{\pi}\sin(mx)\cdot\cos(nx)dx = \begin{cases} 0 & m,n\ \text{integers and}\ m+n\ \text{odd} \\ 2m/(m^2-n^2) & m,n\ \text{integers and}\ m+n\ \text{even} \end{cases}$
4. $\int_0^{\pi/2}\sin^{2m}x\,dx = \int_0^{\pi/2}\cos^{2m}x\,dx = \dfrac{1.3.5....2m-1}{2.4.6....2m}\dfrac{\pi}{2}$

Definite integrals of exponential functions

1. $\int_0^{\infty} e^{-ax}\cos bx\,dx = \dfrac{a}{a^2+b^2}$	2. $\int_0^{\infty} e^{-ax}\sin bx\,dx = \dfrac{b}{a^2+b^2}$
3. $\int_0^{\infty} e^{-ax^2}dx = \dfrac{1}{2}\sqrt{\dfrac{\pi}{a}}$	4. $\int_0^{\infty} x^n e^{-ax}dx = \dfrac{\Gamma(n+1)}{a^{n+1}}$

5. $\int_0^\infty x^m e^{-ax^2}\,dx = \dfrac{\Gamma\left(\dfrac{m+1}{2}\right)}{2a^{(m+1)/2}}$	6. $\int_0^\infty \dfrac{x\,dx}{e^x - 1} = \dfrac{\pi^2}{6}$
7. $\int_0^\infty \dfrac{x^{n-1}}{e^x - 1}\,dx = \Gamma(n)\left(\dfrac{1}{1^n} + \dfrac{1}{2^n} + \dfrac{1}{3^n} + \dots\right)$	8. $\int_0^\infty \dfrac{x\,dx}{e^x + 1} = \dfrac{\pi^2}{12}$
9. $\int_0^\infty \dfrac{x^{n-1}}{e^x + 1}\,dx = \Gamma(n)\left(\dfrac{1}{1^n} - \dfrac{1}{2^n} + \dfrac{1}{3^n} - \dots\right)$	10. $\int_0^\infty \dfrac{e^{-ax} - e^{-bx}}{x\sec(px)}\,dx = \dfrac{1}{2}\ln\left(\dfrac{b^2 + p^2}{a^2 + p^2}\right)$
11. $\int_0^\infty \dfrac{e^{-ax} - e^{-bx}}{x\csc(px)}\,dx = \arctan\dfrac{b}{p} - \arctan\dfrac{a}{p}$	12. $\int_0^\infty \dfrac{e^{-ax}(1 - \cos x)}{x^2}\,dx = \operatorname{arccot}a - \dfrac{a}{2}\ln(a^2 + 1)$

Solved Examples

JEE Main/Boards

Example 1: Evaluate:

(i) $\displaystyle\int_0^a \dfrac{dx}{\sqrt{(a^2/4) - \left(x - (a/2)\right)^2}}$ 　(ii) $\displaystyle\int_{-a}^a \sqrt{\dfrac{a - x}{a + x}}\,dx$

Sol: (i) As we know $\displaystyle\int \dfrac{dx}{\sqrt{a^2 - x^2}} = \sin^{-1}\dfrac{x}{a}$, therefore by using this formula we can solve the given problem.

(ii) Put $x = a\cos\theta : \theta \in [0, p]$ and solve it using the appropriate formula.

(i) $\displaystyle\int_0^a \dfrac{dx}{\sqrt{(a^2/4) - \left(x - (a/2)\right)^2}}$

$= \left(\sin^{-1}\dfrac{x - (a/2)}{(a/2)}\right)_0^a; = \left(\sin^{-1}\dfrac{2x - a}{a}\right)_0^a$

$= [\sin^{-1} 1 - \sin^{-1}(-1)] = 2\sin^{-1}(1) = 2 \times \dfrac{\pi}{2} = \pi.$ (ii)

Then $dx = -a\sin\theta\,d\theta$. Hence,

$\displaystyle\int_{-a}^a \sqrt{\dfrac{a - x}{a + x}}\,dx = \int_\pi^0 \sqrt{\dfrac{1 - \cos\theta}{1 + \cos\theta}}\,(-a\sin\theta)d\theta$

$= a\displaystyle\int_0^\pi \sqrt{\dfrac{2\sin^2(\theta/2)}{2\cos^2(\theta/2)}} \cdot 2\sin\dfrac{\theta}{2}\cos\dfrac{\theta}{2}\,d\theta$

$= a\displaystyle\int_0^\pi 2\sin^2\dfrac{\theta}{2}\,d\theta = a\int_0^\pi (1 - \cos\theta)d\theta$

$= a(\theta - \sin\theta)_0^\pi = a(\pi) = a\pi.$

Example 2: Evaluate $\displaystyle\int_0^{\pi/2} \dfrac{\sin x}{\sin x + \cos x}\,dx$

Sol: Let $\displaystyle\int_0^a f(x)\,dx = \int_0^a f(a - x)\,dx$.

By using this we can write $\displaystyle\int_0^{\pi/2} \dfrac{\sin x}{\sin x + \cos x}\,dx$

as $\displaystyle\int_0^{\pi/2} \dfrac{\sin[(\pi/2) - x]}{\sin[(\pi/2) - x] + \cos[(\pi/2) - x]}\,dx$ and by adding

we can get the result.

$I = \displaystyle\int_0^{\pi/2} \dfrac{\sin[(\pi/2) - x]}{\sin[(\pi/2) - x] + \cos[(\pi/2) - x]}\,dx$

$= \displaystyle\int_0^{\pi/2} \dfrac{\cos x}{\cos x + \sin x}\,dx$

$\therefore\ 2I = \displaystyle\int_0^{\pi/2} \dfrac{\sin x + \cos x}{\sin x + \cos x}\,dx = \int_0^{\pi/2} dx = \dfrac{\pi}{2}$

$\therefore\quad I = \dfrac{\pi}{4}$

Example 3: Evaluate $\int_0^1 \log\left(\frac{1}{x}-1\right)dx$

Sol: Here $\log\left(\frac{1-x}{x}\right) = \log(1-x) - \log(x)$ and $\int_0^a f(x)dx = \int_0^a f(a-x)dx$ by using these two formulae we can solve it.

$I = \int_0^1 \log\left(\frac{1}{x}-1\right)dx$

(Put $x = \cos^2 t$: $\cos t > 0$; then $dx = -2\cos t \sin t \, dt$)

$= -\int_{\pi/2}^0 \log(\sec^2 t - 1) \cdot 2\cos t \sin t \, dt$

$= \int_0^{\pi/2} \log(\tan^2 t) \cdot \sin 2t \, dt = 2\int_0^{\pi/2} \sin 2t \cdot \log(\tan t)dt$

$= 2\int_0^{\pi/2} \log(\cot t) \cdot \sin 2t \, dt$

$\therefore 2I = 2\int_0^{\pi/2} \log(\tan t \cdot \cot t) \times \sin 2t \, dt = 0$

Example 4: Evaluate:

(i) $I = \int_0^\pi |\cos x| \, dx$

(ii) $I = \int_{-2}^1 |2x+1| \, dx$

(iii) $I = \int_1^4 f(x)dx$, where $f(x) = \begin{array}{l} 4x+3, \quad 1 \le x \le 2 \\ 3x+5, \quad 2 < x \le 4 \end{array}$

Sol: (i) Here $|\cos(\pi - x)| = |\cos x|$ hence $|\cos x| = \cos x$ therefore using the formula $\int \cos x = \sin x$ we can solve it.

(ii) By putting $2x+1 = z$ we can solve it.

(iii) As $\int_a^b f(x)dx = \int_a^c f(x)dx + \int_c^b f(x)dx$ $(a < c < b)$ By using this formula we can obtain the result.

(i) $I = 2\int_0^{\pi/2} |\cos x| \, dx$

$= 2\int_0^{\pi/2} \cos x \, dx = 2(\sin x)_0^{\pi/2} = 2(1) = 2$

(ii) $I = \int_{-2}^1 |2x+1| \, dx$ (put $2x+1 = z$)

$= \frac{1}{2}\int_{-3}^3 |z| \, dz = \int_0^3 |z| \, dz = \frac{9}{2}$.

(iii) $I = \int_1^2 f(x)dx + \int_2^4 f(x)dx$

$= \int_1^2 (4x+3)dx + \int_2^4 (3x+5)dx$

$= (2x^2 + 3x)_1^2 + \left(\frac{3x^2}{2} + 5x\right)_2^4$

$= 9 + 28 = 37$.

Example 5: Evaluate $I = \int_0^{1.7} [x^2]dx$, where $[x]$ is the greatest integer function

Sol: $[x^2]$ takes constant values 0, 1, 2 in intervals $[0, 1)$, $(1, \sqrt{2})$, $(\sqrt{2}, \sqrt{3})$ respectively. By substituting these values we will get the required result.

$I = \int_0^1 [x^2]dx + \int_1^{\sqrt{2}} [x^2]dx + \int_{\sqrt{2}}^{1.7} [x^2]dx$

$= \int_0^1 0 \, dx + \int_1^{\sqrt{2}} 1 \, dx + \int_{\sqrt{2}}^{1.7} 2 \, dx$

$= 0 + (\sqrt{2}-1) + 2(1.7 - \sqrt{2}) = 2.4 - \sqrt{2}$

Example 6: Let $f(x)$ be an odd function in the interval $\left[-\frac{T}{2}, \frac{T}{2}\right]$ with period T, prove that $F(x) = \int_a^x f(t)dt$ is a periodic function with period T.

Sol: As $f(x)$ is an odd function.

$F(x+T) = \int_a^{x+T} f(t)dt = \int_a^x f(t)dt + \int_x^{x+T} f(t)dt = F(x) + I(x)$

where $I(x) = \int_x^{x+T} f(t)dt = \int_{-\frac{T}{2}}^{\frac{T}{2}} f(t)dt = 0$ (since f is an odd function). Hence $F(x)$ is a periodic function with period T.

Example 7: Evaluate $\int_0^\pi \theta \sin^2 \theta \cos^2 \theta \, d\theta$

Sol: As we know, $\int_0^a f(x)dx = \int_0^a f(a-x)dx$, hence by using this formula we can evaluate it.

Let $I = \int_0^\pi \theta \sin^2 \theta \cos^2 \theta \, d\theta$

$$= \int_0^\pi (\pi - \theta) \sin^2(\pi - \theta) \cos^2(\pi - \theta) d\theta$$

$$= \int_0^\pi (\pi - \theta) \sin^2 \theta \cos^2 \theta \, d\theta$$

$$= \pi \int_0^\pi \sin^2 \theta \cos^2 \theta \, d\theta - \int_0^\pi \theta \sin^2 \theta \cos^2 \theta \, d\theta$$

$$= \pi \int_0^\pi \left(\frac{\sin 2\theta}{2} \right)^2 d\theta - I$$

$$\Rightarrow \quad 2I = \frac{\pi}{4} \int_0^\pi \sin^2 2\theta \, d\theta = \frac{\pi}{4} \int_0^\pi \left(\frac{1 - \cos 4\theta}{2} \right) d\theta$$

$$= \frac{\pi}{8} \left[\theta - \frac{\sin 4\theta}{4} \right]_0^\pi = \frac{\pi^2}{8}$$

$$\therefore \quad I = \frac{\pi^2}{16}$$

Example 8: Evaluate $\lim\limits_{n \to \infty} \sum\limits_{r=1}^{n-1} \frac{1}{n} \sqrt{\left(\frac{n+r}{n-r} \right)}$

Sol: Here by using the limit as a sum method we can solve the given problem.

$$\lim_{n \to \infty} \sum_{r=1}^{n-1} \frac{1}{n} \sqrt{\left(\frac{n+r}{n-r} \right)}$$

$$= \lim_{n \to \infty} \sum_{r=1}^{n-1} \frac{1}{n} \sqrt{\frac{1 + r/n}{1 - r/n}} = \int_0^1 \sqrt{\frac{1 + x}{1 - x}} dx$$

$$= \int_0^1 \frac{1 + x}{\sqrt{1 - x^2}} dx = \int_0^1 \frac{dx}{\sqrt{1 - x^2}} + \int_0^1 \frac{x \, dx}{\sqrt{1 - x^2}}$$

$$= [\sin^{-1} x - \sqrt{1 - x^2}]_0^1$$

$$= [\sin^{-1} 1 - 0] - [\sin^{-1} 0 - 1] = \frac{\pi}{2} + 1$$

Example 9: Integrate : $I = \int\limits_{\frac{\pi}{4}}^{\frac{3\pi}{4}} \frac{\theta}{1 + \sin \theta} d\theta$

Sol: As $\int_0^a f(x) dx = \int_0^a f(a - x) dx$ hence we can

write $\int\limits_{\frac{\pi}{4}}^{\frac{3\pi}{4}} \frac{\theta}{1 + \sin \theta} d\theta$ as $\int\limits_{\frac{\pi}{4}}^{\frac{3\pi}{4}} \frac{\pi - \theta}{1 + \sin \theta} d\theta$ and then

by putting $\theta = \frac{\pi}{2} + y$ we can solve the given problem.

$$I = \int\limits_{\frac{\pi}{4}}^{\frac{3\pi}{4}} \frac{\theta}{1 + \sin \theta} d\theta = \int\limits_{\frac{\pi}{4}}^{\frac{3\pi}{4}} \frac{\pi - \theta}{1 + \sin \theta} d\theta$$

$$2I = \pi \int\limits_{\frac{\pi}{4}}^{\frac{3\pi}{4}} \frac{d\theta}{1 + \sin \theta} \quad ; \quad \text{Put } \theta = \frac{\pi}{2} + y$$

$$= \pi \int\limits_{-\frac{\pi}{4}}^{\frac{\pi}{4}} \frac{dy}{1 + \cos y} = 2\pi \int\limits_0^{\frac{\pi}{4}} \frac{dy}{1 + \cos y}$$

$$I = \frac{\pi}{2} \int_0^{\frac{\pi}{4}} \sec^2 \frac{y}{2} dy = \pi \left[\tan \frac{y}{2} \right]_0^{\pi/4} = \pi \tan \frac{\pi}{8}$$

JEE Advanced/Boards

Example 1: Show that $1 < \int_0^1 e^{x^2} dx < e$.

Sol: e^{x^2} is an increasing function in [0, 1]. Further, $e^0 \le e^{x^2} \le e^1 \; \forall \; x \in [0, 1]$

$$\therefore \int_0^1 1 \, dx < \int_0^1 e^{x^2} dx < \int_0^1 e \, dx$$

or $1 < \int_0^1 e^{x^2} dx < e$.

Example 2: If $F(x) = \int_0^{x^2} \frac{t^2 - 5t + 4}{4 + e^{2t}} dt$, find the critical points of $F(x)$.

Sol: By using Leibnitz rule we can write

$$F(x) = \int_0^{x^2} \frac{t^2 - 5t + 4}{4 + e^{2t}} dt,$$

as $F'(x) = \frac{(x^2)^2 - 5x^2 + 4}{4 + e^{2x^2}} \cdot (2x) = 0.$

By Leibnitz Rule,

$$F'(x) = \frac{(x^2)^2 - 5x^2 + 4}{4 + e^{2x^2}} \cdot (2x)$$

$$F'(x) = 0$$

$$\Rightarrow \quad (x^4 - 5x^2 + 4) x = 0$$

$$\Rightarrow \quad (x^2 - 4)(x^2 - 1) x = 0$$

$$\Rightarrow \quad x = 0, \pm 1, \pm 2$$

These are the critical points of $F(x)$.

Example 3: Evaluate: $\int\limits_0^{\pi/2} \log \sin x \, dx$

Sol: We can write $\int\limits_0^{\pi/2} \log \sin x \, dx$

As $\int\limits_0^{\pi/2} \log \sin\left(\dfrac{\pi}{2} - x\right) dx$ and then by adding these two integration we can obtain the result.

$$I = \int\limits_0^{\pi/2} \log \sin x \, dx = \int\limits_0^{\pi/2} \log \sin\left(\dfrac{\pi}{2} - x\right) dx$$

$$= \int\limits_0^{\pi/2} \log \cos x \, dx$$

$$\therefore 2I = \int\limits_0^{\pi/2} (\log \sin x + \log \cos x) dx = \int\limits_0^{\pi/2} \log(\sin x \cos x) dx$$

$$= \int\limits_0^{\pi/2} \log\left(\dfrac{\sin 2x}{2}\right) dx = \int\limits_0^{\pi/2} \log \sin 2x \, dx - \int\limits_0^{\pi/2} \log 2 \, dx$$

$$= -\dfrac{\pi}{2}\log 2 + \int\limits_0^{\pi/2} \log \sin 2x \, dx \text{ (Put } 2x = t)$$

$$= -\dfrac{\pi}{2}\log 2 + \dfrac{1}{2}\int\limits_0^{\pi} \log \sin t \, dt$$

$$= -\dfrac{\pi}{2}\log 2 + \dfrac{1}{2}(2)\int\limits_0^{\pi/2} \log \sin t \, dt.$$

$$\therefore 2I = -\dfrac{\pi}{2}\log 2 + I \implies I = -\dfrac{\pi}{2}\log 2$$

Example 4: Evaluate: (i) $I = \int\limits_1^3 (x^2 + x) dx$

(ii) $I = \int\limits_a^b \sin x \, dx$ as limit of a sum.

Sol: By using the limit as a sum method we can solve the problems above.

(i) $f(x) = x^2 + x$, $a = 1$, $b = 3$, $nh = 3 - 1 = 2$

$$I = \lim_{n\to\infty} h\sum_{r=1}^n f(a+rh)$$

$$= \lim_{n\to\infty} h\sum_{r=1}^n ((a+rh)^2 + (a+rh))$$

$$= \lim_{n\to\infty} h\left(\sum_{r=1}^n r^2 h^2 + rh(2a+1) + (a^2 + a)\right)$$

$$= \lim_{n\to\infty} h\left(\dfrac{n(2+h)(4+h)}{6} + (2a+1)\dfrac{n(2+h)}{2} + n(a^2 + a)\right)$$

$$= \lim_{n\to\infty}\left(\dfrac{2(2+h)(4+h)}{6} + (2a+1)\dfrac{2(2+h)}{2} + 2(a^2 + a)\right)$$

$$= \dfrac{8}{3} + 6 + 4 = \dfrac{38}{3}$$

(ii) $I = \int\limits_a^b \sin x \, dx$

$$nh = b - a; \qquad I = \lim_{h\to 0} h\left(\sum_{r=1}^n \sin(a+rh)\right)$$

$$= \lim_{h\to 0} \dfrac{\frac{h}{2}}{\sin\frac{h}{2}}\left(\sum_{r=1}^n 2\sin\dfrac{h}{2}\sin(a+rh)\right)$$

$$= \lim_{h\to 0} \dfrac{\frac{h}{2}}{\sin\frac{h}{2}}\left(\sum_{r=1}^n \cos\left(a+hr-\dfrac{h}{2}\right) - \cos\left(a+hr+\dfrac{h}{2}\right)\right)$$

$$= \lim_{h\to 0} \dfrac{\frac{h}{2}}{\sin\frac{h}{2}}\left(\cos\left(a+\dfrac{h}{2}\right) - \cos\left(a+nh+\dfrac{h}{2}\right)\right)$$

$$= \cos a - \cos b$$

Example 5: Evaluate $I = \int\limits_{-1}^3 (|x-2| + 2[x]) dx$, where $[x]$ is the greatest integer function.

Sol: By putting $x - 2 = y$ and it is negative in interval -3 to -1 and positive in interval 0 to 1.

$$I_1 = \int\limits_{-1}^3 |x-2| dx; \text{ Put } x - 2 = y$$

$$\int\limits_{-3}^1 |y| dy = \int\limits_{-3}^{-1} -y \, dy + 2\int\limits_0^1 y \, dy$$

$$= -\dfrac{1}{2}[y^2]_{-3}^{-1} + [y^2]_0^1 = 4 + 1 = 5$$

$$I_2 = \int\limits_{-1}^3 [x] dx$$

$$= \int\limits_{-1}^0 -dx + \int\limits_0^1 0 \, dx + \int\limits_1^2 dx + \int\limits_2^3 2 \, dx = -1 + 0 + 1 + 2 = 2$$

$$\therefore I = I_1 + 2I_2 = 9$$

Example 6: Show that $I = \int\limits_0^{\infty} \dfrac{x \log x}{(1+x^2)^2} dx = 0$

Sol: By splitting the given integration into two intervals i.e. from 0 to 1 and then 1 to ∞ we can solve the given problem.

$$\int_0^\infty \frac{x\log x}{(1+x^2)^2}dx = \int_0^1 \frac{x\log x}{(1+x^2)^2}dx + \int_1^\infty \frac{x\log x}{(1+x^2)^2}dx$$

Put $x = 1/y$ in the second integral

$$\therefore \int_1^\infty \frac{x\log x}{(1+x^2)^2}dx = \int_1^0 \frac{y^4 \log y}{y^3(1+y^2)^2}dy = -\int_0^1 \frac{y\log y}{(1+y^2)^2}dy$$

Thus $I = \int_0^1 \frac{x\log x}{(1+x^2)^2}dx - \int_0^1 \frac{y\log y}{(1+y^2)^2}dy = 0$

Example 7: If $I = \int_{\frac{1}{\sqrt 3}}^{\frac{1}{\sqrt 3}} \frac{x^4}{1-x^4}\cos^{-1}\left(\frac{2x}{1-x^2}\right)dx$, then find its value.

Sol: We can write $\cos^{-1}\left(\frac{2x}{1-x^2}\right)$ as $\cos^{-1}\left(\frac{-2x}{1-x^2}\right)$

$= \left(\pi - \cos^{-1}\frac{2x}{1+x^2}\right)$ and then by solving we will get the result.

$$I = \int_{-\frac{1}{\sqrt7}}^{\frac{1}{\sqrt3}} \frac{x^4}{1-x^4}\cos^{-1}\left(\frac{2x}{1-x^2}\right)dx = \int_{-\frac{1}{\sqrt3}}^{\frac{1}{\sqrt3}} \frac{x^4}{1-x^4}\cos^{-1}\left(\frac{-2x}{1-x^2}\right)dx$$

$$= \int_{-\frac{1}{\sqrt3}}^{\frac{1}{\sqrt3}} \frac{x^4}{1-x^4}\left(\pi - \cos^{-1}\frac{2x}{1+x^2}\right)dx$$

$$2I = \pi\int_{-\frac{1}{\sqrt3}}^{\frac{1}{\sqrt3}} \frac{x^4}{1-x^4}dx = 2\pi\int_0^{\frac{1}{\sqrt3}} \frac{x^4}{1-x^4}dx$$

$$I = \pi(-1)\int_0^{\frac{1}{\sqrt3}}\left(1 - \frac{1}{1-x^4}\right)dx$$

$$= -\frac{\pi}{\sqrt3} + \frac{\pi}{2}\int_0^{\frac{1}{\sqrt3}}\frac{1}{1-x^2} + \frac{1}{1+x^2}dx$$

$$= -\frac{\pi}{\sqrt3} + \frac{\pi}{2}\cdot\frac{\pi}{6} + \frac{\pi}{4}\int_0^{\frac{1}{\sqrt3}}\frac{1}{1-x} + \frac{1}{1+x}dx$$

$$= -\frac{\pi}{\sqrt3} + \frac{\pi^2}{12} + \frac{\pi}{4}\left(\log\frac{|1+x|}{|1-x|}\right)_0^{\frac{1}{\sqrt3}}$$

$$= \frac{\pi^2}{12} - \frac{\pi}{\sqrt3} + \frac{\pi}{4}\log\left(\frac{\sqrt3+1}{\sqrt3-1}\right)$$

Example 8: Evaluate $\int_a^b (px+q)dx$ as a limit of a sum

Sol: Here as $f(x) = px + q$, therefore using the limit as sum method we can solve the given problem.

$$I = \int_a^b (px+q)dx$$

$$= \lim_{h\to0} h[f(a) + f(a+h) + \ldots + f(a+(n-1)h)]$$

$$= \lim_{h\to0} h[(pa+q) + \{p(a+h)+q\} + \ldots +$$

$$\{p(a+(n-1)h)+q\}]$$

$$= \lim_{h\to0} h[p(a+a+\ldots+a) + ph(1+2+\ldots+(n-1))$$

$$+ q(1+1+\ldots+1)]$$

$$= \lim_{h\to0} h\left[pna + \frac{1}{2}pnh(n-1) + qn\right]$$

$$= \lim_{h\to0}\left[hpna + \frac{1}{2}pnh(hn-h) + qnh\right] \qquad \ldots(i)$$

Since, $h = (b-a)/n$, or $nh = b-a$, we obtain from (i)

$$I = \lim_{h\to0}\left[(pa+q)(b-a) + \frac{p}{2}(b-a)(b-a-h)\right]$$

$$= (pa+q)(b-a) + \frac{p}{2}(b-a)^2$$

$$= \frac{p}{2}(b-a)(2a+b-a) + q(b-a)$$

$$= \frac{p}{2}(b^2-a^2) + q(b-a).$$

Example 9: If $U_n = \int_0^\pi \frac{1-\cos n\pi}{1-\cos x}dx$ where n is a positive integer or zero, then show that $U_{n+2} + U_n = 2U_{n+1}$.

Hence show that $\int_0^{\pi/2}\frac{\sin^2 n\theta}{\sin^2\theta}d\theta = \frac{n\pi}{2}$

Sol: Here $U_n = U_{n+2} - U_{n+1}$ therefore by substituting $n+2$ and $n+1$ in place of n and solving we will get the required result.

$$\because U_n = \int_0^\pi \frac{1-\cos nx}{1-\cos x}dx$$

$$\therefore U_{n+2} - U_{n+1}$$

$$= \int_0^\pi \frac{\{(1-\cos(n+2)x)\} - \{1-\cos(n+1)x\}}{(1-\cos x)}dx$$

$$= \int_0^\pi \frac{\cos(n+1)x - \cos(n+2)x}{(1-\cos x)}dx$$

$$= \int_0^\pi \frac{2\sin(n+(3/2))x \sin(x/2)}{2\sin^2(x/2)}$$

$$\Rightarrow U_{n+2} - U_{n+1} = \int_0^\pi \frac{\sin(n+(3/2))x}{\sin(x/2)}dx \qquad \ldots(i)$$

Similarly

$$\Rightarrow U_{n+1} - U_n = \int_0^\pi \frac{\sin(n+(1/2))x}{\sin(x/2)}dx \qquad \ldots(ii)$$

from (1) and (2), we get

$$(U_{n+2} - U_{n+1}) - (U_{n+1} - U_n)$$

$$= \int_0^\pi \frac{\sin(n+(3/2))x - \sin(n+(1/2))x}{\sin(x/2)}$$

$$= \int_0^\pi \frac{2\cos(n+1)x \sin(x/2)}{\sin(x/2)}dx = 2\left\{\frac{\sin(n+1)x}{(n+1)}\right\}_0^\pi = 0$$

$$\therefore U_{n+2} + U_n = 2U_{n+1}$$

Hence proved

Now $U_{n+2} - U_{n+1} = U_{n+1} - U_n$.

Similarly implies

$$U_{n+2} - U_{n+1} = U_{n+1} - U_n = U_n - U_{n-1} = \ldots\ldots = U_1 - U_0$$

$$\therefore U_n - U_{n-1} = U_1 - U_0 = \pi - 0$$

$$\Rightarrow U_n = \pi + U_{n-1}$$

$$= \pi + \pi + U_{n-2}$$

$$= 2\pi + U_{n-2}$$

$$U_n = n\pi + U_0 \quad \ldots\ldots(3) \; [\because U_0 = 0]$$

$$U_n = np$$

Hence $\therefore \int_0^{\pi/2} \frac{\sin^2 n\theta}{\sin^2 \theta}d\theta = \int_0^{\pi/2} \frac{1-\cos 2n\theta}{1-\cos 2\theta}d\theta$

Put $2\theta = x \therefore d\theta = \frac{dx}{2}$

Hence $\int_0^{\pi/2} \frac{\sin^2 n\theta}{\sin^2 \theta}d\theta = \frac{1}{2}\int_0^\pi \frac{1-\cos nx}{1-\cos x}dx$

$$= \frac{1}{2}U_n = \frac{1}{2}n\pi \qquad \{\text{from (1)}\}$$

Example 10: Solve $\int_{-\pi/4}^{\pi/4} \frac{x+(\pi/4)}{2-\cos 2x}dx$.

Sol: By splitting $\int_{-\pi/4}^{\pi/4} \frac{x+(\pi/4)}{2-\cos 2x}dx$

$$= \int_{-\pi/4}^{\pi/4} \frac{x}{2-\cos 2x}dx + \frac{\pi}{4}\int_{-\pi/4}^{\pi/4} \frac{1}{2-\cos 2x}dx \text{ and as we}$$

know $\left(\frac{x}{2-\cos 2x}\right)$ is an odd function

therefore $\int_{-\pi/4}^{\pi/4} \frac{x}{2-\cos 2x}dx = 0$.

Therefore $0 + \frac{\pi}{4}2\int_0^{\pi/4} \frac{1}{2-\cos 2x}dx$

This is because $\left(\frac{x}{2-\cos 2x}\right)$ is an odd function,

whereas $\left(\frac{1}{2-\cos 2x}\right)$ is an even function

$$= \frac{\pi}{2}\int_0^{\pi/4} \frac{dx}{2-((1-\tan^2 x)/(1+\tan^2 x))}$$

$$= \frac{\pi}{2}\int_0^{\pi/4} \frac{(1+\tan^2 x)dx}{2(1+\tan^2 x)-(1-\tan^2 x)} = \frac{\pi}{2}\int_0^{\pi/4} \frac{\sec^2 x \, dx}{1+3\tan^2 x}$$

Now let $\tan x = t \quad \therefore \sec^2 x \, dx = dt$

$$\Rightarrow \frac{\pi}{2}\int_0^1 \frac{dt}{1+3t^2} = \frac{\pi}{2\sqrt{3}}\left(\tan^{-1}\sqrt{3}t\right)_0^1 = \frac{\pi^2}{6\sqrt{3}}$$

Example 11: Show that

$$\frac{\pi}{6} < \int_0^1 \frac{dx}{\sqrt{4-x^2-x^3}} < \frac{\pi}{4\sqrt{2}}$$

Sol: Since $0 < x < 1$

so $\frac{1}{\sqrt{4-x^2}} < \frac{1}{\sqrt{4-x^2-x^3}} < \frac{1}{\sqrt{4-2x^2}}$

Hence by using the property:

If $f(x) \le g(x)$ on $[a, b]$, then $\int_a^b f(x)dx \le \int_a^b g(x)dx$ we

can solve the given problem.

Integrate the above relation

$$\int_0^1 \frac{dx}{\sqrt{4-x^2}} < \int_0^1 \frac{dx}{\sqrt{4-x^2-x^2}} < \int_0^1 \frac{dx}{\sqrt{4-2x^2}}$$

$$\left(\sin^{-1}\frac{x}{2}\right)_0^1 < \int_0^1 \frac{dx}{\sqrt{4-x^2-x^3}} < \frac{1}{\sqrt{2}}\left(\sin^{-1}\frac{x}{\sqrt{2}}\right)_0^1$$

$$\frac{\pi}{6} < \int_0^1 \frac{dx}{\sqrt{4-x^2-x^3}} < \frac{\pi}{4\sqrt{2}}.$$

Hence proved.

Exercise 1

Q.1 $\int\limits_{1/4}^{1/2} \dfrac{dx}{\sqrt{x-x^2}}$

Q.2 $\int\limits_{0}^{\pi/2} \dfrac{dx}{(4\sin^2 x + 5\cos^2 x)}$

Q.3 $\int\limits_{0}^{\pi/2} \dfrac{\sin^2 x}{1+\sin x \cos x}dx$

Q.4 $\int\limits_{0}^{1} |5x-3| \, dx$

Q.5 $\int\limits_{1}^{3} f(x)dx$, where $f(x) = \begin{bmatrix} 2x+1, \ 1 \le x \le 2 \\ x^2+1, \ 2 \le x \le 3 \end{bmatrix}$

Q.6 $\int\limits_{-\pi/4}^{\pi/4} |\sin x| \, dx$

Q.7 $\int\limits_{0}^{\pi} \dfrac{x}{(1+\sin^2 x)}dx$

Q.8 Evaluate using limit of a sum: $\int\limits_{0}^{2}(x^2+1)dx$

Q.9 Evaluate: $\int\limits_{0}^{\pi/2} |\sin x - \cos x| \, dx$

Q.10 If f and g are continuous function on [0, a] satisfying $f(x) = f(a-x)$ and $g(x) + g(a-x) = 2$ then, show that

$$\int\limits_{0}^{a} f(x)g(x)dx = \int f(x)dx.$$

Q.11 Evaluate: $\int\limits_{0}^{100\pi} \sqrt{1-\cos 2x} \, dx$

Q.12 (i) Show that if f(t) is an odd function then $\int\limits_{0}^{a} f(t)dx$ is an even function w.r.t. x.

(ii) Can $\int\limits_{a}^{x} f(t)dt$ be an odd function if f(t)dt is an even function?

Q.13 If $f(x) = \int\limits_{0}^{x^2} \sqrt{1+t^2}dt$, then find the value of f'(x).

Q.14 Evaluate $\int\limits_{-\pi/2}^{\pi/2} \dfrac{\pi + 4x^2}{-\cos(|x|+(\pi/3))}dx$.

Q.15 If $f(x) = \int\limits_{1}^{x} \dfrac{\log t}{1+t}dt$ then prove that

$f(x) + f\left(\dfrac{1}{x}\right) = \dfrac{1}{2}(\log x)^2$.

Q.16 $\int\limits_{\frac{1}{x}}^{2t} |\log x| \, dt$

Q.17 $\int\limits_{0}^{x} \dfrac{\sin(n+(1/2))x}{2\sin(x/2)}dx, \ n \in N.$

Q.18 If $F(x) = \int\limits_{\frac{5x}{4}}^{x}(3\sin t + 4\cos t)dx$. Find the least value of F(x) on the interval $\left[\dfrac{5\pi}{4}, \dfrac{4\pi}{3}\right]$.

Q.19 If $I_A = \int\limits_{0}^{\frac{\pi}{4}} \tan^n \theta \, d\theta$, $n \in N$, then find $n(I_{n-1} + I_{n+1})$ and I_B.

Q.20 If "a" is a positive integer, solve for "a"

$$\int\limits_{0}^{a}\left(a^2\left(\dfrac{\cos 3x}{4} + \dfrac{3}{4}\cos x\right) + a\sin x - 20\cos x\right)dx \le \dfrac{-a^3}{3}.$$

Q.21 If $f(x) = \sin x$, then find its mean value on (−2, 0).

Q.22 Evaluate $I = \int\limits_{0}^{\pi} \dfrac{1}{x+\sqrt{a^2-x^2}}dx$.

Q.23 Show that $I = \int\limits_{0}^{a^2}\left[\sqrt{x}\right]dx = \dfrac{n(n-1)(4\pi+1)}{6}$, where [x] is the greatest integer function.

Q.24 Show that $I = \int\limits_{0}^{nx+\lambda} |\sin x| \, dx = 2n+1-\cos\lambda, \ n \in N, \ 0 \le \lambda < \pi$.

Q.25 Show that $I = \int_0^\pi \dfrac{\pi \sin 2x, \sin\left((\pi/2)\cos x\right)}{2x - \pi} dx = \dfrac{8}{x^2}$.

Q.26 Let f and g be function satisfying the following conditions:

(i) $f(0) = 1$ (ii) $f(x) = g(x)$, $g'(x) = f(x)$

(iii) $g(0) = 0$ (iv) $g(x) \geq 0 \;\forall\; x \in R$

Find $f(1)$.

Q.27 Show that

(i) $\displaystyle\int_0^\pi \log(1 + \cos x)dx = \pi\log(1/2)$;

(ii) $\displaystyle\int_{\pi/6}^{\pi/3} \dfrac{dx}{1 + \sqrt{\cot x}} = \dfrac{\pi}{12}$

Q.28 Prove that

$\displaystyle\int_0^\pi \dfrac{dx}{1 - 2a\cos x + x^2} - \dfrac{\pi}{1 - a^2}$ or $\dfrac{\pi}{a^2 - 1}$; $a > 0$,

According as $a < 1$ or $a > 1$.

Q.29 (i) Evaluate $\displaystyle\lim_{n \to 0} \dfrac{\int_0^a x\, dx}{\alpha \sin \alpha}$

(ii) If $y = x\displaystyle\int_x^a \log dx$, Find $\dfrac{dy}{dx}$ at $x = e$.

Q.30 Find the intervals of increase of $f(x)$ defined by $f(x) = \displaystyle\int_0^\alpha (t^2 + 2t)(t^2 - 1)dt$.

Exercise 2

Single Correct Choice Type

Q.1 $\displaystyle\int_{-1}^1 f(x)\,dx$ is equal to where $f(x) = \begin{cases} 1 - 2x, & x < 0 \\ 1 + 2x, & x \geq 0 \end{cases}$

(A) 4   (B) –4   (C) 2   (D) –2

Q.2 $\displaystyle\int_{-1}^1 e^{|x|}\,dx$ equals

(A) 2e   (B) 2e – 1   (C) 2e – 2   (D) e – 2

Q.3 $\displaystyle\int_0^1 [x]dx$ equals ; where $[\cdot]$ is G.I.F.

(A) 0   (B) 2   (C) 3   (D) 1

Q.4 $\displaystyle\int_0^x |\cos x|dx$ equals

(A) 1   (B) 2   (C) 3   (D) 4

Q.5 $\displaystyle\int_{-2}^2 |2x + 3|dx$ equals

(A) $\dfrac{25}{2}$ (B) 0   (C) $\dfrac{25}{4}$ (D) $\dfrac{25}{3}$

Q.6 $\displaystyle\int_{-2}^2 |1 - x^2|\,dx =$

(A) 2   (B) 4   (C) 6   (D) 8

Q.7 The point of extremum of $\displaystyle\int_0^{x^2} \dfrac{t^2 - 5t + 4}{2 + e^t} dt$ are

(A) $x = -2$ (B) $x = 1$

(C) $x = 0$ (D) All of the above

Q.8 The point of intersection

$F_1(x) = \displaystyle\int_2^x (2t - 5)\,dt$ and $F_2(x) = \displaystyle\int_0^x 2t\,dt$, are -

(A) $\left(\dfrac{6}{5}, \dfrac{36}{25}\right)$ (B) $\left(\dfrac{2}{3}, \dfrac{4}{9}\right)$ (C) $\left(\dfrac{1}{3}, \dfrac{1}{9}\right)$ (D) $\left(\dfrac{1}{5}, \dfrac{1}{25}\right)$

Q.9 If f and g are continuous function on [0, a) satisfying $f(x) = f(a - x)$ and $g(x) + g(a-x) = 2$, then $I = \displaystyle\int_0^a f(x)g(x)dx =$

(A) $\displaystyle\int_0^a f(x)dx$ (B) $\displaystyle\int_a^0 f(x)dx$

(C) $2\displaystyle\int_0^a f(x)dx$ (D) None of these

Q.10 The value of integral $\displaystyle\int_0^{\log 5} \dfrac{e^x \sqrt{e^x - 1}}{e^x + 3}dx =$

(A) 3 + 2p   (B) 4 – p

(C) 2 + p   (D) None of these

Q.11 The value of the integral $\displaystyle\int_{-\alpha}^\pi \sin mx \sin nx\, dx$ for m $\neq$ n (m, n $\in$ I), is -

(A) 0   (B) p   (C) $\pi/2$ (D) 2p

Q.12 $\int_{1/e}^{e} |\log x| dx =$

(A) $1 - \dfrac{1}{e}$

(B) $2\left(1 - \dfrac{1}{e}\right)$

(C) $e^{-1} - 1$

(D) None of these

Q.13 $\int_{0}^{\pi} \dfrac{dx}{1 - 2a\cos x + a^2} =$

(A) $\dfrac{\pi}{2(1 - a^2)}$

(B) $\pi(1 - a^2)$

(C) $\dfrac{\pi}{1 - a}$

(D) None of these

Q.14 $\int_{0}^{1} (1 - x)^9 dx =$

(A) x

(B) $\dfrac{1}{10}$

(C) $\dfrac{11}{10}$

(D) 2

Q.15 $\int_{0}^{\pi} \dfrac{dx}{\left(x + \sqrt{x^2 + 1}\right)^3} =$

(A) $\dfrac{3}{8}$

(B) $\dfrac{1}{8}$

(C) $-\dfrac{3}{8}$

(D) None of these

Q.16 If [x] denotes the greatest integer less than or equal to x, then the value $\int_{1}^{5} [|x - 3|] dx$ is -

(A) 1

(B) 2

(C) 4

(D) 8

Q.17 $\int_{-\pi/2}^{\pi/2} \dfrac{\sin x}{1 + \cos^2 x} e^{-\cos^2 x} dx$ is equal to -

(A) $2e^{-1}$

(B) 1

(C) 0

(D) None of these

Q.18 The value of

$\int_{-1/2}^{1/2}\left[\left(\dfrac{x+1}{x-1}\right)^2 + \left(\dfrac{x-1}{x+1}\right)^2 - 2\right]^{1/2} dx$ equal

(A) log (4/3)

(B) 2 log (4/3)

(C) 4 log (4/3)

(D) –4 log (4/3)

Q.19 Let f(x) = x – [x], for every real number x, where [x] is integral pat of x. Then $\int_{-1}^{1} f(x) dx$ is

(A) 1

(B) 2

(C) 0

(D) ½

Q.20 If [x] stands for the greatest integer function, the value of $\int_{4}^{10} \dfrac{[x^2]}{[x^2 - 28x + 196] + [x^2]} dx$ is

(A) 0

(B) 1

(C) 3

(D) None of these

Q.21 The value of $\int_{-1}^{3} (|x - 2| + [x]) dx$ is ([x] stands for greatest integer less than or equal to x)

(A) 7

(B) 5

(C) 4

(D) 3

Q.22 $\int_{0}^{\pi/2} \dfrac{\sin^2 x}{\sin x + \cos x} dx$ is equal to

(A) $\dfrac{\pi}{2}$

(B) $\sqrt{2}\log(\sqrt{2} + 1)$

(C) $\dfrac{1}{\sqrt{2}}\log(\sqrt{2} + 1)$

(D) None of these

Q.23 If $u_{10} - \int_{0}^{\pi/2} x^{10} \sin x \, dx$ then the value of $u_{10} + 90\,u_{8}$ is

(A) $9\left(\dfrac{\pi}{2}\right)^8$

(B) $\left(\dfrac{\pi}{2}\right)^9$

(C) $10\left(\dfrac{\pi}{2}\right)^9$

(D) $9\left(\dfrac{\pi}{2}\right)^9$

Q.24 For any integer n, the integral $\int_{0}^{\pi} e^{\sin^2 x} \cos^3(2n + 1)x \, dx$ has the value

(A) π

(B) 1

(C) 0

(D) None of these

Q.25 The value of $\int_{-\pi/2}^{\pi/2} \sin(\log(x + \sqrt{x^2 + 1})) dx$ is

(A) 1

(B) –1

(C) 0

(D) None of these

Q.26 The value of $\alpha \in (-\pi, 0)$ satisfying $\sin \alpha + \int_{\alpha}^{2\alpha} \cos 2x \, dx = 0$ is

(A) $-\pi/2$

(B) $-p$

(C) $-\pi/3$

(D) 0

Q.27 If $f(x) = \int_{x^2}^{x^4} \sin \sqrt{t} \, dt$, then $f'(x)$ equals

(A) $\sin x^2 - \sin x$

(B) $4x^3 \sin x^2 - 2x \sin x$

(C) $x^4 \sin x^2 - x \sin x$

(D) None of these

Q.28 $\int_0^\pi x \sin x \cos^4 x \, dx =$

(A) $\dfrac{\pi}{10}$ (B) $\dfrac{\pi}{5}$ (C) $-\dfrac{\pi}{5}$ (D) None of these

Q.29 If $f(x) = ae^{2x} + be^x + cx$, satisfies the conditions $f(0) = -1$, $f'(\log 2) = 31$, $\int_0^{\log 4}(f(x) - cx)dx = \dfrac{39}{2}$, then

(A) $a = 5, b = 6, c = 3$ (B) $a = 5, b = -6, c = 3$
(C) $a = -5, b = 6, c = 3$ (D) None of these

Q.30 $\int_{-\pi/4}^{\pi/4} \dfrac{e^x \sec^2 x \, dx}{e^{2x} - 1}$ is equal to

(A) 0   (B) 2   (C) e   (D) None of these

Q.31 $\int_{-1}^{a} \log_a\left(x + \sqrt{1 + x^2}\right)dx$ is equal to

(A) $2\log_a a$ (B) 0
(C) $\log_a 2 + \log a$ (D) None of these

Q.32 The value of $\int_{-2}^{2} \dfrac{\sin^2 x}{\left[(x/\pi)\right] + (1/2)}dx$, where [x] = the greatest integer less than or equal to x, is

(A) 1   (B) 0   (C) $4 - \sin 4$ (D) None of these

Q.33 If $f(x) = \int_0^x \log(1 + t^2)dt$ then the value of $f'(1)$ is equal to

(A) 2   (B) 0   (C) 1   (D) None of these

Q.34 $\int_0^x \dfrac{dx}{1 + 3^{\cos x}}$ is equal to

(A) π (B) 0   (C) $\dfrac{\pi}{2}$ (D) None of these

Previous Years' Questions

Q.1 The value of the integral
$\int_0^{\pi/2} \dfrac{\sqrt{\cot x}}{\sqrt{\cot x} + \sqrt{\tan x}}dx$ is   *(1983)*

(A) $\pi/4$ (B) $\pi/2$ (C) p   (D) None of these

Q.2 For any integer n, the integral
$\int_0^x e^{\cos^2 x} \cos^3(2r + 1)x \, dx$ has the value   *(1985)*

(A) π (B) 1   (C) 0   (D) None of these

Q.3 Let $f: R \to R$ be a differentiable function and $f(1) = 4$. Then, the value of $\lim_{x \to 1}\int_4^{f(x)} \dfrac{2t}{x - 1}dt$ is   *(1990)*

(A) $8f'(1)$ (B) $4f'(1)$ (C) $2f'(1)$ (D) $f'(1)$

Q.4 The value of $\int_0^{\pi/2} \dfrac{dx}{1 + \tan^3 x}$ is   *(1993)*

(A) 0   (B) 1   (C) $\pi/2$ (D) $\pi/4$

Q.5 The value of $\int_0^{2\pi}[2\sin x]dx$ where [·] represents the greatest integral function, is   *(1995)*

(A) $-\dfrac{5\pi}{3}$ (B) $-\pi$ (C) $\dfrac{5\pi}{3}$ (D) $-2p$

Q.6 $\int_0^x f(t)dt + x - \int_x^1 t f(t)dt$, then the value of $f(1)$ is   *(1998)*

(A) $\dfrac{1}{2}$ (B) 0   (C) 1   (D) $-\dfrac{1}{2}$

Q.7 $\int_{\pi/4}^{3\pi/4} \dfrac{dx}{1 + \cos x}$ is equal to   *(1999)*

(A) 2   (B) -2 (C) $\dfrac{1}{2}$ (D) $-\dfrac{1}{2}$

Q.8 If for a real number y, [y] is the greatest integer less than or equal to y, then the value of the integral $\int_{\pi/2}^{3\pi/2}[2\sin x]dx$ is   *(1999)*

(A) $-\pi$ (B) 0   (C) $-\dfrac{\pi}{2}$ (D) $\dfrac{\pi}{2}$

Q.9 The value of $\int_{-\pi}^{\pi} \dfrac{\cos^2 x}{1 + a^x}dx, a > 0$, is   *(2001)*

(A) π (B) $a\pi$ (C) $\dfrac{\pi}{2}$ (D) 2π

Q.10 Let $f: (0, \infty) \to R$ and $F(x) = \int_0^x f(t)dt$, If $F(x^2) = x^2(1 + x)$, then $f(4)$ equals   *(2001)*

(A) $\dfrac{5}{4}$ (B) 7   (C) 4   (D) 2

Q.11 Let $f(x) = \int_1^x \sqrt{2 - t^2}dt$. Then, the real value of x if it satisfies $x^2 - f'(x) = 0$ are   *(2002)*

(A) ± 1 (B) $\pm\dfrac{1}{\sqrt{2}}$ (C) $\pm\dfrac{1}{2}$ (D) 0 and 1

Q.12 Let T > 0 be a fixed real number. Suppose, f is a continuous function such that for all $x \in R$, f(x + T) = f(x). If $I = \int_0^T f(x)dx$, then the value of $\int_3^{3+3T} f(2x)dx$, *(2002)*

(A) $\frac{3}{2}I$ (B) I   (C) 3I   (D) 6I

Q.13 If $f(x) = \int_{x^2}^{x^2+1} e^{-t^2}dt$, then f(x) increases in *(2003)*

(A) (2, 2)   (B) No value of x   (C) (0, ∞)   (D) (– ∞, 0)

Q.14 The value of the integral $\int_0^1 \sqrt{\frac{1-x}{1+x}}dx$ is   *(2004)*

(A) $\frac{\pi}{2}+1$ (B) $\frac{\pi}{2}-1$ (C) –1   (D) 1

Q.15 Match the conditions expressions in column I with statement in column II   *(2007)*

Column I	Column II
(A) $\int_{-1}^1 \frac{dx}{1+x^2}$	(p) $\frac{1}{2}\log\left(\frac{2}{3}\right)$
(B) $\int_0^1 \frac{dx}{\sqrt{1-x^2}}$	(q) $2\log\left(\frac{2}{3}\right)$
(C) $\int_2^3 \frac{dx}{1-x^2}$	(r) $\frac{\pi}{3}$
(D) $\int_1^2 \frac{dx}{x\sqrt{x^2-1}}$	(s) $\frac{\pi}{2}$

Q.16 The value of $\int_{-2}^2 |1-x^2|\, dx$ is..... *(1989)*

Q.17 The value of $\int_{\pi/4}^{3\pi/4} \frac{x}{1+\sin x}dx$..... *(1993)*

Q.18 The value of $\int_2^3 \frac{\sqrt{x}}{\sqrt{5-x}+\sqrt{x}}dx$ is..... *(1994)*

Q.19 Let $= \int_0^1 \frac{\sin x}{\sqrt{x}}dx$ and $J = \int_0^1 \frac{\cos x}{\sqrt{x}}dx$.

Then which one of the following is true?   *(2008)*

(A) $I > \frac{2}{3}$ and $J > 2$ (B) $I < \frac{2}{3}$ and $J < 2$

(C) $I < \frac{2}{3}$ and $J > 2$ (D) $I > \frac{2}{3}$ and $J < 2$

Q.20 $\int_0^x [\cot x]dx, [.]$ denotes the greatest integer function, is equal to   *(2009)*

(A) $\frac{\pi}{2}$ (B) 1   (C) -1   (D) $-\frac{\pi}{2}$

Q.21 Let $p(x)$ be a function defined on R such that $p(x) = p(1-x)$ for all $p(0) = 1$ $p(1) = 41$. Then $\int_0^1 p(x)dx$ equals.   *(2010)*

(A) 21   (B) 41   (C) 42   (D) $\sqrt{41}$

Q.22 The value of $\int_0^1 \frac{8\log(1+x)}{1+x^2}dx$. is   *(2011)*

(A) $\frac{\pi}{8}\log 2$ (B) $\frac{\pi}{2}\log 2$ (C) $\log 2$ (D) $\pi\log 2$

Q.23 If $g(x) = \int_0^x \cos 4t\, dt$, then $g(x+\pi)$ equals *(2012)*

(A) $\frac{g(x)}{g(\pi)}$ (B) $g(x)+g(\pi)$

(C) $g(x)-g(\pi)$ (D) $g(x).g(\pi)$

Q.24 Statement-I: The value of the integral $\int_{\pi/6}^{\pi/3} \frac{dx}{1+\sqrt{\tan x}}$ is equal to $\pi/6$.

Statement-II: $\int_a^b f(x)dx = \int_a^b f(a+b-x)dx$ *(2013)*

(A) Statement-I is true; statement-II is true; statement-II is a correct explanation for statement-I.

(B) Statement-I is true; statement-II is true; statement-II is not a correct explanation for statement-I.

(C) Statement-I is true; statement-II is false.

(D) Statement-I is false; statement-II is true.

Q.25 The integral $\int \left(1+x-\frac{1}{x}\right)e^{x+\frac{1}{x}}dx$ is equal to *(2014)*

(A) $(x+1)e^{x+\frac{1}{x}}+c$ (B) $-xe^{x+\frac{1}{x}}+c$

(C) $(x-1)e^{x+\frac{1}{x}}+c$ (D) $xe^{x+\frac{1}{x}}+c$

Q.26 The integral $\int_2^4 \frac{\log x^2}{\log x^2 + \log(36-12x+x^2)}dx$ is equal to   *(2015)*

(A) 1   (B) 4   (C) 1   (D) 6

Exercise 1

Q.1 $\int_0^1 e^{\tan^{-1}x} \sin^{-1}(\cos x)dx.$

Q.2 Prove that :

(i) $\int_0^1 \sqrt{(x-\alpha)(\beta-x)}dx - \dfrac{(\beta-\alpha)^2 x}{8}$

(ii) $\int_0^a \sqrt{\dfrac{x-\alpha}{\beta-x}}dx = (\beta-\alpha)\dfrac{\pi}{2}$

(iii) $\int_0^a \dfrac{dx}{x\sqrt{(x-\alpha)(\beta-x)}} = \dfrac{\pi}{\sqrt{\alpha\beta}}$ where $\alpha, \beta > 0$

(iv) $\int_0^b \dfrac{x\,dx}{\sqrt{(x-\alpha)(\beta-x)}} = (\alpha+\beta)\dfrac{\pi}{2}$ where $\alpha < \beta$

Q.3 (i) Let $\beta(\Pi) = \int_0^{n\pi} \sqrt{1-\sin t}\,dt.$
Find the value of $\beta(2) - \beta(1)$.

(ii) Determine a positive integer $n \leq 5$, such that $\int_0^1 e^x(x-1)^n\,dx = 16 - 6e.$

Q.4 (i) $\int_0^{\pi/2} e^x\left[\cos(\sin x)\cos^2\dfrac{x}{2} + \sin(\sin x)\sin^2\dfrac{x}{2}\right]dx$

(ii) $\int_0^\pi \{(1+x)e^x + (1-x)e^{-x}\}\ln x\,dx.$

Q.5 If $P = \int_0^\infty \dfrac{x^2}{1+x^4}dx; Q = \int_0^\infty \dfrac{x\,dx}{1+x^4}$ and $R = \int_0^\infty \dfrac{dx}{1-x^4}$
then prove that

(i) $Q = \dfrac{\pi}{4}$,

(ii) $P = R$

(iii) $P - \sqrt{2}\,Q + R = \dfrac{\pi}{2\sqrt{2}}$

Q.6 $\int_1^2 \dfrac{(x^2-1)dx}{x^2\sqrt{2x^4-2x^2+1}} = \dfrac{u}{v}$ where u and v are
in their lowest form. Find the value of $\dfrac{(1000)u}{v}$

Q.7 Let $h(x) = (fog)(x) + K$ where K is any constant. If $\dfrac{d}{dx}(h(x)) = -\dfrac{\sin x}{\cos^{-2}(\cos x)}$ then
compute the value of j(0) where j(x)

$\int_{g(x)}^{f(x)} \dfrac{f(t)}{g(t)}dt,$ where f and g are trigonometric functions.

Q.8 $\int_0^{\pi/2} \sqrt{\dfrac{1-\sin 2x}{1+\sin 2x}}dx$

Q.9 If the value of the definite integral $I = \int_0^2 (3x^2 - 3x + 1)$
$\cos(x^3 - 3x^3 + 4x - 2)\,dx$ can be expressed in the form as
$p(\sin q$ where $p, q \in N$, then find $(p + q)$.

Q.10 $\int_{-\sqrt{2}}^{\sqrt{3}} \dfrac{2x^7 + 3x^6 - 10x^5 - 7x^3 - 12x^2}{x^2+2}dx.$

Q.11 For $a \geq 2$, if the value of the definite integral
$\int_0^a \dfrac{dx}{a^2 + (x-(1x))^2}$ equals $\dfrac{x}{5050}$. Find the value of a.

Q.12 $\int_{-2}^2 \dfrac{x^2 - x}{\sqrt{x^2+4}}.$

Q.13 Let $u = \int_0^{\pi/4}\left(\dfrac{\cos x}{\sin x - \cos x}\right)^2 dx$ and
$v = \int_0^{\pi/4}\left(\dfrac{\sin x + \cos x}{\cos x}\right)^2 dx.$ Find the value of $\dfrac{v}{u}$

Q.14 $\int_0^{\pi/4} \dfrac{x\,dx}{\cos x(\cos x + \sin x)}.$

Q.15 $\int_0^1 \dfrac{\sin^{-1}\sqrt{x}}{x^2 - x + 1}dx$

Q.16 $\int_1^{\frac{1+\sqrt{5}}{2}} \dfrac{x^2+1}{x^4 - x^2 + 1}\ln\left(1 + x - \dfrac{1}{x}\right)dx$

Q.17 $\displaystyle\lim_{x\to 0} n^2 \int_{-1\pi}^{1\pi} (2010\sin x + 2012\cos x)\,|x|\,dx$.

Q.18 Find the value of the definite integral

$$\int_0^\pi |\sqrt{2}\sin x + 2\cos x|\,dx .$$

Q.19 If $\displaystyle\int_0^\pi \sqrt{(\cos x + \cos 2x + \cos 3x)^2 + (\sin x + \sin 2x + \sin 3x)^2}\,dx$

has the value equal to $\left(\dfrac{\pi}{k} + \sqrt{w}\right)$. w are positive integer.
Find the value of $(k^2 + w^2)$.

Q.20 $\displaystyle\int_0^1 \frac{1-x}{1+x}\,\frac{dx}{\sqrt{x+x^2+x^3}}$

Q.21 $\displaystyle\int_0^{\pi/2} \frac{a\sin x + b\cos x}{\sin\left(\dfrac{\pi}{4}+x\right)}\,dx$.

Q.22 A continuous real function f satisfies $f(2x) = 3\,f(x)$ $\forall\, x \in R$.

If $\displaystyle\int_0^\pi f(x)\,dx = 1$, then compute the value of definite integral $\displaystyle\int_1^2 f(x)\,dx$.

Q.23 The value of $\displaystyle\int_{-1}^3 \{|x-2| + [x]\}\,dx$, where [x] denotes the greatest integer less than or equal to x is.

Q.24 $\displaystyle\int_1^0 \sin^{-1}\frac{2x}{1+x^2}\,dx$.

Q.25 $\displaystyle\int_0^1 \frac{(ax+b)\sec x\tan x}{4+\tan^2 x}\,dx\ (a,b>0)$

Q.26 $\displaystyle\int_0^\pi \frac{(2x+3)\sin x}{(1+\cos^2 x)}\,dx$.

Q.27 Evaluate $\displaystyle\int_0^{\pi/2} \frac{\sqrt{\cos x}}{\sqrt{\cos x}+\sqrt{\sin x}}$

Q.28 If $\displaystyle\int_0^{n\pi} \frac{x\,|\sin x|}{1+|\cos x|}\,dx\ (n \in N)$ is equal to $100\,\pi\,\log 2$, then the value of n.

Q.29 Evaluate $\displaystyle\int_0^{\pi/2} \frac{\cos x}{1+\cos x+\sin x}\,dx$.

Q.30 $\displaystyle\int_0^\pi \frac{\ln(I+ax)}{1+x^2}\,dx,\ a \in N.$

Q.31 $\displaystyle\int_0^{\frac{\ln 3}{2}} \frac{e^x+1}{e^{2x}+1}\,dx$.

Q.32 If $\displaystyle\int_{a+1}^a \sqrt{x}\,dx = 2a\int_0^{\pi/2} \sin^3 x\,dx$, find the value of $\displaystyle\int_a^{a+1} x\,dx$.

Q.33 Let $\alpha,\ \beta$ be the distinct positive roots of the equation $\tan x = 2x$ then evaluate $\displaystyle\int_0^1 (\sin_\alpha x \cdot \sin_\beta x)\,dx$, independent of α and β.

Q.34 Show that $\displaystyle\int_0^{p+q} |\cos x|\,dx = 2q + \sin p$ where $q \in N$ & $-\dfrac{\pi}{2} < p < \dfrac{\pi}{2}$.

Q.35 Show that the sum of the two integrals

$$\int_{-1}^{-\pi} e^{(x+1)^2}\,dx + 3\int_{1/3}^{2/3} e^{(x-2x)^2}\,dx\ \text{is zero.}$$

Q.36 Let $F(x) = \max(\sin px, \cos px)$. Find the value of $\dfrac{\pi}{4\sqrt{2}} \displaystyle\int_{-10}^{10} F(x)\,dx$.

Q.37 $\displaystyle\int_0^{\pi/2} \tan^{-1}\left[\frac{\sqrt{1+\sin x}+\sqrt{1-\sin x}}{\sqrt{1+\sin x}-\sqrt{1-\sin x}}\right]dx$.

Q.38 Comment upon the nature of roots of the quadratic equation $x^2 + 2x = k + \displaystyle\int_0^1 |t+k|\,dt$ dependent on the value of $k \in R$.

Q.39 $\displaystyle\int_{-1}^1 \frac{(2x^{232}+x^{998}+4x^{1668}\sin x^{691})}{1+x^{666}}$

Q.40 $\pi\displaystyle\int_0^\pi \frac{x^2\sin 2x \cdot \sin\left(\dfrac{\pi}{2}\cos x\right)}{2x-\pi}\,dx$

Q.41 Evaluate $\displaystyle\int_{1/3}^{1} \frac{\left(x - x^3\right)^{1/3}}{x^4} dx$

Q.42 $\displaystyle\lim_{x\to\infty} \frac{1}{n^2} \sum_{k=0}^{n-1} \left[k \int_{k}^{k-1} \sqrt{(x-k)(k+1-x)} dx \right]$

Q.43 Let $I = \displaystyle\int_{0}^{\pi/2} \frac{\cos x + 4}{3\sin x + 4\cos x + 25} dx$ anc

$I = \displaystyle\int_{0}^{\pi/2} \frac{\sin x + 3}{3\sin x + 4\cos x + 25} dx$.

If $25\,I = a\pi + b\ln\dfrac{c}{d}$ where a, b, c and d $\in$ N and $\dfrac{c}{d}$ is not a perfect square of a rational then find the value of (a + b + c + d).

Q.44 Let y = f(x) be a quadratic function with f(2) = 1. Find the value of the integral

$\displaystyle\int_{2-\pi}^{2+\pi} f(x) . \sin\left(\frac{x-2}{2}\right) dx$.

Exercise 2

Single Correct Choice Type

Q.1 $\displaystyle\int_{0}^{2} | x^2 + 2x - 3 | dx$ equals

(A) 5/3 (B) 7/3 (C) 4 (D) 0

Q.2 The correct evaluation of $\displaystyle\int_{0}^{\pi/2} \left| \sin\left(x - \frac{\pi}{4}\right) \right| dx$ is -

(A) $2 + \sqrt{2}$ (B) $2 - \sqrt{2}$ (C) $-2 + \sqrt{2}$ (D) 0

Q.3 The correct evaluation of $\displaystyle\int_{0}^{\pi} | \sin^4 x | dx$ is -

(A) $\dfrac{8\pi}{3}$ (B) $\dfrac{2\pi}{3}$ (C) $\dfrac{4\pi}{3}$ (D) $\dfrac{3\pi}{8}$

Q.4 $\displaystyle\int_{0}^{1.5} [x^2]dx$, where [·] denotes the greatest integer function, equals -

(A) $2 + \sqrt{2}$ (B) $2 - \sqrt{2}$
(C) $-2 + \sqrt{2}$ (D) $-2 - \sqrt{2}$

Q.5 Solve $\displaystyle\int_{0}^{\pi} \frac{x}{a^2\cos^2 x + b^2\sin^2 x} dx$

(A) $\dfrac{\pi^2}{2ab}$ (B) $\dfrac{\pi^2}{4ab}$

(C) $\dfrac{\pi^2}{3ab}$ (D) $\dfrac{\pi}{5ab}$

Q.6 $\displaystyle\int_{0}^{\pi/4} \frac{\sec x}{1 + 2\sin^2 x}$ is equal to -

(A) $\dfrac{1}{3}\left| \log(\sqrt{2}+1) + \dfrac{\pi}{2\sqrt{2}} \right|$ (B) $\dfrac{1}{3}\left| \log(\sqrt{2}+1) - \dfrac{\pi}{2\sqrt{2}} \right|$

(C) $3\left| \log(\sqrt{2}+1) - \dfrac{\pi}{2\sqrt{2}} \right|$ (D) $3\left| \log(\sqrt{2}+1) + \dfrac{\pi}{2\sqrt{2}} \right|$

Q.7 If $\displaystyle\int_{0}^{1} e^{x^2}(x - \alpha)dx = 0$, then

(A) $1 < \alpha < 2$ (B) $\alpha < 0$
(C) $0 < \alpha < 1$ (D) None of these

Q.8 $\displaystyle\int_{b}^{\pi/2} \{x - [\sin x]\}dx$ is equal to -

(A) $\dfrac{\pi^2}{8}$ (B) $\dfrac{\pi^2}{8} - 1$ (C) $\dfrac{\pi^2}{8} - 2$ (D) None of these

Q.9 The value of the integral $\displaystyle\int_{b}^{100} \sin\{x - [x]\}\pi dx$ is -

(A) $\dfrac{100}{\pi}$ (B) $\dfrac{200}{\pi}$ (C) 100π (D) 200π

Q.10 The value of the integral $\displaystyle\int_{0}^{\pi} \frac{x\log x}{(1+x^2)^2} dx$ is -

(A) 1 (B) 0 (C) 2 (D) None of these

Q.11 $\displaystyle\lim_{n\to\infty} \left[\left(1 + \frac{1}{n}\right)\left(1 + \frac{2}{n}\right)\left(1 + \frac{3}{n}\right)......\left(1 + \frac{n}{n}\right) \right]^{1/n}$ is equal to -

(A) e/4 (B) 4/e (C) 2/e (D) None of these

Q.12 The solution of the equation

$\displaystyle\int_{\log 2}^{x} \frac{1}{\sqrt{e^x - 1}} dx = \frac{\pi}{6}$

(A) x = log 4 (B) x = log 2
(C) $x = \log\left(\dfrac{1}{4}\right)$ (D) None of these

Q.13 The value of

$$\lim_{n\to\infty}\left(\frac{1}{1+n^3}+\frac{4}{8+n^3}+\frac{9}{27+n^3}+\dots+n\text{ terms}\right)\text{ is -}$$

(A) $\frac{1}{3}\log 2$ (B) 0 (C) $\frac{1}{3}\log 3$ (D) None of these

Q.14 $\lim_{n\to\infty}$

$$\left\{\frac{n^2}{(n^2+1^2)^{3/2}}+\frac{n^2}{(n^2+2^2)^{3/2}}+\dots+\frac{n^2}{[n^2+(n-1)^2]^{3/2}}\right\}\text{ is}$$

equal -

(A) $-\frac{1}{\sqrt{2}}$ (B) $\frac{1}{\sqrt{2}}$ (C) $\sqrt{2}$ (D) None of these

Q.15 $\lim_{n\to\infty}\left\{\frac{\sqrt{n}}{\sqrt{n^2}}+\frac{\sqrt{n}}{\sqrt{(n+4)^3}}+\frac{\sqrt{n}}{\sqrt{(n+8)^3}}+\dots+\frac{\sqrt{n}}{\sqrt{[n+4(n-1)]^3}}\right\}$

is equal -

(A) $\frac{1}{10}\left[5-\sqrt{5}\right]$ (B) $\left[5-\sqrt{5}\right]$ (C) $\frac{1}{5}\left[5-\sqrt{5}\right]$ (D) 0

Q.16 $\lim_{n\to\infty}\left\{\tan\left(\frac{\pi}{2n}\right)\tan\left(\frac{2\pi}{2n}\right)\tan\left(\frac{3\pi}{2n}\right)\dots\tan\left(\frac{n\pi}{2n}\right)\right\}^{\frac{1}{6}}$

is equal -

(A) 0 (B) 1 (C) −1 (D) 2

Previous Years' Questions

Q.1 The integral $\int_{1/2}^{1/2}\left([x]+\log\left(\frac{1+x}{1-x}\right)\right)dx$ equals *(2002)*

(A) $-\frac{1}{2}$ (B) 0 (C) 1 (D) $\log\left(\frac{1}{2}\right)$

Q.2 If $I(m,n)=\int_0^1 t^m(1+t)^n dt$, then the expression for $I(m,n)$ in terms of $I(m+1, n-1)$ is *(2003)*

(A) $\frac{2^n}{m+1}-\frac{n}{m+1}I(m+1,n-1)$

(B) $\frac{n}{m+1}I(m+1,n-1)$

(C) $\frac{2^n}{m+1}+\frac{n}{m+1}I(m+1,n-1)$

(D) $\frac{m}{m+1}I(m+1,n-1)$

Q.3 Let f be a non-negative function defined on the interval [0, 1] . If $\int_0^x\sqrt{1-(f'(t))^2}\,dt=\int_0^x f(t)dt,\ 0\le x\le 1$ and $f(0)=0$, then *(2009)*

(A) $f\left(\frac{1}{2}\right)<\frac{1}{2}$ and $f\left(\frac{1}{3}\right)>\frac{1}{3}$

(B) $f\left(\frac{1}{2}\right)>\frac{1}{2}$ and $f\left(\frac{1}{3}\right)>\frac{1}{3}$

(C) $f\left(\frac{1}{2}\right)<\frac{1}{2}$ and $f\left(\frac{1}{3}\right)<\frac{1}{3}$

(D) $f\left(\frac{1}{2}\right)>\frac{1}{2}$ and $f\left(\frac{1}{3}\right)<\frac{1}{3}$

Q.4 The value of

$$\int_{\sqrt{\log 2}}^{\sqrt{\log 3}}\frac{x\sin x^2}{\sin x^2+\sin(\log 6-x^2)}dx\text{ is -}\qquad(2011)$$

(A) $\frac{1}{4}\log\frac{3}{2}$ (B) $\frac{1}{2}\log\frac{3}{2}$ (C) $\log\frac{3}{2}$ (D) $\frac{1}{6}\log\frac{3}{2}$

Q.5 Let $S_n=\sum_{k=0}^{n}\frac{n}{n^2+kn+k^2}$ and

$T_n=\sum_{k=0}^{n-1}\frac{n}{n^2+kn+k^2}$, for n = 1, 2, 3 ....., then *(2008)*

(A) $S_n<\frac{\pi}{3\sqrt{3}}$ (B) $S_n>\frac{\pi}{3\sqrt{3}}$

(C) $T_n<\frac{\pi}{3\sqrt{3}}$ (D) $T_n>\frac{\pi}{3\sqrt{3}}$

Q.6 If $I_n=\int_{-\pi}^{\pi}\frac{\sin nx}{(1+\pi^x)\sin x}dx$, n = 0, 1, 2,..., then *(2009)*

(A) $I_n=I_{n+2}$ (B) $\sum_{m=1}^{10}I_{2m+1}=10\pi$

(C) $\sum_{m=1}^{10}I_{2m}=0$ (D) $I_n=I_{n+1}$

Q.7 The value(s) of $\int_0^1\frac{x^4(1-x)^4}{1+x^2}dx$ is (are) *(2010)*

(A) $\frac{22}{7}-\pi$ (B) $\frac{2}{105}$ (C) 0 (D) $\frac{71}{15}-\frac{3\pi}{2}$

Paragraph for Q.8

Read the following passage and answer the questions. For every function f(x) which is twice differentiable, these will be good approximation of

$$\int_a^b f(x)dx = \left(\frac{b-a}{2}\right)\{f(a) + f(b)\},$$

for more accurate results for $c\in(a, b)$,

$$F(c) = \frac{c-a}{2}[f(a) - f(c)] + \frac{b-c}{2}[f(b) - f(c)]$$

When $c = \frac{a+b}{2}$

$$\int_a^b f(x)dx = \frac{b-a}{4}\{f(a) + f(b) + 2f(c)\}dx \qquad (2006)$$

Q.8 Good approximation of $\int_0^{\pi/2} \sin x\,dx$, is $\qquad$ **(2003)**

(A) $\pi/4$

(B) $\pi(\sqrt{2}+1)/4$

(C) $\pi(\sqrt{2}+1)/8$

(D) $\pi/8$

Q.9 If $f''(x) < 0$, "$x\in$ (a, b), and $(c,f(c))$ is point of maxima where $c\in(a, b)$, then $f'(c)$ is - $\qquad$ **(2009)**

(A) $\dfrac{f(b)-f(a)}{b-a}$

(B) $3\left(\dfrac{f(b)-f(a)}{b-a}\right)$

(C) $2\left(\dfrac{f(b)-f(a)}{b-a}\right)$

(D) 0

Q.10 If $\displaystyle\lim_{t\to a}\frac{\int_a^1 f(x)dx - ((t-a)/2)\{f(t) + f(a)\}}{(t-a)^3} = 0$, then degree of polynomial function f(x) at most is - **(2002)**

(A) 0

(B) 1

(C) 3

(D) 2

Q.11 For any real number x, let [x] denotes the largest integer less than or equal to x. Let f be a real valued function defined on the interval

[−10, 10] by $f(x) = \begin{cases} x - [x] & \text{if } [x] \text{ is odd} \\ 1 + [x] - x & \text{if } [x] \text{ is even} \end{cases}$

Then the value of $\dfrac{\pi^2}{10}\int_{-10}^{10} f(x)\cos px\,dx$ is $\quad$ .... **(2010)**

Q.12 For x > 0, let $f(x) = \int_1^x \dfrac{\log t}{1+t}dt$. Find the function $f(x) + f(1/x)$ and show that $f(e) + f(1/e) = 1/2$. Here, $\ln t = \log_e t$ **(2000)**

Q.13 If f is an even function, then prove that

$$\int_0^{\pi/2} f(\cos 2x)\cos x\,dx = \sqrt{2}\int_0^{\pi/4} f(\sin 2x)\cos x\,dx \quad (2003)$$

Q.14 Evaluate $\displaystyle\int_{-\pi/3}^{\pi/3} \frac{\pi + 4x^3}{2 - \cos(|x| + (\pi/3))}dx$. **(2004)**

Q.15 Evaluate

$$\int_0^\pi e^{|\cos x|}\left(2\sin\left(\frac{1}{2}\cos x\right) + 3\cos\left(\frac{1}{2}\cos x\right)\right)\sin x\,dx \quad (2005)$$

Q.16 The value of $\dfrac{(5050)\int_0^1 (1-x^{50})^{100}dx}{\int_0^1 (1-x^{50})^{101}dx}$ **(2006)**

Q.17 Let $g(x) = \int_0^x \dfrac{f'(t)}{1+t^2}dt$, then which of the following is true? **(2008)**

(A) $g(x)$ is positive on $(-\infty,0)$ and negative on $(\infty,0)$

(B) $g(x)$ is negative on $(-\infty,0)$ and positive or $(0,\infty)$

(C) $g(x)$ changes sign on both $(-\infty,0)$ and $\therefore$

(D) $g(x)$ does not change sign on (∞,∞)

Q.18 $\displaystyle\int_{-1}^1 g'(x)dx =$ **(2008)**

(A) $2g(-1)$

(B) 0

(C) $-2g(1)$

(D) $2g(1)$

Q.19 The total number of distinct $x \in [0,1]$ for which $\displaystyle\int_0^x \frac{t^2}{1+t^4}dt = 2x - 1$ is **(2016)**

Q.20 Let $f R \to R$ be a continuous function which satisfies $f(x) = \int_0^x f(t)dt$. Then the value of f(ln5) is **(2009)**

Q.21 Let f be a non-negative function defined on the interval $[0,1]\int_0^x \sqrt{1-(f'(t))^2}\,dt = \int_0^x f(t)dt\; 0 \le x \le 1$ and $f(0) = 0$ then **(2009)**

(A) $f\left(\dfrac{1}{2}\right) < \dfrac{1}{2}$ and $f\left(\dfrac{1}{3}\right) > \dfrac{1}{3}$

(B) $f\left(\dfrac{1}{2}\right) > \dfrac{1}{2}$ and $f\left(\dfrac{1}{3}\right) > \dfrac{1}{3}$

(C) $f\left(\dfrac{1}{2}\right) < \dfrac{1}{2}$ and $f\left(\dfrac{1}{3}\right) < \dfrac{1}{3}$

(D) $f\left(\dfrac{1}{2}\right) > \dfrac{1}{2}$ and $f\left(\dfrac{1}{3}\right) < \dfrac{1}{3}$

Q.22 Match the statements/expressions in column I with the open intervals in column II.   *(2009)*

Column I	Column II
(A) Interval contained in the domain of definition of non-zero	(p) $\left(-\dfrac{\pi}{2},\dfrac{\pi}{2}\right)$
(B) Interval containing the value of the integral $\int\limits_{1}^{5}(x-1)(x-2)(x-3)(x-4)\,dx$	(q) $\left(0,\dfrac{\pi}{2}\right)$
(C) Interval in which at least one of the points of local maximum of $\cos^2 x+\sin$ lies	(r) $\left(\dfrac{\pi}{8},\dfrac{\pi}{2}\right)$
(D) Interval in which $\tan^{-1}(\sin x+\cos x)$ is increasing	(s) $\left(0,\dfrac{\pi}{2}\right)$
	(t) $(-\pi,\pi)$

Q.23 Match the statements in column I with those in column II.   *(2010)*

Column I	Column II
(A) A line from the origin meets the lines $\dfrac{x-2}{1}=\dfrac{y-1}{-2}=\dfrac{z+1}{1}$ and $\dfrac{x-\frac{8}{3}}{2}=\dfrac{y+3}{-1}=\dfrac{z-1}{1}$ at P and Q respectively. If length $PQ=d$ Then d^2 is	(p) -4
(B) The values of x satisfying $\tan^{-1}(x+3)-\tan^{-1}(x-3)=\sin^{-1}\left(\dfrac{3}{5}\right)$	(q) 0
(C) Non-zero vectors $\vec{a},\vec{b}$ and $\vec{c}$ satisfy $\vec{a}.\vec{c}=0\,\left(\vec{b}-\vec{a}\right).\left(\vec{b}-\vec{c}\right)=0$ and possible values of are	(r) 4
(D) Let f be the function on $[-\pi,\pi]$ given by $f(0)=9$ and $f(x)=\sin\left(\dfrac{9x}{2}\right)/\sin\left(\dfrac{x}{2}\right)\,x\neq 0$. The value of $\dfrac{2}{\pi}\int\limits_{-\pi}^{\pi}f(x)\,dx$ is	(r) 5
	(s) 6

Q.24 The value of $\lim\limits_{x\to 0}\dfrac{1}{x^3}\int\limits_{0}^{x}\dfrac{t\log(1+n)}{t^4+4}\,dt$ is   *(2010)*

(A) 0   (B) $\dfrac{1}{12}$ (C) $\dfrac{1}{24}$ (D) $\dfrac{1}{64}$

Q.25 The value (s) of $\int\limits_{0}^{1}\dfrac{x^4(1-x)^4}{1+x^2}\,dx$ is (are)   *(2010)*

(A) $\dfrac{22}{7}-\pi$ (B) $\dfrac{2}{105}$

(C) 0   (D) $\dfrac{71}{15}-\dfrac{3\pi}{2}$

Q.26 The value of $\int\limits_{\sqrt{\log 2}}^{\sqrt{\log 3}}\dfrac{x\sin x^2}{\sin x^2+\sin(\log 6-x^2)}\,dx$ is   *(2011)*

(A) $\dfrac{1}{4}\log\dfrac{3}{2}$ (B) $\dfrac{1}{2}\log\dfrac{3}{2}$ (C) $\log\dfrac{3}{2}$ (D) $\dfrac{1}{6}\log\dfrac{3}{2}$

Q.27 The value of the integral $\int\limits_{-\pi/2}^{\pi/2}\left(x^2+\log\dfrac{\pi+x}{\pi-x}\right)\cos x\,dx$ is   *(2012)*

(A) 0   (B) $\dfrac{\pi^2}{2}-4$ (C) $\dfrac{\pi^2}{2}+4$ (D) $\dfrac{\pi^2}{2}$

Q.28 The following integral $\int\limits_{\pi/4}^{\pi/2}(2\cos ecx)^{17}\,dx$ is equal to

(2014)

(A) $\int\limits_{0}^{\log\left(1+\sqrt{2}\right)}2\left(e^{u}+e^{-u}\right)^{16}du$

(B) $\int\limits_{0}^{\log\left(1+\sqrt{2}\right)}2\left(e^{u}+e^{-u}\right)^{16}du$

(C) $\int\limits_{0}^{\log\left(1+\sqrt{2}\right)}\left(e^{u}-e^{-u}\right)^{16}du$

(D) $\int\limits_{0}^{\log\left(1+\sqrt{2}\right)}2\left(e^{u}-e^{-u}\right)^{16}du$

Q.29 Match the following: *(2014)*

List I	List II
(i) The number of polynomials $f(x)$ with non negative integer coefficients of degree ≤ 2 satisfying $f(0)=0$ and $\int\limits_{0}^{1}f(x)dx=1$, is	(p) 8
(ii) The number of points in the interval $\left[-\sqrt{13},\sqrt{13}\right]$ at which $f(x)=\sin\left(x^{2}\right)+\cos\left(x^{2}\right)$ attains its maximum value, is	(q) 2
(iii) $\int\limits_{-2}^{2}\dfrac{3x^{2}}{\left(1+e^{x}\right)}dx$ equals	(r) 4
(iv) $\dfrac{\left(\int\limits_{-1/2}^{1/2}\cos 2.x.\log\left(\dfrac{1+x}{1-x}\right)dx\right)}{\int\limits_{0}^{1/2}\cos 2x.\log\left(\dfrac{1+x}{1-x}\right)dx}$ equals	(s) 0

Codes:

	i	ii	iii	iv
(A)	r	q	s	p
(B)	q	r	s	p
(C)	r	q	p	s
(D)	q	r	p	s

Q.30 The value of $\int\limits_{0}^{1}4x^{3}\left\{\dfrac{d^{2}}{dx^{2}}\left(1-x^{2}\right)\right\}dx$ is _____

(2014)

Q.31 If $\alpha=\int\limits_{0}^{1}\left(e^{\varepsilon x+3\tan^{-1}x}\right)\left(\dfrac{12+9x^{2}}{1+x^{2}}\right)dx$

Where $\tan^{-1}$ takes only principal values, then the value of $\left(\log_{e}\left|1+\alpha\right|-\dfrac{3\pi}{4}\right)$ is *(2015)*

Q.32 The option's) with the values of a and L that satisfy the following equation is(are)

$$\dfrac{\int\limits_{0}^{4\pi}e^{t}\left(\sin^{6}at+\cos^{4}at\right)}{\int\limits_{0}^{\pi}e^{t}\left(\sin^{6}at+\cos^{4}at\right)}=L?$$

(2015)

(A) $a=2,L=\dfrac{e^{4\pi}-1}{e^{\pi}-1}$ (B) $a=2,L=\dfrac{e^{4\pi}+1}{e^{\pi}+1}$

(C) $a=4,L=\dfrac{e^{4\pi}-1}{e^{\pi}-1}$ (D) $a=4,L=\dfrac{e^{4\pi}+1}{e^{\pi}+1}$

Q.33 The correct statement(s) is(are) *(2015)*

(A) $f'(1)<0$

(B) $f'(2)<0$

(C) $f'(x)\ne 0$ for any $x\in(1,3)$

(D) $f(x)=0$ for some $x\in(1,3)$

Q.34 Let $f:R\to R$ be a function defined by

$f(x)=\begin{Bmatrix}[x] & x\le 2 \\ 0 & x>2\end{Bmatrix}$ where $[x]$ is the greatest integer less than or equal to x.

If $I=\int\limits_{-1}^{2}\dfrac{xf\left(x^{2}\right)}{2+f\left(x+1\right)}dx\,dx$, then the value of $(4I-1)$ is

(2015)

Q.35 The value of $\int\limits_{-\frac{\pi}{2}}^{\frac{\pi}{2}}\dfrac{x^{2}\cos x}{1+e^{x}}dx$ is equal to *(2016)*

(A) $\dfrac{\pi^{2}}{4}-2$ (B) $\dfrac{\pi^{2}}{4}+2$

(C) $\pi^{2}-e^{\frac{\pi}{2}}$ (D) $\pi^{2}+e^{\frac{\pi}{2}}$

Important Questions

JEE Main/Boards

Exercise 1

Q.3	Q.8	Q.12
Q.17	Q.21	Q.23
Q.26	Q.28	

Exercise 2

Q.9	Q.12	Q.17
Q.20	Q.23	Q.29
Q.32	Q.34	

Previous Years' Questions

Q.4	Q.8	Q.11

JEE Advanced/Boards

Exercise 1

Q.2	Q.7	Q.10
Q.15	Q.22	Q.27
Q.32	Q.34	Q.44

Exercise 2

Q.2	Q.7	Q.10
Q.12	Q.15	

Previous Years' Questions

Q.1	Q.4	Q.6
Q.7	Q.10	Q.15

Answer Key

JEE Main/Boards

Exercise 1

Q.1 $\dfrac{\pi}{6}$

Q.2 $\dfrac{\pi}{4\sqrt{5}}$

Q.3 $\dfrac{\pi}{3\sqrt{3}}$

Q.4 $\dfrac{13}{10}$

Q.5 $\dfrac{34}{3}$

Q.6 $2-\sqrt{2}$

Q.7 $\dfrac{\pi^2}{2\sqrt{2}}$

Q.8 $\dfrac{14}{3}$

Q.9 $2(\sqrt{2}-1)$

Q.11 $200\sqrt{2}$

Q.12 (ii) Not necessary

Q.13 $2x\sqrt{1+x^4}$

Q.14 $\dfrac{4\pi}{\sqrt{3}}\tan^{-1}\left(\dfrac{1}{2}\right)$

Q.16 $2-\dfrac{2}{e}+2e\log 2$

Q.17 $\dfrac{\pi}{2}$

Q.18 $\dfrac{3}{2}-2\sqrt{3}+\dfrac{1}{\sqrt{2}}$

Q.19 $1,\dfrac{5}{12}-\dfrac{1}{2}\log 2$

Q.20 a = 1, 2, 3 or 4

Q.21 –1   **Q.22** $\dfrac{\pi}{4}$ **Q.26** $\dfrac{e^2+1}{2e}$

Q.29 (i) $\dfrac{1}{2}$; (ii) 1+e   **Q.30** $(-\infty,-2)\cup(-1,0)\cup(1,\infty)$

Exercise 2

Single Correct Choice Type

Q.1 A   **Q.2** C   **Q.3** C   **Q.4** B   **Q.5** A   **Q.6** B

Q.7 D   **Q.8** A   **Q.9** A   **Q.10** B   **Q.11** A   **Q.12** B

Q.13 C   **Q.14** B   **Q.15** A   **Q.16** B   **Q.17** C   **Q.18** C

Q.19 A   **Q.20** C   **Q.21** A   **Q.22** C   **Q.23** C   **Q.24** C

Q.25 C   **Q.26** C   **Q.27** B   **Q.28** B   **Q.29** B   **Q.30** A

Q.31 B   **Q.32** C   **Q.33** C   **Q.34** C

Previous Years' Questions

Q.1 A   **Q.2** C   **Q.3** A   **Q.4** D   **Q.5** A   **Q.6** A

Q.7 A   **Q.8** C   **Q.9** C   **Q.10** C   **Q.11** A   **Q.12** C

Q.13 D   **Q.14** B   **Q.15** A $\to$ s ; B $\to$ s ; C $\to$ p ; D $\to$ r   **Q.16** 4   **Q.17** $\pi(\sqrt{2}-1)$

Q.18 $\dfrac{1}{2}$ **Q.19** B   **Q.20** D   **Q.21** A   **Q.22** D   **Q.23** B C

Q.24 D   **Q.25** D   **Q.26** C

JEE Advanced/Boards

Exercise 1

Q.1 $\dfrac{\pi^2}{8}-\dfrac{\pi}{4}(1+\log 2)+\dfrac{1}{2}$ **Q.3** (a) 4 (b) n = 3

Q.4 (i) $\dfrac{1}{2}[e^{\pi/2}(\cos 1+\sin 1)-1]$ (ii) $e^{1+e}+e^{1-e}+e^{-e}-e^e+e-e^{-1}$ **Q.6** 125

Q.7 $1-\sec(1)$ **Q.8** ln 2   **Q.9** 4

Q.10 $\dfrac{\pi}{2\sqrt{2}}-\dfrac{16\sqrt{2}}{5}$ **Q.11** 2525   **Q.12** $4\sqrt{2}-4\ln\left|1+\sqrt{2}\right|$

Q.13 4   **Q.14** $\dfrac{\pi}{8}\log 2$ **Q.15** $\dfrac{\pi^2}{6\sqrt{3}}$

Q.16 $\dfrac{\pi}{8}\log 2$ **Q.17** 2012   **Q.18** $2\sqrt{6}$

Q.19 153   **Q.20** $\dfrac{\pi}{3}$ **Q.21** $\dfrac{\pi(a+b)}{2\sqrt{2}}$

Q.22 5

Q.23 90

Q.24 $\dfrac{\pi}{\sqrt{3}}$

Q.25 $\dfrac{(a\pi + 2b)\pi}{3\sqrt{3}}$

Q.26 $\dfrac{\pi(\pi+3)}{2}$

Q.27 $= \dfrac{1}{2}\left[x\right]_0^{\pi/2} = \dfrac{1}{2}\left(\dfrac{\pi}{2}-0\right) = \dfrac{\pi}{4}$

Q.28 10

Q.29 $\dfrac{1}{2}\left[\dfrac{\pi}{2}-\log 2\right]$

Q.30 $\tan^{-1}(a)\,.\,\log\sqrt{1+a^2}$

Q.31 $\dfrac{1}{2}\left[\dfrac{\pi}{6}+\log 3 - \log 2\right]$

Q.32 $\dfrac{9}{2}$

Q.33 0

Q.36 5

Q.37 $\dfrac{3\pi^2}{16}$

Q.38 Real and distinct $\forall\, k \in R$

Q.39 $\dfrac{\pi+4}{666}$

Q.40 8

Q.41 6

Q.42 $\dfrac{\pi}{16}$

Q.43 62

Q.44 I = 8 as $\displaystyle\int_0^{\pi/2} y\sin y\,dy = 1$

Exercise 2

Single Correct Choice Type

Q.1 C
Q.2 B
Q.3 D
Q.4 B
Q.5 A
Q.6 A

Q.7 C
Q.8 A
Q.9 B
Q.10 B
Q.11 B
Q.12 A

Q.13 A
Q.14 B
Q.15 A
Q.16 B

Previous Years' Questions

Q.1 A
Q.2 A
Q.3 C
Q.4 A
Q.5 A, D
Q.6 A, B, C

Q.7 A
Q.8 C
Q.9 A
Q.10 B
Q.11 4

Q.12 $f(e)+f\left(\dfrac{1}{e}\right) = \dfrac{1}{2}(\log e)^2 = \dfrac{1}{2}$

Q.13 $I = \sqrt{2}\displaystyle\int_0^{\pi/4} f(\sin 2t)\cos t\, dt$

Q.14 $\dfrac{4\pi}{\sqrt{3}}\tan^{-1}\left(\dfrac{1}{2}\right)$

Q.15 $\dfrac{24}{5}\left(e\cos\left(\dfrac{1}{2}\right)+\dfrac{e}{2}\sin\left(\dfrac{1}{2}\right)-1\right)$

Q.16 5051
Q.17 B
Q.18 D
Q.19 A

Q.20 0
Q.21 C
Q.22 A → p, q, s; B → p, t; C → p, q, r, t; D → s
Q.23 A

Q.24 B
Q.25 A
Q.26 A
Q.27 B
Q.28 A

Q.29 D
Q.30 2
Q.31 9
Q.32 A, C
Q.33 A, B, C
Q.34 –2

Q.35 A

Solutions

JEE Main/Boards

Exercise 1

Sol 1: $\displaystyle\int_{1/4}^{1/2}\frac{dx}{\sqrt{x-x^2}} = \int_{1/4}^{1/2}\frac{dx}{\sqrt{\frac{1}{4}-\left(x-\frac{1}{2}\right)^2}}$

$= \sin^{-1}\dfrac{\left(x-\frac{1}{2}\right)}{1/2}\Bigg|_{1/4}^{1/2} = \sin^{-1}0 - \sin^{-1}\left(\dfrac{-1/4}{1/2}\right)$

$= \sin^{-1}\left(\dfrac{1}{2}\right) = \dfrac{\pi}{6}$

Sol 2: $\displaystyle\int_0^{\pi/2}\frac{dx}{4\sin^2 x+5\cos^2 x} = \int_0^{\pi/2}\frac{dx}{4+\cos^2 x}$

$= \displaystyle\int_0^{\pi/2}\frac{dx}{\frac{9}{2}+\frac{\cos 2\theta}{2}} = \int_0^{\pi/2}\frac{2dx}{9+\frac{1-\tan^2\theta}{1+\tan^2\theta}}$

$= 2\displaystyle\int_0^{\pi/2}\frac{\sec^2\theta d\theta}{10+8\tan^2\theta}$

$= \displaystyle\int_0^{\pi/4}\frac{2\sec^2\theta d\theta}{10+8\tan^2\theta}+\int_{\pi/4}^{\pi/2}\frac{2\csc^2\theta}{10\cot^2\theta+8}d\theta$

$= \displaystyle\int_0^{\pi/4}\frac{\sec^2\theta d\theta}{5+4\tan^2\theta}+\int_{\pi/4}^{\pi/2}\frac{\csc^2\theta}{5\cot^2\theta+4}d\theta$

$= \displaystyle\int_0^1\frac{dt}{5+4t^2}+\int_1^0\frac{-dt}{5t^2+4} = \int_0^1\frac{dt}{5+4t^2}+\int_0^1\frac{dt}{5t^2+4}$

$= \dfrac{1}{4}\times\dfrac{1}{\sqrt{5}/2}\tan^{-1}\dfrac{t}{\sqrt{5}/2}\Bigg|_0^1+\dfrac{1}{5}\times\dfrac{1}{2/\sqrt{5}}\tan^{-1}\dfrac{t}{2/\sqrt{5}}\Bigg|_0^1$

$= \dfrac{1}{2\sqrt{5}}\tan^{-1}\dfrac{2}{\sqrt{5}}+\dfrac{1}{2\sqrt{5}}\tan^{-1}\dfrac{\sqrt{5}}{2}$

$= \dfrac{1}{2\sqrt{5}}\left[\tan^{-1}\dfrac{2}{\sqrt{5}}+\cot^{-1}\dfrac{2}{\sqrt{5}}\right]$

$= \dfrac{1}{2\sqrt{5}}\times\dfrac{\pi}{2} = \dfrac{\pi}{4\sqrt{5}}$

Sol 3: $\displaystyle\int_0^{\pi/2}\frac{\sin^2 x}{1+\sin x\cos x}dx$

$I = \displaystyle\int_0^{\pi/2}\frac{\sin^2\left(\frac{\pi}{2}-x\right)}{1+\sin x\cos x}dx$ or $I = \int_0^{\pi/2}\frac{\cos^2 x}{1+\sin x\cos x}dx$

$\therefore 2I = \displaystyle\int_0^{\pi/2}\frac{1}{1+\sin x\cos x}dx = \int_0^{\pi/2}\frac{\sec^2 x dx}{1+\tan^2 x+\tan x}$

$= \displaystyle\lim_{x\to\infty}\int_0^x\frac{dt}{1+t^2+t} = \lim_{x\to\infty}\int_0^x\frac{dt}{\left(t+\frac{1}{2}\right)^2+\left(\frac{\sqrt{3}}{2}\right)^2}$

$= \dfrac{2}{\sqrt{3}}\tan^{-1}\dfrac{2t+1}{\sqrt{3}}\Bigg|_0^\infty$

$= \dfrac{2}{\sqrt{3}}\left[\dfrac{\pi}{2}-\dfrac{\pi}{6}\right] = \dfrac{2\pi}{3\sqrt{3}}$

$\therefore I = \dfrac{\pi}{3\sqrt{3}}$

Sol 4: $\displaystyle\int_0^{3/5}(3-5x)dx+\int_{3/5}^1(5x-3)dx$

$= 3x-\dfrac{5}{2}x^2\Big|_0^{3/5}+\dfrac{5x^2}{2}-3x\Big|_{3/5}^1$

$= \dfrac{9}{5}-\dfrac{5}{2}\times\dfrac{9}{25}+\left(\dfrac{5}{2}-3\right)-\left(\dfrac{5}{2}\times\dfrac{9}{25}-\dfrac{9}{5}\right)$

$= \dfrac{9}{5}-\dfrac{9}{10}-\dfrac{1}{2}-\dfrac{9}{10}+\dfrac{9}{5}$

$= \dfrac{18}{5}-\dfrac{9}{5}-\dfrac{1}{2} = \dfrac{9}{5}-\dfrac{1}{2} = \dfrac{13}{10}$

Sol 5: $\displaystyle\int_1^2(2x+1)dx+\int_2^3(x^2+1)dx$

$= x^2+x\Big|_1^2+\dfrac{x^3}{3}+x\Big|_2^3$

$= (4+2-2)+(9+3)-\left(\dfrac{8}{3}+2\right)$

$= 4+12-2-\dfrac{8}{3} = 14-\dfrac{8}{3} = \dfrac{34}{3}$

Sol 6: $\displaystyle\int_{-\pi/4}^{\pi/4} \sin x\,dx = 2\int_{0}^{\pi/4}\sin x\,dx = 2\left[-\cos x\big|_{0}^{\pi/4}\right]$

$= 2\left[-\left(\dfrac{1}{\sqrt{2}}-1\right)\right] = 2\left[1-\dfrac{1}{\sqrt{2}}\right] = 2-\sqrt{2}$

Sol 7: $\displaystyle\int_{0}^{\pi}\dfrac{x}{(1+\sin^2 x)}dx = \int_{0}^{\pi}\dfrac{\pi-x}{(1+\sin^2 x)}dx = I$

$\therefore 2I = \pi\displaystyle\int_{0}^{\pi}\dfrac{1}{1+\sin^2 x}dx = 2\pi\int_{0}^{\pi/2}\dfrac{1}{1+\sin^2 x}dx$

$\therefore I = \pi\displaystyle\int_{0}^{\pi/2}\dfrac{1}{1+\dfrac{1-\cos 2x}{2}}dx = \pi\int_{0}^{\pi/2}\dfrac{2}{3-\cos 2x}dx$

$= 2\pi\displaystyle\int_{0}^{\pi/2}\dfrac{dx}{3-\dfrac{(1-\tan^2 x)}{1+\tan^2 x}} = 2\pi\int_{0}^{\pi/2}\dfrac{\sec^2 x\,dx}{2+4\tan^2 x}$

$= \dfrac{\pi}{2}\displaystyle\int_{0}^{\pi/2}\dfrac{\sec^2 x\,dx}{\dfrac{1}{2}+\tan^2 x} = \dfrac{\pi}{2}\left[\int_{0}^{1}\dfrac{dt}{\dfrac{1}{2}+t^2} - \int_{\pi/4}^{\pi/2}\dfrac{\mathrm{cosec}^2 x\,dx}{\dfrac{1}{2}\cot^2 x+1}\right]$

$= \dfrac{\pi}{2}\left[\left[\dfrac{1}{\dfrac{1}{\sqrt{2}}}\tan^{-1}\dfrac{t}{\dfrac{1}{\sqrt{2}}}\right]_{0}^{1} - \int_{1}^{0}\dfrac{dt}{\dfrac{1}{2}t^2+1}\right]$

$= \dfrac{\pi}{2}\left[\sqrt{2}\tan^{-1}\sqrt{2} + 2\displaystyle\int_{0}^{1}\dfrac{dt}{t^2+2}\right]$

$= \dfrac{\pi}{2}\left[\sqrt{2}\tan^{-1}\sqrt{2} + \dfrac{2}{\sqrt{2}}\tan^{-1}\dfrac{t}{\sqrt{2}}\right]_{0}^{1}$

$= \dfrac{\pi}{2}\left[\sqrt{2}\tan^{-1}\sqrt{2} + \sqrt{2}\tan^{-1}\dfrac{1}{\sqrt{2}}\right]$

$= \dfrac{\pi}{2}\times\sqrt{2}\times\dfrac{\pi}{2} = \dfrac{\pi^2}{2\sqrt{2}}$

Sol 8: $\displaystyle\int_{0}^{2}(x^2+1)dx$

$h = \dfrac{b-a}{n} = \dfrac{2-0}{n} = \dfrac{2}{n}$

$\therefore I = \displaystyle\lim_{n\to 0}\sum_{r=1}^{n} hf(a+rh) = \lim_{n\to 0}\sum_{r=1}^{n} h((rh)^2+1)$

$= \displaystyle\lim_{n\to 0}\sum_{r=1}^{n}(r^2 h^3 + h)$

$= \displaystyle\lim_{n\to\infty}\left[h^3\times\dfrac{n(n+1)(2n+1)}{6}+hn\right]$

$= \displaystyle\lim_{n\to\infty}\left[\dfrac{8}{n^3}\dfrac{(n)(n+1)(2n+1)}{6}+\dfrac{2}{n}\times n\right]$

$= \dfrac{8\times 1\times 2}{6}+2 = \dfrac{14}{3}$

Sol 9: $\displaystyle\int_{0}^{\pi/2}\left|\sin x-\cos x\right|dx$

$= \displaystyle\int_{0}^{\pi/4}(\cos x-\sin x)dx + \int_{\pi/4}^{\pi/2}(\sin x-\cos x)dx$

$= \sin x + \cos x\big|_{0}^{\pi/4} + (-\cos x-\sin x)\big|_{\pi/4}^{\pi/2}$

$= \left(\dfrac{1}{\sqrt{2}}+\dfrac{1}{\sqrt{2}}-1\right) - \left(1-\left(\dfrac{1}{\sqrt{2}}+\dfrac{1}{\sqrt{2}}\right)\right)$

$= \sqrt{2}-1-1+\sqrt{2} = 2\sqrt{2}-2$

Sol 10: $f(x) = f(a-x)$

$g(x) + g(a-x) = 2$

$\displaystyle\int_{0}^{a}f(x)g(x)dx = \int_{0}^{a}f(a-x)g(a-x)dx = I$

$\therefore 2I = \displaystyle\int_{0}^{a}\left[f(x)g(x)+f(a-x)g(a-x)\right]dx = \int_{0}^{a}\left[f(x)\times 2\right]dx$

$I = \displaystyle\int_{0}^{a}f(x)dx$

Sol 11: $\displaystyle\int_{0}^{100\pi}\sqrt{1-\cos 2x}\,dx$

$\displaystyle\int_{0}^{100\pi}\sqrt{2\sin^2 x}\,dx = \sqrt{2}\int_{0}^{100\pi}\sqrt{\sin^2 x}\,dx$

$Q\sin^2(\pi-x) = \sin^2 x$

$\because I = 100\sqrt{2}\displaystyle\int_{0}^{\pi}\sqrt{\sin^2 x}\,dx = 100\sqrt{2}\int_{0}^{\pi}\left|\sin x\right|dx$

Also $\left|\sin(\pi-x)\right| = \left|\sin x\right|$

$\therefore I = 200\sqrt{2}\displaystyle\int_{0}^{\pi/2}\left|\sin x\right|dx$

$= 200\sqrt{2}\left(-\left|\cos x\right|\right)_{0}^{\pi/2} = 200\sqrt{2}$

Sol 12: (i) $f(-t) = -f(t)$

$$g(x) = \int_a^x f(t)\,dt$$

$$g(-x) = \int_a^{-x} f(t)\,dt = \int_a^{-a} f(t)\,dt + \int_{-a}^{-x} f(t)\,dt$$

$Qf(t)$ = odd function

So $\int_a^{-a} f(t)\,dt = 0$

$$\therefore \int_a^{-x} f(t)\,dt = \int_{-a}^{-x} f(t)\,dt$$

Put $t = -p$

$$= -\int_a^x f(-p)\,dp \quad \because f(-p) = -f(p)$$

$$= \int_a^x f(p)\,dp$$

$$\therefore g(-x) = g(x)$$

(ii) $f(t) = f(-t)$

$$g(x) = \int_a^x f(t)\,dt \; ; \quad g(-x) = \int_a^{-x} f(t)\,dt$$

Put $t = -p$

$$= -\int_{-a}^x f(-p)\,dp = -\int_{-a}^x f(p)\,dp = -\int_{-a}^x f(t)\,dt$$

$$\therefore g(-x) = \int_x^{-a} f(t)\,dx$$

$$\therefore g(x) + g(-x) = \int_a^x f(t)\,dt + \int_x^{-a} f(t)\,dt = \int_a^{-a} f(t)\,dt$$

$\therefore$ It is not necessary that if $f(t)$ is even then $\int_a^x f(t)\,dt$ is odd

Sol 13: $f(x) = \int_a^{x^2} \sqrt{1+t^2}\,dt$

$f'(x) = \sqrt{1+x^4}\,dx^2 = 2x\sqrt{1+x^4}$

Sol 14: $I = \int_{-\pi/3}^{\pi/3} \dfrac{\pi + 4x^3 dx}{2 - \cos\left(|x| + \dfrac{\pi}{3}\right)}$

$$I = \int_{-\pi/3}^{\pi/3} \dfrac{\pi - 4x^3}{2 - \cos\left(|x| + \dfrac{\pi}{3}\right)}\,dx$$

$$\therefore 2I = 2\pi \int_{-\pi/3}^{\pi/3} \dfrac{dx}{2 - \cos\left(|x| + \dfrac{\pi}{3}\right)}$$

$$\Rightarrow I = 2\pi \int_0^{\pi/3} \dfrac{dx}{2 - \dfrac{\left(1 - \tan^2\left(\dfrac{x}{2} + \dfrac{\pi}{6}\right)\right)}{\left(1 + \tan^2\left(\dfrac{x}{2} + \dfrac{\pi}{6}\right)\right)}}$$

$$= \dfrac{2\pi}{3} \int_0^{\pi/3} \dfrac{\sec^2\left(\dfrac{x}{2} + \dfrac{\pi}{6}\right)dx}{\dfrac{1}{3} + \tan^2\left(\dfrac{x}{2} + \dfrac{\pi}{6}\right)}$$

Put $\tan\left(\dfrac{x}{2} + \dfrac{\pi}{6}\right) = t$

$$\dfrac{1}{2}\sec^2\left(\dfrac{x}{2} + \dfrac{\pi}{6}\right)dx = dt$$

$$\therefore I = \dfrac{4\pi}{3} \int_{1/\sqrt{3}}^{\sqrt{3}} \dfrac{dt}{\dfrac{1}{3} + t^2} = \dfrac{4\pi}{3} \times \sqrt{3}\,\tan^{-1}\sqrt{3}t\,\Big|_{1/\sqrt{3}}^{\sqrt{3}}$$

$$= \dfrac{4\pi}{\sqrt{3}}\left[\tan^{-1}3 - \tan^{-1}1\right] = \dfrac{4\pi}{\sqrt{3}}\left[\tan^{-1}\left(\dfrac{3-1}{1 + 3\times1}\right)\right]$$

$$= \dfrac{4\pi}{3}\tan^{-1}\dfrac{1}{2}$$

Sol 15: $f(x) = \int \dfrac{\log t}{1 + t}\,dt$

To Prove. $f(x) + f\left(\dfrac{1}{x}\right) = \dfrac{1}{2}(\log x)^2$

Put $t = \dfrac{1}{p} \Rightarrow dt = -\dfrac{1}{p^2}\,dp$

$$f\left(\dfrac{1}{x}\right) = \int_1^x \dfrac{\log\dfrac{1}{p}}{1\left(1 + \dfrac{1}{p}\right)}\left(-\dfrac{1}{p^2}\right)dp$$

$$f\left(\dfrac{1}{x}\right) = \int_1^x \dfrac{\log p}{p(1 + p)}\,dp = \int_1^x \dfrac{\log t}{t(t+1)}\,dt$$

$$\therefore f(x) + f\left(\frac{1}{x}\right) = \int\limits_{1}^{x} \frac{\log t}{t} dt$$

$$= (\log)^2 \Big|_{1}^{x} - \int \frac{\log t}{t} dt$$

$$\therefore 2I = (\log x)^2$$

$$\therefore f(x) + f\left(\frac{1}{x}\right) = \frac{1}{2}(\log x)^2$$

Sol 16: $-\int\limits_{1/e}^{1} \log x\, dx + \int\limits_{1}^{2e} \log x\, dx$

$$= -\left(x\log x - x\right)\Big|_{1/e}^{1} + \left(x\log x - x\right)\Big|_{1}^{2e}$$

$$= -\left[0 - 1 - \left[\frac{1}{e}\log\frac{1}{e} - \frac{1}{e}\right]\right]$$

$$+ \left[(2e\log 2e - 2e) - (0 - 1)\right]$$

$$= 1 - \frac{1}{e} - \frac{1}{e} + 2e\log 2 + 2e - 2e + 1$$

$$= 2 - \frac{2}{e} + 2e\log 2$$

Sol 17: $\int\limits_{0}^{\pi} \dfrac{\sin\left(n + \dfrac{1}{2}\right)x}{2\sin\dfrac{x}{2}} dx \quad n \in N$

$$2\sin\left(n + \frac{1}{2}\right)x \cos\frac{x}{2} = \sin(nx + 2) + \sin(nx)$$

$$= \frac{1}{2}\int\limits_{0}^{\pi} \frac{\sin nx + \sin(nx + x)}{\sin x} dx$$

$$= \frac{1}{2}\int\limits_{0}^{\pi} \frac{\sin(n+1)x + \sin nx}{\sin x} dx$$

If n is odd

$$I = \frac{1}{2}\int\limits_{0}^{\pi} \frac{\sin nx - \sin(n+1)x}{\sin x} dx$$

$$\therefore 2I = \int\limits_{0}^{\pi} \frac{\sin nx}{\sin x} dx = \pi \qquad \Rightarrow I = \frac{\pi}{2}$$

If n is even

$$2I = \int\limits_{0}^{\pi} \frac{\sin(n+1)x}{\sin x} dx = \pi; \quad I = \frac{\pi}{2}$$

Sol 18: $F(x) = \int\limits_{5\pi/4}^{x} (3\sin t + 4\cos t)dt$

$$= 3(-\cos t)\Big|_{5\pi/4}^{x} + 4\sin t\Big|_{5\pi/4}^{x}$$

$$= 3\left[-\left(\cos x - \frac{1}{\sqrt{2}}\right)\right] + 4\left[\sin x - \frac{1}{\sqrt{2}}\right]$$

$$= -3\cos x + \frac{3}{\sqrt{2}} + 4\sin x - \frac{4}{\sqrt{2}}$$

$$= \left(\frac{4\sin x - 3\cos x}{5}\right)5 - \frac{1}{\sqrt{2}}$$

From interval $\left[\dfrac{5\pi}{4}, \dfrac{4\pi}{3}\right]$ $\sin x < \cos x$

$\therefore$ We get min value of $x = \dfrac{4\pi}{3}$

$\therefore$ Min value $= -4 \times \dfrac{\sqrt{3}}{2} + \dfrac{3}{2} - \dfrac{1}{\sqrt{2}} = \dfrac{3}{2} - 2\sqrt{3} - \dfrac{1}{\sqrt{2}}$

Sol 19: $I_n = \int\limits_{0}^{\pi/4} \tan^n\theta\, d\theta$

$$I_{n-1} + I_{n+1} = \int\limits_{0}^{\pi/4} \left(\tan^{n-1}\theta + \tan^{n+1}\theta\right)d\theta$$

$$= \int\limits_{0}^{\pi/4} (\tan^{n-1}\theta)\sec^2\theta\, d\theta$$

$\tan\theta = t \Rightarrow \sec^2 q\, d\theta = dt$

$$= \int\limits_{0}^{1} t^{n-1}dt = \frac{t^n}{n}\Big|_{0}^{1} = \frac{1}{n}$$

$$\therefore \quad n(I_{n-1} + I_{n+1}) = n \times \frac{1}{n} = 1$$

$$I_7 = \int\limits_{0}^{\pi/4} \tan^7\theta\, d\theta$$

$$= \int\limits_{0}^{\pi/4} \tan^5\theta\sec^2\theta\, d\theta - \int\limits_{0}^{\pi/4} \tan^3\theta\sec^2\theta\, d\theta$$

$$+ \int\limits_{0}^{\pi/4} \tan\theta\sec^2\theta\, d\theta - \int\limits_{0}^{\pi/4} \tan\theta\, d\theta$$

$$= \frac{1}{6} - \frac{1}{4} + \frac{1}{2} - \log\sqrt{2} = \frac{5}{12} - \frac{1}{2}\log 2$$

Sol 20: $\displaystyle\int_0^{\pi/2}\left\{a^2\left(\dfrac{\cos 3x}{4}+\dfrac{3}{4}\cos x\right)+a\sin x-20\cos x\right\}dx\le -\dfrac{a^2}{3}$

$\displaystyle\int_0^{\pi/2}\left\{a^2\cos^3 x+a\sin x-20\cos x\right\}dx\le -\dfrac{a^2}{3}$

$=a^2\left[+\dfrac{1}{12}\sin^3 x\Big|_0^{\pi/2}+\dfrac{3}{4}(+\sin x)\Big|_0^{\pi/2}\right]$

$+\,a(-\cos x)\Big|_0^{\pi/2}-20\sin x\Big|_0^{\pi/2}\le -\dfrac{a^2}{3}$

$=a^2\left[-\dfrac{1}{12}+\dfrac{3}{4}\right]+a-20\le -\dfrac{a^2}{3}$

$\Rightarrow a^2+a-20\le 0$

$(a+5)(a-4)\le 0\ \therefore\ a\in[-5,4]$

$\therefore a$ is +ve interger

So $a=1,2,3$ or 4

Sol 21: $f(x)=\sin x$

Mean value of $\sin x$ from $[-2,0)$

$\therefore\ \displaystyle\int_{-2}^{0}\dfrac{\sin x}{2}dx=\dfrac{-1[0+2]}{2}=-1$

Sol 22: $I=\displaystyle\int_0^a\dfrac{1}{x+\sqrt{a^2-x^2}}dx$

$x=a\cos\theta$

$dx=-a\sin q\,d\theta$

$\displaystyle\int_{\pi/2}^{a}\dfrac{-a\sin\theta\,d\theta}{a\cos\theta+a\sin\theta}=\int_0^{\pi/2}\left(\dfrac{\sin\theta}{\cos\theta+\sin\theta}\right)d\theta$

$=\displaystyle\int_0^{\pi/2}\left(\dfrac{\cos\theta}{\cos\theta+\sin\theta}\right)d\theta$

$\Rightarrow I=\dfrac{1}{2}\left[\displaystyle\int_0^{\pi/2}d\theta\right]=\dfrac{1}{2}\left[\dfrac{\pi}{2}\right]=\dfrac{\pi}{4}$

Sol 23: $I=\displaystyle\int_0^1 0\,dx+\int_1^4 1\,dx+\int_4^9 2\,dx\ldots+\int_{(n-1)^2}^{n^2}(n-1)dx$

$\therefore\ \displaystyle\sum_{n=0}^{n}\int_{(n-1)^2}^{n^2}(n-1)dx$

$=\displaystyle\sum(n-1)x\Big|_{(n-1)^2}^{n^2}=\sum(n-1)\left(n^2-(n-1)^2\right)$

$=\displaystyle\sum(n-1)(2n-1)=\sum(2n^2-3n+1)$

$=\dfrac{2n(n+1)(2n+1)}{6}-\dfrac{3n(n+1)}{2}+n$

$=\dfrac{n(n+1)}{2}\left[\dfrac{4n+2}{3}-3\right]+n=\dfrac{n(n+1)(4n-7)}{6}+n$

$=n\left(\dfrac{4n^2-3n-7+6}{6}\right)=\dfrac{n(4n^2-3n-1)}{6}$

$=\dfrac{n(n-1)(4n+1)}{6}$

Sol 24: $\displaystyle\int_0^{n\pi+\lambda}|\sin x|\,dx=2n+1-\cos\lambda$

$n\in N,\ 0\le\lambda<p$

LHS $=\displaystyle\int_0^{\lambda}|\sin x|\,dx+\int_{\lambda}^{n\pi+\lambda}|\sin x|\,dx$

$=-\cos x\Big|_0^{\lambda}+n\displaystyle\int_0^{\pi}|\sin x|\,dx$

$=-(\cos\lambda-1)+2n\displaystyle\int_0^{\pi/2}\sin x\,dx$

$=2n+1-\cos\lambda$

Sol 25: $I=\displaystyle\int_0^{\pi}\dfrac{x\sin 2x\cdot\sin\left(\dfrac{\pi}{2}\cos x\right)}{2x-\pi}dx$ ..... (i)

$I=\displaystyle\int_0^{\pi}\dfrac{(\pi-x)(-\sin 2x)\sin\left(\dfrac{\pi}{2}(-\cos x)\right)dx}{2(\pi-x)-\pi}$ ..... (ii)

On adding (i) and (ii)

$2I=\displaystyle\int_0^{\pi}\dfrac{(2x-\pi)\sin 2x\sin\left(\dfrac{\pi}{2}\cos x\right)}{(2x-\pi)}dx$

$=\displaystyle\int_0^{\pi}\sin 2x\sin\left(\dfrac{\pi}{2}\cos x\right)dx$

or $I=\displaystyle\int_0^{\pi/2}\sin 2x\,s\,n\left(\dfrac{\pi}{2}\cos x\right)dx$

Let $\dfrac{\pi}{2}\cos x=t$

$-\dfrac{\pi}{2}\sin x\,dx=dt$ or $\sin x\,dx=-\dfrac{2}{\pi}dt$

$$I = -\frac{2}{\pi}\int_{\pi/2}^{0} 2\times\frac{2t}{\pi}\sin t\,dt$$

$$= \frac{8}{\pi^2}\int_0^{\pi/2} t\sin t\,dt = \frac{8}{\pi^2}\left[t(-\cos t)\Big|_0^{\pi/2} + \int_0^{\pi/2}\cos t\,dt\right]$$

$$= \frac{8}{\pi^2}\left[0 + \sin t\Big|_0^{\pi/2}\right] = \frac{8}{\pi^2}$$

Sol 26: Let $f(x) = K_1 e^x + K_2 e^{-x}$

$g(x) + f'(x) = K_1 e^x - K_2 e^{-x}$

$\therefore g'(x) = K_1 e^x + K_2 e^{-x} = f(x)$

$\therefore f(0) = 1 \Rightarrow K_1 + K_2 = 1$

Also $g(0) = 0 \Rightarrow K_1 - K_2 = 0$

$$K_1 = K_2 = \frac{1}{2}$$

$$\therefore f(x) = \frac{e^x + e^x}{2}$$

$$\therefore f(1) = \frac{e + \dfrac{1}{e}}{2} = \frac{e^2 + 1}{2e}$$

Sol 27: (i) $\displaystyle\int_0^\pi \log(1+\cos x)\,dx$

$$= \int_0^\pi \log(1-\cos x)\,dx = I$$

$$\therefore 2I = \int_0^\pi \log(1-\cos^2 x)\,dx$$

$$= 2\int_0^\pi \log\sin x\,dx = 4\int_0^{\pi/2}\log\sin x\,dx$$

$$= 4\times\left(\frac{-\pi}{2}\right)\log 2$$

$$2I = -2\pi\log 2$$

$$\therefore I = -\pi\log 2 = \pi\log\frac{1}{2}$$

(ii) $\displaystyle\int_{\pi/6}^{\pi/3}\frac{dx}{1+\sqrt{\cot x}} = \int_{\pi/6}^{\pi/3}\frac{dx}{1+\sqrt{\cot\left(\dfrac{\pi}{2}-x\right)}}$

$$= \int_{\pi/6}^{\pi/3}\frac{dx}{1+\sqrt{\tan x}}$$

$$\therefore 2I = \int_{\pi/6}^{\pi/3}\left(\frac{\sqrt{\tan x}+1}{1+\sqrt{\tan x}}\right)dx = \frac{\pi}{3}-\frac{\pi}{6} = \frac{\pi}{6}$$

$$\therefore I = \frac{\pi}{12}$$

Sol 28: $\displaystyle\int_0^\pi \frac{dx}{1-2a\cos x + a^2}$

$$= \int_0^\pi \frac{dx}{1+a^2 - \dfrac{2a\left(1-\tan^2\dfrac{x}{2}\right)}{1+\tan^2\dfrac{x}{2}}}$$

$$= \int_0^\pi \frac{\sec^2\dfrac{x}{2}\,dx}{(1+a^2)\left(1+\tan^2\dfrac{x}{2}\right)-2a\left(1-\tan^2\dfrac{x}{2}\right)}$$

$$= \int_0^\pi \frac{\sec^2\dfrac{x}{2}\,dx}{(1+a^2-2a)+\tan^2\dfrac{x}{2}(1+a^2+2a)}$$

Putting $\dfrac{x}{2} = t$

$$\sec^2\frac{x}{2}\,dx = 2dt$$

$$\int_0^\infty \frac{dt}{(1+a)^2 t^2 + (1-a)^2} = \frac{1}{1+a^2}\int_0^\infty \frac{dt}{t^2 + \left(\dfrac{1-a}{1+a}\right)^2}$$

$$= \frac{1}{(1+a^2)}\times\frac{1+a}{|1-a|}\tan^{-1}\frac{t}{\left|\dfrac{1-a}{1+a}\right|}\Bigg|_0^\infty$$

$$= \frac{1}{1-a^2}\frac{\pi}{2} \quad\text{If } a < 1$$

$$= \frac{1}{a^2-1}\frac{\pi}{2} \quad\text{if } a > 1$$

Sol 29: (i) $\displaystyle\lim_{\alpha\to 0}\frac{\displaystyle\int_0^\alpha x\,dx}{\alpha\sin x} = \lim_{\alpha\to 0}\frac{\alpha^2}{2\alpha\sin\alpha}$

$$= \frac{1}{2}\lim_{\alpha\to 0}\frac{1}{\left(\dfrac{\sin\alpha}{\alpha}\right)} = \frac{1}{2}$$

(ii) $y = x^{\displaystyle\int_1^x \ell n t\,dt}$

$$\log y = \left(\int_1^x \log t\,dt\right)\log x$$

$$\frac{1}{y}\frac{dy}{dx} = \frac{1}{x}\int_1^x \log t\,dt + (\log x)(\log x)$$

2.42

$$\frac{dy}{dx} = x^{\int_1^x \log t\, dt} \left[\log^2 x + \frac{1}{x} \int_1^x \log t\, dt \right]$$

$$\frac{dy}{dx}\Big|_{x=e} = e^{\int_1^e \log t\, dt} \left[\log^2 e + \frac{1}{e} \int_1^e \log t\, dt \right]$$

$$= e(e\log e - e - (-1)) \left[\log^2 e + \frac{1}{e}(e\log e - e(-1)) \right]$$

$$= e\left[\frac{1}{e} + 1\right] = e + 1$$

Sol 30: $f(x) = \int_1^x (t^2 + 2t)(t^2 - 1)dt$

$f'(x) = (x^2 + 2x)(x^2 - 1) > 0$

$x(x + 2)(x - 1)(x + 1) > 0$

$\therefore x \in (-\infty, -2) \cup (-1, 0) \cup (1, \infty)$

Exercise 2

Single Correct Choice Type

Sol 1: (A) $\int_{-1}^0 (1 - 2x)dx + \int_0^1 (1 + 2x)dx$

$$= x - x^2\Big|_{-1}^0 + x + x^2\Big|_0^1 = 0 - [-1 - 1] + [1 + 1] = 4$$

Sol 2: (C) $\int_{-1}^0 e^{-x}dx + \int_1^1 e^x dx$

$$= -e^{-x}\Big|_{-1}^0 + e^x\Big|_0^1 - [1 - e^{+1}] + [e^1 - 1]$$

$$= e^{+1} + e^1 - 2 = 2e - 2$$

Sol 3: (C) $\int_0^3 [x]dx = \int_0^1 0\,dx + \int_1^2 1\,dx + \int_2^3 2\,dx$

$$= 0 + 1 + 2 = 3$$

Sol 4: (B) $\int_0^{\pi/2} \cos x\, dx + \int_{\pi/2}^{\pi} -\cos x\, dx$

$$= \sin x\Big|_0^{\pi/2} - \sin x\Big|_{\pi/2}^{\pi} = 1 - [0 - 1] = 2$$

Sol 5: (A) $\int_{-2}^{-3/2} -(2x + 3)dx + \int_{-3/2}^2 (2x + 3)dx$

$$= -\left[x^2 + 3x\right]_{-2}^{-3/2} + x^2 + 3x\Big|_{-3/2}^2$$

$$= -\left[\frac{9}{4} - \frac{9}{2} - (4 - 6)\right] + \left[4 + 6 - \left(\frac{9}{4} - \frac{9}{2}\right)\right]$$

$$= \frac{9}{4} - 2 + 10 + \frac{9}{4} = \frac{9}{2} + 8 = \frac{25}{2}$$

Sol 6: (B) $\int_{-2}^2 |1 - x^2|\, dx = 2\int_0^2 |1 - x^2|\, dx$

$$= 2\left[\int_0^1 (1 - x^2)dx + \int_1^2 (x^2 - 1)dx \right]$$

$$= 2\left[x - \frac{x^3}{3}\Big|_0^1 + \frac{x^3}{3} - x\Big|_1^2 \right]$$

$$= 2\left[1 - \frac{1}{3} + \left(\frac{8}{3} - 2\right) - \left(\frac{1}{3} - 1\right) \right]$$

$$= 2\left[\frac{2}{3} + \frac{2}{3} + \frac{2}{3} \right] = 4$$

Sol 7: (D) $f'(x) = \left(\frac{x^4 - 5x^2 + 4)}{2 + e^{x^2}} \right) \times 2x = 0$

$x = 0$ or $(x^2 - 4)(x^2 - 1) = 0$

$\therefore x = 0, x = \pm 2, x = \pm 1$

Sol 8: (A) $F_1(x) = \int_2^x (2t - 5)dt = t^2 - 5t\Big|_2^x$

$$= x^2 - 5x - (4 - 10) = x^2 - 5x + 6$$

$F_2(x) = \int_0^x 2t\, dt = x^2$

$\therefore x^2 = x^2 - 5x + 6 \Rightarrow x = \frac{6}{5}, y = \frac{36}{25}$

Sol 9: (A) $f(x) = f(a - x)$

$g(x) = 2 - g(a - x)$

$I = \int_0^a f(x)\, g(x)\, dx = \int_0^a f(a - x) \cdot (2 - g(a - x))\, dx$

$\Rightarrow 2\int_0^a f(a - x)dx - \int_0^a f(a - x) \cdot g(a - x) \cdot dx$

Put $a - x = t$

$- dx = dt$

$\Rightarrow -2 \int_a^0 f(t)\, dt - \int_a^0 -f(t) \cdot g(t)dt$

$$\Rightarrow -2\int_a^0 f(t).dt + \int_a^0 f(t).g(t).dt$$

$$\Rightarrow I = 2\int_0^a f(t).dt - \int_0^a f(t).g(t)\,dt$$

$$I = 2\int_0^a f(t).dt - I \Rightarrow 2I = 2\int_0^a f(t).dt \quad I = \int_0^a f(x)dx$$

Sol 10: (B) $\displaystyle\int_0^{\log 5} \frac{e^x\sqrt{e^x-1}}{e^x+3}dx$

$e^x + 3 = t$

$e^x dx = dt$

$$\int_4^8 \frac{\sqrt{t-4}}{t}dt$$

$t = 4\sec^2\theta; \qquad dt = 8\sec^2\theta\tan\theta dq$

$$\int_0^{\pi/4} \frac{2\tan\theta \times 8\sec^2\theta\tan\theta}{4\sec^2\theta}d\theta = \int_0^{\pi/4} 4(\sec^2\theta - 1)d\theta$$

$$= 4\tan\theta\Big|_0^{\pi/4} - 4\theta\Big|_0^{\pi/4} = 4 - p$$

Sol 11: (A)

$$\int_{-\pi}^{\pi} \sin mx\,\sin nx\,dx = 2\int_0^{\pi} \sin mx\,\sin nx\,dx$$

$$= \int_0^{\pi}[\cos(m-n)x - \cos(m+n)x]dx = 0$$

Sol 12: (B) $-\displaystyle\int_{1/e}^1 \log x\,dx + \int_1^e \log x\,dx$

$$= -\left[x\log x - x\Big|_{1/e}^1\right] + x\log x - x\Big|_1^e$$

$$= -\left[(-1) - \left(\frac{1}{e}\log\frac{1}{e} - \frac{1}{e}\right)\right] + \left[(e\log e - e) - (-1)\right]$$

$$= 1 + \left(-\frac{1}{e}\right) - \frac{1}{e} + e - e + 1$$

$$= 2 - \frac{2}{e} = 2\left(1 - \frac{1}{e}\right)$$

Sol 13: (C) $\displaystyle\int_0^{\pi} \frac{dx}{1 - 2a\cos x + a^2}$

$$\Rightarrow \text{Put } \cos x = \frac{1 - \tan^2(x/2)}{1 + \tan^2(x/2)}$$

$$\Rightarrow \int_0^{\pi} \frac{\sec^2(x/2).dx}{(1+a^2)(1+\tan^2(x/2)) - 2a(1 - \tan^2 x/2)}$$

Put $\tan(x/2) = t$

$$\frac{1}{2}.\sec^2(x/2)\,dx = dt \Rightarrow 2\int_0^{\infty} \frac{dt}{(1-a^2) + (1+a^2)t^2}$$

$$\Rightarrow \frac{2}{1-a}.\tan^{-1}\cdot\frac{(1+a)t}{(1-a)}\Big|_0^{\infty} \Rightarrow \frac{2}{1-a}\left(\tan^{-1}\infty - \tan^{-1}0\right)$$

$$\Rightarrow \frac{2}{1-a}.\frac{\pi}{2} \Rightarrow \frac{\pi}{1-a}$$

Sol 14: (B) $\displaystyle\int_0^{\pi/4} \frac{\sec x}{1 + 2\sin^2 x}dx = \int_0^1 (1 - (1-x)^9)dx$

$$= \int_0^1 x^9 dx = \frac{1}{10}$$

Sol 15: (A) $\displaystyle\int_0^{\infty} \frac{dx}{\left(x + \sqrt{x^2+1}\right)^3}$

$x = \tan\theta$

$$\int_0^{\pi/2} \frac{\sec^2\theta d\theta}{(\tan\theta + \sec\theta)^3} = \int_0^{\pi/2} \frac{\cos\theta}{(1+\sin\theta)^3}d\theta$$

$1 + \sin\theta = t$

$\cos\theta d\theta = dt$

$$\int_1^2 \frac{dt}{t^3} = -\frac{1}{2t^2}\Big|_1^2 = -\frac{1}{2}\left[\frac{1}{4} - 1\right] = \frac{3}{8}$$

Sol 16: (B) $\displaystyle\int_1^5 \left[|x-3|\right]dx$

$$\int_1^2 1dx + \int_2^3 0dx + \int_3^4 0dx + \int_4^5 1dx$$

$1 + 1 = 2$

Sol 17: (C) $I = \displaystyle\int_{-\pi/2}^{\pi/2} \frac{\sin x}{1 + \cos^2 x}e^{-\cos^2 x}dx$

$$I = \int_{-\pi/2}^{\pi/2} \frac{-\sin x}{1 + \cos^2 x}e^{-\cos^2 x}dx$$

$\Rightarrow 2I = 0 \Rightarrow I = 0$

Sol 18: (C) $I = \displaystyle\int_{-1/2}^{1/2}\left[\left(\frac{x+1}{x-1}\right)^2 + \left(\frac{x-1}{x+1}\right)^2 - 2\right]^{1/2} dx$

$$I = 2 \int_0^{1/2} \left| \frac{x+1}{x-1} - \frac{x-1}{x+1} \right|^{2/2} dx$$

$$= 2 \int_0^{1/2} \left| \left(\frac{x+1}{x-1} - \frac{x-1}{x+1} \right) \right| dx = 2 \int_0^{1/2} \left| \frac{4x}{x^2-1} \right| dx$$

$$= -4 \log \left| (x^2-1) \right|_0^{1/2} = -4 \log \frac{3}{4}$$

Sol 19: (A) $\int_{-1}^{1} \{x - [x]\} dx$

$$= \frac{x^2}{2} \Big|_{-1}^{1} - \left[\int_{-1}^{0} (-1) dx + \int_0^1 0 dx \right]$$

$$= 0 - [-(0+1) + 0] = 1$$

Sol 20: (C) $I = \int_4^{10} \frac{[x^2]}{[(x-14)^2] + [x^2]} dx$

$$I = \int_4^{10} \frac{[(14-x)^2]}{[x^2] + [(x-14)^2]} dx$$

$$\therefore 2I = \int_4^{10} dx = 10 - 4 = 6 \Rightarrow I = 3$$

Sol 21: (A) $\int_{-1}^{3} \left(|x-2| + [x] \right) dx$

$$= \int_{-1}^{2} (2-x) dx + \int_2^3 (x-2) dx + \int_{-1}^{0} (-1) dx$$

$$+ \int_0^1 0 dx + \int_1^2 1 dx + \int_2^3 2 dx$$

$$= 2x - \frac{x^2}{2} \Big|_{-1}^{2} + \frac{x^2}{2} - 2x \Big|_2^3 - 1(0+1) + 1 + 2$$

$$= [4-2] - \left(-2 - \frac{1}{2} \right) + \left(\frac{9}{2} - 6 \right) - (2-4) + 2$$

$$= 2 + \frac{5}{2} - \frac{3}{2} + 4 = 6 + 1 = 7$$

Sol 22: (C) $\int_0^{\pi/2} \frac{\sin^2 x}{\sin x + \cos x} dx = \int_0^{\pi/2} \frac{\cos^2 x}{\sin x + \cos x} dx$

$$\therefore 2I = \int_0^{\pi/2} \frac{1}{\sin x + \cos x} dx = \frac{1}{\sqrt{2}} \int_0^{\pi/2} \frac{dx}{\cos \left(x - \frac{\pi}{4} \right)}$$

$$= \frac{2}{\sqrt{2}} \int_0^{\pi/4} \sec \left(x - \frac{\pi}{4} \right) dx$$

$$= \frac{1}{\sqrt{2}} \log \left[\sec \left(x - \frac{\pi}{4} \right) + \tan \left(x - \frac{\pi}{4} \right) \right] \Big|_0^{\pi/4}$$

$$2I = \frac{2}{\sqrt{2}} \left[\log \left(\sqrt{2} + 1 \right) \right] \Rightarrow I = \frac{1}{\sqrt{2}} \log \left(\sqrt{2} + 1 \right)$$

Sol 23: (C) $\mu_{10} = \int_0^{\pi/2} x^{10} \sin x \, dx$

$$\mu_8 = \int_0^{\pi/2} x^8 \sin x \, dx$$

$$\mu_{10} = x^{10} (-\cos x) \Big|_0^{\pi/2} - \int_0^{\pi/2} (-\cos x) 10 x^9 dx$$

$$\mu_{10} = \left(\frac{\pi}{2} \right)^{10} (0) + \int_0^{\pi/2} \cos x \cdot 10 \cdot x^9 dx$$

$$= 10 \left[\int_0^{\pi/2} \cos x \cdot x^9 dx \right]$$

$$= 10 \left[x^9 \sin x \Big|_0^{\pi/2} - \int_0^{\pi/2} 9 x^8 \sin x \, dx \right]$$

$$= 10 \times \left(\frac{\pi}{2} \right)^9 - 90 \mu_8$$

$$\therefore \mu_{10} + 90 \mu_8 = 10 \left(\frac{\pi}{2} \right)^9$$

Sol 24: (C) $\int_0^{\pi} e^{\sin^2 x} \cos^3 (2n+1) x \, dx$

$$I = \int_0^{\pi} e^{\sin^2 x} \cos^3 \left((2n+1)\pi - (2n+1)x \right) dx$$

$$= - \int_0^{\pi} \sin^2 x \cos^3 \left((2n+1)x \right) dx$$

$$\therefore 2I = 0 \Rightarrow I = 0$$

Sol 25: (C) $I = \int_{-\pi/2}^{\pi/2} \sin \log \left(x + \sqrt{x^2+1} \right) dx$

$$= \int_{-\pi/2}^{\pi/2} \sin \log \left(\sqrt{x^2+1} - x \right) dx$$

$$= \int_{-\pi/2}^{\pi/2} \sin \log \frac{1}{\sqrt{x^2+1} + x} dx$$

$$= -\int_{-\pi/2}^{\pi/2} \sin\log\left(\sqrt{x^2+1}+x\right)dx = -I$$

$$\therefore 2I = 0 \quad \Rightarrow \quad I = 0$$

Sol 26: (C) $\sin\alpha + \int_{\alpha}^{2\alpha} \cos 2x\, dx = 0$

$$\Rightarrow \sin\alpha + \frac{1}{2}\sin 2x\Big|_{\alpha}^{2\alpha} = 0$$

$$\Rightarrow \sin\alpha + \frac{1}{2}[\sin 4\alpha - \sin 2\alpha] = 0$$

$$\Rightarrow \sin\alpha + \cos 3\alpha\, \sin\alpha = 0$$

$$\Rightarrow \sin\alpha = 0 \text{ or } \cos 3\alpha = -1$$

$$\Rightarrow \alpha = n\pi, \text{ or } 3\alpha = (2n+1)\,\pi$$

$$\therefore \alpha = -\frac{\pi}{3}$$

Sol 27: (B) $f'(x) = \sin\sqrt{x^4}\,dx^4 - \sin\sqrt{x^2}\,dx^2$

$$= 4x^3 \sin x^2 - 2x \sin x$$

Sol 28: (B) $\int_{0}^{\pi} x\sin x\cos^4 x\, dx = \int_{0}^{\pi} (\pi-x)\sin x\cos^4 x\, dx$

$$\therefore 2I = \pi\int_{0}^{\pi} \sin x\cos^4 x\, dx$$

Let $\cos x = t$

$-\sin x\, dx = dt$

$$2I = -\pi\int_{1}^{-1} t^4\, dt$$

$$2I = \pi\int_{-1}^{1} t^4\, dt = \frac{\pi}{5}\,[1+1]$$

$$\therefore I = \frac{\pi}{5}$$

Sol 29: (B) $f(0) = a + b = -1$

$f'(x) = 2ae^{2x} + be^x + c$ ..... (i)

$f'(\log 2) = 8a + 2b + c = 31$ ..... (ii)

$$\int_{0}^{\log 4} \left(f(x)-(x)\right)dx = \int_{0}^{\log 4} \left(ae^{2x}+be^x\right)dx$$

$$= \frac{a}{2}e^{2x} + be^x\Big|_{0}^{\log 4} = 8a + 4b - \left(\frac{a}{2}+b\right) = \frac{39}{2}$$

$= 15a + 6b = 39$ ..... (iii)

$\Rightarrow 9a = 45$

$a = 5; b = -6; c = 3$

Sol 30: (A) $I = \int_{-\pi/4}^{\pi/4} \frac{e^{-x}\sec^2 x\, dx}{e^{-2x}-1} = \int_{-\pi/4}^{\pi/4} \frac{e^x \sec^2 x}{1-e^{2x}}dx$

$$\therefore 2I = \int_{-\pi/4}^{\pi/4} \left(\frac{e^x \sec^2 x}{e^{2x}-1} - \frac{e^x \sec^2 x}{e^{2x}-1}\right)dx = 0$$

$I = 0$

Sol 31: (B) $I = \int_{-1}^{a} \log\left(x+\sqrt{1+x^2}\right)$

$$= \log\left(\frac{-1}{x-\sqrt{1+x^2}}\right) = \log\left(\frac{1}{\sqrt{1+x^2}-x}\right)$$

$$= \int_{-1}^{a} -\log\left(\sqrt{1+x^2}-x\right)$$

$$\therefore 2I = \int_{-1}^{a} \log(x^2+1-x^2) = \int_{-1}^{a} \log 1 = 0$$

Sol 32: (C) $\int_{-2}^{0} \frac{\sin^2 x}{-\dfrac{1}{2}}dx + \int_{0}^{2} \frac{\sin^2 x}{\dfrac{1}{2}}dx$

$$= -\frac{2}{2}\int_{-2}^{0} \frac{1-\cos 2x}{-\dfrac{1}{2}}dx + \frac{2}{2}\int_{0}^{2}\left(1-\cos 2x\right)dx$$

$$= -1[2] + \frac{1}{2}\sin 2x\Big|_{-2}^{0} + 1[2] - \frac{1}{2}\sin 2x\Big|_{0}^{2}$$

$$= -2 + \frac{1}{2}[0+\sin 4] + 2 - \frac{1}{2}[\sin 4] = 4 - \sin 4$$

Sol 33: (C) $f(x) = \int_{0}^{x} \log(1+t^2)dt$

$f'(x) = \log(1+x^2)$

$$f''(x) = \frac{1}{1+x^2} \times 2x$$

$$\therefore f''(1) = \frac{2}{2} = 1$$

Sol 34: (C) $\int_{0}^{\pi} \frac{dx}{1+3^{\cos x}}$

$$= \int_{0}^{\pi} \frac{dx}{1+3\cos^{(\pi-x)}} = \int_{0}^{\pi} \frac{dx}{1+3^{-\cos x}}$$

$$= \int_0^{\pi} \frac{3^{\cos x}}{1 + 3^{\cos x}} dx$$

$$\therefore 2I = \int_0^{\pi} dx = \pi$$

$$\therefore I = \frac{\pi}{2}$$

Previous Years' Questions

Sol 1: (A) Let $I = \int_0^{\pi/2} \frac{\sqrt{\cot x}}{\sqrt{\cot x} + \sqrt{\tan x}} dx$ ....(i)

$$\Rightarrow I = \int_0^{\pi/2} \frac{\sqrt{\tan x}}{\sqrt{\cot x} + \sqrt{\tan x}} dx$$ ....(ii)

On adding Eqs. (i) and (ii), we get

$$2I = \int_0^{\pi/2} I \, dx$$

$$\therefore I = \frac{\pi}{4}$$

Sol 2: (C) Let $I = \int_0^{\pi} e^{\cos^2 x} . \cos^3\{(2n+1)x\} dx$

Using $\int_0^a f(x) dx$

$$= \begin{cases} 0, & f(a-x) = -f(x) \\ 2\int_0^{a/2} f(x) dx, & f(a-x) = f(x) \end{cases}$$

Again, let $f(x) = e^{\cos^2 x} . \cos^3\{(2n+1)x\}$

$\therefore f(\pi - x) = (e^{\cos^2 x}) \{- \cos^3(2n+1)x\} = -f(x)$

$\therefore I = 0$

Sol 3: (A) $\lim_{x \to 1} \int_4^{f(x)} \frac{2t}{x-1} dt = \lim_{x \to 1} \frac{\int_4^{f(x)} 2t \, ct}{x-1}$

(using L' Hospital's rule)

$$= \lim_{x \to 1} \frac{2f(x).f'(x)}{1} = 2f(1) . f'(1)$$

$= 8f'(1) [\because f(1) = 4]$

Sol 4: (D) Let $I = \int_0^{\pi/2} \frac{1}{1 + \tan^3 x} dx$

$$= \int_0^{\pi/2} \frac{1}{1 + \frac{\sin^3 x}{\cos^3 x}} dx$$

$$\Rightarrow I = \int_0^{\pi/2} \frac{\cos^3 x}{\cos^3 x + \sin^3 x} dx$$ .....(i)

$$\Rightarrow I = \int_0^{\pi/2} \frac{\cos^3\left(\frac{\pi}{2} - x\right)}{\cos^3\left(\frac{\pi}{2} - x\right) + \sin^3\left(\frac{\pi}{2} - x\right)} dx$$

$$\Rightarrow I = \int_0^{\pi/2} \frac{\sin^3 x}{\sin^3 x + \cos^3 x} dx$$ .....(ii)

On adding Eqs (i) and (ii), we get $2I = \int_0^{\pi/2} 1 \, dx$

$$2I = [x]_0^{\pi/2} = \frac{\pi}{2} \Rightarrow I = \frac{\pi}{4}$$

Now, $\int_0^1 f(x) dx = \frac{2A}{\pi}$

$$\Rightarrow \int_0^1 \left\{ A \sin\left(\frac{\pi x}{2}\right) + B \right\} dx = \frac{2A}{\pi}$$

$$\Rightarrow \left[-\frac{2A}{\pi} \cos\frac{\pi x}{2} + Bx \right]_0^1 = \frac{2A}{\pi}$$

$$\Rightarrow B + \frac{2A}{\pi} = \frac{2A}{\pi} \Rightarrow B = 0$$

Sol 5: (A) It is a questions of greatest integer function. We have subdivide the interval π to 2π as under keeping in view that we have to evaluate [2 sin x]

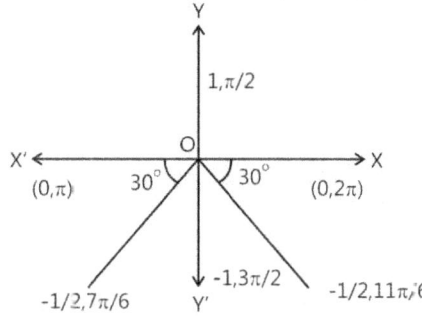

We known that, $\sin\frac{\pi}{6} = \frac{1}{2}$,

$$\sin\left(\pi + \frac{\pi}{6}\right) = \sin\frac{7\pi}{6} = -\frac{1}{2}$$

$$\sin\frac{11\pi}{6} = \sin\left(2\pi - \frac{\pi}{6}\right) = -\sin\frac{\pi}{6} = -\frac{1}{2}$$

$$\sin\frac{9\pi}{6} = \sin\frac{3\pi}{6} = -1$$

Hence, we divide the interval π to 2π as

$\left(\pi, \dfrac{7\pi}{6}\right), \left(\dfrac{7\pi}{6}, \dfrac{11\pi}{6}\right), \left(\dfrac{11\pi}{6}, 2\pi\right)$

$\sin x = \left(0, -\dfrac{1}{2}\right), \left(-1, -\dfrac{1}{2}\right), \left(-\dfrac{1}{2}, 0\right)$

$2\sin x = (0, -1), (-2, -1), (-1, 0)$

$[2\sin x] \, x = -1$

$= \displaystyle\int_\pi^{7\pi/6} [2\sin x]dx + \int_{7\pi/6}^{11\pi/6} [2\sin x]dx$

$\quad + \displaystyle\int_{11\pi/6}^{2\pi} [2\sin x]dx$

$= \displaystyle\int_\pi^{7\pi/6} -1 \, dx + \int_{7\pi/6}^{11\pi/6} -2 \, dx + \int_{11\pi/6}^{2\pi} -1 \, dx$

$= -\dfrac{\pi}{6} - 2\left(\dfrac{4\pi}{6}\right) - \dfrac{\pi}{6} = -\dfrac{10\pi}{6} = -\dfrac{5\pi}{3}$

Sol 6: (A) Given, $\displaystyle\int_0^x f(t)dt = x + \int_x^1 t\, f(t)dt$

On differentiating both sides w.r.t. x, we get

$f(x)\,1 = 1 - x f(x)\, . \, 1$

$\Rightarrow (1 + x)\, f(x) = 1$

$\Rightarrow f(x) = \dfrac{1}{1 + x}$

$\Rightarrow f(1) = \dfrac{1}{1 + 1} = \dfrac{1}{2}$

Sol 7: (A) Let $I = \displaystyle\int_{\pi/4}^{3\pi/4} \dfrac{dx}{1 + \cos x}$ .....(i)

$\Rightarrow I = \displaystyle\int_{\pi/4}^{3\pi/4} \dfrac{dx}{1 + \cos(\pi - x)}$

$I = \displaystyle\int_{\pi/4}^{3\pi/4} \dfrac{dx}{1 - \cos x}$ ....(ii)

On adding Eqs. (i) and (ii), we get

$2I = \displaystyle\int_{\pi/4}^{3\pi/4} \left(\dfrac{1}{1 + \cos x} + \dfrac{1}{1 - \cos x}\right)dx$

$\Rightarrow 2I = \displaystyle\int_{\pi/4}^{3\pi/4} \left(\dfrac{2}{1 - \cos^2 x}\right)dx$

$\Rightarrow I = \displaystyle\int_{\pi/4}^{3\pi/4} \text{cosec}^2 x \, dx = [-\cot x]_{\pi/4}^{3\pi/4}$

$= \left[-\cot\dfrac{3\pi}{4} + \cot\dfrac{\pi}{4}\right] = -(-1) + 1 = 2$

Sol 8: (C) The graph of $y = 2\sin x$ for $\pi/2 \le x \le 3\pi/2$ is given in figure. From the graphs, it is clear that

$[2\sin x] = \begin{cases} 2, & \text{if} \quad x = \pi/2 \\ 1, & \text{if} \quad \pi/2 < x \le 5\pi/6 \\ 0, & \text{if} \quad 5\pi/6 < x \le \pi \\ -1, & \text{if} \quad \pi < x \le 7\pi/6 \\ -2, & \text{if} \quad 7\pi/6 < x \le 3\pi/2 \end{cases}$

Therefore, $\displaystyle\int_{\pi/2}^{3\pi/2} [2\sin x]dx$

$= \displaystyle\int_{\pi/2}^{5\pi/2} dx + \int_{5\pi/6}^{\pi} 0 dx + \int_{\pi}^{7\pi/6} (-1)dx$

$\quad + \displaystyle\int_{7\pi/6}^{3\pi/2} (-2)dx$

$= \left[x\right]_{\pi/2}^{5\pi/6} + \left[-x\right]_{\pi}^{7\pi/6} + \left[-2x\right]_{7\pi/6}^{3\pi/2}$

$= \left(\dfrac{5\pi}{6} - \dfrac{\pi}{2}\right) + \left(-\dfrac{7\pi}{6} + \pi\right)$

$\quad + \left(\dfrac{-2.3\pi}{2} + \dfrac{2.7\pi}{6}\right)$

$= \pi\left(\dfrac{5}{6} - \dfrac{1}{2}\right) + \pi\left(1 - \dfrac{7}{6}\right) + \pi\left(\dfrac{7}{3} - 3\right)$

$= \pi\left(\dfrac{5-3}{6}\right) + \pi\left(-\dfrac{1}{6}\right) + \pi\left(\dfrac{7-9}{3}\right) = -\dfrac{\pi}{2}$

Sol 9: (C) Let $I = \displaystyle\int_{-\pi}^{\pi} \dfrac{\cos^2 x}{1 + a^x}dx$ ....(i)

$= \displaystyle\int_{\pi}^{-\pi} \dfrac{\cos^2(-x)}{1 + a^{-x}}d(-x)$

$\Rightarrow I = \displaystyle\int_{-\pi}^{\pi} a^x \dfrac{\cos^2 x}{1 + a^x}dx$ ....(ii)

On adding Eqs. (i) and (ii), we get

2.48

$$2I = \int_{-\pi}^{\pi}\left(\frac{1+a^x}{1+a^x}\right)\cos^2 x\,dx$$

$$= \int_{-\pi}^{\pi}\cos^2 x\,dx = 2\int_0^{\pi}\frac{1+\cos 2x}{2}\,dx$$

$$= \int_0^{\pi}(1+\cos 2x)dx$$

$$= \int_0^{\pi}1\,dx = \int_0^{\pi}\cos 2x\,dx$$

$$= \left[x\right]_0^x + 2\int_0^{\pi/2}\cos 2x\,dx$$

$$= \pi + 0$$

$$\Rightarrow 2I = \pi \Rightarrow I = \pi/2$$

Sol 10: (C) Given, $F(x) = \int_0^x f(t)dt$

By Leibnitz rule,

$F'(x) = f(x)$ ...(i)

But $F(x^2) = x^2(1+x) = x^2 + x^3$ (given)

$\Rightarrow F(x) = x + x^{3/2}$

$\Rightarrow F'(x) = 1 + \dfrac{3}{2}x^{1/2}$

$\Rightarrow f(x) = F'(x) = 1 + \dfrac{3}{2}x^{1/2}$ [from Eq. (i)]

$\Rightarrow f(4) = 1 + \dfrac{3}{4}(4)^{1/2}$

$\Rightarrow f(4) = 1 + \dfrac{3}{2}\times 2 = 4$

Sol 11: (A) Given, $f(x) = \int_1^x \sqrt{2-t^2}\,dt$

$\Rightarrow f'(x) = \sqrt{2-x^2}$

Also $x^2 - f'(x) = 0$

$\therefore x^2 = \sqrt{2-x^2}$

$\Rightarrow x^4 = 2 - x^2 \Rightarrow x^4 + x^2 - 2 = 0$

$\Rightarrow x = \pm 1$

Sol 12: (C) $\int_3^{3+3T} f(2x)dx$ put $2x = y \Rightarrow dx = \dfrac{1}{2}dy$

$\therefore \dfrac{1}{2}\int_6^{6+6T} f(y)dy = \dfrac{6I}{2} = 3I$

Sol 13: (D) Given, $f(x) = \int_{x^2}^{x^2+1} e^{-t^2}\,dt$

On differentiating both sides using Newton's Leibnitz formula, we get

$$f'(x) = e^{-(x^2+1)^2}\cdot\left\{\frac{d}{dx}(x^2+1)\right\} - e^{-(x^2)^2}\left\{\frac{d}{dx}(x^2)\right\}$$

$$= e^{-(x^2+1)^2}\cdot 2x - e^{-(x^2)^2}\cdot 2x$$

$$= 2xe^{-(x^4+2x^2+1)}(1 - e^{2x^2+1})$$

[where, $e^{2x^2+1} > 1, \forall x$ and $e^{-(x^4+2x^2+1)} > 0 \forall x$]

$\therefore f'(x) > 0$

which shows $2x < 0$ or $x < 0$

$\Rightarrow x \in (-\infty, 0)$

Sol 14: (B) $I = \int_0^1 \sqrt{\dfrac{1-x}{1+x}}\,dx = \int_0^1 \dfrac{1-x}{\sqrt{1-x^2}}\,dx$

$$\int_0^1 \frac{1}{\sqrt{1-x^2}}\,dx - \int_0^1 \frac{x}{\sqrt{1-x^2}}\,dx$$

$$I = \left[\sin^{-1}x\right]_0^1 + \int_1^0 \frac{t}{t}\,dt$$

(where, $t^2 = 1 - x^2 \Rightarrow t\,dt = -x\,dx$)

$$I = (\sin^{-1}1 - \sin^{-1}0) + \left[t\right]_1^0 = \frac{\pi}{2} - 1$$

Sol 15: (A) Let $I = \int_{-1}^1 \dfrac{dx}{1+x^2}$

Put $x = \tan\theta \Rightarrow dx = \sec^2\theta\,d\theta$

$\therefore I = 2\int_0^{\pi/4} d\theta = \dfrac{\pi}{2}$

(B) Let $I = \int_0^1 \dfrac{dx}{\sqrt{1-x^2}}$

Put $x = \sin\theta \Rightarrow dx = \cos\theta\,d\theta$

$\therefore I = \int_0^{\pi/2} 1\,d\theta = \dfrac{\pi}{2}$

(C) $\int_2^3 \dfrac{dx}{1-x^2} = \dfrac{1}{2}\left[\log\left(\dfrac{1+x}{1-x}\right)\right]_2^3$

$$= \frac{1}{2}\left[\log\left(\frac{4}{-2}\right) - \log\left(\frac{3}{-1}\right)\right] = \frac{1}{2}\left[\log\left(\frac{2}{3}\right)\right]$$

(D) $\int_1^2 \dfrac{dx}{x\sqrt{x^2-1}} = \left[\sec^{-1}x\right]_1^2 = \dfrac{\pi}{3} - 0 = \dfrac{\pi}{3}$

Sol 16: $\int_{-2}^{2}|1-x^2|\,dx$

$=\int_{-2}^{-1}(x^2-1)dx+\int_{-1}^{1}(1-x^2)dx$

$+\int_{1}^{2}(x^2-1)dx$

$=\left[\dfrac{x^3}{3}-x\right]_{-2}^{-1}+\left[x-\dfrac{x^3}{3}\right]_{-1}^{1}+\left[\dfrac{x^3}{3}-x\right]_{1}^{2}$

$=\left(-\dfrac{1}{3}+1+\dfrac{8}{3}-2\right)+\left(1-\dfrac{1}{3}+1-\dfrac{1}{3}\right)$

$+\left(\dfrac{8}{3}-2-\dfrac{1}{3}+1\right)$

$=4$

Sol 17: Let $I=\int_{\pi/4}^{3\pi/4}\dfrac{x}{1+\sin x}\,dx$ ... (i)

$\Rightarrow I=\int_{\pi/4}^{3\pi/4}\dfrac{\left(\dfrac{\pi}{4}+\dfrac{3\pi}{4}-x\right)}{1+\sin\left(\dfrac{\pi}{4}+\dfrac{3\pi}{4}-x\right)}\,dx$

$\left[\because \int_{a}^{b}f(x)dx=\int_{a}^{b}f(a+b-x)dx\right]$

$\Rightarrow I=\int_{\pi/4}^{3\pi/4}\dfrac{\pi-x}{1+\sin(\pi-x)}\,dx$

$\Rightarrow I=\int_{\pi/4}^{3\pi/4}\dfrac{\pi}{1+\sin x}\,dx-\int_{\pi/4}^{3\pi/4}\dfrac{x}{1+\sin x}\,dx$

$\Rightarrow 1=\pi\int_{\pi/4}^{3\pi/4}\dfrac{dx}{1+\sin x}-1$ [from Eq. (i)]

$\Rightarrow 1=\dfrac{\pi}{2}\int_{\pi/4}^{3\pi/4}\dfrac{dx}{(1+\sin x)}$

$\Rightarrow 1=\dfrac{\pi}{2}\int_{\pi/4}^{3\pi/4}\dfrac{(1-\sin x)}{(1+\sin x)(1-\sin x)}\,dx$

$=\dfrac{\pi}{2}\int_{\pi/4}^{3\pi/4}\dfrac{(1-\sin x)}{1-\sin^2 x}\,dx$

$=\dfrac{\pi}{2}\int_{\pi/4}^{3\pi/4}\left(\dfrac{1}{\cos^2 x}-\dfrac{\sin x}{\cos^2 x}\right)dx$

$=\dfrac{\pi}{2}\int_{\pi/4}^{3\pi/4}(\sec^2 x-\sec x.\tan x)dx$

$=\dfrac{\pi}{2}\left[\tan x-\sec x\right]_{\pi/4}^{3\pi/4}$

$=\dfrac{\pi}{2}[-1-1-(-\sqrt{2}-\sqrt{2})]$

$=\dfrac{\pi}{2}(-2+2\sqrt{2})$

$=\pi(\sqrt{2}-1)$

Sol 18: Let $I=\int_{2}^{3}\dfrac{\sqrt{x}}{\sqrt{5-x}+\sqrt{x}}\,dx$ ... (i)

$\Rightarrow I=\int_{2}^{3}\dfrac{\sqrt{2+3-x}}{\sqrt{(2+3)-(5-x)}+\sqrt{2+3-x}}\,dx$

$\Rightarrow I=\int_{2}^{3}\dfrac{\sqrt{5-x}}{\sqrt{x}+\sqrt{5-x}}\,dx$ ... (ii)

On adding Eqs. (i) and (ii), we get

$\Rightarrow 2I=\int_{2}^{3}\dfrac{\sqrt{x}+\sqrt{5-x}}{\sqrt{5-x}+\sqrt{x}}\,dx$

$\Rightarrow 2I=\int_{2}^{3}1\,dx=1 \Rightarrow I=\dfrac{1}{2}$

Sol 19: (B)

$I=\int_{0}^{1}\dfrac{\sin x}{\sqrt{x}}\,dx<\int_{0}^{1}\dfrac{x}{\sqrt{x}}\,dx=\int_{0}^{2}\sqrt{x}\,dx=\dfrac{2}{3}x^{3/2}\Big|_{2}^{2}=\dfrac{2}{3}$

$\Rightarrow I<\dfrac{2}{3}$

$J=\int_{0}^{1}\dfrac{\cos x}{\sqrt{x}}\,dx<\int_{0}^{1}\dfrac{1}{\sqrt{x}}\,dx\,|_{0}^{1}=2$

$\therefore J\le 2.$

Sol 20: (D) Let $\int_{0}^{x}[\cot x]dx$ ... (i)

$=\int_{0}^{x}[\cot(\pi-x)]dx,\int_{0}^{x}[-\cot x]dx$ ... (ii)

Adding (1) and (2)

$2I=\int_{0}^{x}[\cot x]dx+\int_{0}^{x}[-\cot x]dx=\int_{0}^{x}(-1)dx$

$\left[\begin{array}{l}\because [x]+[-x]=-1\,\text{if}\,x\notin Z\\ \qquad\qquad =0\,\text{if}\,x\in Z\end{array}\right]$

$=[-x]\,_{0}^{x}=-\pi$

$\therefore =-\dfrac{\pi}{2}$

Sol 21: (A) $p'(x) = p'(1-x)$

$\Rightarrow p(x) = -p(1-x) + c$

at $x = 0$

Now $p(0) = -p(1-x) + 42$

$\Rightarrow p(x) + p(1-x) = 42$

$I = \int_0^1 p(x)dx \int_0^1 p(1-x)dx$

$2I = \int_0^1 (42)dx \quad \Rightarrow I = 21.$

Sol 22: (D) $I = 8\int_0^1 \dfrac{\log(1+x)}{1+x^2}dx$

$= 8\int_0^{\frac{\pi}{4}} \dfrac{\log(1+\tan\theta)}{1+\tan^2\theta} \sec^2\theta \, d\theta \, (\text{let } x = \tan\theta)$

$= \int_0^{\frac{\pi}{4}} \log\left(1+\tan\left(\dfrac{\pi}{4}-\theta\right)\right)d\theta = \int_0^{\frac{\pi}{4}} \log 2 \, d\theta - \int_0^{\frac{\pi}{4}} \log(1+\tan\theta)d\theta$

$= 8\log 2 \dfrac{\pi}{4} - 1$

$2I = 2\pi\log 2$

$I = \pi\log 2$

Sol 23: (B, C) $g(x) = \int_0^x \cos 4t \, dt$

$\Rightarrow g'(x) = \cos 4x \quad \Rightarrow g(x) = \dfrac{\sin 4x}{4} + k$

$\Rightarrow g(x) = \dfrac{\sin 4x}{4} \left[\because g(0) = 0\right]$

$= g(x) + g(\pi) = g(x) - g(\pi) \; (\because g(\pi) = 0)$

Sol 24: (D) $I = \int_{\pi/6}^{\pi/3} \dfrac{dx}{1+\sqrt{\tan x}}$

$= \int_{\pi/6}^{\pi/3} \dfrac{dx}{1+\sqrt{\tan\left(\dfrac{\pi}{2}-x\right)}}$

$= \int_{\pi/6}^{\pi/3} \dfrac{\sqrt{\tan x}\,dx}{1+\sqrt{\tan x}}$

$= \int_{\pi/6}^{\pi/3} \dfrac{\sqrt{\tan x}\,dx}{1+\sqrt{\tan x}}$

$2I = \int_{\pi/6}^{\pi/3} dx$

$\Rightarrow I = \dfrac{1}{2}\left[\dfrac{\pi}{3} - \dfrac{\pi}{6}\right] = \dfrac{\pi}{12}$, statement-1 is false

$\int_a^b f(x)dx = \int_a^b f(a+b-x)dx$ it is property

Sol 25: (D) $\int\left\{e^{\left(x-\frac{1}{x}\right)} + \left(1-\dfrac{1}{x^2}\right)e^{e+\frac{1}{x}}\right\}dx$

$= x.e^{x+\frac{1}{x}} + c$

As $\int\left(xf'(x) + f(x)\right)dx(x) + c$

Sol 26: (C) $I = \int_2^4 \dfrac{\log x^2}{\log x^2 + \log(36-12x+x^2)}dx$

$I = \dfrac{2}{2}\int_2^4 \dfrac{\log x}{\log x + \log(6-x)}dx \qquad \text{... (i)}$

$I = \int_2^4 \dfrac{\log(6-x)}{\log x(6-x) - \log}dx \quad \left\{\int_a^b f(x)dx = \int_a^b f(a+b-x)dx\right\} \text{ ...(ii)}$

Equation (i) & (ii) gives

$= \int_2^4 \dfrac{\log x + \log(6-x)}{\log x + \log(6-x)}dx = \int_2^4 dx = 2$

Hence $I = 1$

JEE Advanced/Boards

Exercise 1

Sol 1: $\int_0^1 e^{\tan^{-1}x} \sin^{-1}(\cos x)dx.$

$\int_0^1 (\tan^{-1}x)\sin^{-1}\left(\sin\left(\dfrac{\pi}{2}-x\right)\right)dx$

$= \int_0^1 \left(\dfrac{\pi}{2}\tan^{-1}x - x\tan^{-1}x\right)dx$

$$= \frac{\pi}{2}\left[x\tan^{-1}x\Big|_0^1 - \frac{1}{2}\int_0^1\frac{2x}{1+x^2}dx\right] - \int_0^1 x\tan^{-1}x\,dx$$

$$= \frac{\pi}{2}\left[\frac{\pi}{4} - \frac{1}{2}\log 2\right] - \left[(\tan^{-1}x)\frac{x^2}{2}\Big|_0^1 - \frac{1}{2}\int_0^1\frac{x^2+1-1}{1+x^2}dx\right]$$

$$= \frac{\pi^2}{8} - \frac{\pi}{4}\log - \frac{1}{2}\times\frac{\pi}{4} + \frac{1}{2}\left[(x)_0^1 - \tan^{-1}\Big|_0^1\right]$$

$$= \frac{\pi^2}{8} - \frac{\pi}{4}\log - \frac{\pi}{8} + \frac{1}{2} - \frac{1}{2}\times\frac{\pi}{4}$$

$$= \frac{\pi^2}{8} - \frac{\pi}{4}((\log 2)+1) + \frac{1}{2}$$

Sol 2: (i) Put $x = \alpha\cos^2\theta + \beta\sin^2\theta$

$dx = 2(\beta - \alpha)\sin\theta\cos\theta\,d\theta$

$$I = \int_\alpha^b \sqrt{(x-\alpha)(\beta-x)}dx$$

$$= \int_0^{\pi/2}\sqrt{(\beta-\alpha)\cos^2\theta(\beta-a)\sin^2\theta}$$

$$\times (\beta-\alpha)\sin 2\theta\,d\theta$$

$$= \frac{(\beta-\alpha)^2}{2}\int_0^{\pi/2}\sin^2 2\theta\,d\theta$$

$$= \frac{(\beta-\alpha)^2}{4}\int_0^{\pi/2}(1-\cos 4\theta)d\theta$$

$$= \frac{(\beta-\alpha)^2}{4}\times\frac{\pi}{2} = \frac{(\beta-\alpha)^2\pi}{8}$$

(ii) $I = \int_\alpha^\beta\sqrt{\frac{(x-\alpha)}{(\beta-\alpha)}}dx$

$$= \int_0^{\pi/2}\sqrt{\frac{(\beta-\alpha)\cos^2\theta}{(\beta-\alpha)\sin^2\theta}}\times(\beta-\alpha)\sin 2\theta\,d\theta$$

$$= 2(\beta-\alpha)\int_0^{\pi/2}\frac{\cos\theta}{\sin\theta}\times\sin\theta\cos\theta\,d\theta$$

$$= 2(\beta-\alpha)\int_0^{\pi/2}\left(\frac{1+\cos 2\theta}{2}\right)d\theta$$

$$= (\beta-\alpha)\frac{\pi}{2}$$

(iii) $I = \int_0^{\pi/2}\frac{2(\beta-\alpha)\sin\theta\cos\theta\,d\theta}{(\alpha\cos^2\theta+\beta\sin^2\theta)\times(\beta-\alpha)\cos\theta\sin\theta}$

$$= \int_0^{\pi/2}\frac{2d\theta}{\alpha\cos^2\theta+\beta\sin^2\theta} = \frac{1}{\beta}\int_0^{\pi/2}\frac{2\sec^2\theta\,d\theta}{\frac{\alpha}{\beta}+\tan^2\theta}$$

Put $\tan\theta = t$

$$\frac{2}{\beta}\int_0^\infty\frac{dt}{\frac{\alpha}{\beta}+t^2} = \frac{2}{\beta}\times\frac{1}{\sqrt{\frac{\alpha}{\beta}}}\tan^{-1}\frac{t}{\sqrt{\frac{\alpha}{\beta}}}\Bigg|_0^\infty$$

$$= \frac{2}{\sqrt{\alpha\beta}}\times\frac{\pi}{2} = \frac{\pi}{\sqrt{\alpha\beta}}$$

(iv) $I = \int_0^{\pi/2}\frac{(\alpha\cos^2\theta+\beta\sin^2\theta)\times(\beta-\alpha)\sin 2\theta\,d\theta}{(\beta-\alpha)\sin\theta\cos\theta}$

$$= 2\int_0^{\pi/2}(\alpha\cos^2\theta+\beta\sin^2\theta)d\theta$$

$$= 2\int_0^{\pi/2}\left((\beta-\alpha)\sin^2\theta+\alpha\right)d\theta$$

$$= 2\alpha\times\frac{\pi}{2} + 2\int_0^{\pi/2}\frac{(\beta-\alpha)}{2}(1-\cos 2\theta)d\theta$$

$$= \alpha\pi + 2\left(\frac{\beta-\alpha}{2}\right)\int_0^{\pi/2}(1-\cos 2\theta)d\theta$$

$$= \alpha\pi + \frac{\beta-\alpha}{2}\times\frac{\pi}{2}\times 2 = \left(\frac{\beta}{2}+\frac{\alpha}{2}\right)\pi = (\alpha+\beta)\frac{\pi}{2}$$

Sol 3: (i) $\int_0^{2\pi}\sqrt{1-\sin t}\,dt - \int_0^\pi\sqrt{1-\sin t}\,dt$

$$= \frac{1}{2}\int_0^{2\pi}\left(\sqrt{1-\sin t}+\sqrt{1+\sin t}\right)dt - \int_0^\pi\sqrt{1-\sin t}\,dt$$

$$= \int_0^\pi\sqrt{1-\sin t}+\sqrt{1+\sin t}\,dt - \int_0^\pi\sqrt{1-\sin t}\,dt$$

$$= \int_0^\pi\sqrt{1-\sin t}\,dt$$

$$= \int_0^\pi\left|\sin\frac{t}{2}+\cos\frac{t}{2}\right|dt$$

$$= -2\cos\frac{t}{2}+2\sin\frac{t}{2}\Bigg|_0^\pi$$

$$= -2[0-1] + 2[1-0] = 4$$

(ii) $\int_0^1 e^x(x-1)^n$

$= \left[(x-1)^n e^x \Big|_0^1 - n\int_0^1 (x-1)^{n-1} e^x dx \right]$

$= -(-1)^n - n\left[\begin{array}{l} (x-1)^{n-1} e^x \Big|_0^1 \\ -(n-1)\int_0^1 (x-1)^{n-2} e^x dx \end{array} \right]$

$= -(-1)^n + n(-1)^{n-1}$

$+ n(n-1)\left[\begin{array}{l} (x-1)^{n-2} e^x \Big|_0^1 \\ -(n-2)\int_0^1 (x-1)^{n-3} e^x dx \end{array} \right]$

$= -(-1)^n + n(-1)^{n-1} - n(n-1)(-1)^{n-2}$

$\qquad - n(n-1)(n-2)\int_0^1 (x-1)^{n-3} e^x dx$

Taking n = 3

$= -(-1)^3 + 3(-1)^2 - 3(3-1)(-1)^1 - 3(2)(1)\int_0^1 e^x dx$

$= +1 + 3 + 6 - 6(e^1 - 1)$

$= 16 - 6e$

Sol 4: (i) $\int_0^{\pi/2} e^x \left\{ \begin{array}{l} \cos(\sin x)\cos^2 \frac{x}{2} \\ + \sin(\sin x)\sin^{-2}\frac{x}{2} \end{array} \right\} dx$

$\frac{1}{2}\int_0^{\pi/2} e^x \left\{ \begin{array}{l} \cos(\sin x)[\cos x + 1] \\ + \sin(\sin x)[1 - \cos x] \end{array} \right\} dx$

$\frac{1}{2}\int_0^{\pi/2} e^x \left\{ \begin{array}{l} [\cos(\sin x) + \sin(\sin x)] \\ + \cos x[\cos(\sin x) - \sin(\sin x)] \end{array} \right\}$

Put $\cos(\sin x) + \sin(\sin x) = t$

$(-\sin(\sin x)\cos x + \cos(\sin x)\cos x)dx = dt$

$\frac{1}{2}\int_0^{\pi/2} e^x \{f(x) + f'(x)\}dx$

$= \frac{1}{2}e^x f(x)\Big|_0^{\pi/2} = \frac{1}{2}e^x\{\cos(\sin x) + \sin(\sin x)\}\Big|_0^{\pi/2}$

$= \frac{1}{2}\left[e^{\pi/2}(\cos 1 + \sin 1) - e^0(\cos 0) \right]$

$= \frac{1}{2}\left[e^{\pi/2}(\cos 1 + \sin 1) - 1 \right]$

(ii) $\int_1^e \left\{ (1+x)e^x + (1-x)e^{-x} \right\} \log x \, dx$

$\log x = t$

$x = e^t \Rightarrow dx = e^t dt$

$\int_0^1 \left\{ \left\{ 1+e^t \right\} e^{e^t} + \left\{ 1 - e^t \right\} e^{-e^t} \right\} dt$

$= \int_0^1 \left\{ \left(e^{e^t} + e^{-e^t} \right) t + \left(e^t e^{e^t} - e^t e^{-e^t} \right) t \right\} e^t dt$

Sol 5: $R = \int_0^\infty \frac{dx}{1+x^4}$

Put $x = \frac{1}{t} \Rightarrow dx = -\frac{1}{t^2} dt$

$= \int_\infty^0 \frac{-t^2}{1+t^2} dt = \int_0^\infty \frac{x^2}{1+x^2} = P$

$\therefore 2I = 2P = \int_0^\infty \frac{1+x^2}{1+x^4} dx$

$= \int_0^\infty \frac{1+x^{-2}}{\left(1-\frac{1}{x}\right)^2 + 2} dx$

$= \int_{-\infty}^\infty \frac{dt}{t^2 + 2}$ (Put $x - \frac{1}{x} = t$)

$\therefore 2I = \frac{1}{\sqrt 2}\tan^{-1}\frac{1}{\sqrt 2}\Big|_{-\infty}^\infty$

$= \frac{1}{\sqrt 2}\left[\frac{\pi}{2} + \frac{\pi}{2} \right] = \frac{\pi}{\sqrt 2}$

$\therefore I = \frac{\pi}{2\sqrt 2}$

$\int_0^\infty \frac{x\,dx}{1+x^4}$

Put $x^2 = t \Rightarrow 2x\,dx = dt$

$= \frac{1}{2}\int_0^\infty \frac{dt}{1+t^2} = \frac{1}{2}\tan^{-1}t\Big|_0^\infty = \frac{\pi}{2}$

$\therefore P + R - \sqrt 2 Q = \frac{\pi}{2\sqrt 2} + \frac{\pi}{2\sqrt 2} - \frac{\sqrt 2 \pi}{4}$

$= \frac{\pi}{\sqrt 2} - \frac{\pi}{2\sqrt 2} = \frac{\pi}{2\sqrt 2}$

Sol 6: $\displaystyle\int_1^2 \frac{(x^2-1)}{x^5\sqrt{2-\dfrac{2}{x^2}+\dfrac{1}{x^4}}}dx$

Put $2-\dfrac{2}{x^2}+\dfrac{1}{x^4}=t \Rightarrow \left(\dfrac{4}{x^3}-\dfrac{4}{x^5}\right)dx=dt$

$\dfrac{1}{4}\displaystyle\int_1^{\frac{25}{16}}\frac{dt}{\sqrt{t}} = \dfrac{1}{4}\times 2\sqrt{t}\Big|_1^{\frac{25}{16}} = \dfrac{1}{2}\left[\dfrac{5}{4}-1\right]=\dfrac{1}{8}$

$\therefore \dfrac{1000}{8}=125$

Sol 7: $h(x)=fog(x)+k$

$\dfrac{dh(x)}{dx}=f'\{g(x)\}g'(x)=\dfrac{-\sin x}{\cos^2(\cos x)}$

$j(x)=\displaystyle\int_{g(x)}^{f(x)}\frac{f(t)}{g(t)}dt$

$h(x)=-\displaystyle\int\frac{\sin x}{\cos^2(\cos x)}dx = \int\frac{dt}{\cos^2 t}=\tan t$

$=\tan(\cos x)+c$

$\therefore f(x)=\tan x,\ g(x)=\cos x$

$J(x)=\displaystyle\int_{\cos x}^{\tan x}\frac{\tan t}{\cos t}dt$

$j(0)=\displaystyle\int_1^0\frac{\sin t}{\cos^2 t}dt$

$\cos t=u \Rightarrow -\sin t\,dt=du$

$=\displaystyle\int_0^1\frac{du}{u^2}=\dfrac{-1}{u}\Big|_1^{\cos 1}=-\left[\dfrac{1}{\cos 1}-1\right]=1-\sec 1$

Sol 8: $\displaystyle\int_0^{\pi/2}\sqrt{\frac{1-\sin 2x}{1+\sin 2x}}dx = \int_0^{\pi/2}\sqrt{\left(\frac{1-\tan x}{1+\tan x}\right)^2}dx$

$=\displaystyle\int_0^{\pi/2}\left|\tan\left(x-\frac{\pi}{4}\right)\right|dx = 2\log\sec\left(x-\frac{\pi}{4}\right)\Big|_0^{\pi/4}$

$=2\ell n\sqrt{2}=\log 2$

Sol 9: $2I=\displaystyle\int_0^2(3x^2-3x+1)\cos(x^3-3x^2+4x-2)dx$

$\qquad +\displaystyle\int_0^2(3x^2-9x+7)\cos(x^3-3x^2+4x-2)dx$

$2I=2\displaystyle\int_0^2(3x^2-6x+4)\cos(x^3-3x^2+4x-2)dx$

Put $x^3-3x^2+4x-2=t$

$I=\displaystyle\int_{-2}^2\cos t\,dt = \sin t\Big|_{-2}^2 = \sin 2+\sin 2 = 2\sin 2$

$\therefore p=q=2 \Rightarrow p+q=4$

Sol 10: $I=\displaystyle\int_{-\sqrt{2}}^{\sqrt{2}}\frac{3x^6-12x^2+1}{x^2+2}dx$

$=\displaystyle\int_{-\sqrt{2}}^{\sqrt{2}}\frac{3x^6+6x^4-6x^4-12x^2+1}{x^2+2}dx$

$=\displaystyle\int_{-\sqrt{2}}^{\sqrt{2}}\frac{3x^4(x^2+2)-6x^2(x^2+2)+1}{x^2+2}dx$

$=\displaystyle\int_{-\sqrt{2}}^{\sqrt{2}}3x^4-6x^2+\frac{1}{x^2+2}dx$

$=2\times 3\left[\dfrac{x^5}{5}-\dfrac{2}{3}x^3\right]_0^{\sqrt{2}}+2\times\dfrac{1}{\sqrt{2}}\tan^{-1}\dfrac{x}{\sqrt{2}}\Big|_0^{\sqrt{2}}$

$=6\left[\dfrac{4\sqrt{2}}{5}-\dfrac{4\sqrt{2}}{3}\right]+\dfrac{\pi}{2\sqrt{2}}=\dfrac{-16\sqrt{2}}{5}+\dfrac{\pi}{2\sqrt{2}}$

Sol 11: $\displaystyle\int_0^\infty\frac{dx}{a^2+x^2}=\dfrac{1}{a}\tan^{-1}\dfrac{x}{a}\Big|_0^\infty$

$\dfrac{1}{a}\dfrac{\pi}{2}=\dfrac{\pi}{5050} \Rightarrow a=2525$

Sol 12: $\displaystyle\int_{-2}^2\frac{x^2-x}{\sqrt{x^2+4}}dx = \int_{-2}^2\frac{x^2+x}{\sqrt{x^2+4}}dx$

$\therefore I=\displaystyle\int_{-2}^2\frac{x^2}{\sqrt{x^2+4}}dx = 2\int_0^2\frac{x^2}{\sqrt{x^2+4}}dx$

$=2\displaystyle\int_0^2\left(\sqrt{x^2+4}-\frac{4}{\sqrt{x^2+4}}\right)dx$

$=2\left[\dfrac{x}{2}\sqrt{x^2+4}+2\log\left|x+\sqrt{x^2+4}\right|\right.$
$\qquad\left.-8\log\left|x+\sqrt{x^2+4}\right|\right]_0^2$

$=2\sqrt{8}-4\log\left|2+2\sqrt{2}\right|+4\log 2$

$=4\sqrt{2}-4\log\left|1+\sqrt{2}\right|$

Sol 13: $u = \dfrac{1}{2}\displaystyle\int_0^{\pi/4}\left(\dfrac{\cos x}{\sin\left(x+\dfrac{\pi}{4}\right)}\right)^2 dx$

$= \dfrac{1}{2}\displaystyle\int_0^{\pi/4}\left(\dfrac{\cos\left(\dfrac{\pi}{4}-x\right)}{\cos x}\right)^2 dx$

$v = 2\displaystyle\int_0^{\pi/4}\left(\dfrac{\cos\left(\dfrac{\pi}{4}-x\right)}{\cos x}\right)^2$

$\therefore \dfrac{v}{u} = \dfrac{2}{1/2} = 4$

Sol 14: $\dfrac{1}{\sqrt{2}}\displaystyle\int_0^{\pi/4}\dfrac{x\,dx}{\cos x\cos\left(\dfrac{\pi}{4}-x\right)}$

$= \dfrac{1}{\sqrt{2}}\displaystyle\int_0^{\pi/4}\dfrac{\left(\dfrac{\pi}{4}-x\right)dx}{\cos\left(\dfrac{\pi}{4}-x\right)\cos x}$

$\therefore 2I = \dfrac{1}{\sqrt{2}}\displaystyle\int_0^{\pi/4}\dfrac{\dfrac{\pi}{4}dx}{\cos x\cos\left(\dfrac{\pi}{4}-x\right)}$

$I = \dfrac{\pi}{8\sqrt{2}}\displaystyle\int_0^{\pi/4}\dfrac{dx}{\cos x\cos\left(\dfrac{\pi}{4}-x\right)}$

$= \dfrac{\pi}{8}\displaystyle\int_0^{\pi/4}\dfrac{dx}{\cos^2 x+\cos x\sin x}$

$= \dfrac{\pi}{8}\displaystyle\int_0^{\pi/4}\dfrac{\sec^2 x\,dx}{1+\tan x}$

$= \dfrac{\pi}{8}\displaystyle\int_0^1\dfrac{dt}{1+t} = \dfrac{\pi}{8}\log(1+t)\Big|_0^1 = \dfrac{\pi}{8}\log 2$

Sol 15: $\displaystyle\int_0^1\dfrac{\sin^{-1}\sqrt{x}}{x^2-x+1}dx$

Let $\sin^{-1}\sqrt{x} = t$

$\dfrac{1}{\sqrt{1-x}}\times\dfrac{1}{2\sqrt{x}}dx = dt$

$\Rightarrow dx = \left(\sqrt{1-\sin^2 t}\right)2\sin t\,dt = \sin 2t\,dt$

$\displaystyle\int_0^{\pi/2}\dfrac{t\sin 2t\,dt}{\sin^4 t-\sin^2 t+1}$

$= \displaystyle\int_0^{\pi/2}\dfrac{t\sin 2t\,dt}{1-\sin^2 t\cos^2 t} = \displaystyle\int_0^{\pi/2}\dfrac{\left(\dfrac{\pi}{2}-t\right)\sin 2t\,dt}{1-\sin^2 t\cos^2 t}$

$\therefore 2I = \dfrac{\pi}{2}\displaystyle\int_0^{\pi/2}\dfrac{\sin 2t\,dt}{1-\sin^2 t\cos^2 t}$

$2I = \dfrac{\pi}{2}\displaystyle\int_0^{\pi/2}\dfrac{\sin 2t\,dt}{1-\dfrac{\sin^2 2t}{2}} = 4\cdot\dfrac{\pi}{2}\displaystyle\int_0^{\pi/2}\dfrac{\sin 2t\,dt}{3+\cos^2 2t}$

$2I = \left[2\pi\displaystyle\int_1^{-1}\left(\dfrac{dt}{3+t^2}\right)\right]-\dfrac{1}{2}$

$\therefore I = \dfrac{\pi}{2}\displaystyle\int_{-1}^1\dfrac{dt}{3+t^2} = \dfrac{\pi}{2}\times\dfrac{1}{\sqrt{3}}\tan^{-1}\dfrac{t}{\sqrt{3}}\Big|_{-1}^1$

$= \dfrac{1}{2}\dfrac{\pi}{\sqrt{3}}\left[\dfrac{\pi}{6}+\dfrac{\pi}{6}\right] = \dfrac{\pi^2}{6\sqrt{3}}$

Sol 16: $\displaystyle\int_{\frac{1+\sqrt{5}}{2}}^{\frac{1+\sqrt{5}}{2}}\dfrac{x^2+1}{x^2\left(x^2+\dfrac{1}{x^2}-1\right)}\log\left(1+x-\dfrac{1}{x}\right)dx$

$= \displaystyle\int_1^{\frac{1+\sqrt{5}}{2}}\dfrac{1+x^2}{\left(\left(x-\dfrac{1}{x}\right)^2+1\right)}\log\left(1+\left(x-\dfrac{1}{x}\right)\right)dx$

$x-\dfrac{1}{x} = t \Rightarrow \left(1+\dfrac{1}{x^2}\right)dx = dt$

$= \displaystyle\int_0^1\dfrac{\log(1+t)}{(t^2+1)}dt$

$t = \tan\theta$

$= \displaystyle\int_0^{\pi/4}\log(1+\tan\theta)d\theta$

$= \displaystyle\int_0^{\pi/4}\log\left(1+\dfrac{1-\tan\theta}{1+\tan\theta}\right)d\theta$

$= \displaystyle\int_0^{\pi/4}\left[\log(2)-\log(1+\tan\theta)\right]d\theta$

2.55

$$\therefore 2I = \int_0^{\pi/4} \log 2\, d\theta = \frac{\pi}{4}\log 2 \Rightarrow I = \frac{\pi}{8}\log 2$$

Sol 17: $\displaystyle \lim_{n\to\infty} n^2 \int_{-1/n}^{1/n} (2010\sin x + 2012\cos x)\,|x|\,dx$

$$= \lim_{n\to\infty} n^2 \int_{-1/n}^{1/n} (2012\cos x)\,|x|\,dx$$

$$= 2012 \lim_{n\to\infty} 2n^2 \int_0^{1/n} x\cos x\,dx$$

$$= 2012 \times 2 \lim_{n\to\infty} n^2 \left[x\sin x \Big|_0^{1/n} - \int_0^{1/n} \sin x\,dx \right]$$

$$= 2012 \times 2 \lim_{n\to\infty} n^2 \left[\frac{1}{n}\sin\frac{1}{n} + \cos\frac{1}{n} - 1 \right]$$

$$= 2012 \times 2 \lim_{n\to\infty} \left[\frac{\sin\frac{1}{n}}{\frac{1}{n}} + \frac{\cos\frac{1}{n} - 1}{\frac{1}{n^2}} \right]$$

$$= 2012 \times 2 \left[1 - \frac{1}{2} \right] = 2012$$

Sol 18: $\displaystyle \int_0^{\pi} \left| \sqrt{2}\sin x + 2\cos x \right| dx$

$\sqrt{2}\sin x + 2\cos x > 0$

$\Rightarrow \tan x > -\sqrt{2}$

$\therefore x < \pi - \tan^{-1}\sqrt{2}$

$$= \int_0^{\pi - \tan^{-1}\sqrt{2}} \left(\sqrt{2}\sin x + 2\cos x \right) dx$$

$$+ \int_{\pi}^{\pi - \tan^{-1}\sqrt{2}} \left(\sqrt{2}\sin x + 2\cos x \right) dx$$

$$= -\sqrt{2}\cos x \Big|_0^{\pi - \tan^{-1}\sqrt{2}} + 2\sin x \Big|_0^{\pi - \tan^{-1}\sqrt{2}}$$

$$+ 2\sin x \Big|_{\pi}^{\pi - \tan^{-1}\sqrt{2}} - \sqrt{2}\cos x \Big|_{\pi}^{\pi - \tan^{-1}\sqrt{2}}$$

$$= -\sqrt{2}[-\cos\tan^{-1}\sqrt{2} - 1] + 2\sin\tan^{-1}\sqrt{2}$$

$$+ 2[\sin\tan^{-1}\sqrt{2}] - \sqrt{2}\left[-\cos\tan^{-1}\sqrt{2} + 1 \right]$$

$$= 2\sqrt{2} \times \frac{1}{\sqrt{3}} + 4 \times \frac{\sqrt{2}}{\sqrt{3}} = \frac{6\sqrt{2}}{\sqrt{3}} = 2\sqrt{6}$$

Sol 19: $\cos x + \cos 3x = 2\cos 2x\cos x$

$\sin x + \sin 3x = 2\sin 2x\cos x$

$$\therefore I = \int_0^{\pi} \sqrt{(2\cos x + 1)^2 \left[\cos^2 2x + \sin^2 2x \right]}\,dx$$

$$= \int_0^{\pi} |2\cos x + 1|\,dx$$

$$= \int_0^{2\pi/3} (2\cos x + 1)dx + \int_{2\pi/3}^{\pi} (-2\cos x - 1)dx$$

$$= 2\sin x \Big|_0^{2\pi/3} + \frac{2\pi}{3} - 2\sin x \Big|_{2\pi/3}^{\pi} - \left(n - \frac{2\pi}{3} \right)$$

$$= 2\left[\frac{\sqrt{3}}{2} \right] + \left(\frac{4\pi}{3} - \pi \right) - 2\left(0 - \frac{\sqrt{3}}{2} \right)$$

$$= \sqrt{3} + \sqrt{3} + \frac{\pi}{3} = 2\sqrt{3} + \frac{\pi}{3} = \sqrt{12} + \frac{\pi}{3}$$

$\therefore w = 12,\ k = 3$

$\Rightarrow k^2 + w^2 = 9 + 144 = 153$

Sol 20: $\displaystyle \int_0^1 \frac{(1-x)(1+x)}{x(1+x)(1+x)} \frac{dx}{\sqrt{\frac{1}{x} + 1 + x}}$

$$= \int_0^1 \frac{1 - x^2}{x^2} \left(\frac{1}{\left(\frac{1}{x} + 1 \right)(1+x)} \right) \frac{dx}{\sqrt{\frac{1}{x} + x + 1}}$$

$$= \int_0^1 \frac{x^{-2} - 1}{\left(\frac{1}{x} + x + 1 \right)} \frac{dx}{\sqrt{\frac{1}{x} + x + 1}}$$

Put $\dfrac{1}{x} + x + 1 = t \Rightarrow -\displaystyle\int_{\infty}^{3} \frac{dt}{(t+1)\sqrt{t}}$

Put $t = \tan^2\theta \Rightarrow dt = 2\tan\theta\sec^2\theta\,d\theta$

$$= -\int_{\pi/2}^{\pi/3} \frac{2\tan\theta\sec^2\theta}{\sec^2\theta\tan}\,d\theta = 2\int_{\pi/3}^{\pi/2} d\theta$$

$$= 2 \times \left(\frac{\pi}{2} - \frac{\pi}{3} \right) = \frac{\pi}{3}$$

Sol 21: $\displaystyle \int_0^{\pi/2} \left(\frac{a\sin x + b\cos x}{\sin x + \cos x} \right)\sqrt{2}\,dx$

$$= \int_0^{\pi/2} \frac{(a\cos x + b\sin x)\sqrt{2}}{\sin x + \cos x}\,dx$$

2.56

$$\therefore 2I = \sqrt{2}(a+b)\frac{\pi}{2} \Rightarrow I = \frac{(a+b)\pi}{2\sqrt{2}}$$

Sol 22: $\int_0^1 f(x)dx = 1$

$$\Rightarrow \int_0^1 \frac{f(2x)}{3}dx = 1 \Rightarrow \int_0^1 f(2x)dx = 3$$

$$\therefore 2I \int_0^{\pi/2} \frac{(a+b)(\sin x + \cos x)}{(\sin x + \cos x)}\sqrt{2}dx$$

Put $2x = t \Rightarrow dx = \frac{dt}{2}$

$$\int_1^2 f(t)\frac{dt}{2} = 3 \Rightarrow \int_1^2 f(t)dt = 6$$

$$\therefore \int_1^2 f(t)dt = 6 - 1 = 5$$

Sol 23: $\int_{-1}^3 \{|x-2| + [x]\}dx = \int_{-1}^0 \{|x-2| + [x]\}dx +$

$$\int_0^{-1} \{|x-2| + [x]\}dx = \int_1^2 \{|x-2| + [x]\}dx$$

$$+ \int_2^3 \{|x-2| + [x]\}dx$$

$$\int_{-1}^0 (2-x-1)dx + \int_0^1 (2-x+0)dx +$$

$$\int_1^2 (2-x+1)dx + \int_2^3 (x-2+2)dx +$$

$$= x - \frac{x^2}{2}\Big|_{-1}^0 + 2x - \frac{x^2}{2}\Big|_0^1 + 3x - \frac{x^2}{2}\Big|_1^2 + \frac{x^2}{2}\Big|_2^3$$

$$= -\left(-1-\frac{1}{2}\right) + \left(2-\frac{1}{2}\right) + (6-2) - \left(3-\frac{1}{2}\right) + \frac{9}{2} - 2$$

$$= -\left(-1-\frac{1}{2}\right) + \left(2-\frac{1}{2}\right) + (6-2) - \left(3-\frac{1}{2}\right) + \frac{9}{2} - 2$$

$$= 7$$

Sol 24: $x = \tan\theta$

$dx = \sec\theta d\theta$

$$\int_0^{\pi/3} \left(\sin^{-1}\frac{2\tan\theta}{1+\tan\theta}\right)\sec^2\theta d\theta$$

$$= \int_0^{\pi/4} 2\theta\sec^2\theta d\theta + \int_{\pi/4}^{\pi/3} (\pi - 2\theta)\sec^2\theta d\theta$$

$$= -2\left[\theta\tan\theta\Big|_{\pi/4}^{\pi/3} - \int_{\pi/4}^{\pi/3} \tan\theta d\theta\right] + \pi\tan\theta\Big|_{\pi/4}^{\pi/3}$$

$$+2\left[\theta\tan\theta\Big|_0^{\pi/4} - \int_0^{\pi/4} \tan\theta d\theta\right]$$

$$= -2\left[\left(\frac{\pi}{3}\times\sqrt{3} - \frac{\pi}{4}\right)\right] + \left[\log 2 - \frac{1}{2}\log 2]^2\right]$$

$$+ \pi(\sqrt{3} - 1) + 2\left[\frac{\pi}{4}\right] - \left[\frac{1}{2}\log 2\right]2$$

$$= -\frac{2\pi}{\sqrt{3}} + \frac{\pi}{2} + \log 2 + \sqrt{3}\pi - \pi + \frac{\pi}{2} - \log 2$$

$$= \frac{\pi}{\sqrt{3}}$$

Sol 25: $\int_0^\pi \frac{(ax + b\sec x\tan x)}{4 + \tan^2 x}dx$

$$\frac{2I}{a\pi + 2b} = \left[\int_0^\pi \frac{\sec x\tan x}{4 + \tan^2 x}dx\right]$$

$$= \int_1^{-1} \frac{dt}{3 + t^2} = \frac{1}{\sqrt{3}}\tan^{-1}\frac{t}{\sqrt{3}}\Big|_1^{-1}$$

$$= \frac{1}{\sqrt{3}}\left[\pi - \frac{\pi}{6} - \frac{\pi}{6}\right] = \frac{1}{\sqrt{3}}\times\frac{2\pi}{3} \quad \therefore I = \frac{(a\pi + 2b)\pi}{3\sqrt{3}}$$

Sol 26: $\int_0^\pi \frac{(2x - 3)\sin x}{(1 + \cos^2 x)}dx$

$$2I = \int_0^\pi \frac{(2\pi + 6)\sin x}{1 + \cos^2 x}dx$$

$$\frac{I}{\pi + 3} = \int_0^\pi \frac{\sin x}{1 + \cos^2 x}dx = \int_1^{-1} \frac{-dt}{1 + t^2} = \int_1^1 \frac{dt}{1 + t^2}$$

$$= \tan^{-1}t\Big|_{-1}^1 = \frac{\pi}{4} + \frac{\pi}{4} \Rightarrow I = (\pi + 3)\frac{\pi}{2}$$

Sol 27: Let $f(x) = \frac{\sqrt{\cos x}}{\sqrt{\cos x} + \sqrt{\sin x}}$ ....(i)

Then, $f\left(\frac{\pi}{2} - x\right) = \frac{\sqrt{\cos\left(\frac{\pi}{2} - x\right)}}{\sqrt{\cos\left(\frac{\pi}{2} - x\right)} + \sqrt{\sin\left(\frac{\pi}{2} - x\right)}}$

$$= \frac{\sqrt{\sin x}}{\sqrt{\sin x} + \sqrt{\cos x}} \qquad \qquad \ldots(ii)$$

Now, $f(x) + f\left(\frac{\pi}{2} - x\right) = \frac{\sqrt{\cos x} + \sqrt{\sin x}}{\sqrt{\cos x} + \sqrt{\sin x}} = 1$

$$\therefore I = \frac{1}{2} \int_0^{\pi/2} \left[f(x) + f\left(\frac{\pi}{2} - x\right) \right] dx = \frac{1}{2} \int_0^{\pi/2} x \, dx$$

$$= \frac{1}{2} \left[x \right]_0^{\pi/2} = \frac{1}{2} \left(\frac{\pi}{2} - 0 \right) = \frac{\pi}{4}$$

Sol 28: $2I = \int_0^{n\pi} \frac{n\pi \, |\sin x|}{1 + (\cos x)} dx$

$$2I = n^2 \pi \int_0^{\pi} \frac{\sin x}{1 + \cos x} dx = 2n^2 \pi \int_0^{\pi/2} \frac{\sin x}{1 + \cos x} dx$$

$$= 2n^2 \pi \int_1^0 \frac{-dt}{1 + t} = 2n^2 \pi \log(t+1) \big|_0^1 = 2n^2 \pi \log 2$$

$\therefore I = n^2 \pi \log 2 = 100\pi \log 2$

$\therefore n = 10$

Sol 29: $\int_0^{\pi/2} \frac{\cos x}{1 + \cos x + \sin x} dx$

$$= \int_0^{\pi/2} \frac{\cos x}{(1 + \cos x) + \sin x} dx$$

$$= \int_0^{\pi/2} \frac{\cos^2 \frac{x}{2} - \sin^2 \frac{x}{2}}{2\cos^2 \frac{x}{2} + 2\sin \frac{x}{2} \cos \frac{x}{2}} dx$$

$$= \int_0^{\pi/2} \frac{1 - \tan^2 \frac{x}{2}}{2 + 2\tan \frac{x}{2}} dx \text{ s}$$

[Dividing numerator and denominator By $\cos^2 \frac{x}{2}$]

$$= \frac{1}{2} \int_0^{\pi/2} \frac{\left(1 - \tan \frac{x}{2}\right)\left(1 + \tan \frac{x}{2}\right)}{1 + \tan \frac{x}{2}} dx$$

$$= \frac{1}{2} \int_0^{\pi/2} \left(1 - \tan \frac{x}{2}\right) dx$$

$$= \frac{1}{2} \left[x + 2 \log \cos \frac{x}{2} \right]_0^{\pi/2}$$

$$= \frac{1}{2} \left[\left(\frac{\pi}{2} + 2 \log \cos \frac{\pi}{4} \right) - (0 + 2 \log 1) \right]$$

$$= \frac{1}{2} \left[\frac{\pi}{2} + 2 \log \frac{1}{\sqrt{2}} \right] = \frac{1}{2} \left[\frac{\pi}{2} + \log \frac{1}{2} \right] = \frac{1}{2} \left[\frac{\pi}{2} - \log 2 \right]$$

Sol 30: $\int_0^a \frac{\log(1 + ax)}{1 + x^2} dx$

$ax = \tan\theta \Rightarrow abx = \sec^2\theta \, d\theta$

$$\int_0^{\tan^{-1} a^2} \frac{a\log(1 + \tan\theta)}{(a^2 + \tan^2 \theta)} \times \sec^2 \theta = d\theta$$

$$a \left[\log(1 + \tan\theta) \int_0^{\tan^{-1} a^2} \frac{\sec^2 \theta}{a^2 + \tan^2 \theta} d\theta \Big|_0^{\tan^{-1} a^2} \right.$$
$$\left. - \int_0^{\tan^{-1} a^2} \frac{\sec^2 \theta}{(1 + \tan\theta)} \int_0^{\tan^{-1} \theta} \frac{\sec^2 \theta}{a^2 + \tan^2 \theta} d\theta \right]$$

$$a \left[\log(1 + \tan\theta) \times \frac{1}{a} \tan^{-1} \frac{\tan\theta}{\alpha} \Big|_0^{\tan^{-1} a^2} \right.$$
$$\left. - \int_0^{\tan^{-1} a^2} \left(\frac{\sec^2 \theta}{1 + \tan\theta} \times \frac{1}{a} \tan^{-1} \frac{\tan\theta}{a} \right) d\theta \right]$$

$2I = \log(1 + a^2) \tan^{-1} a$

$I = \tan^{-1} a \log \sqrt{1 + a^2}$

Sol 31: $\int_0^{\frac{\ell n 3}{2}} \frac{e^x + 1}{e^{2x} + 1} dx = \int_0^{\frac{\ell n 3}{2}} \frac{e^x dx}{e^{2x} + 1} + \int_0^{\frac{\ell n 3}{2}} \frac{1}{e^{2x} + 1} dx$

$e^{2x} = t$

$2e^{2x} dx = dt$

$dx = \frac{1}{2t} dt$

$$\tan^{-1} e^x \Big|_0^{\frac{\log 3}{2}} + \frac{1}{2} \int_1^3 \frac{1}{(t+1)t} dt$$

$$= \tan^{-1} \sqrt{3} - \frac{\pi}{4} + \frac{1}{2} \int_1^3 \left(\frac{1}{t} - \frac{1}{t+1} \right) dt$$

$$= \frac{1}{2}\left[\frac{\pi}{6} + \log 3 - \log 2\right]$$

Sol 32: Given, $\int_a^a \sqrt{x}\, dx = 2a \int_0^{\pi/2} \sin^3 x\, dx$

$$\Rightarrow \left[\frac{x^{3/2}}{3/2}\right]_0^a = 2a \int_0^{\pi/2} \frac{3\sin x - \sin 3x}{4}\, dx$$

$$\left[\because \sin 3x = 3\sin x - 4\sin^3 x\right]$$

$$\Rightarrow \frac{2}{3}\left[a^{3/2} - 0\right] = \frac{a}{2}\left[3(-\cos x) - \left(-\frac{\cos 3x}{3}\right)\right]_0^{\pi/2}$$

$$\Rightarrow \frac{2}{3}a^{3/2} = \frac{a}{2}\left[-3\left(\cos\frac{\pi}{2} - \cos 0\right) + \frac{1}{3}\left(\cos\frac{3\pi}{2} - \cos 0\right)\right]$$

$$\Rightarrow \frac{2}{3}a^{3/2} = \frac{a}{2}\left(-3(0-1) + \frac{1}{3}(0-1)\right)$$

$$\Rightarrow \frac{2}{3}a^{3/2} = \frac{4a}{3} \Rightarrow a\sqrt{a} - 2a = 0$$

$$\Rightarrow a\left(\sqrt{a} - 2\right) = 0 \Rightarrow a = 0 \text{ or } \sqrt{a} = 2 \Rightarrow a = 0 \text{ or } a = 4$$

When a = 0:

$$\int_a^{a+1} x\, dx = \int_0^1 x\, dx = \left[\frac{x^2}{2}\right]_0^1 = \frac{1}{2}(1-0) = \frac{1}{2}$$

When a = 4:

$$\int_0^{a+1} x\, dx = \int_4^5 x\, dx = \left[\frac{x^2}{2}\right]_4^5 = \frac{1}{2}(25 - 16) = \frac{9}{2}$$

Sol 33: $\tan x = 2x$

$$\frac{1}{2}\int_0^1 [\cos(\alpha - \beta)x - \cos(\alpha + \beta)x]\, dx$$

$$\frac{1}{2}\left[\frac{\sin(\alpha - \beta)x}{(\alpha - \beta)}\Big|_0^1 - \frac{\sin(\alpha + \beta)x}{(\alpha + \beta)}\Big|_0^1\right]$$

$$\frac{1}{2}\left[\frac{\sin(\alpha - \beta)}{\alpha - \beta} - \frac{\sin(\alpha - \beta)}{(\alpha + \beta)}\right]$$

$\sin\alpha = 2\alpha\cos a$

$\sin\beta = 2\beta\cos b$

$$= \frac{1}{2}\left[\left(\frac{\frac{\sin\alpha}{2\beta}\sin\beta - \frac{\sin\alpha}{2\alpha}\sin\beta}{\alpha - \beta}\right) - \left(\frac{\frac{\sin\alpha\, isn\beta}{2\beta} + \frac{\cos\alpha\cos\beta}{2\alpha}}{(\alpha + \beta)}\right)\right]$$

$$= \frac{1}{2}\sin\alpha\sin\beta\left[\frac{1}{2\alpha\beta} - \frac{1}{2\alpha\beta}\right] = 0$$

Sol 34: $\int_0^{p+q\pi} |\cos x|\, dx$

$$\int_0^p \cos x + \int_p^{p+q\pi} |\cos x|\, dx$$

$$= \sin x\Big|_0^p + \int_0^{q\pi} (\cos x)\, dx$$

$$= \sin p + q \times 2 \int_0^{q\pi} (\cos x)\, dx$$

$$= 2q + \sin p$$

Sol 35: $\int_{-4}^{-5} e^{(x+5)^2}\, dx + 3\int_{1/3}^{2/3} e^{9\left(x-\frac{2}{3}\right)^2}\, dx$

Let $x + 5 = t$ and $3\left(x - \frac{2}{3}\right) = t$

$$= \int_1^0 e^{t^2}\, dt + (-1)\int_1^0 e^{t^2}\, dt = 0$$

Put t = −2

$$\int_0^1 e^{t^2}\, dt + (-1)\int_1^0 e^{z^2}\, dz = 0$$

Sol 36: $\sin\pi x > \cos\pi x$

$$2n\pi + \frac{\pi}{4} < \pi x < 2n\pi + \frac{\pi}{4}$$

$$2n + \frac{1}{4} < x < 2n + \frac{1}{4}$$

$$\therefore \frac{\pi}{4\sqrt{2}}\int_{-10}^{10} F(x)\, dx$$

$$= \frac{\pi}{4\sqrt{2}} \times 2 \times 10 \int f(x)\, dx$$

$$= \frac{5\pi}{\sqrt{2}}\left[\int_0^{1/4} \cos\pi x\, dx + \int_{1/4}^1 \sin\pi x\, dx\right]$$

$$= \frac{5\pi}{\sqrt{2}}\left[\frac{1}{\pi}\left[\sin\frac{\pi}{4} - 0\right] - \frac{1}{\pi}\left[\cos\pi - \cos\frac{\pi}{4}\right]\right]$$

$$= \frac{5\pi}{\sqrt{2}}\left[\frac{1}{\sqrt{2\pi}} + \frac{1}{\sqrt{2\pi}}\right] = 5$$

Sol 37: $\displaystyle\int_0^{\pi/2} \tan^{-1}\left[\frac{1+\sin x - (1-\sin x)}{1+\sin x + (1-\sin x) - 2\sqrt{1-\sin^2 x}}\right]dx$

$$= \int_0^{\pi/2} \tan^{-1}\frac{2\sin x}{2-2\cos x}dx = \int_0^{\pi/2} \tan^{-1}\frac{2\sin x}{2-2\cos x}dx$$

$$= \int_0^{\pi/2} \tan^{-1}\tan\left(\frac{\pi}{2}-\frac{x}{2}\right)dx = \int_0^{\pi/2}\left(\frac{\pi}{2}-\frac{x}{2}\right)dx$$

$$= \frac{\pi^2}{4} - \frac{1}{2}\times\frac{\pi^2}{4}\times\frac{1}{2} = \frac{\pi^2}{4} - \frac{\pi^2}{16} = \frac{3\pi^2}{16}$$

Sol 38: $x^2 + 2x = k + \displaystyle\int_0^1 |t+k|dt$

$t = k = 0 \Rightarrow dt = dU$

$$kt\int_k^{k+1} udu = \frac{1}{2}\left[(k+1)^2 - k^2\right]$$

$$= \frac{2k+1}{2}$$

$$x^2 + 2x = \frac{4k+1}{2} \Rightarrow x^2 + 2x - \left(\frac{4k+1}{2}\right) = 0$$

$$\Rightarrow x = -2 \pm \frac{\sqrt{4 + 2(4k+1)}}{2a}$$

$\Rightarrow x =$ real and distinct

Sol 39: $I = \displaystyle\int_{-1}^{1} \frac{2x^{332} + x^{998} + 4x^{1668}\sin x^{691}}{1+x^{666}}dx$

$$I = \int_{-1}^{1} \frac{2x^{332} + x^{998}}{1+x^{666}}dx = 2\int_0^1 \frac{2x^{332} + x^{998}}{1+x^{666}}dx$$

$$= 2\left[\int_0^1 \left(\frac{x^{332}}{1+x^{666}} + x^{332}\right)dx\right]$$

$$= 2\frac{1}{333} + 2\int_0^1 \frac{x^{332}}{1+(x^{333})^2}dx$$

$$= \frac{2}{333} + 2\left[\int_0^1\left(\frac{dt}{1+t^2}\right)\right]\frac{1}{333}$$

$$= \frac{2}{333} + \frac{2}{333}\tan^{-1}t\Big|_0^1 = \frac{2}{333}\left[1+\frac{\pi}{4}\right] = \frac{\pi+4}{666}$$

Sol 40: $2I = \pi\displaystyle\int_0^{\pi} \frac{[x^2 - (x-\pi)^2]\sin 2x\sin\left(\frac{\pi}{2}\cos x\right)dx}{2x-\pi}$

$$= \pi^2\int_0^{\pi} \sin 2x\sin\left(\frac{\pi}{2}\cos x\right)dx$$

Let $\dfrac{\pi}{2}\cos x = t \Rightarrow -\dfrac{\pi}{2}\sin x\, dx = dt$

$$= -\pi^2 \times \frac{2}{\pi}\int_{\pi/2}^{-\pi/2} 2\cdot\frac{2}{\pi}t\sin t\, dt$$

$$2I = 8\int_{-\pi/2}^{\pi/2} t\sin t\, dt$$

$$I = 4\left[-t\cos t\Big|_{-\pi/2}^{\pi/2} + \int_{-\pi/2}^{\pi/2}\cos t\, dt\right] = 4[+2] = 8$$

Sol 41: $\displaystyle\int_{1/3}^{1} \frac{\left(x-x^3\right)^{1/3}}{x^4}dx = \int_{1/3}^{1}\frac{\left[x^3\left(\frac{1}{x^2}-1\right)\right]^{1/3}}{x^4}$

$$dx = \int_{1/3}^{1}\frac{\left(\frac{1}{x^2}-1\right)^{1/3}}{x^3}dx$$

Put $\dfrac{1}{x^2} - 1 = t$, then $-\dfrac{2}{x^3}dx = dt$ or $\dfrac{1}{x^3}dx = -\dfrac{1}{2}dt$

When $x = 1, t = \dfrac{1}{1^2} - 1 = 0$ and when $x = \dfrac{1}{3}, t = 9 - 1 = 8$

Now, $\displaystyle\int_{1/3}^{1}\frac{\left(x-x^3\right)^{1/3}}{x^4}dx = -\frac{1}{2}\int_8^0 t^{1/3}\, dt = -\frac{1}{2}\left[\frac{t^{4/3}}{\frac{4}{3}}\right]_8^0$

$$= -\frac{3}{8}\left[0 - 8^{4/3}\right] = -\frac{3}{8}\left[-2^4\right] = -\frac{3}{8}(-16) = 6$$

Sol 42: $\displaystyle\lim_{n\to\infty}\frac{1}{n^2}\left[\sum_{k=0}^{n-1}k\int_k^{k+1}\sqrt{(x-k)(k+1-x)}dx\right]$

$x - k = t$

$$\int_0^1 \sqrt{t(1-t)}dt = \int_0^1\sqrt{\left(\frac{1}{2}\right)^2 - (t-1)^2}dx$$

$$= \frac{1}{2}\left(t-\frac{1}{2}\right)\sqrt{\frac{1}{4}-\left(t-\frac{1}{2}\right)^2} + \frac{1}{8}\sin^{-1}\frac{\left(t-\frac{1}{2}\right)}{\frac{1}{2}}\Bigg|_0^1$$

$$= \frac{\pi}{2} \times \frac{1}{8} + \frac{\pi}{2} \times \frac{1}{8} = \frac{\pi}{8}$$

$$\therefore \lim_{n \to \infty} \frac{1}{n^2} \sum_{k=0}^{n-1} k \times \frac{\pi}{8} = \int_0^1 \frac{\pi}{8} x \, dx = \frac{\pi}{8} \times \frac{1}{2} = \frac{\pi}{16}$$

Sol 43: $I = \int_0^{\pi/2} \frac{\sin x + 3}{5 \sin(x + \alpha) + 25} dx \quad \cos \alpha = \frac{3}{5}$

$$4I + 3J = \int_0^{\pi/2} \frac{4\cos x + 3\sin x + 25}{4\cos x + 3\sin x + 25} dx = \frac{\pi}{2}$$

$$3I - 4J = \int_0^{\pi/2} \frac{3\cos x - 4\sin x}{4\cos x + 3\sin x + 25} dx$$

$$= \log(4\cos x + 3\sin x + 25)\Big|_0^{\pi/2}$$

$$= \log(28) - \log(29) = \log \frac{28}{29}$$

$$16I + 9I = 2\pi + 3\log \frac{28}{29} = 2\pi + 3\log \frac{28}{29}$$

$$a + b + c + d = 2 + 3 + 28 + 29 = 62$$

Sol 44: $f(x) = ax^2 + bx + c$

$f'(x) = 2ax + b$

$f'(2) = 4a + b = 1$

$f'(2) = 4a + b = 1$

$$\int_{2-\pi}^{2+\pi} f(x) \sin\left(\frac{x-2}{2}\right) dx = -\int_{2-\pi}^{2+\pi} f(4-x) \sin \frac{(x-2)}{2} dx$$

$$2I = -\int_{2-\pi}^{2+\pi} \left[f(x) - f(4-x) \right] \sin \frac{(x-2)}{2} dx$$

$$= -\int_{2-\pi}^{2+\pi} \left\{ \begin{array}{l} ax^2 + bx + c \\ \quad - \left[a(4-x)^2 + b(4-x) + c \right] \end{array} \right\} \sin\left(\frac{x-2}{2}\right) dx$$

$$= \int_{2-\pi}^{2+\pi} \left[\begin{array}{l} a(x-4+x)(x+4-x) \\ \quad +b(x-4+x) \end{array} \right] \sin\left(\frac{x-2}{2}\right) dx$$

$$= \int_{2-\pi}^{2+\pi} \left(a(2x-4)4 + 2bx - 4b \right) \sin\left(\frac{x-2}{2}\right) dx$$

$$= \int_{2-\pi}^{2+\pi} (8ax + 2bx - 4) \sin\frac{(x-2)}{2} dx$$

$$= 4\int_{2-\pi}^{2+\pi} \frac{(x+2)}{2} \sin\left(\frac{x-2}{2}\right) dx \quad \frac{x-2}{2} = t$$

$$= 4\int_{-\pi/2}^{\pi/2} t\sin t \, dt = 4\left[t(-\cos t)\Big|_{-\pi/2}^{\pi/2} + \int_{-\pi/2}^{\pi/2} \cos t \, dt \right]$$

$$= 4\left[\sin t \Big|_{-\pi/2}^{\pi/2} \right] = 8$$

Exercise 2

Single Correct Choice Type

Sol 1: (C) $\int_0^2 |(x+3)(x-1)| dx$

$$= \int_0^1 (x+3)(1-x)dx + \int_1^2 (x+3)(x-1)dx$$

$$= -\int_1^1 (x^2 + 2x - 3)dx + \int_1^2 (x^2 + 2x - 3)dx$$

$$= -\left[\frac{x^3}{3} + x^2 - 3x \Big|_0^1 \right] + \left[\frac{x^3}{3} + x^2 - 3x \Big|_1^2 \right]$$

$$= -\left[\frac{1}{3} + 1 - 3 \right] + \left[\frac{8}{3} + 4 - 6 - \left(\frac{1}{3} + 1 - 3 \right) \right]$$

$$= \frac{5}{3} + \frac{2}{3} + \frac{5}{3} = 4$$

Sol 2: (B) $\int_0^{\pi/2} \left| \frac{1}{\sqrt{2}} \sin x - \frac{1}{\sqrt{2}} \cos x \right| dx$

$$= \frac{1}{\sqrt{2}} \left[\int_0^{\pi/4} (\cos x - \sin x)dx + \int_{\pi/4}^{\pi/2} (\sin x - \cos x)dx \right]$$

$$= \frac{1}{\sqrt{2}} \left[\sin x + \cos x \Big|_0^{\pi/4} + (-\cos x - \sin x)\Big|_{\pi/4}^{\pi/2} \right]$$

$$= \frac{1}{\sqrt{2}} \left[\left[\frac{1}{\sqrt{2}} + \frac{1}{\sqrt{2}} - 1 \right] - \left[1 - \left(\frac{1}{\sqrt{2}} + \frac{1}{\sqrt{2}} \right) \right] \right]$$

$$= \frac{1}{\sqrt{2}} \left[\sqrt{2} - 1 - 1 + \sqrt{2} \right] = \frac{2\sqrt{2} - 2}{\sqrt{2}} = 2 - \sqrt{2}$$

Sol 3: (D) $2\int_0^{\pi/2} (\sin^4 x)dx = 2 \times \left(\frac{(4-1)(4-3)}{4 \times (4-2)} \right) \times \frac{\pi}{2} = \frac{3\pi}{8}$

Sol 4: (B) $\int_0^1 0dx = \int_1^{\sqrt{2}} 1dx + \int_{\sqrt{2}}^{1.5} 2dx$

$$= \sqrt{2} - 1 + 2(1.5 - \sqrt{2}) = 2 - \sqrt{2}$$

Sol 5: (A) $I = \int\limits_0^\pi \dfrac{x}{a^2 \cos^2 x + b^2 \sin^2 x}\, dx$ ...(i)

Then $I = \int\limits_0^\pi \dfrac{(\pi - x)}{a^2 \cos^2(\pi - x) + b^2 \sin^2(\pi - x)}\, dx$

$$\left[\because \int\limits_0^a f(x)\,dx = \int\limits_0^a f(a-x)\,dx \right]$$

Or $I = \int\limits_0^\pi \dfrac{\pi - x}{a^2 \cos^2 x + b^2 \sin^2 x}\, dx$ ...(ii)

Adding (i) and (ii), we get

$$2I = \int\limits_0^\pi \dfrac{x + \pi - x}{a^2 \cos^2 x + b^2 \sin^2 x}\, dx$$

$$= \pi \int\limits_0^\pi \dfrac{1}{a^2 \cos^2 x + b^2 \sin^2 x}\, dx$$

$$= 2\pi \int\limits_0^{\pi/2} \dfrac{1}{a^2 \cos^2 x + b^2 \sin^2 x}\, dx$$

$$\left[\because \int\limits_0^{2a} f(x)\,dx = 2\int\limits_0^a f(x)\,dx,\ \text{if } f(2a-x) = f(x) \right]$$

$$\therefore 2I = 2\pi \int\limits_0^{\pi/2} \dfrac{\sec^2 x}{a^2 + b^2 \tan^2 x}\, dx$$

[Dividing num. and denom. By $\cos^2 x$]

Or $I = \pi \int\limits_0^{\pi/2} \dfrac{\sec^2 x}{a^2 + b^2 \tan^2 x}\, dx$

Let $\tan x = z$. Then, $\sec^2 x\, dx = dz$

Also $x = 0 \Rightarrow z = \tan 0 = 0$ and $x \to \dfrac{\pi}{2} \Rightarrow z \to \tan \dfrac{\pi}{2}$

or $z \to \infty$

$$\therefore I = \pi \int\limits_0^\infty \dfrac{dz}{a^2 + b^2 z^2} = \dfrac{\pi}{b^2} \int\limits_0^\infty \dfrac{dz}{(a/b)^2 + z^2}$$

$$= \dfrac{\pi}{b^2} \times \dfrac{1}{(a/b)}\left[\tan^{-1}\left(\dfrac{z}{a/b}\right) \right]_0^\infty$$

$$\Rightarrow I = \dfrac{\pi}{ab}\left[\tan^{-1}\left(\dfrac{bz}{a}\right) \right]_0^\infty = \dfrac{\pi}{ab}\left(\tan^{-1}\infty\ \tan^{-1} 0 \right)$$

$$= \dfrac{\pi}{ab}\left(\dfrac{\pi}{2} - 0 \right) = \dfrac{\pi^2}{2ab}$$

Sol 6: (A) $\int\limits_0^{\pi/4} \dfrac{\sec x}{1 + 2\sin^2 x}\, dx$

$$= \int\limits_0^{\pi/4} \dfrac{dx}{\cos x + 2\sin^2 x \cos x}$$

$$= \int\limits_0^{\pi/4} \dfrac{\cos x\ dx}{\cos^2 x + 2\sin^2 x \cos^2 x}$$

$$= \int\limits_0^{\pi/4} \dfrac{\cos x\, dx}{(1 - \sin^2 x)(1 + 2\sin^2 x)}$$

$$= \int\limits_0^{1/\sqrt{2}} \dfrac{dt}{(1 - t^2)(1 + 2t^2)}$$

$$= \dfrac{1}{3} \int\limits_0^{1/\sqrt{2}} \left(\dfrac{1}{1 - t^2} + \dfrac{2}{1 + 2t^2} \right) dt$$

$$= \dfrac{1}{3} \int\limits_0^{1/\sqrt{2}} \dfrac{1}{1 - t^2}\, dt + \dfrac{1}{3} \int\limits_0^{1/\sqrt{2}} \dfrac{1}{\dfrac{1}{2} + t^2}\, dt$$

$$= \dfrac{1}{6} \ell n \dfrac{1+t}{1-t}\bigg|_0^{1/\sqrt{2}} + \dfrac{1}{3} \times \dfrac{1}{1/\sqrt{2}} \tan^{-1}\sqrt{2}t \bigg|_0^{1/\sqrt{2}}$$

$$= \dfrac{1}{6} \ell n \dfrac{\sqrt{2}+1}{\sqrt{2}-1} + \dfrac{\sqrt{2}}{3} \times \dfrac{\pi}{4} = \dfrac{1}{3}\left[\ell n(\sqrt{2}+1) + \dfrac{\pi}{2\sqrt{2}} \right]$$

Sol 7: (C) $\int\limits_0^1 e^{x^2}(x - \alpha)\,dx = 0$

For this integral to be zero

If $\alpha < 0$ then $x - \alpha$ when $x \in (0, 1) > 0$

$\therefore$ It is not possible that integral reduce to zero

If $2 > \alpha > 1$ then $x - a$

when $x \in (0, 1) < 0$ function gives negative value and so cannot reduced zero.

$\therefore$ If $0 < \alpha < 1$, f^n can take both positive and negative valuce and it is possible that integral reduced to zero

Sol 8: (A) $\int\limits_0^{\pi/2} \{x - [\sin x]\}\,dx$

[ ] $\to$ greatest integer function

$[\sin x] = 0$ $\sin x \in [0, 1)$ i.e. $x \in \left[0, \dfrac{\pi}{2}\right)$

$$= \int\limits_0^{\pi/2} (x - 0)\,dx = \dfrac{x^2}{2}\bigg|_0^{\pi/2} = \dfrac{\pi^2}{8}$$

Sol 9: (B) $\int\limits_{0}^{100} \sin\left(x[x]\right)\pi dx$

Since $x - [x]$ has a period of 1

$\therefore I = 100\int\limits_{0}^{1} \sin\pi x dx = \dfrac{100}{\pi}\left(-\cos\pi x\Big|_{0}^{1}\right)$

$= \dfrac{100}{\pi}(-(-1-1)) = \dfrac{200}{\pi}$

Sol 10: (B) $\int\limits_{0}^{\infty} \dfrac{x\log x}{(1+x^2)^2}dx$

$x = \tan\theta \Rightarrow dx = \sec^2\theta d\theta$

$I = \int\limits_{0}^{\pi/2} \dfrac{\tan\theta\log(\tan\theta)\sec^2\theta d\theta}{\sec^4\theta}$

$= \int\limits_{0}^{\pi/2} \tan\theta\log(\tan\theta)\cos^2\theta d\theta$

$= \int\limits_{0}^{\pi/2} \sin\theta\cos\theta\log(\tan\theta)d\theta$

$= \int\limits_{0}^{\pi/2} \cos\theta\sin\theta\log\cot\theta d\theta$

$\therefore 2I = \int\limits_{0}^{\pi/2} \sin\theta\cos\theta\left[\log\tan\theta + \log\cot\theta\right]d\theta$

$= \int\limits_{0}^{\pi/2} \sin\theta\cos\theta\left[\log\tan - \log\tan\theta\right]d\theta = 0$

Sol 11: (B) $\log I = \lim\limits_{n\to\infty}\sum\limits_{r=1}^{n} \log\left(1+\dfrac{r}{n}\right)\times\dfrac{1}{n}$

$\log I = \int\limits_{0}^{1} \log(1+x)dx$

$\log I = x\log(1+x)\Big|_{0}^{1} - \int\limits_{0}^{1} \dfrac{x}{1+x}dx$

$= \log 2 - \int\limits_{0}^{1} 1 - \dfrac{1}{1+x}dx = \log 2 - [1] + \log(1+x)\Big|_{0}^{1}$

$= 2\log 2 - \log e = \log\dfrac{4}{e} \Rightarrow I = \dfrac{4}{e}$

Sol 12: (A) $\int\limits_{\log 2}^{x} \dfrac{1}{\sqrt{e^x-1}}dx$

Put $e^x - 1 = t^2 \Rightarrow e^x dx = 2t dt$ or $dx = \dfrac{2t}{1+t^2}dt$

$\int\limits_{1}^{\sqrt{e^x-1}} \dfrac{2}{1+t^2}dt = 2\tan^{-1}t\Big|_{1}^{\sqrt{e^x-1}} = \dfrac{\pi}{6}$

$2\left[\tan^{-1}t - \dfrac{\pi}{4}\right] = \dfrac{\pi}{6}$

$\therefore \tan^{-1}t = \dfrac{\pi}{3} \Rightarrow \therefore t = \tan\dfrac{\pi}{3} = \sqrt{3}$

$\therefore e^x - 1 = 3 \Rightarrow x = \log 4$

Sol 13: (A) $\lim\limits_{n\to\infty}\sum\limits_{r=1}^{n} \dfrac{r^2}{r^3+n^3} = \lim\limits_{n\to\infty}\sum\limits_{r=1}^{n} \dfrac{\left(\dfrac{r}{n}\right)^2}{\left(\dfrac{r}{n}\right)^3+1}\times\dfrac{1}{n}$

$= \int\limits_{0}^{1} \dfrac{x^2}{x^3+1}dx = \dfrac{1}{3}\ell n(x^3+1)\Big|_{0}^{1} = \dfrac{\ell n2}{3}$

Sol 14: (B) $\lim\limits_{n\to\infty}\left[\left(\sum\limits_{r=1}^{n} \dfrac{n^2}{(n^2+r^2)^{3/2}}\right) - \dfrac{1}{2n}\right]$

$= \lim\limits_{n\to\infty}\sum\limits_{r=1}^{n} \dfrac{1}{\left(1+\left(\dfrac{r}{n}\right)^2\right)^{3/2}}\times\dfrac{1}{n} = \int\limits_{0}^{1} \dfrac{1}{\left(1+x^2\right)^{3/2}}dx$

Put $x = \tan\theta \Rightarrow dx = \sec^2\theta d\theta$

$\dfrac{\pi}{4}\int \dfrac{\sec^2\theta d\theta}{\sec^3\theta} = \int\limits_{0}^{\pi/4} \cos\theta d\theta = \sin\theta\Big|_{0}^{\pi/4} = \dfrac{1}{\sqrt{2}}$

Sol 15: (A) $\lim\limits_{n\to\infty}\sum\limits_{r=1}^{n} \dfrac{\sqrt{n}}{\sqrt{[n+4(r-1)]^3}} - \dfrac{\sqrt{n}}{\sqrt{(n+4n)^3}}$

$\lim\limits_{n\to\infty}\sum\limits_{r=1}^{n} \dfrac{1}{\sqrt{\left(1+\dfrac{4r}{n}\right)^3}}\times\dfrac{1}{n} - \dfrac{1}{(5)^{3/2}\times n^{1/2}}$

$= \int\limits_{0}^{1} \dfrac{1}{\sqrt{(1+4x)^3}}dx = \int\limits_{0}^{1} (1+4x)^{-3/2}dx$

$= \dfrac{(1+4x)^{-1/2}}{-1/2}\times\dfrac{1}{4}\Big|_{0}^{1} = -\dfrac{1}{2}\left(\dfrac{1}{1+4x}\right)^{1/2}\Big|_{0}^{1}$

$= -\dfrac{1}{2}\left[\dfrac{1}{\sqrt{5}}-1\right] = \dfrac{1}{-0}\left(5-\sqrt{5}\right)$

Sol 16: (B) $\log I = \lim\limits_{n\to\infty}\left(\sum\limits_{r=1}^{n}\log\tan\left(\dfrac{\pi r}{2n}\right)\right)\dfrac{1}{n}$

$= \displaystyle\int_{0}^{1}\log\tan\left(\dfrac{\pi}{2}x\right)dx$

$\dfrac{\pi}{2}x = t \Rightarrow dx = \dfrac{2}{\pi}dt$

$\Rightarrow \displaystyle\int_{0}^{\pi/2}\log\tan t\,dt = 0$

$\therefore I = e^{0} = 1$

Previous Years' Questions

Sol 1: (A) $\displaystyle\int_{-1/2}^{1/2}\left([x]+\log\left(\dfrac{1+x}{1-x}\right)\right)dx$

$= \displaystyle\int_{-1/2}^{1/2}[x]dx + \int_{-1/2}^{1/2}\log\left(\dfrac{1+x}{1-x}\right)dx$

$= \displaystyle\int_{-1/2}^{1/2}[x]dx + 0$

$\left[\because \log\left(\dfrac{1+x}{1-x}\right)\text{is an odd function}\right]$

$= \displaystyle\int_{-1/2}^{0}[x]dx + \int_{0}^{1/2}[x]dx$

$= \displaystyle\int_{-1/2}^{0}(-1)dx + \int_{0}^{1/2}(0)dx$

$= \Big[x\Big]_{-1/2}^{0}$

$= -\left(0+\dfrac{1}{2}\right) = -\dfrac{1}{2}$

Sol 2: (A) Here, $I(m,n) = \displaystyle\int_{0}^{1}t^{m}(1+t)^{n}dt$ reduce into $I(m+1, n-1)$ [we apply integration by parts taking $(1+t)^{n}$ as first and t^{m} as second function]

$\therefore I(m,n) = \left[(1+t)^{n}\cdot\dfrac{t^{m+1}}{m+1}\right]_{0}^{1}$

$\qquad -\displaystyle\int_{0}^{1}n(1+t)^{(n-1)}\cdot\dfrac{t^{m+1}}{m+1}dt$

$= \dfrac{2^{n}}{m+1} - \dfrac{n}{m+1}\displaystyle\int_{0}^{1}(1+t)^{(n-1)}\cdot t^{m+1}dt$

$\therefore I(m,n) = \dfrac{2^{n}}{m+1} - \dfrac{n}{m+1}\cdot I(m+1, n-1)$

Sol 3: (C) Given $\displaystyle\int_{0}^{x}\sqrt{1-(f'(t))^{2}}\,dt$

$= \displaystyle\int_{0}^{x}f(t)dt,\ 0\le x\le 1$

Differentiating both sides w.r.t.x by using Leibnitz rule, we get

$\sqrt{1-(f'(x))^{2}} = f(x)$

$\Rightarrow f'(x) = \pm\sqrt{1-(f(x))^{2}}$

$\Rightarrow \displaystyle\int\dfrac{f'(x)}{\sqrt{1-(f(x))^{2}}}dx = \pm\int dx$

$\Rightarrow \sin^{-1}(f(x)) = \pm x + c$

Put $x = 0$

$\Rightarrow \sin^{-1}(f(0)) = c$

$\Rightarrow c = \sin^{-1}(0) = 0 \qquad (\because f(0) = 0)$

$\therefore f(x) = \pm\sin x$

but $f(x) \ge 0,\ \forall\ x \in [0, 1]$

$\therefore f(x) = \sin x$

As we know that,

$\sin x < x\ \forall\ x > 0$

$\therefore \sin\left(\dfrac{1}{2}\right) < \dfrac{1}{2}$ and $\sin\left(\dfrac{1}{3}\right) < \dfrac{1}{3}$

$\Rightarrow f\left(\dfrac{1}{2}\right) < \dfrac{1}{2}$ and $f\left(\dfrac{1}{3}\right) < \dfrac{1}{3}$

Sol 4: (A) $x^{2} = t \Rightarrow 2x\,dx = dt$

$\therefore I = \dfrac{1}{2}\displaystyle\int_{\log 2}^{\log 3}\dfrac{\sin t}{\sin t + \sin(\log 6 - t)}dt$

$I = \dfrac{1}{2}\displaystyle\int_{\log 2}^{\log 3}\dfrac{\sin(\log 6 - t)}{\sin(\log 6 - t) + \sin t}dt$

$2I = \dfrac{1}{2}\displaystyle\int_{\log 2}^{\log 3}1\,dt \Rightarrow I\,\dfrac{1}{4}\log\dfrac{3}{2}$

Sol 5: (A, D) Gives, $S_n = \sum_{k=0}^{n} \dfrac{n}{n^2 + kn + k^2}$

$= \sum_{k=0}^{n} \dfrac{1}{n} \left(\dfrac{1}{1 + \dfrac{k}{n} + \dfrac{k^2}{n^2}} \right) < \lim_{n \to \infty} \sum_{k=0}^{n} \dfrac{1}{n}$

$\left(\dfrac{1}{1 + \dfrac{k}{n} + \left(\dfrac{k}{n}\right)^2} \right)$

$= \int_0^1 \dfrac{1}{1+x+x^2}\, dx = \left[\dfrac{2}{\sqrt{3}} \tan^{-1}\left(\dfrac{2}{\sqrt{3}}\left(x+\dfrac{1}{2}\right)\right) \right]_0^1$

$= \dfrac{2}{\sqrt{3}} \cdot \left(\dfrac{\pi}{3} - \dfrac{\pi}{6}\right) = \dfrac{\pi}{3\sqrt{3}}$

i.e. $S_n < \dfrac{\pi}{3\sqrt{3}}$

Similarly, $T_n > \dfrac{\pi}{3\sqrt{3}}$

Sol 6: (A, B, C) Given $I_n = \int_{-\pi}^{\pi} \dfrac{\sin nx}{(1+\pi^x)\sin x}\, dx$ ...(i)

Using $\int_a^b f(x)\,dx = \int_a^b f(b+a-x)\,dx$

we get $I_n = \int_{-\pi}^{\pi} \dfrac{\pi^x \sin nx}{(1+\pi^x)\sin x}\, dx$ ...(ii)

Adding Eqs. (i) and (ii), we get

$2I_n = \int_{-\pi}^{\pi} \dfrac{\sin nx}{\sin x}\, dx = 2\int_0^{\pi} \dfrac{\sin nx}{\sin x}\, dx$

$(\because f(x) = \dfrac{\sin nx}{\sin x}$ is an even function$)$

$\Rightarrow I_n = \int_0^{\pi} \dfrac{\sin nx}{\sin x}\, dx$

Now, $I_{n+2} - I_n = \int_0^{\pi} \dfrac{\sin(n+2)x - \sin nx}{\sin x}\, dx$

$= \int_0^{\pi} \dfrac{2\cos(n+1)x.\sin x}{\sin x}\, dx$

$= 2\int_0^{\pi} \cos(n+1)x\, dx$

$= 2\left[\dfrac{\sin(n+1)x}{(n+1)} \right]_0^{\pi} = 0$

$\therefore I_{n+2} = I_n$.   ...(iii)

Since, $I_n = \int_0^{\pi} \dfrac{\sin nx}{\sin x}\, dx \Rightarrow I_1 = \pi$ and $I_2 = 0$

$\therefore$ From Eq. (iii) $I_1 = I_3 = I_5 = ... = p$
and $I_2 = I_4 = I_6 = ... = 0$

$\Rightarrow \sum_{m=1}^{10} I_{2m+1} = 10\pi$ and $\sum_{m=1}^{10} I_{2m} = 0$

$\therefore$ Correct options are A, B, C.

Sol 7: (A) Let $I = \int_0^1 \dfrac{x^4(1-x)^4}{1+x^2}\, dx$

$= \int_0^1 \dfrac{(x^4 -1)(1-x)^4 + (1-x)^4}{(1+x^2)}\, dx$

$= \int_0^1 (x^2 - 1)(1-x)^4\, dx + \int_0^1 \dfrac{(1+x^2 - 2x)^2}{(1+x^2)}\, dx$

$= \int_0^1 \left\{ (x^2 - 1)(1-x)^4 + (1+x^2) - 4x + 4 - \dfrac{4x^2}{(1+x^2)} \right\} dx$

$= \int_0^1 \left((x^2 - 1)(1-x)^4 + (1+x^2) - 4x + 4 - \dfrac{4}{1-x^2} \right) dx$

$= \int_0^1 \left(x^6 - 4x_1^5 + 5x^4 - 4x^2 + 4 - \dfrac{4}{1+x^2} \right) dx$

$= \int_0^1 \left[\dfrac{x^7}{7} - \dfrac{4x^6}{6} + \dfrac{5x^5}{5} - \dfrac{4x^3}{3} + 4x - 4\tan^{-1} x \right]_0^1$

$= \dfrac{1}{7} - \dfrac{4}{6} + \dfrac{5}{5} - \dfrac{4}{3} + 4 - 4\left(\dfrac{\pi}{4} - 0\right) = \dfrac{22}{7} - \pi$

$= (\text{cosec } x.\cot x + \sec^2 x - \cos x)$

$. (\cos^3 x - \cos x) . \cos x$

$= -\left[\dfrac{\sin^2 x + \cos^3 x - \cos^3 x.\sin^2 x}{\sin^2 x.\cos^2 x} \right]$

$. \cos^2 x . \sin^2 x$

$= -\sin^2 x - \cos^3 x (1 - \sin^2 x)$

$= -\sin^2 x - \cos^5 x$

$\therefore \int_0^{\pi/2} f(x)\,dx = -\int_0^{\pi/2} (\sin^2 x + \cos^5 x)\,dx$

$$\left[\because \int_0^{\pi/2} \sin^m x . \cos^n x\, dx = \frac{\sqrt{\frac{m+1}{2}}\sqrt{\frac{n+2}{2}}}{2\sqrt{\frac{m+n+2}{2}}}\right]$$

$$\therefore \int_0^{\pi/2} f(x)dx = -\left\{\frac{\sqrt{\frac{3}{2}}.\sqrt{\frac{1}{2}}}{2\sqrt{2}} + \frac{\sqrt{\frac{6}{2}}.\sqrt{\frac{1}{2}}}{2\sqrt{\frac{7}{2}}}\right\}$$

Sol 8: (C) $\int_0^{\pi/2} \sin x\, dx$

$$= \frac{\frac{\pi}{2}-0}{4}\left(\sin 0 + \sin\left(\frac{\pi}{2}\right) + 2\sin\left(\frac{0+\frac{\pi}{2}}{2}\right)\right)$$

$$= \frac{\pi}{8}(1+\sqrt{2})$$

Sol 9: (A) $F'(c) = (b-a) f'(c) + f(a) - f(b)$

$F''(c) = f''(c)(b-a) < 0$

$\Rightarrow F'(c) = 0$

$\Rightarrow f'(c) = \dfrac{f(b)-f(a)}{b-a}$

Sol 10: (B) Given, $\displaystyle\lim_{t\to a} \frac{\int_a^t f(x)dx - \frac{(t-a)}{2}\{f(t)+f(a)\}}{(t-a)^3} = 0$

Using L' Hospital's rule

$\Rightarrow \displaystyle\lim_{h\to 0} \frac{\int_a^{a+h} f(x)dx - \frac{h}{2}\{f(a+h)+f(a)\}}{h^3} = 0$

$$f(a+h) - \frac{1}{2}\{f(a+h)+f(a)\}$$

$\Rightarrow \displaystyle\lim_{h\to 0} \frac{-\frac{h}{2}\{f'(a+h)\}}{3h^2} = 0$

Again, using L'Hospital's rule

$$f'(a+h) - \frac{1}{2}f'(a+h) - \frac{1}{2}f'(a+h)$$

$\Rightarrow \displaystyle\lim_{h\to 0} \frac{-\frac{h}{2}f''(a+h)}{6h} = 0$

$\Rightarrow \displaystyle\lim_{h\to 0} \frac{-\frac{h}{2}f''(a+h)}{6h} = 0$

$\Rightarrow f''(a) = 0, \forall a \Rightarrow R$

$\Rightarrow f(x)$ must have maximum degree 1

Sol 11: Given, $f(x) = \begin{cases} x-[x] & \text{if }[x]\text{ is odd} \\ 1+[x]-x & \text{if }[x]\text{ is even} \end{cases}$

$f(x)$ and $\cos \pi x$ both are periodic with period 2 and both are even.

$$\therefore \int_{-10}^{10} f(x)\cos \pi x\, dx = 2\int_0^{10} f(x)\cos \pi x\, dx$$

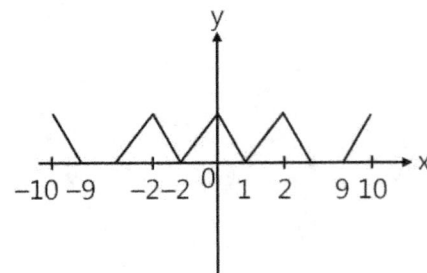

$$= 10\int_0^3 f(x)\cos \pi x\, dx$$

Now, $\displaystyle\int_0^1 f(x)\cos \pi x\, dx$

$$= \int_0^1 (1-x)\cos \pi x\, dx = -\int_0^1 u\cos \pi u\, du \text{ and}$$

$$\int_1^2 f(x)\cos \pi x\, dx = \int_1^2 (x-1)\cos \pi x\, dx = -\int_0^1 u\cos \pi u\, du$$

$$\therefore \int_{-10}^{10} f(x)\cos \pi x\, dx = -20\int_0^1 u\cos \pi u\, du = \frac{40}{\pi^2}$$

$$\Rightarrow \frac{\pi^2}{10}\int_{-10}^{10} f(x)\cos \pi x\, dx = 4$$

Sol 12: $f(x) = \displaystyle\int_1^x \frac{\ln t}{1+t}\, dt$ for $x > 0$ (given)

Now, $f(1/x) = \displaystyle\int_1^{1/x} \frac{\ln t}{1+t}\, dt$

Put $t = 1/u$

$\Rightarrow dt = (-1/u^2)du$

$\therefore f(1/x) = \displaystyle\int_1^x \frac{\ln(1/u)}{1+1/u} . \frac{(-1)}{u^2}\, du$

$$= \int_1^x \frac{\ln u}{u(u+1)}\, du = \int_1^x \frac{\ln t}{t(1+t)}\, dt$$

Now, $f(x) + f\left(\dfrac{1}{x}\right) = \displaystyle\int_1^x \dfrac{\log t}{(1+t)}dt + \int_1^x \dfrac{\log t}{(1+t)}dt$

$= \displaystyle\int_1^x \dfrac{(1+t)\log t}{t(1+t)}dt + \int_1^x \dfrac{x\log t}{t}dt$

$= \dfrac{1}{2}\Big[(\log t)^2\Big]_1^x = \dfrac{1}{2}(\log x)^2$

Put $x = e$

$\therefore\ f(e) + f\left(\dfrac{1}{e}\right) = \dfrac{1}{2}(\log e)^2 = \dfrac{1}{2}$

Hence proved.

Sol 13: Let $I = \displaystyle\int_0^{\pi/2} f(\cos 2x)\cos x\,dx$ ... (i)

$I = \displaystyle\int_0^{\pi/2} f\left(\cos 2\left(\dfrac{\pi}{2}-x\right)\right)\cos\left(\dfrac{\pi}{2}-x\right)dx$

$\left[\text{using}\displaystyle\int_0^a f(x)dx = \int_0^a f(a-x)dx\right]$

$I = \displaystyle\int_0^{\pi/2} f(\cos 2x)\sin x\,dx$ .. (ii)

On adding Eqs. (i) and (ii), we get

$2I = \displaystyle\int_0^{\pi/2} f(\cos 2x)(\sin x + \cos x)dx$

$= \sqrt{2}\displaystyle\int_0^{\pi/2} f(\cos 2x)[\cos(x - \pi/4)]dx$

Put $-x + \dfrac{\pi}{4} = t \Rightarrow -dx = dt$

$\therefore\ 2I = -\sqrt{2}\displaystyle\int_{\pi/4}^{-\pi/4} f\left(\cos\left(\dfrac{\pi}{2}-2t\right)\right)\cos t\,dt$

$\therefore\ 2I = \sqrt{2}\displaystyle\int_{-\pi/4}^{\pi/4} f(\sin 2t)\cot t\,dt$

$\therefore\ I = \sqrt{2}\displaystyle\int_0^{\pi/4} f(\sin 2t)\cos t\,dt$

Sol 14:

Let $I = \displaystyle\int_{-\pi/3}^{\pi/3} \dfrac{\pi\,dx}{2-\cos\left(|x|+\dfrac{\pi}{3}\right)} + 4\int_{-\pi/3}^{\pi/3} \dfrac{x^3\,dx}{2-\cos\left(|x|+\dfrac{\pi}{3}\right)}$

Using $\displaystyle\int_{-a}^a f(x)dx = \begin{cases} 0, & f(-x) = -f(x) \\ 2\displaystyle\int_0^a f(x)dx, & f(-x) = f(x) \end{cases}$

$\therefore\ I = 2\displaystyle\int_0^{\pi/3} \dfrac{\pi\,dx}{2-\cos\left(|x|+\dfrac{\pi}{3}\right)} + 0$

$\left[\dfrac{x^3 dx}{2-\cos\left(|x|+\dfrac{\pi}{3}\right)}\ \text{is odd}\right]$

$I = 2\pi\displaystyle\int_0^{\pi/3} \dfrac{dx}{2-\cos(x+\pi/3)}$

Put $x + \dfrac{\pi}{3} = t \Rightarrow dx = dt$

$\therefore\ I = 2\pi\displaystyle\int_{\pi/3}^{2\pi/3} \dfrac{dt}{2-\cos t} = 2\pi\int_{\pi/3}^{2\pi/3} \dfrac{\sec^2\dfrac{t}{2}dt}{1+3\tan^2\dfrac{t}{2}}$

Put $\tan\dfrac{t}{2} = u \Rightarrow \sec^2\dfrac{t}{2}dt = 2du$

$\Rightarrow I = 2\pi\displaystyle\int_{1/\sqrt{3}}^{\sqrt{3}} \dfrac{2\,du}{1+3u^2} = \dfrac{4\pi}{3}[\sqrt{3}\tan^{-1}\sqrt{3}u]_{\frac{1}{\sqrt{3}}}^{\sqrt{3}}$

$= \dfrac{4\pi}{\sqrt{3}}(\tan^{-1}3 - \tan^{-1}1) = \dfrac{4\pi}{\sqrt{3}}\tan^{-1}\left(\dfrac{1}{2}\right)$

$\therefore\ \displaystyle\int_{-\pi/3}^{\pi/3} \dfrac{\pi+4x^3}{2-\cos\left(|x|+\dfrac{\pi}{3}\right)}dx = \dfrac{4\pi}{\sqrt{3}}\tan^{-1}\left(\dfrac{1}{2}\right)$

Sol 15: Let

$I = \displaystyle\int_0^\pi e^{|\cos x|}\left(2\sin\left(\dfrac{1}{2}\cos x\right) + 3\cos\left(\dfrac{1}{2}\cos x\right)\right)\sin x\,dx$

$\Rightarrow I = \displaystyle\int_0^\pi e^{|\cos x|}\cdot \sin x\cdot 2\sin\left(\dfrac{1}{2}\cos x\right)dx$

$+\displaystyle\int_0^\pi e^{|\cos x|}\cdot 3\cos\left(\dfrac{1}{2}\cos x\right)\cdot \sin x\,dx$ ... (i)

$\Rightarrow I = I_1 + I_2$

$\left(\begin{array}{l}\text{using}\displaystyle\int_0^{2a} f(x)dx \\ = \begin{cases} 0, & f(2a-x) = -f(x) \\ 2\displaystyle\int_0^a f(x)dx, & f(2a-x) = +f(x) \end{cases}\end{array}\right)$

where $I_1 = 0$ [$\because\ f(\pi - x) = -f(x)$] ...(ii)

and

$I_2 = 6\displaystyle\int_0^{\pi/2} e^{\cos x}\cdot \sin x\cdot \cos\left(\dfrac{1}{2}\cos x\right)dx$

Now, $I_2 = 6\int_0^1 e^t \cdot \cos\left(\dfrac{t}{2}\right) dt$

(Put $\cos x = t \Rightarrow -\sin x\, dx = dt$)

$= 6\left[e^t \cos\left(\dfrac{t}{2}\right) + \dfrac{1}{2}\int e^t \sin\dfrac{t}{2}\, dt \right]_0^1$

$= 6\left[e^t \cos\left(\dfrac{t}{2}\right) + \dfrac{1}{2}\left(e^t \sin\dfrac{t}{2} - \int \dfrac{e^t}{2}\cos\dfrac{t}{2}\, dt \right) \right]_0^1$

$= 6\left[e^t \cos\dfrac{t}{2} + \dfrac{1}{2}e^t \sin\dfrac{t}{2} \right]_0^1 - \dfrac{I_2}{4}$

$= \dfrac{24}{5}\left(e\cos\left(\dfrac{1}{2}\right) + \dfrac{e}{2}\sin\left(\dfrac{1}{2}\right) - 1 \right)$ ... (iii)

From Eqs. (i), we get

$I = \dfrac{24}{5}\left(e\cos\left(\dfrac{1}{2}\right) + \dfrac{e}{2}\sin\left(\dfrac{1}{2}\right) - 1 \right)$

Sol 16: Let $I_2 = \int_0^1 (1-x^{50})^{101} dx$,

using integration by parts

$= \left[(1-x^{50})^{101} \cdot x \right]_0^1$

$+ \int_0^1 (1-x^{50})^{100}\, 50 \cdot x^{49} \cdot x\, dx$

$= 0 - \int_0^1 (50)(101)(1-x^{50})^{100}(-x^{50}) dx$

$= -50(101)\int_0^1 (1-x^{50})^{101} dx + (50)(101)\int_0^1 (1-x^{50})^{100} dx$

$= 5050 I_2 + 5050 I_1$

$\therefore I_2 + 5050 I_2 = 5050 I_1$

$\therefore \dfrac{(5050)I_1}{I_2} = 5051$

Sol 17: (B) $g(x) = \dfrac{f'(e^x)e^x}{1 + e^{2x}}$

Hence positive for $(0,\infty)$ and negative for $(-\infty, 0)$

Consider the line

$L_1 : \dfrac{x+1}{3} = \dfrac{y+2}{1} = \dfrac{z+1}{2}, L_2 : \dfrac{x-2}{1} = \dfrac{y+2}{2} = \dfrac{z-3}{3}$

Sol 18: (D) Hence $\int_{-1}^{1} g'(x) = g(1) - g(-1) = 2g(1)$

A circle C of radius 1 is inscribed in an equilateral triangle PQR. The points of contact of C with the sides PQ, QR, RP are D, E, F, respectively. The line PQ is given by the equation $\sqrt{3}x + y - 6 = 0$ and the point D is $\left(\dfrac{3\sqrt{3}}{2}, \dfrac{3}{2} \right)$

Further, it is given that the origin and the centre of C are on the same side of the line PQ.

Sol 19: (A) Let $f(x) = \int_0^x \dfrac{t^2}{1+t^4} dt = 2x - 1$

$f(x) = \dfrac{x^2 1 + x^4 - 2}{x^4 + 1} \Rightarrow \dfrac{-2x^4 + x - 2}{x^4 + 1} < \forall x \in R$

$f(0) > 0, f(1) < 0$

$\therefore$ One solution in (0, 1)

Sol 20: $f(x) = \int_0^x f(t) dt \Rightarrow f(0) = 0$

Also $f(x) = f(x), x > 0 \Rightarrow f(x) = ke^x, x > 0$

$\because f(0) = 0$ and $f(x)$ is continuous

$\Rightarrow f(x) = 0 \forall x > 0$

$\therefore f(\ln 5) = 0$

Sol 21: (C) $f' = \pm\sqrt{1 - f^2}$

$\Rightarrow f(x) = \sin x$ or $f(x) = -\sin x$ (not possible)

$\Rightarrow f(x) = \sin x$

Also $x > \sin x \forall > 0$.

Sol 22: $A \to p, q, s; B \to p, t; C \to p, q, r, t; D \to s$

(A) $(x-3)^2 \dfrac{dy}{dx} + y = 0$

$\int \dfrac{dx}{(x-3)^2} = \int \dfrac{dy}{y} \Rightarrow \dfrac{1}{x-3} = \ln|y| + c$

So domain is $R - \{3\}$.

(B) Put $x = t + 3$

$\int_{-2}^{2}(t+2)(t+1)t(t-1)(t-2)dt = \int_{-2}^{2} t(t^2 - 1)(t^2 - 4)dt = 0$

(C) $f(x) = \dfrac{5}{4} - \left(\sin x - \dfrac{1}{2} \right)^2$

Maximum value occurs when $\sin x = \dfrac{1}{2}$

(D) $f(x) > 0$ if $\cos > \sin x$

Sol 23: (A) Let the line be $\dfrac{x}{a} = \dfrac{y}{b} = \dfrac{z}{c}$ intersects the lines

$\Rightarrow$ S.D $= 0 \Rightarrow a + 3b + 5c = 0$ and

$3a + b - 5c = 0 \Rightarrow a : b : c :: 5r : -5r : 2r$

on solving with given lines we get points of intersection

$P \equiv (5, -5, 2)$ and $Q \equiv \left(\dfrac{10}{3}, -\dfrac{10}{3}, \dfrac{8}{3}\right) \Rightarrow PQ^2 = d^2 = 6$

(B) (p, r)

$\tan^{-1}(x+3) - \tan^{-1}(x-3) = \sin^{-1}(3/5)$

$\Rightarrow \tan^{-1}\dfrac{(x+3)-(x-3)}{1+(x^2-9)} = \tan^{-1}\dfrac{3}{4} \Rightarrow \dfrac{6}{x^2-8} = \dfrac{3}{4}$

$\therefore x^2 - 8 = 8$

Or $x = \pm 4$

(C) (q, s)

As $\bar{a} = \mu\bar{b} + 4\bar{c} \Rightarrow \mu\left(|\bar{b}|\right) = -4\,\bar{b}.\bar{c}$ and $|\bar{b}|^2 = 4\bar{a}.\bar{c}$

and $|\bar{b}|^2 + \bar{b}.\bar{c} - \bar{d}.\bar{c} = 0$

Again, as $2\,|\bar{b} + \bar{c}| = |\bar{b} - \bar{a}|$

Solving and eliminating $\bar{b}.\bar{c}$ and eliminating $|\bar{a}|^2$

We get $\left(2\mu^2 - 10\mu\right)|\bar{b}|^2 = 0 \Rightarrow \mu = 0$ and 5.

(D) $I = \dfrac{2}{\pi} \displaystyle\int_{-x}^{x} \dfrac{\sin 9(x/2)}{\sin(x/2)} dx = \dfrac{2}{\pi} \times 2 \displaystyle\int_{0}^{\pi} \dfrac{\sin 9(x/2)}{\sin(x/2)} dx$

$x/2 = \theta \Rightarrow dx = 2b\theta$

$x = \pi\theta = \pi/2$

$= \dfrac{8}{\pi} \displaystyle\int_{0}^{\pi/2} \dfrac{\sin 9\theta - \sin 7\theta}{\sin\theta} + \dfrac{(\sin 7\theta - \sin 5\theta)}{\sin\theta}$

$+ \dfrac{(\sin 5\theta - \sin 3\theta)}{\sin\theta} + \dfrac{(\sin 3\theta - \sin\theta)}{\sin\theta} + \dfrac{\sin\theta}{\sin} d\theta = \dfrac{16}{\pi}$

$\displaystyle\int_{0}^{\pi/2} (\cos 8\theta + \cos 6\theta + \cos 4\theta + \cos 2\theta + 1) + \dfrac{8}{\pi} \displaystyle\int_{0}^{\pi/2} d\theta$

$= \dfrac{16}{\pi}\left[\dfrac{\sin 8\theta}{8} + \dfrac{\sin 6\theta}{6} + \dfrac{\sin 4\theta}{4} + \dfrac{\sin 2\theta}{2}\right]$

$+ \dfrac{8}{\pi}[\theta]^{\pi/2} = 0\dfrac{8}{\pi} \times \left[\dfrac{\pi}{2} - 0\right] = 4$

Sol 24: (B) $\displaystyle\lim_{x \to 0} \dfrac{x\ln(1+x)}{(x^4+4) \times 3x^2} = \lim_{x \to 0}\dfrac{1}{4} \times \dfrac{1}{3} = \dfrac{1}{12}$

Sol 25: (A)

$\displaystyle\int_{0}^{1} \dfrac{x^4(1-x)^1}{1+x^2} = \int_{0}^{1} \dfrac{x^4\left[(1+x^2)-2x\right]^2}{1+x^2} =$

$\displaystyle\int_{0}^{1} \dfrac{x^4(1-x)^1}{1+x^2} = \int_{0}^{1} \dfrac{x^4\left[(1+x^2)-2x\right]^2}{1+x^2}$

$= \displaystyle\int_{0}^{1} x^4\left[(1+x^2) - 4x + \dfrac{4x^2}{1+x^2}\right] dx$

$= \displaystyle\int \left[x^6 + x^4 - 4x^5 + \dfrac{4x^6}{1+x^2}\right] dx$

Now on polynomial division of x^6 by $1+x^2$, we obtain

$\displaystyle\int \left[x^6 + x^4 - 4x^5 + 4\left[(x^4 - x^2 + 1) - \dfrac{1}{1+x^2}\right]\right]$

$dx = \displaystyle\int\left[(x^6 - 4x^5 + 5x^4 - 4x^2 + 4) - \dfrac{4}{1+x^2}\right] dx$

$= \left[\dfrac{x^7}{7} - \dfrac{4x^6}{6} + \dfrac{5x^5}{5} - \dfrac{4x^3}{3} + 4x\right]_{0}^{1} - 4\left[\tan^{-x} x\right]$

$= \left(\dfrac{1}{7} - \dfrac{4}{6} + 1\dfrac{4}{3} + 4\right) - 4\left(\dfrac{\pi}{4}\right) = \left(\dfrac{1}{7} + 3\right) - \pi = \dfrac{22}{7} - \pi$

Sol 26: (A) $x^2 = t \Rightarrow 2x\,dx = dt$

$I = \dfrac{1}{2}\displaystyle\int_{\ln 2}^{\ln 3} \dfrac{\sin t}{\sin t + \sin(\ln 6 - t)} dt$

and $I = \dfrac{1}{2}\displaystyle\int_{\ln 2}^{\ln 3} \dfrac{\sin(\ln 6 - t)}{\sin t(\ln 6 - t) + \sin t} dt$

$x^2 = t \Rightarrow 2x\,dx = dt$

$2I = \dfrac{1}{2}\displaystyle\int_{\ln 2}^{\ln 3} 1\,dt \Rightarrow I = \dfrac{1}{4}\ln\dfrac{3}{2}$

Sol 27: (B)

$\displaystyle\int_{-\pi/2}^{\pi/2} \left(x^2 + \ln\left(\dfrac{\pi+x}{\pi-x}\right)\right)\cos x\,dx$

$= 2\displaystyle\int_{0}^{\pi/2} x^2\cos x\,dx + 0 \left(\because \ln\left(\dfrac{\pi+x}{\pi-x}\right) \text{is an odd function}\right)$

$= 2\left[(x^2\sin x)\Big|_{0}^{\pi/2} \displaystyle\int_{0}^{\pi/2} 2x\sin x\,dx\right] = 2\left(\dfrac{\pi^2}{4} - 0\right) - 4\displaystyle\int_{0}^{\pi/2} x\sin x\,dx$

$= \dfrac{\pi^2}{2} - 4\left[(-x\cos x)^{\pi/2}\Big|_{0} + \displaystyle\int_{0}^{\pi/2}\cos x\,dx\right] = \dfrac{\pi^2}{2} - 4$

Sol 28: (A) $\int\limits_{\frac{\pi}{4}}^{\frac{\pi}{2}} (2\cosec x)^{17} dx$

Let

$e^n + e^{-n} = 2\cosec x, x = \frac{\pi}{4} \Rightarrow u\log(1+\sqrt{2}), x = \frac{\pi}{2} \Rightarrow u = 0$

$\Rightarrow \cosec x + \cot x = e^n$ and

$x - \cot x = e^{-n} \Rightarrow \cot x = \dfrac{e^u - e^{-u}}{2}$

$\left(e^u - e^{-u}\right) dx = -2\cosec x \cot x\, dx$

$\Rightarrow -\int \left(e^u + e^{-u}\right)^{17} \dfrac{\left(e^u - e^{-u}\right)}{2\cosec x \cot x} du$

$= -2 \int\limits_{\log(1+\sqrt{2})}^{0} \left(e^u + e^{-u}\right) du = \int\limits^{\log(1+\sqrt{2})} 2(e^u + e^u) du$

Sol 29: (D) (p) $f(x) = ax^2 + bx, \int\limits_0^1 f(x) dx = 1$

$\Rightarrow 2a + 3b = 6$

$\Rightarrow (a,b) \equiv (0,2) \text{ and } (3,0)$

(q) $f(x) = \sqrt{2}\cos\left(x^2 - \dfrac{\pi}{4}\right)$

$x^2 - \dfrac{\pi}{4} = 2n\pi \Rightarrow x^2 = 2n\pi + \dfrac{\pi}{4}$

$\Rightarrow x = \pm\sqrt{\dfrac{\pi}{4}}, \pm\sqrt{\dfrac{9\pi}{4}}$ as $x \in \left[-\sqrt{3}, \sqrt{13}\right]$

(r) $\int\limits_0^2 \left(\dfrac{3x^2}{1+e^x} + \dfrac{3x^2}{1+e^{-x}}\right) dx = \int\limits_0^2 3x^2 dx = 8$

(s) $\int\limits_{-1/2}^{1/2} \cos 2x \ln\left(\dfrac{1+x}{1-x}\right) dx = 0$ as it is an odd function

Sol 30: (2) $\int\limits_0^1 4x^3 \dfrac{d^2}{dx^2}\left(1-x^2\right) dx$

$= \left[4x^3 \dfrac{d}{dx}\left(1-x^2\right)^5\right]_0^1 - \int\limits_0^1 12x^2 \dfrac{d}{dx}\left(1-x^2\right) dx$

$= \left[4x^3 \times 5\left(1-x^2\right)^4 (-2x)\right]_0^1 - 12\left[\left[x^2\left(1-x^2\right)^5\right]_0^1 - \int\limits_0^1 2x(1-x)^5 dx\right]$

$= 0 - 0 - 12\left[0-0\right] + 12\int\limits_0^1 2x\left(1-x^2\right) dx$

$= 12 \times \left[-\dfrac{\left(1-x^2\right)^6}{6}\right]_0^1 = 12\left[0 + \dfrac{1}{6}\right] = 2$

Sol 31: $\alpha = \int\limits_0^1 \left(e^{9x+3\tan^{-1}x}\right)\left(\dfrac{12+9x^2}{1+x^2}\right) dx$

Put $9x + 3\tan^{-1}x = t \Rightarrow \left(9 + \dfrac{3}{1+x^2}\right) dx = dt$

$\Rightarrow \alpha = \int\limits_0^4 e^t dt = e^{9+\frac{3\pi}{4}} - 1 \Rightarrow \left(\log_e|1+\alpha| - \dfrac{3\pi}{4}\right) = 9$

Sol 32: (A, C) Let $\int\limits_0^\pi e^t \left(\left(\sin^6 at + \cos^4 at\right)\right) dt = A$

$I = \int\limits_\pi^{2\pi} e^t \left(\sin^6 at + \cos^4 at\right) dt$

Put $t = \pi + x$

$dt = dx$

For a = 2 as well as a = 4

$I = e^x \int\limits_0^\pi e^x \left(\sin^6 ax + \cos^4 at\right) dt = e^{2\pi}A \Rightarrow I = e^\pi A$

Similarly $\int\limits_0^\pi e^\pi e^\pi (\sin^6 at + \cos^4 at) dt = e^{2\pi}A$

$L = \dfrac{A + e^\pi A + e^{2\pi}A + e^{3\pi}A}{A} = \dfrac{e^{4\pi}-1}{e^\pi - 1}$ ∴ For both a = 2, 4

Sol 33: (A, B, C) (A) $f'(x) = F(x) + xF'((x)$

$f(1) = F(1) + F'(1)$

$f'(1) = F(1) < 0$

(B) $f(2) = 2F(2)$

$F(x)$ is decreasing and F(1) = 0

Hence F(2) < 0

$\Rightarrow f'(2) < 0$

(C) $f(x) = F(x) + xF'(x)$

$F(x) < 0 \forall x \in (1,0)$

$F'(x) < 0 \forall x \in (1,3)$

Hence $f(x) < 0 \forall x \in (1,3)$

Sol 34: $I = \displaystyle\int_{-1}^{0} \frac{x.0}{2+0}dx + \int_{0}^{1} \frac{x.0}{2+1}dx + \int_{1}^{\sqrt{2}} \frac{x.1}{2+C}dx + 0 = \frac{1}{4}$

$\Rightarrow 4I - 1 = 0$

Sol 35: (A) $I = \displaystyle\int_{-\pi/2}^{\pi/2} \frac{x^2 \cos x}{1+e^x}dx$ ... (i)

$I = \displaystyle\int_{-\pi/2}^{\pi/2} \frac{x^2 \cos x}{1+\dfrac{1}{e^x}}dx$ ... (ii)

$= \displaystyle\int_{-\pi/2}^{\pi/2} \frac{x^2 \cos x.e^x}{1+e^x}dx$

(i) and (ii)

$2I \displaystyle\int_{-\pi/2}^{\pi/2} x^2 \cos x \, dx$

$I = \displaystyle\int_{0}^{\pi/2} x^2 \cos x \, dx$ (even fn)

$= x^2.\sin x \Big|_{0}^{\pi/2} - \displaystyle\int_{0}^{\pi/2} 2x \sin x \, dx$

$= \dfrac{\pi^2}{4} - 2\left[(-x \cos x)_{0}^{\pi/2} - \displaystyle\int_{0}^{\pi/2} (-\cos x)\, dx \right]$

$= \dfrac{\pi^2}{4} - 2\left[0 + \sin x \Big|_{0}^{\pi/2} \right] = \dfrac{\pi^2}{4} - 2[1] = \dfrac{\pi^2}{4} - 2$

3. AREA UNDER THE CURVE AND LINEAR PROGRAMMING

AREA UNDER THE CURVE

1. INTRODUCTION

In the previous chapters we have studied the process of integration and its physical interpretation. The most important application of integration is finding the area under a curve. In this topic we will discuss different curves and the area bounded by some simple plane curves taken together. In order to find the area, we need to know the basics of plotting a curve and then use integration with appropriate limits to get the answer. The process of finding area of some plane region is called **Quadrature**.

2. CURVE TRACING

Let us now discuss the basics of curve tracing. Curve tracing is a technique which provides a rough idea about the nature and shape of a plane curve. Different techniques are used in order to understand the nature of the curve, but there is no fixed rule which provides all the information to draw the graph of a given function (say f(x)). Sometimes it is also very difficult to draw the exact curve of the given function. However, the following steps can be helpful in trying to understand the nature and the shape of the curve.

Step 1: Check whether the origin lies on the given curve. Also check for other points lying on the curve by putting some values.

Step 2: Check whether the curve is increasing or decreasing by finding the derivative of the function. Also check for the boundary points of the curve.

Step 3: Check whether the curve f(x , y) = 0 is symmetric about

(a) X-axis: If the equation remains same on replacing y by –y i.e. f(x , y) = f(x , –y), or, if all the powers of "y" are even, then the graph is symmetric about the X-axis.

(b) Y-axis: If the equation remains same on replacing x by –x i.e. f(x , y) = f(–x , y), or, if all the powers of "x" are even, then the graph is symmetric about the Y-axis.

(c) Origin: If f(–x, –y) = –f(x, y), then the graph is symmetric about the Origin.

For example, the curve given by $x^2 = y+2$ is symmetrical about y-axis, $y^2 = x+2$ is symmetrical about x-axis and the curve $y = x^3$ is symmetrical about the origin.

Step 4: Find out the points of intersection of the curve with the x-axis and y-axis by substituting y = 0 and x = 0 respectively.

For example, the curve $\dfrac{x^2}{9} + \dfrac{y^2}{4} = 1$ intersects the axes at points (± 3, 0) and (0, ±2).

Step 5: Identify the domain of the given function and the region in which the graph can be drawn. For example, the curve $xy^2 = (8 - 4x)$ or $y = 2\sqrt{\dfrac{2-x}{x}}$.

Therefore the value of y is defined only when $\dfrac{2-x}{x} \geq 0$ i.e. $0 < x \leq 2$. Hence, the graph lies between the lines x = 0 and x = 2.

Step 6: Check the behaviour of the graph as $x \rightarrow +\infty$ and as $x \rightarrow -\infty$. Find all the horizontal, vertical and oblique asymptotes, if any.

Step 7: Determine the critical points, the intervals on which the function (f) is concave up or concave down and the inflection points.

The information obtained from the Steps 1 to 7 are used to trace the curve

Illustration 1: Trace the curve $y^2(2a - x) = x^3$ 　　　　　　　　　　　　　　　　　　　**(JEE MAIN)**

Sol: By using curve tracing method as mentioned above.

Given curve: $y^2 = x^3/(2a - x)$ 　　　　　　　　　　　　　　　　　　　　　　　　　　　...(i)

(a) Origin: The point (0 , 0) satisfies the given equation, therefore, it passes through the origin.

(b) Symmetrical about x-axis: On replacing y by –y, the equation remains same, therefore, the given curve is symmetrical about x-axis.

(c) Tangent at the origin: Equation of the tangent is obtained by equating the lowest degree terms to zero.

$\Rightarrow 2ay^2 = 0$ 　　$\Rightarrow y^2 = 0 \Rightarrow y=0$

(d) Asymptote parallel to y-axis: Equation of asymptote is obtained by equating the coefficient of lowest degree of y to 0. The given equation can be written as $(2a - x)y^2 = x^3$

$\therefore$ Equation of asymptote is $2a - x = 0$ or $x = 2a$

(e) Region of absence of curve: The given equation is

$y^2(2a - x) = x^3 \qquad \Rightarrow y^2 = \dfrac{x^3}{(2a - x)}$.

For x < 0 and x > 2a, RHS becomes negative, therefore the curve exists only for $0 \leq x < 2a$.

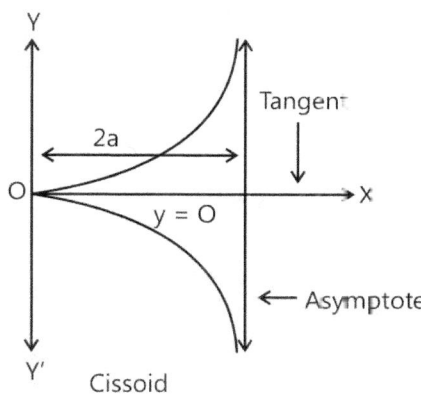

Cissoid

Figure 25.1

Hence the graph of $y^2(2a - x) = x^3$ is as shown in Fig. 25.1. Such a curve is known as a Cissoid.

Illustration 2: Sketch the curve $\dfrac{x^2}{4} + \dfrac{y^2}{9} = 1$ 　　　　　　　　　　　　　**(JEE MAIN)**

Sol: Same as above illustration.

We have, $\dfrac{x^2}{4} + \dfrac{y^2}{9} = 1$ 　　　　　　　　　　　　　　　　　　　　　　　...(i)

(a) Origin: The point (0,0) does not satisfy the equation, hence, the curve does not pass through O.

(b) Symmetry: The equation of the curve contains even powers of x and y so it is symmetric about both x and y axes.

(c) Intercepts: Putting y = 0, we get x = ± 2 i.e. the curve passes through the points (2 , 0) and (-2 , 0). Similarly, on substituting x = 0, we get y = ± 3 i.e. the curve passes through the points (0 , 3) and (0 , -3).

(d) Region where the curve does not exist: If $x^2 > 4$, y becomes imaginary. So the curve does not exist for x > 2 and x < –2. Similarly, if $y^2 > 9$, x becomes imaginary. So, the curve does not exist for y > 3 and y < –3.

(e) Table:

x	– 2	0	1	2
y	0	±3	±2.6	0

Hence the graph of $\dfrac{x^2}{4} + \dfrac{y^2}{9} = 1$ is as shown in Fig. 25.2.

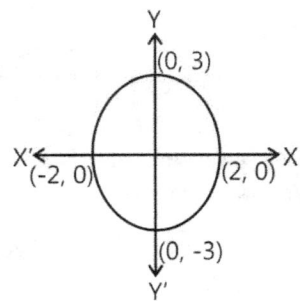

Figure 25.2

NOMORECLASS CONCEPTS

Using the above rules try to trace the Witch of Agnesi

$xy^2 = a^2\,(a - x)$.

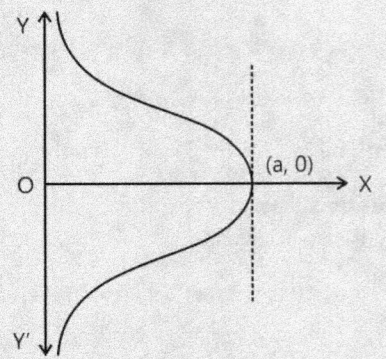

Figure 25.3

3. AREA BOUNDED BY A CURVE

3.1 The Area Bounded by a Curve with X-axis

The area bound the curve y=f(x) with the x-axis between the ordinates

x= a and x=b is given by $\text{Area} = \int\limits_{a}^{b} y\, dx = \int\limits_{a}^{b} f(x)\, dx$

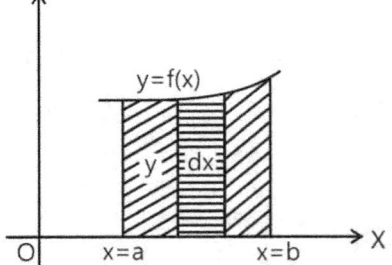

Figure 25.4: Area Bounded By a curve y=f(x) with x-axis

Illustration 3: Find the area bounded by the curve $y = x^3$, x-axis and ordinates x = 1 and x = 2. **(JEE MAIN)**

Sol: By using above formula, we can find out the area under given curve.

$$\text{Required Area} = \int\limits_{1}^{2} y\, dx = \int\limits_{1}^{2} x^3\, dx = \left(\frac{x^4}{4}\right)_{1}^{2} = \frac{15}{4}$$

Illustration 4: Find the area bounded by the curve y = mx x-axis and ordinates x = 1 and x = 2. **(JEE MAIN)**

Sol: Same as above.

$$\text{Required area} = \int\limits_{1}^{2} y\, dx = \int\limits_{1}^{2} mx\, dx = \left(\frac{mx^2}{2}\right)_{1}^{2} = \frac{m}{2}(4-1) = \frac{3}{2}m$$

3.3

Illustration 5: Find the area included between the parabola $y^2 = 4ax$ and its latus rectum ($x = a$).

(JEE ADVANCED)

Sol: Here the curve is $y^2 = 4ax$, latus rectum is $x = a$, and the curve is symmetrical about the x-axis.

(a) The latus rectum is the line perpendicular to the axis of the parabola and passing through the focus S (a, 0).

(b) The parabola is symmetrical about the x-axis.

∴ The required area AOBSA = 2 × area AOSA

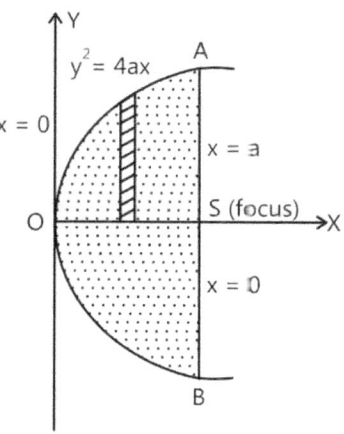

$$= 2\int_0^a y\,dx = 2\int_0^a 2\sqrt{ax}\,dx \quad \left[y^2 = 4ax \Rightarrow y = 2\sqrt{ax} \right]$$

$$= 4\sqrt{a}.\frac{2}{3}\left[x^{3/2} \right]_0^a = \frac{8}{3}\sqrt{a}.a^{3/2} = \frac{8}{3}a^2 \,.$$

Figure 25.5

Illustration 6: Sketch the region $\{(x, y): 4x^2 + 9y^2 = 36\}$ and find its area using integration. (JEE ADVANCED)

Sol: The given curve is an ellipse, where a = 3 and b = 2. The X and Y axis divides this ellipse into four equal parts.

Region $\{(x, y): 4x^2 + 9y^2 = 36\}$ = Region bounded by $\left(\dfrac{x^2}{9} + \dfrac{y^2}{4} = 1 \right)$

Limits for the shaded area are $x = 0$ and $x = 3$.

∴ The required area of the ellipse

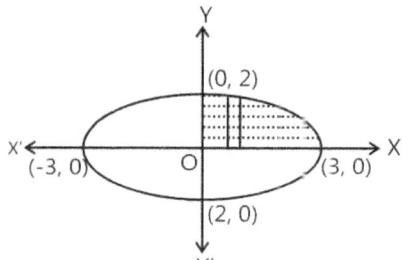

Figure 25.6

$$= 4\int_0^a y\,dx = 4\int_0^3 2\sqrt{1-\frac{x^2}{9}}\,dx \quad \left[\because \frac{x^2}{9} + \frac{y^2}{4} = 1 \Rightarrow \frac{y^2}{4} = 1 - \frac{x^2}{9} \Rightarrow y = 2\sqrt{1-\frac{x^2}{9}} \right]$$

$$= \frac{8}{3}\int_0^3 \sqrt{3^2 - x^2}\,dx = \int_0^3 \frac{8}{3}\left[\frac{x}{2}\sqrt{9-x^2} + \frac{9}{2}\sin^{-1}\left(\frac{x}{3}\right) \right]_0^3 \quad \left[using \int \sqrt{a^2 - x^2}\,dx = \frac{x}{2}\sqrt{a^2 - x^2} + \frac{a^2}{2}\sin^{-1}\frac{x}{a} \right]$$

$$= \frac{8}{3}\left[0 + \frac{9}{2}\sin^{-1}1 - 0 - 0 \right] = \frac{8}{3}\times\frac{9}{2}\times\frac{\pi}{2} = 6\pi \ \text{sq. units.}$$

3.2 The Area Bounded by a Curve with y-Axis

The area bound the curve $y=f(x)$ with y-axis between the ordinates
$y = a$ and $y = b$ is given by

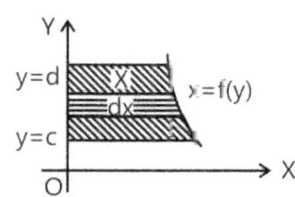

$$\text{Area} = \int_c^d x\,dy = \int_c^d f(y)\,dy$$

Figure 25.7: Area bounded by a curve with y-axis

Illustration 7: Find the area bounded by the curve $x^2 = \dfrac{1}{4}y$, y-axis and between the lines $y = 1$ and $y = 4$.

(JEE MAIN)

Sol: As we know, area bounded by curve with y – axis is given by $\int_c^d x\,dy = \int_c^d f(y)\,dy$.

Required Area $= \int\limits_{1}^{4} x \, dy = 2\int\limits_{1}^{4} \frac{1}{2}\sqrt{y}\,dy = \frac{2}{3}\left[y^{3/2}\right]_{1}^{4} = \frac{2}{3}(8-1) = \frac{14}{3}$ sq. units

Illustration 8: Find the area of the region bounded by the curve $y^2 = 4x$, y-axis and the line $y = 3$.　　**(JEE MAIN)**

Sol: Same as above illustration. $\left(\begin{array}{l} \because y^2 = 4x \\ \dfrac{y^2}{4} = x \end{array}\right)$

Area of region is $A = \int\limits_{y=0}^{y=3} x \, dy = \int\limits_{0}^{3} \frac{y^2}{4} dy$

$= \frac{1}{4}\left[\frac{y^3}{3}\right]_{0}^{3} = \frac{1}{4}\left[\frac{3^3}{3} - \frac{0}{3}\right] = \frac{1}{4}[9] = \frac{9}{4}$ sq. units

Hence, the required area is $\frac{9}{4}$ sq. units.

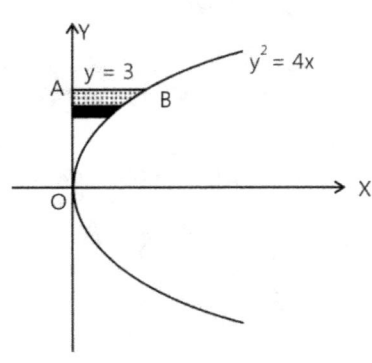

Figure 25.8

3.3 Area of a Curve in Parametric Form

If the given curve is in parametric form say $x = f(t)$, $y = g(t)$, then the area bounded by the curve with x-axis is equal to $\int\limits_{a}^{b} y \, dx = \int\limits_{t_2}^{t_1} g(t)f'(t)dt$　$\left[\because dx = d(f(t)) = f'(t)dt\right]$ Where t_1 and t_2 are the values of t corresponding to the values of a and b of x.

Illustration 9: Find the area bounded by the curve $x = a\cos t$, $y = b\sin t$ in the first quadrant.　　**(JEE MAIN)**

Sol: Solve it using formula of area of a curve in parametric form.

The given equation is the parametric equation of ellipse, on simplifying we get $\dfrac{x^2}{a^2} + \dfrac{y^2}{b^2} = 1$.

$\therefore$ Required area $= \int\limits_{0}^{a} y \, dx = \int\limits_{\pi/2}^{0} (b\sin t(-a\sin t)dt) = ab\int\limits_{0}^{\pi/2} \sin^2 t \, dt = \left(\dfrac{\pi ab}{4}\right)$.

3.4 Symmetrical Area

If the curve is symmetrical about a line or origin, then we find the area of one symmetrical portion and multiply it by the number of symmetrical portions to get the required area.

Illustration 10: Find the area bounded by the parabola $y^2 = 4x$ and its latus rectum.　　**(JEE MAIN)**

Sol: Here the given parabola is symmetrical about x – axis.

Hence required area $= 2\int_0^1 y \, dx$.

Since the curve is symmetrical about x-axis,

$\therefore$ The required Area $= 2\int_0^1 y \, dx = 2\int_0^1 \sqrt{4x} \, dx = 4 \cdot \frac{2}{3}\left[x^{3/2}\right]_0^1 = \frac{8}{3}$

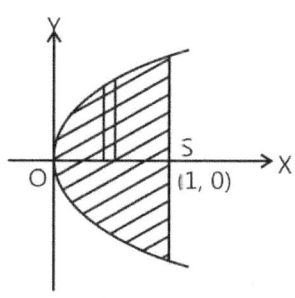

Figure 25.9

3.5 Positive and Negative Area

The area of a plane figure is always taken to be positive. If some part of the area lies above x-axis and some part lies below x-axis, then the area of two parts should be calculated separately and then add the numerical values to get the desired area.

If the curve crosses the x-axis at c (see Fig. 25.10), then the area bounded by the curve y = f(x) and the ordinates x = a and x = b, (b > a) is given by

$$A = \left|\int_a^c f(x)\,dx\right| + \left|\int_c^b f(x)\,dx\right|; \qquad A = \int_a^c f(x)\,dx - \int_c^b f(x)\,dx$$

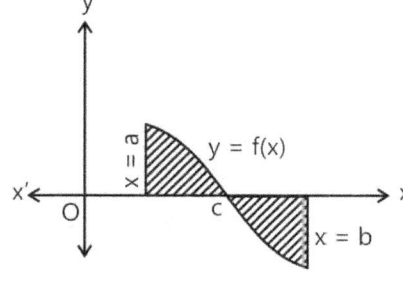

Figure 25.10

Illustration 11: Find the area between the curve y = cos x and x-axis when π/4 < x < π **(JEE MAIN)**

Sol: Here some part of the required area lies above x-axis and some part lies below x-axis,. Hence by using above mentioned method we can obtain required area.

$\therefore$ Required area $= \int_{\pi/4}^{\pi/2} \cos x \, dx + |\int_{\pi/2}^{\pi} \cos x \, dx|$

$= [\sin x]_{\pi/4}^{\pi/2} + |[\sin x]_{\pi/2}^{\pi}| = \left(1 - 1/\sqrt{2}\right) + |0 - 1| = \frac{2\sqrt{2} - 1}{\sqrt{2}}$

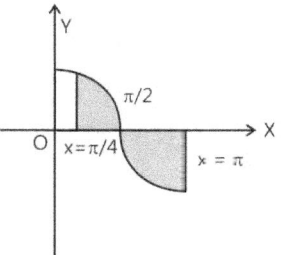

Figure 25.11

Illustration 12: Using integration, find the area of the triangle ABC, whose vertices are A (4, 1), B (6, 6) and C (8, 4) **(JEE ADVANCED)**

Sol: Here by using slope point form we can obtain respective equation of line by which given triangle is made. And after that by using integration method we can obtain required area.

Equation of line AB: $y - 1 = \frac{5}{2}(x - 4) \Rightarrow y = \frac{5x}{2} - 9$

Equation of line AC: $y - 1 = \left(\frac{3}{4}\right)(x - 4) \Rightarrow y = \frac{3x}{4} - 2$

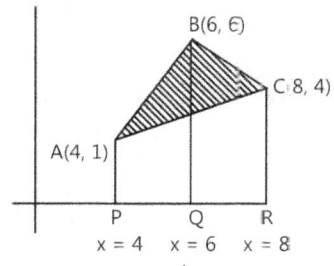

Figure 25.12

Equation of line BC: $(y-6) = \left(\dfrac{-2}{2}\right)(x-6) \Rightarrow y = -x + 12$

∴ The required area = Area of trapezium ABQP + Area of trapezium BCRQ − Area of trapezium ACRP

$$= \int_4^6 \left(\frac{5}{2}x - 9\right) dx + \int_6^8 (-x + 12)\,dx - \int_4^8 \left(\frac{3}{4}x - 2\right) dx$$

$$= \left(\frac{5}{4}x^2 - 9x\right)_4^6 + \left(12x - \frac{x^2}{2}\right)_6^8 - \left(\frac{3}{8}x^2 - 2x\right)_4^8 = 7 + 10 - 10 = 7 \text{ sq. units.}$$

3.6 Area between Two Curves

(a) Area enclosed between two curves.

If $y = f_1(x)$ and $y = f_2(x)$ are two curves (where $f_1(x) > f_2(x)$), which intersect at two points, A $(x = a)$ and B$(x = b)$, then the area enclosed by the two curves between A and B is

Common area $= \int_a^b (y_1 - y_2)\,dx = \int_a^b [f_1(x) - f_2(x)]\,dx$

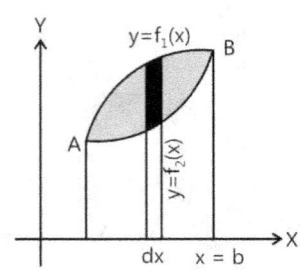

Figure 25.13

Illustration 13: Find the area between two curves $y^2 = 4ax$ and $x^2 = 4ay$. **(JEE MAIN)**

Sol: By using above mentioned formula of finding the area enclosed between two curves, we can obtain required area.

Given, $\quad y^2 = 4ax$ ... (i)

$\qquad\qquad x^2 = 4ay$ ... (ii)

Solving (i) and (ii), we get $x = 4a$ and $y = 4a$.

So required area $= \int_0^{4a} \left(\sqrt{4ax} - \dfrac{x^2}{4a}\right) dx = \left(2\sqrt{a}\,\dfrac{x^{3/2}}{3/2} - \dfrac{x^3}{12a}\right)_0^{4a}$

$= \dfrac{4\sqrt{a}}{3}\,|\,4a\,|^{3/2} - \dfrac{64a^3}{12a} = \dfrac{16}{3}a^2$

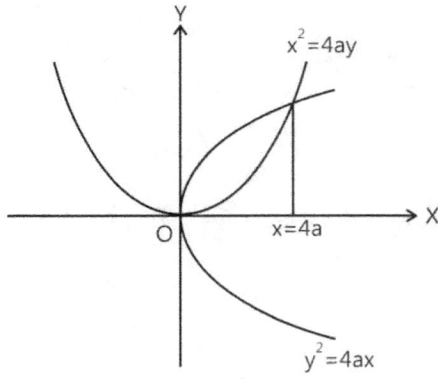

Figure 25.14

(b) Area enclosed by two curves intersecting at one point and the X-axis.

If $y = f_1(x)$ and $y = f_2(x)$ are two curves which intersect at a point P (α, β) and meet x-axis at A $(a, 0)$ and B $(b, 0)$ respectively, then the area enclosed between the curves and x-axis is given by

$$\text{Area} = \int_a^{\alpha} f_1(x)\,dx + \int_{\alpha}^{b} f_2(x)\,dx$$

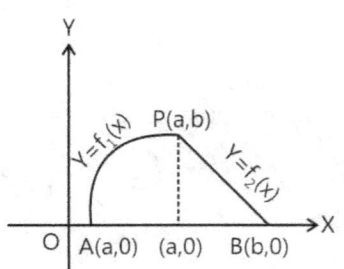

Figure 25.15

(c) Area bounded by two intersecting curves and lines parallel to y-axis.

The area bounded by two curves $y = f(x)$ and $y = g(x)$ (where $a \le x \le b$), when they intersect at $x = c \in (a, b)$, is given

by $A = \int_a^b | f(x) - g(x) | \, dx \Rightarrow A = \int_a^c (f(x) - g(x)) \, dx + \int_c^b (g(x) - f(x)) \, dx$

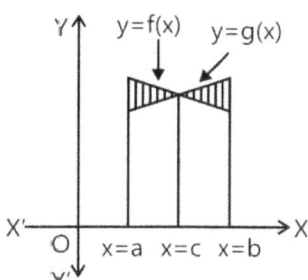

Figure 25.16

Illustration 14: Draw a rough sketch of the region enclosed between the circles $x^2 + y^2 = 4$ and $(x - 2)^2 + y^2 = 4$. Using method of integration, find the area of this enclosed region **(JEE ADVANCED)**

Sol: By solving given equations simultaneously, we will be get intersection points of circles and then by using integration method we can obtain required area.

The figure shown alongside is the sketch of the circles

$x^2 + y^2 = 4$ ... (i)

and, $(x - 2)^2 + y^2 = 4$ ... (ii)

From (i) and (ii), we have $(x - 2)^2 - x^2 = 0$

$\Rightarrow (x - 2 - x)(x - 2 + x) = 0 \qquad \Rightarrow x = 1$ ... (iii)

Solving (i) and (iii), we get $\qquad y = \pm\sqrt{3}$

Therefore, the circles (i) and (ii) intersect at $A(1, \sqrt{3})$ and $B(1, -\sqrt{3})$.

Area of enclosed region = Area OACBO = 2 Area OACO

= 2 [Area OAD + Area ACD]

$= 2\int_0^1 \sqrt{4 - (x - 2)^2} \, dx + 2\int_1^2 \sqrt{4 - x^2} \, dx$

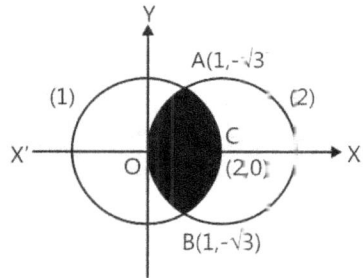

Figure 25.17

$= 2\int_1^2 \sqrt{4 - x^2} \, dx + 2\int_0^1 \sqrt{4 - (x - 2)^2} \, dx$

$= 2\left[\frac{x\sqrt{4 - x^2}}{2} + \frac{4}{2}\sin^{-1}\frac{x}{2} \right]_1^2 + 2\left[\frac{(x - 2)\sqrt{4 - (x - 2)^2}}{2} + \frac{4}{2}\sin^{-1}\left(\frac{x - 2}{2}\right) \right]_0^1 \left[\because \sqrt{a^2 - x^2} \, dx \Rightarrow \frac{x}{2}\sqrt{a^2 - x^2} + \frac{a^2}{2}\sin^{-1}\frac{x}{a} \right]$

$= 2\left(\pi - \frac{\sqrt{3}}{2} - 2\left(\frac{\pi}{6}\right) \right) + 2\left(-\frac{\sqrt{3}}{2} - 2\left(\frac{\pi}{6}\right) + \pi \right) = \frac{8\pi}{3} - 2\sqrt{3}$ sq. units

Illustration 15: Using integration, find the area of the region given below:

$\{(x, y): 0 \le y \le x^2 + 1, 0 \le y \le x + 1, 0 \le x \le 2\}$ **(JEE ADVANCED)**

Sol: Same as above illustration, by solving given equation $y = x^2 + 1$ and $y = x + 1$ we will be get their points of intersection and after that using integration method and taking these points as limit we can obtain required area.

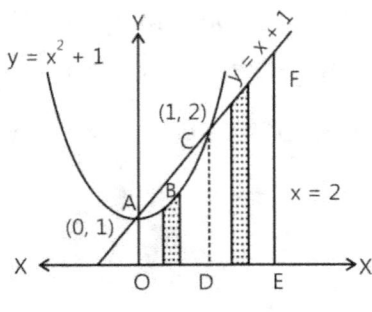

The region is shaded as shown in the Fig. 25.18.

Given, $\quad y = x^2 + 1$ ... (i)

$\quad\quad\quad y = x + 1$ ... (ii)

On solving (i) and (ii), we have $x^2 + 1 = x + 1$

$\Rightarrow x = 0, 1$ and $y = 1, 2$

Figure 25.18

$\therefore$ The shaded region can be divided into two parts OABCDO and CDEFC.

Limits for the area OABEO are $x = 0$ and $x = 1$.

Limits for the area EBDFE are $x = 1$ and $x = 2$.

Area of the shaded region = Area OABEO + Area EBDFE.

$$= \int_0^1 (x^2 + 1)\,dx + \int_1^2 (x+1)\,dx = \left[\frac{x^3}{3} + x\right]_0^1 + \left[\frac{x^2}{2} + x\right]_1^2 = \left(\frac{1}{3} + 1\right) + \left(\frac{4}{2} + 2 - \frac{1}{2} - 1\right) = \frac{23}{6} \text{ sq. units}$$

Illustration 16: Find the area of the following region: $[(x, y): y^2 \leq 4x,\ 4x^2 + 4y^2 \leq 9]$ **(JEE ADVANCED)**

Sol: Similar to above problem, Here the required area is equal to Area AOBA + Area ACBA.

Given $\quad y^2 = 4x$ ... (i)

$$4x^2 + 4y^2 = 9 \quad \Rightarrow \quad x^2 + y^2 = \left(\frac{3}{2}\right)^2 \quad\quad\quad\text{... (ii)}$$

Curves (i) and (ii) intersect at $A\left(\frac{1}{2}, \sqrt{2}\right)$ and $B\left(\frac{1}{2}, -\sqrt{2}\right)$

Limits for the area OAB are $x = 0$, $x = \frac{1}{2}$

Limits for the area ACB are $x = \frac{1}{2}$, $x = \frac{3}{2}$.

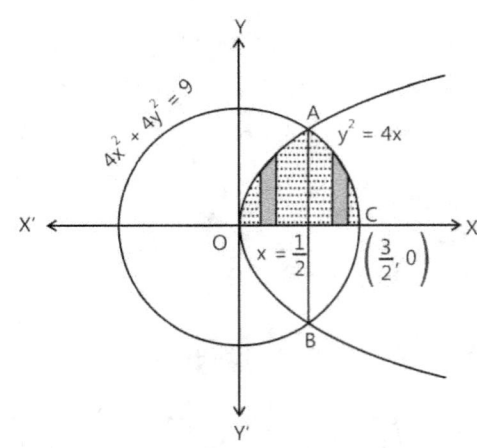

The required area = Area AOBA + Area ACBA

$$= 2\left[\int_0^{1/2} y_1\,dx + \int_{1/2}^{3/2} y_2\,dx\right] = 2\left[\int_0^{1/2} \sqrt{4x}\,dx + \int_{1/2}^{3/2} \sqrt{\frac{9}{4} - x^2}\,dx\right]$$

$$= 4\left[\frac{2}{3} x^{3/2}\right]_0^{1/2} + 2\left[\frac{x}{2} \cdot \sqrt{\frac{9}{4} - x^2} + \frac{9}{8}\sin^{-1}\left(\frac{x}{3/2}\right)\right]_{1/2}^{3/2}$$

Figure 25.19

$$= \frac{8}{3} \cdot \frac{1}{2\sqrt{2}} + \left[0 - \frac{1}{\sqrt{2}} + \frac{9}{4}\left(\sin^{-1} 1 - \sin^{-1}\frac{1}{3}\right)\right] = \frac{4}{3\sqrt{2}} - \frac{1}{\sqrt{2}} + \frac{9}{4}\left(\frac{\pi}{2} - \sin^{-1}\frac{1}{3}\right) = \frac{1}{3\sqrt{2}} + \frac{9}{4}\cos^{-1}\frac{1}{3}.$$

Illustration 17: Draw a rough sketch and find the area of the region bounded by the two parabolas $y^2 = 8x$ and $x^2 = 8y$, by using method of integration.   **(JEE MAIN)**

Sol: As the given two equation is the equation of parabola which intersect at O(0, 0) and A(8, 8), and the required area is equal to Area OBADO – Area OADO.

Given parabolas are $y^2 = 8x$ ... (i)

and, $x^2 = 8y$ ... (ii)

The curves (i) and (ii) intersect at O(0, 0) and A(8, 8).

∴ Required Area = Area OBADO – Area OADO

$$= \int_0^8 (y_1 - y_2)\,dx$$

$$= \int_0^8 \left(\sqrt{8x} - \frac{x^2}{8} \right) dx = \left[2\sqrt{2} \cdot \frac{x^{3/2}}{3/2} - \frac{1}{8}\frac{x^3}{3} \right]_0^8 = \frac{64}{3} \text{ sq. units.}$$

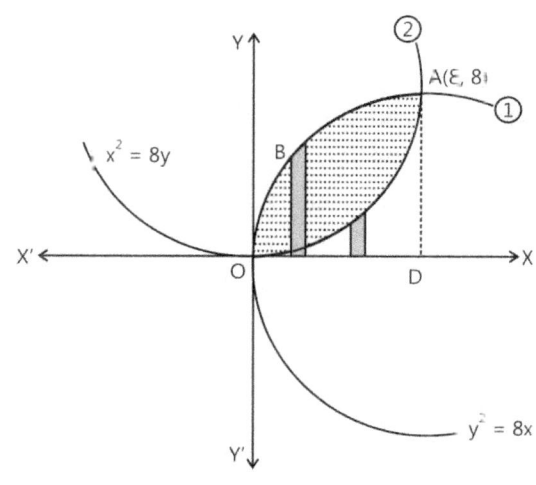

Figure 25.20

Illustration 18: Find the area between the curves $y = 2x$, $x + y = 1$ and x-axis. **(JEE MAIN)**

Sol: Here $y = 2x$ and $x + y = 1$ is a two line intersect at $p\left(\frac{1}{3}, \frac{2}{3} \right)$, therefore using integration method we can obtain required area.

Given $y = 2x$ ... (i)

and, $x + y = 1$ ... (ii)

Solving (i) and (ii), we get $x + 2x = 1$ ⟹ $x = 1/3$.

Line (i) intersects with the x – axis at the origin and the line (ii) intersects with the x – axis at x = 1.

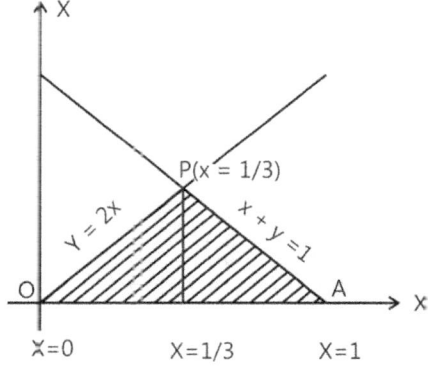

Figure 25.21

So required area $= \int_0^{1/3} 2x\,dx + \int_{1/3}^1 (1-x)\,dx = \left[x^2 \right]_0^{1/3} + \left(x - \frac{x^2}{2} \right)_{1/3}^1$

$= \frac{1}{9} + \left(\frac{1}{2} \right) - \left(\frac{1}{3} - \frac{1}{18} \right) = \frac{1}{3}$ sq. units

Illustration 19: Using the method of integration, find the area of the region bounded by lines: $2x + y = 4$, $3x - 2y = 6$ and $x - 3y + 5 = 0$ **(JEE ADVANCED)**

Sol: Same as above problem.

Given equation of the lines are $2x + y = 4$ ... (i)

$3x - 2y = 6$ ... (ii)

$x - 3y + 5 = 0$ ... (iii)

Solving (i) and (ii), we get (2, 0)

Solving (ii) and (iii), we get (4, 3)

Solving (i) and (iii), we get (1, 2)

$$\therefore \text{ Required Area } = \int_1^4 \left(\frac{x+5}{3}\right) dx - \int_1^2 (4-2x) dx - \int_2^4 \left(\frac{3x-6}{2}\right) dx$$

$$= \frac{1}{3}\left[\frac{x^2}{2} + 5x\right]_1^4 - [4x - x^2]_1^2 - \frac{1}{2}\left[\frac{3x^2}{2} - 6x\right]_2^4$$

$$= \frac{1}{3}\left[(8+20) - \left(\frac{1}{2} + 5\right)\right] - [(8-4) - (4-1)] - \frac{1}{2}[(24-24) - (6-12)]$$

$$= = \frac{7}{2} \text{ sq. units.}$$

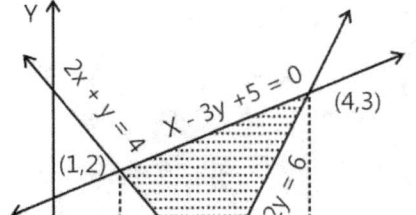

Figure 25.22

SKETCH OF STANDARD CURVES

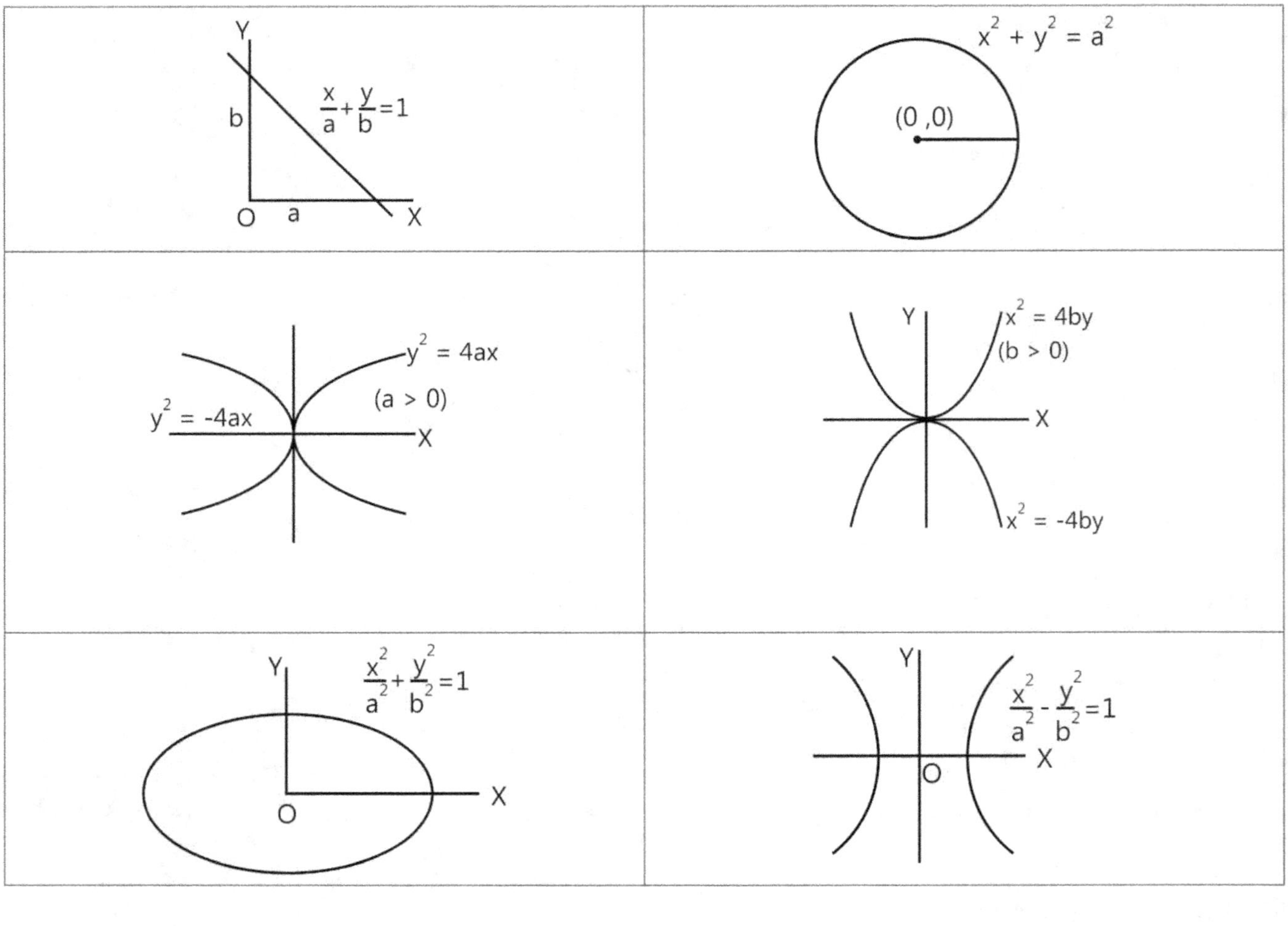

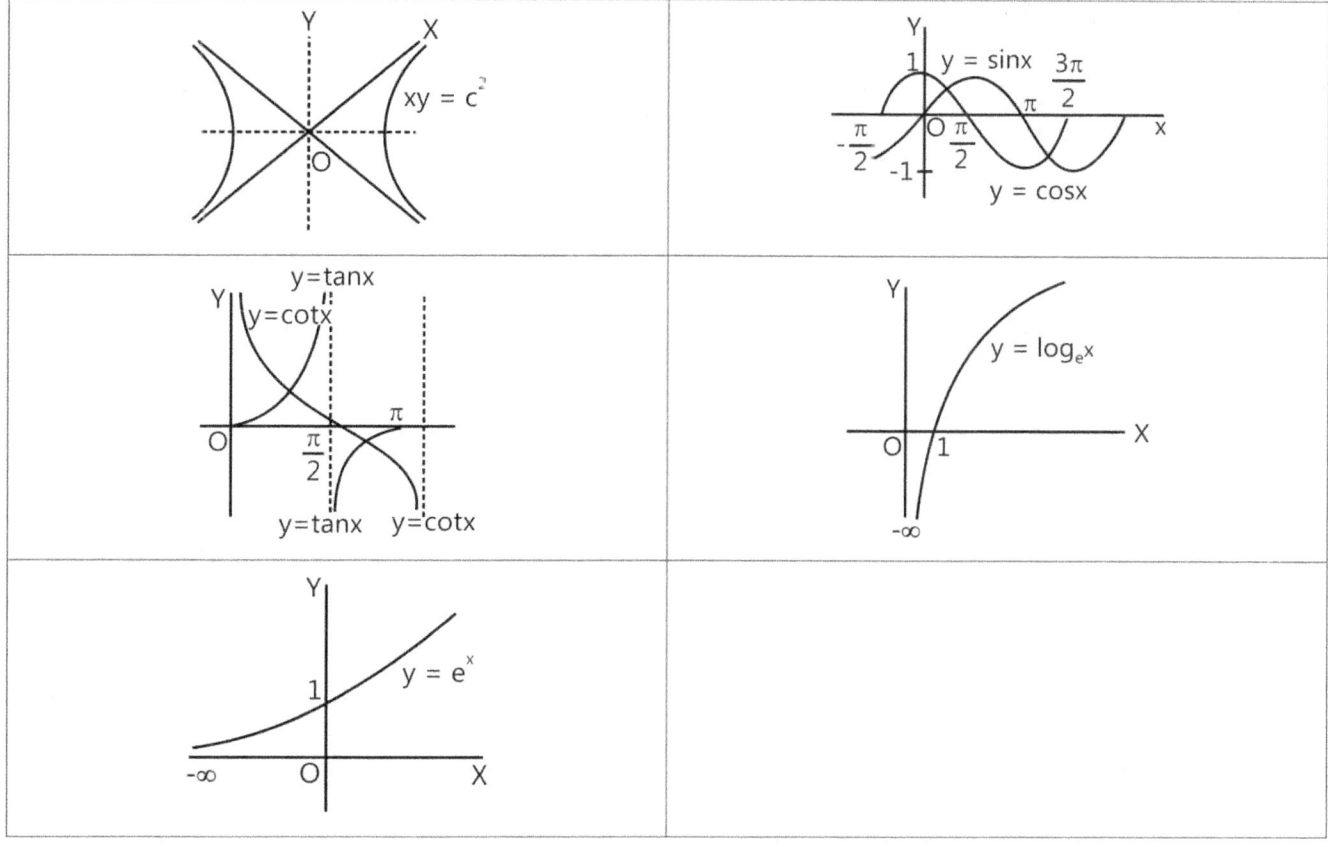

4. STANDARD AREAS

4.1 Area Bounded by Two Parabolas

Area between the parabolas $y^2 = 4ax$ and $x^2 = 4$; $a > 0$, $b > 0$, is

$$|A| = \frac{16ab}{3}$$

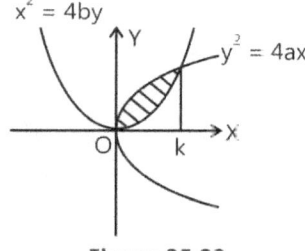

Figure 25.23

Illustration 20: Find the area bounded by $y = \sqrt{x}$ and $x = \sqrt{y}$.

(JEE MAIN)

Sol: By using above mentioned formula.

Area bounded is shaded in the figure

Here, $a = \dfrac{1}{4}$ and $b = \dfrac{1}{4}$

$\therefore$ Using the above formula, Area = (16 ab)/ 3

$$= \frac{16 \times (1/4) \times (1/4)}{3} = \frac{1}{3}$$

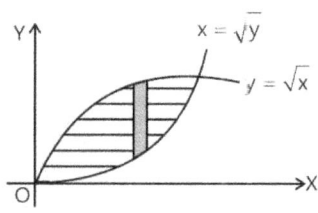

Figure 25.24

4.2 Area Bounded By Parabola and a Line

Area bounded by $y^2 = 4ax$ and $y = mx$; $a > 0$, $m > 0$ is $A = \dfrac{8a^2}{3m^3}$

Area bounded by $x^2 = 4ay$ and $y = mx$; $a > m > 0$

is $y = mx$; $a > m > 0$ $A = \dfrac{8a^2}{3m^3}$

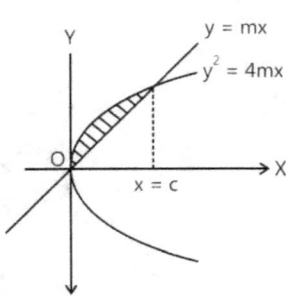

Figure 25.25

(JEE MAIN)

Illustration 21: Find the area bounded by, $x^2 = y$ and $y = |x|$.

Sol: Using above formula, i.e. $A = \dfrac{8a^2}{3m^3}$

Area bounded is shaded in the Fig. 25.26.

Here, $a = 1/4$, $m = 1$

∴ Using the above formula, Area $= 2\left(\dfrac{8a^2}{3m^3}\right) = \dfrac{2 \times 8 \times \left(\dfrac{1}{4}\right)^2}{3 \times (1)^3} = \dfrac{1}{3}$

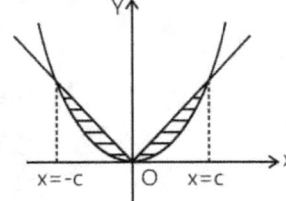

Figure 25.26

(JEE MAIN)

Illustration 22: Find the area bounded by $y^2 = x$ and $x = |y|$.

Sol: Here, $a = 1/4$, $m = 1$, and required area is divided in to two equal parts at above and below x – axis.

Hence required area will be $2\left(\dfrac{8a^2}{3m^3}\right)$.

∴ Using the above formula, Area $= 2\left(\dfrac{8a^2}{3m^3}\right) = \dfrac{2 \times 8 \times (1/4)^2}{3 \times (1)^3} = \dfrac{1}{3}$

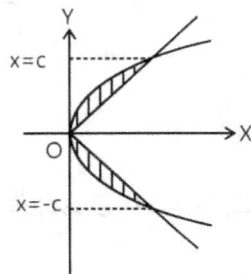

Figure 25.27

4.3 Area Enclosed by Parabola and It's Chord

Area between $y^2 = 4ax$ and its double ordinate at $x = a$ is

Area of AOB $= \dfrac{2}{3}$ (area $\square$ABCD)

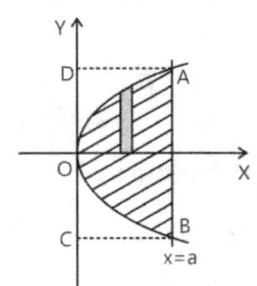

Figure 25.28

(JEE MAIN)

Illustration 23: Find the area bounded by $y = 2x - x^2$, $y + 3 = 0$.

Sol: Here first obtain area of rectangle ABCD and after that by using above mentioned formula we will be get required area.

Solving $y = 2x - x^2$, $y + 3 = 0$, we get $x = -1$ or 3

Area (ABCD) $= 4 \times 4 = 16$.

∴ Required area $= \dfrac{2}{3} \times 16 = \dfrac{32}{3}$

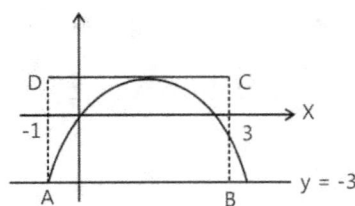

Figure 25.29

4.4 Area of an Ellipse

For an ellipse of the form $\dfrac{x^2}{a^2} + \dfrac{y^2}{b^2} = 1$ is $\qquad$ A = πab

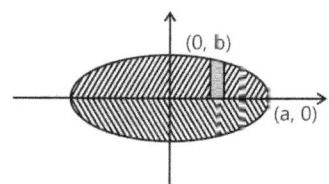

Figure 25.30

> ### NOMORECLASS CONCEPTS
>
> Try to remember some standard areas like for ellipse, parabola. These results are sometimes very helpful.

5. SHIFTING OF ORIGIN

Area remains unchanged even if the coordinate axes are shifted or rotated or both. Hence shifting of origin / rotation of axes in many cases proves to be very convenient in finding the area.

For example: If we have a circle whose centre is not origin, we can find its area easily by shifting circle's centre.

Illustration 24: The line 3x + 2y = 13 divides the area enclosed by the curve $9x^2 + 4y^2 - 18x - 16y - 11 = 0$ into two parts. Find the ratio of the larger area to the smaller area. **(JEE ADVANCED)**

Sol: Given $9x^2 + 4y^2 - 18x - 16y - 11 = 0$ $\qquad\qquad$ ... (i)

$\qquad$ and, 3x + 2y = 13 $\qquad\qquad$ ... (ii)

$9(x^2 - 2x) + 4(y^2 - 4y) = 11;$

$\Rightarrow 9[(x-1)^2 - 1] + 4[(y-2)^2 - 4] = 11$

$\Rightarrow 9(x-1)^2 + 4(y-2)^2 = 36$

$\Rightarrow \dfrac{(x-1)^2}{4} + \dfrac{(y-2)^2}{9} = 1 \Rightarrow \dfrac{X^2}{4} + \dfrac{Y^2}{9} = 1$ $\qquad$ (where X = x − 1 and Y = y − 2)

Hence 3x + 2y = 13

$\Rightarrow 3(X + 1) + 2(Y + 2) = 13$

$\Rightarrow 3X + 2Y = 6$

$\Rightarrow \dfrac{X}{2} + \dfrac{Y}{3} = 1$

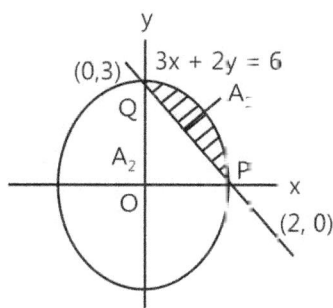

$\therefore$ Area of triangle OPQ = 1/2 × 2 × 3 = 3

Also area of ellipse = π (semi major axes) (semi minor axis) = $\pi . 3 . 2 = 6\pi$

Figure 25.31

$A_1 = \dfrac{6\pi}{4} - \text{area of } \triangle OPQ = \dfrac{3\pi}{2} - 3$

$A_2 = 3\left(\dfrac{6\pi}{4}\right) + \text{area of } \triangle OPQ = \dfrac{9\pi}{2} + 3$

Hence, $\dfrac{A_2}{A_1} = \dfrac{\dfrac{9\pi}{2} + 3}{\dfrac{3\pi}{2} - 3} = \dfrac{3\pi + 2}{\pi - 2}$

6. DETERMINATION OF PARAMETERS

In this type of questions, you will be given area of the curve bounded between some axes or points, and some parameter(s) will be unknown either in equation of curve or a point or an axis. You have to find the value of the parameter by using the methods of evaluating area.

Illustration 25: Find the value of c for which the area of the figure bounded by the curves $y = \dfrac{4}{x^2}$; x = 1 and $y = c$ is equal to $\dfrac{9}{4}$. **(JEE MAIN)**

Sol: By using method of evaluating area we can find out the value of c.

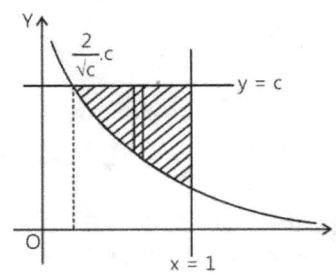

$$A = \int_{2/\sqrt{c}}^{1} \left(c - \frac{4}{x^2} \right) dx = \frac{9}{4}; \quad \left(cx + \frac{4}{x} \right)_{\frac{2}{\sqrt{c}}}^{1} = \frac{9}{4}$$

$$(c+4) - (2\sqrt{c} + 2\sqrt{c}) = \frac{9}{4}; \quad c - 4\sqrt{c} + 4 = \frac{9}{4}$$

$$\Rightarrow \ (\sqrt{c} - 2)^2 = \frac{9}{4} \Rightarrow (\sqrt{c} - 2) = \frac{3}{2} \text{ or } -\frac{3}{2}$$

Figure 25.32

Hence c = (49/4) or (1/4)

Illustration 26: Consider the two curves:

$C_1 : y = 1 + \cos x$, and $C_2 : y = 1 + \cos(x - \alpha)$ for $\alpha \in (0, \pi/2)$ and $x \in [0, p]$.

Find the value of α, for which the area of the figure bounded by the curves C_1, C_2 and x = 0 is same as that of the area bounded by C_2, y = 1 and x = π. For this value of α, find the ratio in which the line y = 1 divides the area of the figure by the curves C_1, C_2 and x = π. **(JEE ADVANCED)**

Sol: Solve C_1 and C_2 to obtain the value of x, after that by following given condition we will be obtain required value of α.

Solving C_1 and C_2 , we get

$$1 + \cos x = 1 + \cos(x - \alpha) \Rightarrow \ x = \alpha - x \quad \Rightarrow x = \frac{\alpha}{2}$$

According to the question,

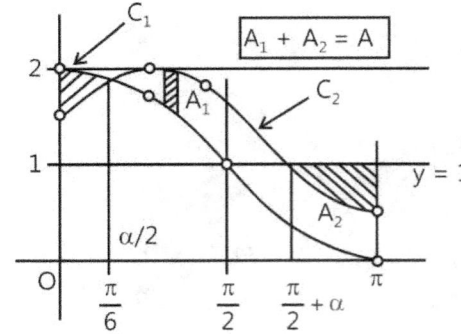

$$\int_{0}^{\alpha/2} (\cos x - \cos(x - \alpha)) dx = - \int_{\frac{\pi}{2}+\alpha}^{\pi} (\cos(x - \alpha)) dx$$

$$\Rightarrow \left[\sin x - \sin(x - \alpha) \right]_{0}^{\alpha/2} = \left[\sin(x - \alpha) \right]_{\pi}^{\frac{\pi}{2}+\alpha}$$

$A_1 + A_2 = A$

Figure 25.33

$$\Rightarrow \left[\sin\frac{\alpha}{2} - \sin\left(-\frac{\alpha}{2} \right) \right] - [0 - \sin(-\alpha)] = \sin\left(\frac{\pi}{2} \right) - \sin(\pi - \alpha)$$

$$\Rightarrow 2\sin\frac{\alpha}{2} - \sin\alpha = 1 - \sin\alpha . \text{ Hence, } 2\sin\frac{\alpha}{2} = 1 \quad \Rightarrow \alpha = \frac{\pi}{3}$$

7. AREA BOUNDED BY THE INVERSE FUNCTION

The area of the region bounded by the inverse of a given function can also be calculated using this method. The graph of inverse of a function is symmetric about the line y = x. We use this property to calculate the area. Hence, area of the function between x = a to x = b, is equal to the area of inverse function from f(a) to f(b).

Illustration 27: Find the area bounded by the curve g(x), the x-axis and the lines at y = –1 and

y = 4, where g(x) is the inverse of the function $f(x) = \dfrac{x^3}{24} + \dfrac{x^2}{8} + \dfrac{13x}{12} + 1$. **(JEE MAIN)**

Sol: Here f(x) is a strictly increasing function therefore required area will be

$$A = \int_0^2 (4 - f(x))dx + \int_{-2}^0 (f(x) + 1)dx$$

Given $f(x) = \dfrac{x^3}{24} + \dfrac{x^2}{8} + \dfrac{13x}{12} + 1$

$\Rightarrow$ f(0) = 1; f(2) = 4 and f(–2) = –1

Also, $f'(x) = \dfrac{x^2}{8} + \dfrac{x}{4} + \dfrac{13}{12}$,

i.e. f(x) is a strictly increasing function.

$$\therefore A = \int_0^2 (4 - f(x))dx + \int_{-2}^0 (f(x) + 1)dx$$

$$A = \int_0^2 \left(4 - \dfrac{x^3}{24} - \dfrac{x^2}{8} - \dfrac{13x}{12} - 1\right)dx + \int_{-2}^0 \left(\dfrac{x^3}{24} + \dfrac{x^2}{8} + \dfrac{13x}{12} + 1 + 1\right)dx$$

$$\therefore A = \left[\left(3.2 - \dfrac{2^4}{24.4} - \dfrac{2^3}{8.3} - \dfrac{13.2^2}{12.2}\right) - (0)\right] + \left[(0) - \left(\dfrac{2^4}{24.4} - \dfrac{2^3}{8.3} + \dfrac{13.2^2}{12.2} - 2.2\right)\right] = \dfrac{16}{3}$$

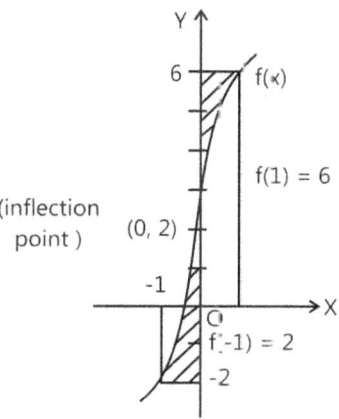

Figure 25.34

Illustration 28: Let f(x) = x³ + 3x + 2 and g(x) is the inverse of it. Find the area bounded by g(x), the x-axis and the ordinate at x = –2 and x = 6. **(JEE ADVANCED)**

Sol: Let $A = \int_{-2}^6 \left|f^{-1}(x)\right| dx$

Substitute $x = f(u)$ or $u = f^{-1}(x)$

$$= \int_{f^{-1}(2)}^{f^{-1}(6)} |u| \, f^{-1}(u) \, du$$

$$= \int_{f^{-1}(2)}^{f^{-1}(6)} |4| \left(3u^2 + 3\right) du$$

We have, $f(-1) = 2$ and $f(1) = 6$

$$= \int_{-1}^1 |u| \left(3u^2 + 3\right) du = 2 \int_0^1 \left(3u^3 + 3u\right) du$$

$$= \left[\dfrac{3}{2} u^4 + 3u^2\right]_0^1 = \dfrac{9}{2} \text{ Sq. units.}$$

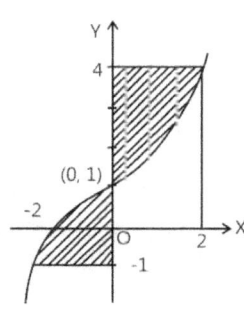

Figure 25.35

8. VARIABLE AREA

If $y = f(x)$ is a monotonic function in (a, b), then the area of the function $y = f(x)$ bounded by the lines at $x = a$, $x = b$, and the line $y = f(c)$, [where $c \in (a, b)$] is minimum when $c = \dfrac{a+b}{2}$.

Proof: $A = \displaystyle\int_a^c f(c) - f(x) dx + \int_c^b (f(x) - f(c)) dx$

$= f(c)(c-a) - \displaystyle\int_a^c (f(x)) dx + \int_c^b (f(x)) dx - f(c)(b-c)$

$= \{(c-a) - (b-c)\} f(c) + \displaystyle\int_c^b (f(x)) dx - \int_a^c (f(x)) dx$

$A = [2c - (a+b)] f(c) + \displaystyle\int_c^b (f(x)) dx - \int_a^c (f(x)) dx$

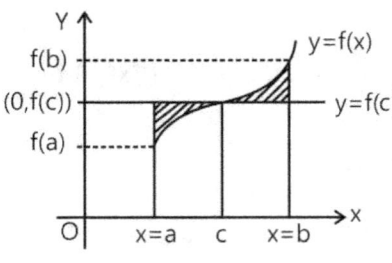

Figure 25.36

For maxima and minima $\dfrac{dA}{dc} = 0 \Rightarrow f'(c) = [2c - (a+b)] = 0$ (as $f'(c) = 0$) hence $c = \dfrac{a+b}{2}$ also $c < \dfrac{a+b}{2}, \dfrac{dA}{dc} < 0$

and $c > \dfrac{a+b}{2}, \dfrac{dA}{dc} > 0$ Hence A is minimum when $c = \dfrac{a+b}{2}$

9. AVERAGE VALUE OF A FUNCTION

In this section, we would study the average of a continuous function. This concept of average is frequently applied in physics and chemistry.

Average of a function $f(x)$ between $x = a$ to $x = b$ is given by $y_{av} = \dfrac{1}{b-a} \displaystyle\int_a^b f(x) dx$

NOMORECLASS CONCEPTS

(a) Average value can be positive, negative or zero .

(b) If the function is defined in $(0, \infty)$, then $y_{av} = \displaystyle\lim_{b \to \infty} \dfrac{1}{b} \int_0^b f(x) dx$ provided the limit exists

(c) Root mean square value (RMS) is defined as $\rho = \left[\dfrac{1}{b-a} \displaystyle\int_a^b f^2(x) dx \right]^{\frac{1}{2}}$

(d) If a function is periodic then we need to calculate average of function in particular time period that is its overall mean.

Illustration 29: Find the average value of y^2 w.r.t. x for the curve $ay = b\sqrt{a^2 - x^2}$ between $x = 0$ & $x = a$. Also find the average value of y w.r.t. x^2 for $0 \le x \le a$. **(JEE MAIN)**

Sol: As average of a function $f(x)$ between $x = a$ to $x = b$ is given by $y_{av} = \dfrac{1}{b-a} \displaystyle\int_a^b f(x) dx$

Let $f(x) = y^2 = \dfrac{b^2}{a^2}(a^2 - x^2)$ Now $f(x)\big|_{av} = \dfrac{b^2}{a^2(a-0)}\displaystyle\int_0^a (a^2 - x^2)dx = \dfrac{2b^2}{3}$

Again y_{av} w.r.t. x^2 as $f(x)\big|_{av} = \dfrac{1}{(a^2 - 0)}\displaystyle\int_0^{a^2} y\, d(x^2) = \dfrac{b}{a^2 a}\displaystyle\int_0^{a^2} \sqrt{a^2 - x^2}\, dx^2 = \dfrac{b}{a^3}\displaystyle\int_0^{a^2} 2t^2\, dt = \dfrac{2ba^3}{3}$

10. DETERMINATION OF FUNCTION

Sometimes the area enclosed by a curve s given es a variable function anc we have to find the function. The area function A_a^x satisfies the differential equation $\dfrac{dA_a^x}{dx} = f(x)$ with initial concition $A_a^a = 0$ i.e. derivative of the area function is the function itself. Thus we can easily f nd $f(x)$ by differentiating area function.

NOMORECLASS CONCEPTS

If $F(x)$ is integral of $f(x)$ then, $A_a^x = \displaystyle\int f(x)\,dx = [F(x) + c]$

And since, $A_a^a = 0 = F(a) + c$ $\Rightarrow c = -F(a)$.

$\therefore A_a^x = F(x) - F(a)$. Finally by taking $x = b$ we get, $A_a^b = F(b) - F(a)$

Note that this is true only if the function doesn't have any zeroes between a and b.

If the function has zero at c then area $= |F(b) - F(c)| + |F(c) - F(a)|$

Illustration 30: The area from 0 to x under a certa n graph is given to be $A = \sqrt{1 + 3x} - 1$, $x \geq 0$;

(a) Find the average rate of change of A w.r.t. x and x increases from 1 to 8.

(b) Find the instantaneous rate of change of A w.r.t x at x = 5.

(c) Find the ordinate (height) y of the graph as a function of x.

(d) Find the average value of the ordinate (height) y, w.r.t. x as x increases from 1 to 8.   **(JEE ADVANCED)**

Sol: Here by differentiating given area function we can obtain the main funztion.

(a) $A(1) = 1$, $A(8) = 4$; $\dfrac{A(8) - A(1)}{8 - 1} = \dfrac{3}{7}$

(b) $\dfrac{dA}{dx}\bigg|_{x=5} = \dfrac{1.3}{2\sqrt{1+3x}}\bigg|_{x=5} = \dfrac{3}{8}$

(c) $y = \dfrac{3}{2\sqrt{1+3x}}$

(d) $\dfrac{1}{(8-1)}\displaystyle\int_1^8 \dfrac{3}{2\sqrt{1+3x}}\,dx = \dfrac{1}{7}\displaystyle\int_1^8 \dfrac{3}{2\sqrt{1+3x}}\,dx = \dfrac{3}{7}$

Illustration 31: Let C_1 & C_2 be the graphs of the function $y = x^2$ & $y = 2x$, $0 \le x \le 1$ respectively. Let C_3 be the graphs of a function $y = f(x)$, $0 \le x \le 1$, $f(0) = 0$. For a point P on C_1, let the lines through P, parallel to the axes, meet C_2 & C_3 at Q & R respectively (see figure). If for every position of P(on C_1), the area of the shaded regions OPQ & ORP are equal, determine the function $f(x)$. **(JEE ADVANCED)**

Sol: Similar to the above mentioned method.

$$\int_0^{h^2} \left(\sqrt{y} - \frac{y}{2} \right) dy = \int_0^h (x^2 - f(x)) dx \text{ differentiate both sides w.r.t. h}$$

$$\left(h - \frac{h^2}{2} \right) 2h = h^2 - f(h)$$

$$f(h) = h^2 - \left(h - \frac{h^2}{2} \right) 2h$$

$$= h^2 - h(2h - h^2) = h^2 - 2h^2 + h^3$$

$$f(h) = h^3 - h^2$$

$$f(x) = x^3 - x^2 = x^2(x - 1)$$

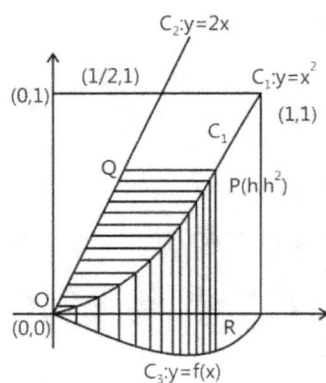

Figure 25.37

11. AREA ENCLOSED BY A CURVE EXPRESSED IN POLAR FORM

$r = a(1 + \cos\theta)$ (Cardioid)

$$A = \frac{1}{2} \int_0^{2\pi} r^2 d\theta = \frac{a^2}{2} \int_0^{2\pi} 4\cos^4 \frac{\theta}{2} d\theta$$

Substitute $\frac{\theta}{2} = t$, $d\theta = 2dt$

$$A = a^2 \int_0^{\pi} 4\cos^4 t\, dt = 8 \times \frac{3\pi a^2}{16}$$

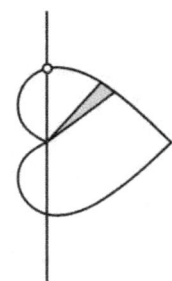

Figure 25.38

Illustration 32: Find the area enclosed by the curves $x = a\sin^3 t$ and $y = a\cos^3 t$. **(JEE MAIN)**

Sol:

$$x^{\frac{2}{3}} + y^{\frac{2}{3}} = a^{\frac{2}{3}} \quad \text{and} \quad dx = 3a\sin^2 t.\cos t.dt$$

$$A = 4\int_0^a y\, dx \; ; \; A = 4a^2 \int_0^{\pi/2} 3\cos^3 t\sin^2 t\cos t dt$$

$$A = 12a^2 \int_0^{\pi/2} \sin^2 t\cos^4 t\, dt = (12a^2) \cdot \frac{1.3.1}{6.4.2} \cdot \frac{\pi}{2} = \frac{12a^2\pi}{32} = \frac{3\pi a^2}{8}$$

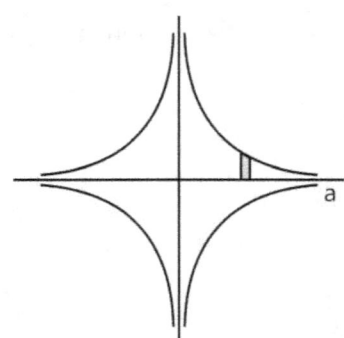

Figure 25.39

Linear Programming

1. INTRODUCTION

Linear Programming was developed during World War II, when a system with which to maximize the efficiency of resources was of utmost importance.

2. LINEAR PROGRAMMING

Linear programming may be defined as the problem of maximising or minimising a linear function subject to linear constraints. The constraints may be equalities or inequalities. Here is an example.

Find numbers x_1 and x_2 that maximize the sum $x_1 + x_2$ subject to the constraints $x_1 \geq 0$, $x_2 \geq 0$, and

$$x_1 + 2x_2 \leq 4$$
$$4x_1 + 2x_2 \leq 12$$
$$-x_1 + x_2 \leq 1$$

Here we have two unknowns and five inequalities (constraints). Notice that these constraints are all linear functions of the variables. The first two constraints, $x_1 \geq 0$ and $x_2 \geq 0$, are special. These are called no negativity constraints and are often found in linear programming problems. The other constraints are called the main constraints. The function to be maximised (or minimized) is called the objective function. In the above example the objective function is $x_1 + x_2$.

3. GRAPHICAL METHOD

As we have only two variables, we can solve this problem by plotting the constraints with x_1 and x_2 as axes. The intersection region of these inequalities is called feasible region for the objective function. This is the region which satisfies all the constraints. Now from this feasible region we have to select point(s) such that objective function is maximized or minimized.

Theorem 1: Let R be the feasible region (convex polygon) for a linear programming problem and let Z=ax + by be the objective function. When Z has an optimal value (maximum or minimum), where the variables x and y are subject to constraints described by linear inequalities, this optimal value must occur at a corner point (vertex) of the feasible region.

Theorem 2: Let R be the feasible region for a linear programming problem and let Z=ax + by be the objective function. If R is bounded, then the objective function Z has both a maximum and a minimum value on R and each of these occurs at corner point (vertex) of R.

Remark: If R is unbounded, then a maximum or a minimum value of the objective function may not exist. However, if it exists it must occur at a corner point of R. (By Theorem 1).

So for the above example

Corner point (x_1, x_2)	Z (= $x_1 + x_2$) value
0,1	1
3,0	3
8/3,2/3	10/3
2/3,5/3	7/3

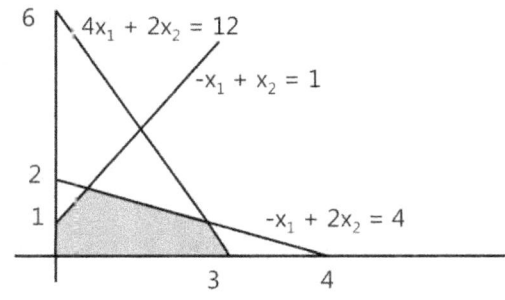

Figure 25.40

Hence (8/3, 2/3) is the optimal solution.

Note that z has also minimum value in the feasible region at (0, 1).

This method of solving is generally called as corner point method. Note that a function can have more than one optimal points.

4. MODELS

There are few important linear programming models which are more frequently used and some of them we encounter in our daily lives.

(a) **Manufacturing/Assignment problems:** In these problems, we determine the number of units of different products which should be produced and sold by a firm when each product requires a fixed manpower, machine hours, labour hours per unit of product, warehouse space per unit of the output. In order to make maximum profit.

Example: There are I persons available for J jobs. The value of person i working 1 day at job j is a_{ij}, for i = 1,......,I, and j = 1,......,J. The problem is to choose an assignment of persons to jobs to maximize the total value.

An assignment is a choice of numbers, x_{ij}, for i = 1,......,I, and j=1,......,J, where x_{ij} represents the proportion of person i's time that is to be spent on job j. Thus,

$$\sum_{j=1}^{J} x_{ij} \leq 1 \quad \text{For i = 1,......, I} \qquad \text{... (i)}$$

$$\sum_{i=1}^{I} x_{ij} \leq 1 \quad \text{For j=1,......, J} \qquad \text{... (ii)}$$

And $x_{ij} \geq 0$ for i = 1,......,I, and j=1,......,J   ... (iii)

Equation (i) reflects the fact that a person cannot spend more than 100% of his time working, (ii) means that only one person is allowed on a job at a time, and (iii) says that no one can work a negative amount of time on any job, Subject to (i), (ii) and (iii) , we wish to maximize the total value of $\sum_{i=1}^{I} \sum_{j=1}^{J} a_{ij} x_{ij}$

(b) **Diet problems:** In these problems, we determine the amount of different kinds of nutrients which should be included in a diet so as to minimise the cost of the desired diet such that it contains a certain minimum amount of each nutrients.

Example: There are m different types of food, $F_1,........,F_m$, that supply varying quantities of the n nutrients , $N_1,........,N_n$, that are essential to good health. Let c_j be the minimum daily requirement of nutrient, N_j contained in one unit of food F_i. The problem is to supply the required nutrients at minimum cost.

Let y_i be the number of units of food F_i to be purchased per day. The cost per day of such a diet is

$$b_1 y_1 + b_2 y_2 + ... + b_m y_m \qquad \text{... (i)}$$

The amount of nutrient N_j contained in this diet is

$$a_{1j} y_1 + a_{2j} y_2 + ... + a_{mj} y_m$$

For j = 1,........, n. We do not consider such a diet unless all the minimum daily requirements are met, that is, unless

$$a_{1j} y_1 + a_{2j} y_2 + ... + a_{mj} y_m \geq c_j \qquad \text{For j = 1,........, n} \qquad \text{... (ii)}$$

Of course, we cannot purchase a negative amount of food, so we automatically have the constraints

$$y_1 \geq 0, y_2 \geq 0,.................,y_m \geq 0 \qquad \text{... (iii)}$$

Our problem is: minimize (i) subject to (ii) and (iii). This is exactly the standard minimum problem.

(c) **Transportation problems:** In these problems, we determine a transportation schedule in order to find the cheapest way of transporting a product from plants/factories situated at different locations to different markets.

Example: There are I ports, or production plants, $P_1, ...P_I$, that supply a certain commodity, and there are J markets, $M_1, ...M_J$, to which this commodity must be shipped. Port P_i possesses an amount s of the commodity (i=1,2,......I), and market M_j must receive the amount r_j of the commodity (j = 1,.........J). Let b_{ij} be the cost of transporting one unit of the commodity from port P_i to market M_j. The problem is to meet the market requirements at minimum transportation cost is

$$\sum_{i=1}^{I}\sum_{j=1}^{J} y_{ij} b_{ij}$$ ... (i)

The amount sent from port P_i is $\sum_{j=1}^{J} y_{ij} \leq y_{ij}$ and since the amount available at port P_i is s_i, we must have

$$\sum_{j=1}^{J} y_{ij} \leq s_i \text{ for i = 1,......,I}$$ ... (ii)

The amount sent to market M_j is $\sum_{i=1}^{I} y_{ij}$, and since the amount required there is r_j, we must have

$$\sum_{i=1}^{I} y_{ij} \leq r_j \text{ for j = 1,......,I}$$ ... (iii)

It is assumed that we cannot send a negative amount from P_I to M_j, we have

$$y_{ij} \geq 0 \quad \text{for I = 1,........I and j =1,........J.}$$ .. (iv)

Our problem is minimize (i) subject to (ii), (iii) and (iv).

FORMULAE SHEET

(a) **Area bounded by a curve with x – axis:** Area $= \int_a^b y\, dx = \int_a^b f(x) dx$

(b) **Area bounded by a curve with y – axis:** Area $= \int_c^d x\, dy = \int_c^d f(y) dy$

(c) **Area of a curve in parametric form:** Area $= \int_a^b y\, dx = \int_{t_2}^{t_1} g(t) f'(t) dt$

(d) **Positive and Negative Area:** $A = \left| \int_a^c f(x) dx \right| + \left| \int_c^b f(x) dx \right|$;

(e) **Area between two curves:**

(i) Area enclosed between two curves intersecting at two different points.

Area $= \int_a^b (y_1 - y_2) dx = \int_a^b [f_1(x) - f_2(x)] dx$

(ii) Area enclosed between two curves intersecting at one point and the x – axis.

Area $= \int_a^\alpha f_1(x) dx + \int_\alpha^b f_2(x) dx$

(iii) Area bounded by two intersecting curves and lines parallel to y – axis.

Area $= \int_a^c (f(x) - g(x)) dx + \int_c^b (g(x) - f(x)) dx$

(a) Standard Areas:

(i) Area bounded by two parabolas $y^2 = 4ax$ and $x^2 = 4by$; $a > 0$, $b > 0$: Area $= \dfrac{16ab}{3}$

(ii) Area bounded by Parabola $y^2 = 4ax$ and Line $y = mx$: Area $= \dfrac{8a^2}{3m^3}$

(iii) Area of an Ellipse $\dfrac{x^2}{a^2} + \dfrac{y^2}{b^2} = 1$: Area $= \pi ab$

Solved Examples

JEE Main/Boards

Example 1: Find area bounded by $y = 4 - x^2$, x-axis and the lines $x = 0$ and $x = 2$.

Sol: By using the formula of Area Bounded by the x – axis, we can obtain

Required Area.

$$= \int_0^2 y\,dx = \int_0^2 (4 - x^2)\,dx$$

$$= \left(4x - \frac{x^3}{3}\right)_0^2 = 8 - \frac{8}{3} = \frac{16}{3} \text{ sq. units}$$

Example 2: Find the area bounded by the curve $y^2 = 2y - x$ and the y-axis.

Sol: Here given equation is the equation of parabola with vertex (1, 1) and curve passes through the origin.

Curve is $y^2 - 2y = -x$ or $(y - 1)^2 = -(x - 1)$

It is a parabola with

Vertex at (1, 1) and the curve passes through the origin. At B, $x = 0$ and $y = 2$

Area

$$= \int_0^2 x\,dy = \int_0^2 (2y - y^2)\,dy = \left(y^2 - \frac{y^3}{3}\right)_0^2 = \frac{4}{3} \text{ sq. units}$$

Example 3: Find the area of the region $\{(x, y) : x^2 \le y \le x\}$

Sol: Consider the function $y = x^2$ and $y = x$ Solving them, we get $x = 0$, $y = 0$ and $x = 1$, $y = 1$; $x^2 \le y \Rightarrow$ area

is above the curve $y = x^2 y \le x \Rightarrow$ area is below the line $y = x$

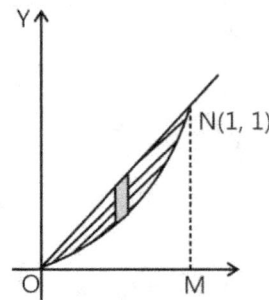

$$\text{Area} = \int_0^1 (x - x^2)\,dx = \left(\frac{x^2}{2} - \frac{x^3}{3}\right)_0^1 = \frac{1}{6} \text{ sq. units}$$

Example 4: Find the area of the region enclosed by $y = \sin x$, $y = \cos x$ and x-axis, $0 \le x \le \dfrac{\pi}{2}$.

Sol: Find point of intersection is P. Therefore after obtaining the co-ordinates of P and then integrating with appropriate limits, we can obtain required Area.

At point of intersection P,

$x = \dfrac{\pi}{4}$ as ordinates of $y = \sin$

x and ; $y = \cos x$ are equal

Hence, P is $\left(\dfrac{\pi}{4}, \dfrac{1}{\sqrt{2}}\right)$ Required area

$$= \int_0^{\pi/4} \sin x\,dx + \int_{\pi/4}^{\pi/2} \cos x\,dx = (-\cos x)_0^{\pi/4} + (\sin x)_{\pi/4}^{\pi/2}$$

$$= \left(-\frac{1}{\sqrt{2}} + 1\right) + \left(1 - \frac{1}{\sqrt{2}}\right) = 2 - \sqrt{2} \text{ sq. units}$$

Example 5: The area bounded by the continuous curve $y = f(x)$, (lying above the x-axis), x-axis and the ordinates $x = 1$ and $x = b$ is $(b - 1) \sin (3b + 4)$. Find $f(x)$

Sol: Using Leibniz rule, we can solve given problem.

$$\int_1^b f(x)\,dx = (b-1)\sin(3b+4)$$

Apply Leibniz Rule: differentiate both sides w.r.t. "b",

$f(b) = \sin(3b+4) + 3(b-1)\cos(3b+4)$

$\Rightarrow f(x) = \sin(3x+4) + 3(x-1)\cos(3x+4)$

Example 6: Find the area bounded by the curve $y = k \sin x$ and $y = 0$ from $x = 0$ to $x = 2\pi$.

Sol: Here the area of OAB is above the x-axis ($y = 0$) and thus it is positive while the area BCD is below x-axis ($y = 0$) and in negative but equal in quantity.

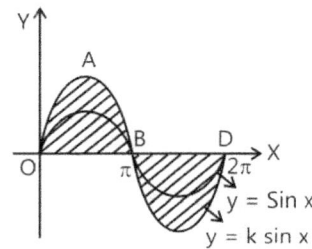

$$\text{Area OAB} = \int_0^\pi y\,dx = \int_0^\pi k\,\sin x\,dx = k[-\cos x]_0^\pi$$

$= k[-\cos\pi] - k[-\cos 0]$

$= k[-(-1)] - k[-(1)] = k + k = 2k$

$\therefore$ Total area = 4k sq. units.

Example 7: Find the area bounded by the curve

$x = a(\theta - \sin\theta),\ y = a(1 - \cos\theta),\ 0 \le \theta \le 2\pi$, with x-axis.

Sol: Substitute the value of y and dx and integrate.

$$\text{Area} = \int_{\theta=0}^{\theta=2\pi} y\,dx = \int_{\theta=0}^{\theta=2\pi} y\frac{dx}{d\theta}\,d\theta$$

$$= \int_0^{2\pi} a(1-\cos\theta)a(1-\cos\theta)\,d\theta$$

$$= a^2 \int_0^{2\pi} (1 - 2\cos\theta + \cos^2\theta)\,d\theta$$

$$= a^2 \int_0^{2\pi} \left[1 - 2\cos\theta + \frac{1}{2}(1 + \cos 2\theta)\right]$$

$$= a^2 \left[\frac{3}{2}\theta - 2\sin\theta + \frac{1}{4}\sin 2\theta\right]_0^{2\pi} = 3\pi a^2 \text{ sq. units.}$$

Example 8: Find the area bounded by the curves $\{(x, y) : y \ge x^2,\ y \le |x|\}$

Sol: Here the region is symmetric about y-axis, the required area is 2[area of shaded region in first quadrant].

The curves intersect each other at $x = 0$ and $x = \pm 1$ as shown in figure. The points of intersection are $(-1, 1)$, $(0, 0)$ and $(1, 1)$.

Since, the region is symmetric about y-axis, the required area is 2[area of shaded region]

Hence, Area $= 2\int_0^1 (x - x^2)\,dx = 2\left[\frac{1}{2}x^2 - \frac{1}{3}x^3\right]_{-0}^{-1} = \frac{1}{3}$ sq. units.

Example 9: Draw a rough sketch of the curve $y = \sin^2 x$, $x \in \left[0, \frac{\pi}{2}\right]$. Find the area enclosed between the curve, x-axis and the line $x = \frac{\pi}{2}$.

Sol: Here by substituting $x = 0, \frac{\pi}{6}, \frac{\pi}{4}, \frac{\pi}{3}, \frac{\pi}{2}$ we will get respective values of y. hence by plotting these values we can draw the given curve.

Some points on the $\sin^2 x$ graph are :

x	0	$\frac{\pi}{6}$	$\frac{\pi}{4}$	$\frac{\pi}{3}$	$\frac{\pi}{2}$
y	0	0.25	0.5	0.75	1

By plotting points and joining them, we trace the curve.

Area bounded by curve $y = \sin^2 x$ between $x = 0$ and $x = \frac{\pi}{2}$

$$= \int_0^{\frac{\pi}{2}} y\,dx = \int_0^{\frac{\pi}{2}} \sin^2 x\,dx = \frac{1}{2}\int_0^{\frac{\pi}{2}} (1 - \cos 2x)\,dx$$

$$= \frac{1}{2}\left[x - \frac{\sin 2x}{2}\right]_0^{\frac{\pi}{2}} = \frac{1}{2}\left[\left(\frac{\pi}{2} - 0\right) - (0 - 0)\right]$$

$$= \frac{\pi}{4} \text{ sq. units}$$

JEE Advanced/Boards

Example 1: A tangent is drawn to $x^2 + 2x - 4ky + 3 = 0$ at a point whose abscissa is 3. The tangent is perpendicular to $x + 3 = 2y$. Find the area bounded by the curve, this tangent, x-axis and line $x = -1$

Sol: As we know multiplication of slopes of two perpendicular line is -1, by using this, we can obtain the value of k and will get standard equation. After that using integration with respective limit, we will be get required area.

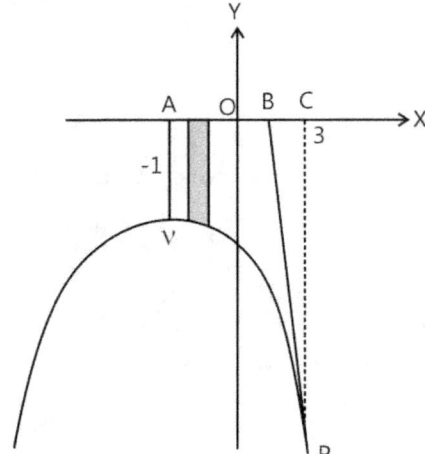

$x^2 + 2x - 4ky + 3 = 0$; $\dfrac{dy}{dx} = \dfrac{x+1}{2k}$ Tangent is perpendicular to $x + 3 = 2y$

$\therefore \dfrac{x+1}{2k}\left(\dfrac{1}{2}\right) = -1$ at $x = 3$

$\Rightarrow 1/k = -1 \Rightarrow k = -1$

$\therefore$ Curve becomes $(x + 1)^2 = -4(y + 1/2)$ which is a parabola with vertex at $V(-1, -1/2)$.

Coordinates of P are $(3, -9/2)$.

Equation of tangent at P is $y + 9/2 = -2(x - 3)$

B is $(3/4, 0)$, C is $(3, 0)$

Required Area = Area (ACPV) – Area of triangle BPC.

$$= \left| \int_{-1}^{3} \dfrac{x^2 + 2x + 3}{-4} dx \right| - \dfrac{1}{2}(BC)(CP)$$

$$= \dfrac{1}{4}\left| \left(\dfrac{x^3}{3} + x^2 + 3x \right)_{-1}^{3} \right| - \dfrac{1}{2}\left(3 - \dfrac{3}{4} \right)\left(\dfrac{9}{2} \right)$$

$$= \dfrac{1}{4}\left(27 + \dfrac{1}{3} - (1 - 3) \right) - \dfrac{81}{16} = \dfrac{109}{48} \text{ sq. units.}$$

Example 2: Let A_n be the area bounded by the curve $y = (\tan x)^n : n \in N$ and the lines $x = 0$, $y = 0$ and $x = \dfrac{\pi}{4}$. Prove that for $n > 2$, $A_n + A_{n-2} = \dfrac{1}{n-1}$ and deduce $\dfrac{1}{2n+2} < A_n < \dfrac{1}{2n-2}$.

Sol: We can write $(\tan x)^n$ as $\tan^{n-2} x (\sec^2 x - 1)$. Therefore by solving $A_n = \displaystyle\int_0^{\pi/4} \tan^{n-2} x (\sec^2 x - 1) dx$ we can prove given equation.

$$A_n = \int_0^{\pi/4} \tan^n x \, dx : n > 2$$

$$= \int_0^{\pi/4} \tan^{n-2} x (\sec^2 x - 1) dx$$

$$\text{or } A_n = \left[\dfrac{\tan^{n-1} x}{n-1} \right]_0^{\pi/2} - A_{n-2}$$

$$\therefore A_n + A_{n-2} = \dfrac{1}{n-1} \qquad \ldots \text{(i)}$$

$\tan^n x \le \tan^{n-2} x$

(as $0 \le \tan x \le 1$ for $0 \le x \le \dfrac{\pi}{4}$)

$\Rightarrow A_n < A_{n-2}$

$\therefore A_n + A_n < A_n + A_{n-2} = \dfrac{1}{n-1}$ by (1)

$$\therefore A_n < \dfrac{1}{2(n-1)} \qquad \ldots \text{(ii)}$$

Similarly $A_{n+2} < A_n$

$\Rightarrow A_{n+2} + A_n < A_n + A_n$

$$\text{or } \dfrac{1}{(n+2)-1} < 2A_n \text{ by (1)}$$

$$\Rightarrow \dfrac{1}{2n+2} < A_n \qquad \ldots \text{(iii)}$$

$$\Rightarrow \dfrac{1}{2n+2} < A_n < \dfrac{1}{2n-2}$$

Example 3: $A(a, 0)$ and $B(0, b)$ are points on the ellipse $\dfrac{x^2}{a^2} + \dfrac{y^2}{b^2} = 1$. Show that the area between the arc AB and chord AB of the ellipse is $\dfrac{1}{4} ab (\pi - 2)$.

Sol: Area between the chord and ellipse = Area bounded by curve AB - Area of $\triangle$ OAB.

Equation of line AB is : $y = -\dfrac{b}{a}(x - a)$

Equation of curve AB is $y = \dfrac{b}{a}\sqrt{a^2 - x^2}$

Area of bounded region is

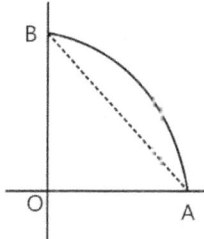

$$\int_0^a \left[\frac{b}{a}\sqrt{a^2 - x^2} - \left(-\frac{b}{a}(x-a)\right)\right] dx$$

$$= \frac{b}{a}\left[0 + \frac{a^2\pi}{4} - \frac{a^2}{2}\right] = \frac{(\pi - 2)}{4}ab$$

Alternate method:

Area between the chord and ellipse = Area bounded by curve AB - Area of $\triangle$ OAB

$$= \frac{1}{4}\pi ab - \frac{1}{2}ab = \frac{(\pi - 2)ab}{4} \text{ sq. units}$$

Example 4: Find the area of the region bounded $y = \dfrac{1}{x}+1, x = 1$ and tangent drawn at the point $P(2, 3/2)$ to the curve $y = \dfrac{1}{x} + 1$.

Sol: Here first obtain equation of tangent and then use the formula for area.

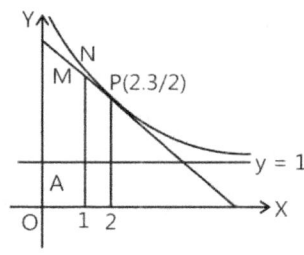

Equation of tangent at $P(2, 3/2)$ to $y = \dfrac{1}{x}+1$ is $y - \dfrac{3}{2} = -\dfrac{1}{4}(x-2)$ or $x + 4y = 8$.

Required area is area of region PMN

$$\text{Area} = \int_1^2 \left(\left(\frac{1}{x}+1\right) - \frac{8-x}{4}\right) dx$$

$$= \left(\ln x - x + \frac{1}{4}\frac{x^2}{2}\right)_1^2 = \ln 2 - \frac{5}{8} \text{ sq. units}$$

Example 5: Find the area of the region bounded by the x-axis and the curve $y = \dfrac{1}{2}(2 - 3x - 2x^2)$.

Sol: Here the curve will intersect the x-axis when $y = 0$, therefore by substituting $y = 0$ in the above equation we will get the points of intersection of curve and x – axis.

$\Rightarrow 2 - 3x - 2x^2 = 0$ or $(2+x)(1-2x) = 0$ or $x = -2$, $x = \dfrac{1}{2}$

Thus, the curve passes through the points $(-2, 0)$ and

$\left(\dfrac{1}{2}, 0\right)$ on the x-axis.

It will have a turning points where $\dfrac{dy}{dx} = 0$

$$\therefore \frac{dy}{dx} = \frac{1}{2}(-3 - 4x) = 0 \ \triangleright x = -\frac{3}{4}$$

Also $\dfrac{d^2y}{dx^2} = -4$. That is, it is a max. at $x = \dfrac{3}{4}$

Also it cuts y-axis where $x = 0$, then $y = 1$. Thus the shape of the curve is as shown in the figure.

The required area is ABC. It is given by

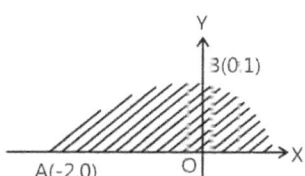

$$\int_{-2}^{1/2} y \, dx = \int_{-2}^{1/2} \frac{1}{2}(2 - 3x - 2x^2) dx$$

$$= \frac{1}{2}\left[2x - \frac{3}{2}x^2 - \frac{2x^3}{3}\right]_{-2}^{1/2}$$

$$= \frac{1}{2}\left[2\left(\frac{1}{2}\right) - \frac{3}{2}\left(\frac{1}{2}\right)^2 - \frac{2}{3}\left(\frac{1}{2}\right)^3\right] -$$

$$\frac{1}{2}\left[2(-2) - \frac{3}{2}(-2)^2 - \frac{2}{3}(-2^3)\right]$$

$$= \frac{1}{2}\left(\frac{13}{24}\right) - \frac{1}{2}\left(-\frac{14}{3}\right) = \frac{125}{48} \text{ sq. units.}$$

Example 6: Find the area of the region bounded by the curve $x^2 = 4y$ and $x = 4y - 2$.

Sol: Solving given equation simultaneously, we will get the point of intersection. Using these points as the limits of integration, we calculate the required area.

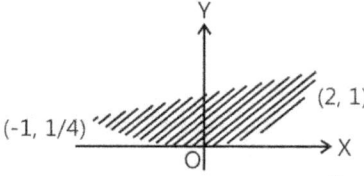

The curve intersect each other, where $\dfrac{x^2}{4} = \dfrac{x+2}{4}$, or $x^2 - x - 2 = 0$, or $x = -1, 2$

Hence, the points of intersection are $(-1, 1/4)$ and $(2, 1)$. The region is plotted in figure. Since, the straight line $x = 4y - 2$ is always above the parabola $x^2 = 4y$ in the interval $[-1, 2]$, the required area is given by

$$\text{Area} = \int_{-1}^2 [f(x) - g(x)] dx$$

$$\text{Area} = \int_{-1}^{2}\left[\frac{x+2}{4} - \frac{x^2}{4}\right]dx = \frac{1}{4}\left[\frac{1}{2}x^2 + 2x - \frac{1}{3}x^3\right]_{-1}^{2}$$

$$= \frac{1}{4}\left[\left(2 + 4 - \frac{8}{3}\right) - \left(\frac{1}{2} - 2 + \frac{1}{3}\right)\right] = \frac{9}{8} \text{ sq. units.}$$

Example 7: Find by using integration, the area of the ellipse $ax^2 + 2hxy + by^2 = 1$.

Sol: The equation can be put in the form $by^2 + 2hxy + (ax^2 - 1) = 0$

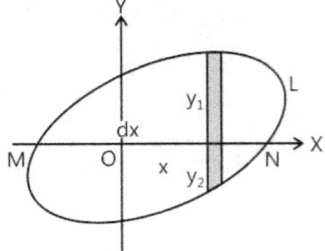

Cut an elementary strip.

Let the thickness of strip = dx

If y_1, y_2 be the values of y corresponding to any value at x.

Length of strip = $y_1 - y_2$

$$= \frac{2}{b}\sqrt{h^2x^2 - b(ax^2 - 1)} = \frac{2}{b}\sqrt{b - (ab - h^2)x^2}$$

$ab - h^2$ being positive here, since the conic is an ellipse.

The extreme values of x, are given by

$$y_1 - y_2 = 0, \text{ i.e., } x = \pm\sqrt{\frac{b}{ab - h^2}}$$

Hence, the area required $= \displaystyle\int_{-\sqrt{b/ab - h^2)}}^{\sqrt{b/ab - h^2)}} (y_1 - y_2)dx$

$$= \int_{-\sqrt{b/(ab-h^2)}}^{\sqrt{b/(ab-h^2)}} \frac{2}{b}\sqrt{b - (ab - h^2)x^2}\, dx$$

and putting $\sqrt{(ab - h^2)}\,x = \sqrt{b}\sin\theta$, this becomes

$$\frac{2}{\sqrt{(ab-h^2)}}\int_{-\pi/2}^{\pi/2}\cos^2\theta\, d\theta = \frac{\pi}{(ab-h^2)} \text{ sq. units.}$$

Example 8: Find the area of region lying above x-axis, and included between the circle $x^2 + y^2 = 2ax$ and the parabola $y^2 = ax$.

Sol: By solving these two equation simultaneously, we can obtain their intersection points and then by subtracting area of parabola from area of circle we will be get the result.

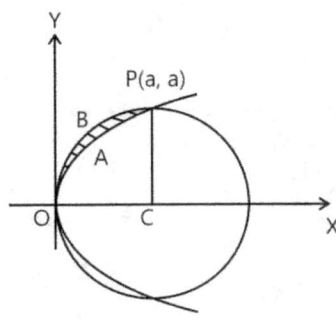

Solving the two equation, simultaneously we see that the two curves intersect at $(0, 0)$, (a, a) and $(a, -a)$. We have to find the area of the region OAPBO, where P is the point of intersection (a, a)

Required area $= \displaystyle\int_{0}^{a}[f(x) - g(x)]dx$

$$= \int_{0}^{a}\sqrt{(2ax - x^2)}\, dx - \int_{0}^{a}\sqrt{ax}\, dx$$

Now, $\displaystyle\int_{0}^{a}\sqrt{(2ax - x^2)}\, dx = \int_{0}^{a}\sqrt{[a^2 - (a - x)^2]}\, dx$

To evaluate this integral, we substitute $a - x = a\sin\theta$ and obtain

$$\int_{0}^{a}\sqrt{(2ax - x^2)}\, dx = \int_{\pi/2}^{0}(a\cos\theta)(-a\cos\theta)d\theta$$

$$= \int_{0}^{\pi/2}a^2\cos^2\theta\, d\theta = a^2\frac{1}{2}\frac{\pi}{2} = \frac{\pi a^2}{4}$$

Also $\displaystyle\int_{0}^{a}\sqrt{(ax)}\, dx = \left|\sqrt{a}\frac{2}{3}x^{3/2}\right|_{0}^{a} = \frac{2a^2}{3}$

$\therefore$ Required area $= a^2\left(\dfrac{\pi}{4} - \dfrac{2}{3}\right)$ sq. units

Example 9: Prove that the area of the region bounded by the curve $a^4y^2 = x^5(2a - x)$, is $\dfrac{5}{4}$ times to that of the circle whose radius is a.

Sol: The curve is a loop lying between the line $x = 0$ and $x = 2a$ and is symmetrical about the x-axis. Therefore the required area

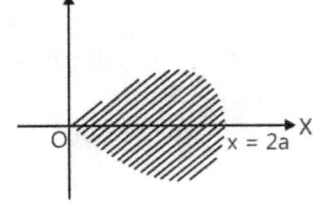

$$= 2\int_{0}^{2a} y\, dx$$

$$= \frac{2}{a^2}\int_{0}^{2a} x^{5/2}\sqrt{2a - x}\, dx$$

To evaluate this integral, we put $x = 2a \sin^2 q$. When, $x = 0, \theta = 0$ and when $x = 2a, \theta = \frac{1}{2}\pi$ $\therefore$ Required area

$$= \frac{2}{a^2} \int_0^{\pi/2} (2a)^{5/2} \sin^5 \theta . \sqrt{2a} \cos\theta . 4a \sin\theta \cos\theta \, d\theta$$

$$= 64a^2 \int_0^{\pi/2} \sin^6\theta \cos^2\theta \, d\theta = 64a^2 \frac{5.3.1.1}{8.6.4.2} . \frac{\pi}{2} = \frac{5a^2\pi}{4}$$

$$= \frac{5}{4} \times \text{ area of the circle whose radius is a.}$$

Example 10: Find the area bounded by the curves $x^2 + y^2 = 25$, $4y = |4 - x^2|$ and $y = 0$.

Sol: Here $x^2 + y^2 = 25$ represent circle with centre at origin and radius 5 unit. Therefore the required area $= 2$ area ABC

$$= 2\left[\int_2^4 \frac{1}{4}(4 - x^2)dx + \int_4^5 \sqrt{25^2 - x^2}\, dx \right]$$

Note: Here the portion is also bounded by two curves but we do not apply

$A = \int [f(x) - g(x)]dx$ rule.

Reason: Range of integration of both the

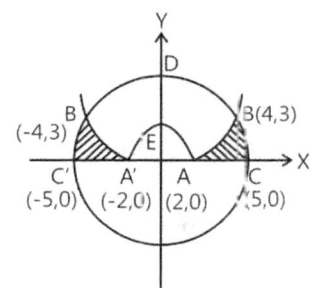

curves is not same.

$$= 2\left[\int_2^4 \frac{1}{4}(x^2 - 4)dx + \int_4^5 \sqrt{(5^2 - x^2)}\, dx \right]$$

$$= \frac{2}{4}\left[\frac{x^3}{3} - 4x \right]_2^4 + 2\left[\frac{x}{2}\sqrt{25 - x^2} + \frac{25}{2}\sin^{-1}\frac{x}{5} \right]_4^5$$

$$= \frac{1}{2}\left[\left(\frac{64}{3} - 16\right) - \left(\frac{8}{3} - 8\right) \right] +$$

$$2\left[\left(0 + \frac{25}{2}\sin^{-1}1\right) - \left(6 + \frac{25}{2}\sin^{-1}\frac{4}{5}\right) \right]$$

$$= \frac{1}{2}\left[\frac{32}{3} \right] + 25\sin^{-1}1 - 12 - 25\sin^{-1}\frac{4}{5}$$

$$= \left(\frac{16}{3} - 12\right) + 25\frac{\pi}{2} - 25\sin^{-1}\frac{4}{5}$$

$$= -\frac{20}{3} + 25\left(\frac{\pi}{2} - \sin^{-1}\frac{4}{5}\right) = \left(25\cos^{-1}\frac{4}{5} - \frac{20}{3}\right) \text{ sq. units.}$$

JEE Main/Boards

Exercise 1

Q.1 Write an expression for finding the area of the shaded portion.

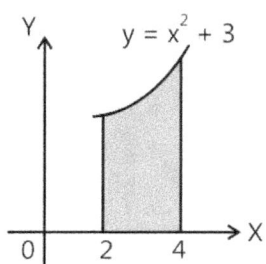

Q.2 Find the area bounded by the curve $y = \cos x$, x-axis and between $x = 0$, $x = \pi$.

Q.3 Find the area bounded by the curve $y = \sin x$, x-axis and between $x = 0$, $x = \pi$.

Q.4 Write an expression for finding the area of the shaded portion.

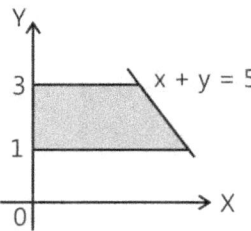

Q.5 Write an expression for finding the area bounded by the curve $x^2 = y$ and the line $y = 2$.

Q.6 Write an expression for finding the area of a circle $x^2 + y^2 = a^2$, above x-axis.

Q.7 On sketching the graph of $y = |x - 2|$ and evaluating $\int_{-1}^{3} |x - 2|\, dx$, what does $\int_{-1}^{3} |x - 2|\, dx$, represent on the graph?

Q.8 Draw the rough sketch of the curve $y = \sqrt{3x + 4}$ and find the area under the curve above x-axis and between $x = 0$ and $x = 4$.

Q.9 Find the area under the curve $y = \dfrac{3}{(1 - 2x)^3}$ above x-axis and between $x = -4$ an $x = -1$.

Q.10 Find the area bounded by the curves $y = 6x - x^2$ and $y = x^2 - 2x$.

Q.11 Draw a rough graph of $f(x) = \sqrt{x} + 1$ in the interval [0, 4] and find the area of the region enclosed by the curve, x-axis and the lines $x = 0$ and $x = 4$.

Q.12 Find the area of the region bounded by the curve $xy - 3x - 2y - 10 = 0$; x-axis and the lines $x = 3$, $x = 4$.

Q.13 Find the area bounded by the curve $y = x \sin x^2$, x-axis and between $x = 0$ and $x = \sqrt{\dfrac{\pi}{2}}$.

Q.14 Using integration, find the area of the region bounded by the following curves, after making a rough sketch:
$y = |x + 1| + 1$, $x = -2$, $x = 3$, $y = 0$.

Q.15 Draw the rough sketch of $y = \sin 2x$ and determine the area enclosed by the curve, the x-axis and the lines $x = \pi/4$ and $x = 3\pi/4$.

Q.16 Find the area of the following region: $\{(x, y) : x^2 + y^2 \leq 2ax, y^2 \geq ax, x \geq 0, y \geq 0\}$.

Q.17 Find the area bounded by the curve $y^2 = 4a^2(x - 3)$ and the lines $x = 3$, $y = 4a$.

Q.18 Make a rough sketch of the region given below and find its area using integration. $\{(x,y): 0 \leq y \leq x^2 + 3$; $0 \leq y \leq 2x + 3, 0 \leq x \leq 3\}$.

Q.19 Determine the area enclosed between the curve $y = 4x - x^2$ and the x-axis.

Exercise 2

Single Correct Choice Type

Q.1 The area of the figure bounded by the curve $y = e^x$, $y = e^{-x}$ and the straight line $x = 1$ is

(A) $e + \dfrac{1}{e}$

(B) $e - \dfrac{1}{e}$

(C) $e + \dfrac{1}{e} - 2$

(D) None of these

Q.2 The area bounded in the first quadrant by the normal at (1, 2) on the curve $y = 4x$, x-axis & the curve is given by

(A) $\dfrac{10}{3}$ (B) $\dfrac{7}{3}$ (C) $\dfrac{4}{3}$ (D) $\dfrac{9}{2}$

Q.3 The area of the figure bounded by the curves $y = \ln x$ and $y = (\ln x)^2$ is

(A) $e + 1$ (B) $e - 1$ (C) $3 - e$ (D) 1

Q.4 The area bounded by the curves $y = x^2 + 1$ & the tangents to it drawn from the origin is:

(A) 2/3 (B) 4/3 (C) 1/3 (D) 1

Q.5 The area bounded by $x^2 + y^2 - 2x = 0$ & $y = \sin\dfrac{\pi x}{2}$ in the upper half of the circle is

(A) $\dfrac{\pi}{2} - \dfrac{4}{\pi}$ (B) $\dfrac{\pi}{4} - \dfrac{2}{\pi}$ (C) $\pi - \dfrac{8}{\pi}$ (D) $\dfrac{\pi}{2} - \dfrac{2}{\pi}$

Q.6 Consider the region formed by the lines $x = 0$, $y = 0$, $x = 2$, $y = 2$. Area enclosed by the curves $y = e^x$ and $y = \ln x$, within this region is removed, then the area of the remaining region is

(A) $2(1 + 2\ell n^2)$ (B) $2(2\ell n2 - 1)$

(C) $(2\ell n2 - 1)$ (D) $1 + 2\ell n^2$

Q.7 The area bounded by the curves $y = x(1 - \ln x)$; $x = e^{-1}$ and positive x-axis between $x = e^{-1}$ and $x = e$ is

(A) $\left(\dfrac{e^2 - 4e^{-2}}{5}\right)$ (B) $\left(\dfrac{e^2 - 5e^{-2}}{4}\right)$

(C) $\left(\dfrac{4e^2 - e^{-2}}{5}\right)$ (D) $\left(\dfrac{5e^2 - e^{-2}}{4}\right)$

Q.8 The positive values of the parameter 'a' for which the area of the figure bounded by the curve $y = \cos ax$, $y = 0$, $x = \dfrac{\pi}{6a}$, $x = \dfrac{x\pi}{2a}$ is greater than 3 are

(A) f

(B) (0, 1/3)

(C) (3, ∞)

(D) None of these

Q.9 The value of 'a' (a > 0) for which the area bounded by the curves $y = \dfrac{x}{6} + \dfrac{1}{x^2}$, $y = 0$, $x = a$ and $x = 2a$ has the least value, is

(A) 2
(B) $\sqrt{2}$
(C) $2^{1/3}$
(D) 1

Q.10 The ratio in which the area enclosed by the curve $y = \cos x$ $\left(0 \leq 0 \leq \dfrac{\pi}{2}\right)$ in the first quadrant is divided by the curve $y = \sin x$, is

(A) $(\sqrt{2} - 1) : 1$
(B) $(\sqrt{2} + 1) : 1$
(C) $\sqrt{2} : 1$
(D) $\sqrt{2} + 1 : \sqrt{2}$

Q.11 The area bounded by the curve $y = f(x)$, the co-ordinate axes & the line $x = x_1$ is given by $x_1 . e^{x_1}$. Therefore f(x) equals

(A) e^x
(B) $x\, e^x$
(C) $xe^x - e^x$
(D) $xe^x - e^x$

Q.12 The area bounded by the curves $y = -\sqrt{-x}$ and $x = -\sqrt{-y}$ where $x, y \leq 0$

(A) Cannot be determined

(B) Is 1/3

(C) Is 2/3

(D) Is same as that of the figure bounded by the curves $y = \sqrt{-x}$; $x \leq 0$ and $x = \sqrt{-y}$; $y \leq 0$

Q.13 The area from 1 to x under a certain graph is given by $A = (1 + 3x)^{1/2} - 1$, $x \geq 0$.. The average value of y w.r.t. x as x increases from 1 to 8 is

(A) 3/7
(B) 1/2
(C) 3/8
(D) 4/7

Q.14 The slope of the tangent to a curve $y = f(x)$ at $(x, f(x))$ is $2x + 1$. If the curve passes through the point (1, 2) then the area of the region by the curve, the x-axis and the line x = 1 is

(A) 5/6
(B) 6/5
(C) 1/6
(D) 1

Q.15 The area of the region for which $0 < y < 3 - 2x - x^2$ & x > 0 is

(A) $\displaystyle\int_1^3 (3 - 2x - x^2)\,dx$
(B) $\displaystyle\int_0^3 (3 - 2x - x^2)\,dx$

(C) $\displaystyle\int_0^1 (3 - 2x - x^2)\,dx$
(D) None of these

Q.16 The graphs of $f(x) = x^2$ and $g(x) = cx^3$ (c > 0) intersect at the points (0, 0) & $\left(\dfrac{1}{c}, \dfrac{1}{c^2}\right)$. If the region which lies between these graphs & over the interval [0, 1/c] has the area equal to 2/3 then the value of c is

(A) 1
(B) 1/3
(C) 1/2
(D) 2

Q.17 The curv linear trapezoid is bounded by the curve $y = x^2 + 1$ and the straight lines x = 1 and x = 2. The co-ordinates of the point (on the given curve) with abscissa xÎ[1, 2] where tangent drawn cut off from the curvilinear trapezoid are ordinary trapezium of the greatest area, is

(A) (1, 2)
(B) (2, 5)

(C) $\left(\dfrac{3}{2}, \dfrac{13}{4}\right)$
(D) None of these

Q.18 In the given figure, if A_1 is the area of the △AOB and A_2 is the area of the parabolic region AOB then the ratio $\dfrac{A_1}{A_2}$ as a → 0 is

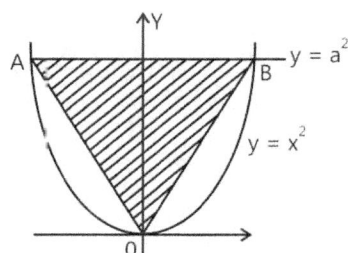

(A) 1
(B) 8/9
(C) 3/4
(D) 2/3

Previous Years' questions

Q.1 The area of the quadrilateral formed by the tangents at the end points of latus rectum to the ellipse $\dfrac{x^2}{9} + \dfrac{y^2}{5} = 1$, is **(2003)**

(A) 27/4 sq. unit
(B) 9 sq. unit

(C) 27/2 sq. unit
(D) 27 sq. unit

Q.2 The area bounded by the curves $y = \sqrt{x}, 2y + 3 = x$ and x-axis in the 1st quadrant is *(2003)*

(A) 9 sq. unit   (B) 27/4 sq. unit

(C) 36 sq. unit   (D) 18 sq. unit

Q.3 The area enclosed between the curves $y = ax^2$ and $x = ay^2 (a > 0)$ is 1 sq unit. Then, the value of a is *(2004)*

(A) $\dfrac{1}{\sqrt{3}}$ (B) $\dfrac{1}{2}$ (C) 1   (D) $\dfrac{1}{3}$

Q.4 The area of the equilateral triangle, in which three coins of radius 1 cm are placed, as shown in the figure, is *(2005)*

(A) $(6 + 4\sqrt{3})$ sq. cm

(B) $(4\sqrt{3} - 6)$ sq. cm

(C) $(7 + 4\sqrt{3})$ sq. cm

(D) $4\sqrt{3}$ sq. cm

Q.5 The area enclosed within the curve $|x| + |y| = 1$ is .... *(1981)*

Q.6 The area of the triangle formed by the positive x-axis and the normal and the tangent to the circle $x^2 + y^2 = 4$ at $(1, \sqrt{3})$ is .... *(1989)*

Q.7 The area bounded by the curves $y = (x - 1)^2$, $y = (x + 1)^2$ and $y = \dfrac{1}{4}$ is *(2005)*

(A) $\dfrac{1}{3}$ sq. unit   (B) $\dfrac{2}{3}$ sq. unit

(C) $\dfrac{1}{4}$ sq. unit   (D) $\dfrac{1}{5}$ sq. unit

Q.8 The area (in sq. units) of the region $\{(x,y) : y^2 \geq 2x \text{ and } x^2 + y^2 \leq 4x, x \geq 0, y \geq 0\}$ is: *(2016)*

(A) $\pi - \dfrac{8}{3}$ (B) $\pi - \dfrac{4\sqrt{2}}{3}$

(C) $\dfrac{\pi}{2} - \dfrac{2\sqrt{2}}{3}$ (D) $\pi - \dfrac{4}{3}$

Q.9 The area (in sq. units) of the region described by $\{(x,y) : y^2 \leq 2x \text{ and } y \geq 4x - 1\}$ is : *(2015)*

(A) $\dfrac{5}{64}$ (B) $\dfrac{15}{64}$ (C) $\dfrac{9}{32}$ (D) $\dfrac{7}{32}$

Q.10 The area (in square units) bounded by the curves $y = \sqrt{x}, 2y - x + 3 = 0, x\text{-axis}$, and lying in the first quadrant is *(2013)*

(A) 36   (B) 18   (C) $\dfrac{27}{4}$ (D) 9

Q.11 The area of the region described by A $= \{(x,y) : x^2 + y^2 \leq 1 \text{ and } y^2 \leq 1 - x\}$ is *(2014)*

(A) $\dfrac{\pi}{2} + \dfrac{4}{3}$ (B) $\dfrac{\pi}{2} - \dfrac{4}{3}$

(C) $\dfrac{\pi}{2} - \dfrac{2}{3}$ (D) $\dfrac{\pi}{2} + \dfrac{2}{3}$

Q.12 If the lines $\dfrac{x-1}{2} = \dfrac{y+1}{3} = \dfrac{z-1}{4}$ and $\dfrac{x-3}{1} = \dfrac{y-k}{2} = \dfrac{z}{1}$ intersect, then k is equal to *(2012)*

(A) -1   (B) $\dfrac{2}{9}$ (C) $\dfrac{9}{2}$ (D) 0

Q.13 The area bounded between the parabolas $x^2 = \dfrac{y}{4}$ and $x^2 = 9y$ and the straight line $y = 2$ is *(2012)*

(A) $20\sqrt{2}$ (B) $\dfrac{10\sqrt{2}}{3}$ (C) $\dfrac{20\sqrt{2}}{3}$ (D) $10\sqrt{2}$

Q.14 The area bounded by the curves $y = \cos x$ and $y = \sin x$ between the ordinates $x = 0$ and $x = \dfrac{3\pi}{2}$ is *(2010)*

(A) $4\sqrt{2} + 2$ (B) $4\sqrt{2} - 1$

(C) $4\sqrt{2} + 1$ (D) $4\sqrt{2} - 2$

Q.15 The area of the region enclosed by the curves $y = x, x = e, y = \dfrac{1}{x}$ and the positive x-axis is *(2011)*

(A) 1 sq. unit   (B) $\dfrac{3}{2}$ sq. units

(C) $\dfrac{5}{2}$ sq. units   (D) $\dfrac{1}{2}$ sq. units

Exercise 1

Q.1 In the adjacent figure, graphs of two functions $y = f(x)$ and $y = \sin x$ are given. $y = \sin x$ intersects, $y = f(x)$ at $A(a, f(a))$; $B(\pi, 0)$ and $C(2\pi, 0)$. A_i (i = 1, 2, 3) is the area bounded by the curve $y = f(x)$ and $y = \sin x$ between $x = 0$ and $x = a$; i = 1, between $x = a$ and $x = \pi$; i = 2, between $x = \pi$ and $x = 2\pi$; i = 3.

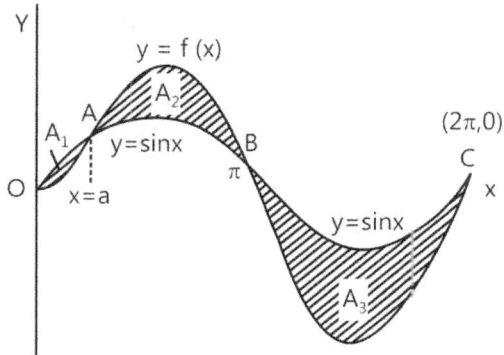

If $A_1 = 1 - \sin a + (a - 1) \cos a$, determine the function $f(x)$. Hence determine 'a' and A_1. Also calculate A_2 and A_3.

Q.2 The figure shows two regions in the first quadrant.

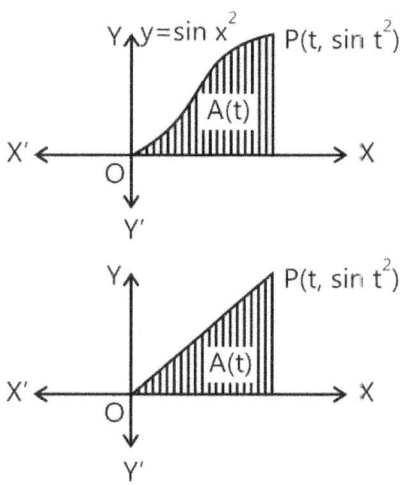

$A(t)$ is the area under the curve $y = \sin x^2$ from 0 to t and $B(t)$ is the area of the triangle with vertices O, P and $M(t, 0)$. Find $\displaystyle \lim_{t \to 0} \frac{A(t)}{B(t)}$.

Q.3 A polynomial function f(x) satisfies the condition $f(x + 1) = f(x) + 2x + 1$. Find f(x) if f(0) = 1. Find also the equations of the pair of tangents from the origin on the curve $y = f(x)$ and compute the area enclosed by the curve and the pair of tangents.

Q.4 Show that the area bounded by the curve $y = \dfrac{\log x - c}{x}$, the x-axis and the vertical line through the maxima point of the curve is independent of the constant c.

Q.5 Consider the curve $y = x^n$ where n > 1 in the 1st quadrant. If the area bounded by the curve, the x-axis and the tangent line to the graph of $y = x^n$ at the point (1, 1) is maximum then find the value of n.

Q.6 For what value of 'a' is the area of the figure bounded by the lines $y = \dfrac{1}{x}, y = \dfrac{1}{2x-1}$, x = 2 & x = a equal to $\ln \dfrac{4}{\sqrt{5}}$?

Q.7 For the curve $f(x) \dfrac{1}{1+x^2}$ let two points on it are $A(\alpha, f(\alpha)), B\left(-\dfrac{1}{\alpha}, f'\left(-\dfrac{1}{\alpha}\right)\right)$ (α > 0). Find the minimum area bounded by the line segments OA, OB and f(x), where 'O' is the origin.

Q.8 Find the area bounded by the curve $y = \sin^{-1} x$ and the lines x = 0, $|y| = \dfrac{\pi}{2}$.

Q.9 If f(x) is monotonic in (a, b) then prove that the area bounded by the ordinates at x = a ; x = b ; y = f(x) and y = f(c), c ∈ (a, b) is minimum when $c = \dfrac{a+b}{2}$. Hence if the area bounded by the graph of $f(x) = \dfrac{x^3}{3} - x^2 + a$, the straight lines x = 0, x = 2 and the x-axis is minimum then find the value of 'a'.

Q.10 Let 'c' be the constant number such that c > 1. If the least area of the figure given by the line passing through the point (1, c) with gradient 'm' and the parabola $y = x^2$ s 36 sq. units find the value of $(c^2 + m^2)$.

Q.11 Let C_1 & C_2 be two curves passing through the origin as shown in the figure. A curve C is said to "bisect the area" the region between C_1 & C_2, if for each point P of C, the two shaded regions A & B shown in the figure have equal area. Determine the upper curve C_2, given that the bisecting curve C has the equation $y = x^2$ & that the lower curve C_1 has the equation $y = x^2/2$.

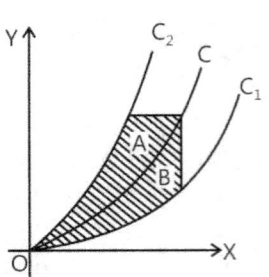

Q.12 Consider one side AB of a square ABCD, (read in order) on the line $y = 2x - 17$ and the other two vertices C, D on the parabola $y = x^2$.

(i) Find the minimum intercept of the line CD on y-axis.

(ii) Find the maximum possible area of the square ABCD.

(iii) Find the area enclosed by the line CD with minimum y-intercept and the parabola $y = x^2$. Consider the two curves $C_1 : y = 1 + \cos x$ & $C_2 : y = 1 + \cos(x - \alpha)$ for $\alpha \in (0, \pi/2)$; $x \in [0, p]$. Find the value of α, for which the area of the figure bounded by the curves C_1, C_2 & $x = 0$ is same as that of the figure bounded by C_2, $y = 1$ & $x = \pi$. For this value of α, find the ratio in which the line $y = 1$ divides the area of the figure by the curves C_1, C_2 & $x = \pi$.

Q.13 Draw the rough sketch of $y^2 = x + 1$ and $y^2 = -x + 1$ and determine area enclosed by the two curves.

Q.14 Let $f : (0, \infty) \to R$ be a continuous and strictly increasing function such that $f^3(x) = \int_0^x t f^2(t)dt$, $\forall\ x > 0$. Find the area enclosed by $y = f(x)$, the x-axis and the ordinate at $x = 3$.

Q.15 For what values of a $\hat{I}[0, 1]$ does the area of the figure bounded by the graph of the function $y = f(x)$ and the straight lines $x = 0$, $x = 1$ & $y = f(a)$ is at a minimum & for what values it is at a maximum if $f(x) = \sqrt{1 - x^2}$. Find also the maximum & the minimum area.

Q.16 Let $f(x) = \sin x\ \forall\ x \in \left[0, \dfrac{\pi}{2}\right]$

$f(x) + f(\pi - x) = 2\ \forall\ x \in \left[\dfrac{\pi}{2}, \pi\right]$ and $f(x) = f(2\pi - x)\ \forall\ x \in (\pi, 2\pi)$.

If the area enclosed by $y = f(x)$ and x-axis is $a\pi + b$, then find the value of $(a^2 + b^2)$.

Q.17 Find the values of $m(m > 0)$ for which the area bounded by the line $y = mx + 2$ and $x = 2y - y^2$ is, (i) 9/2 square units & (ii) minimum. Also find the minimum area.

Q.18 Find the area bounded by the curve $y = x\ e^{-x}$; $xy = 0$ and $x = c$ where c is the x-coordinate of the curve's inflection point.

Q.19 Find the area of the region
$$\left\{(x,y) : y^2 \le 4x, 4x^2 + 4y^2 \le 9\right\}$$

Q.20 For what value of 'a' is the area bounded by the curve $y = a^2x^2 + ax + 1$ and the straight line $y = 0$, $x = 0$ & $x = 1$ the least?

Q.21 Consider two curves $C_1 : y = \dfrac{1}{x}$ and $C_2 : y = \ln x$ on the xy plane. Let D_1 denotes the region surrounded by C_1, C_2 and the line $x = 1$ and D_2 denotes the region surrounded by C_1, C_2 and the line $x = a$. If $D_1 = D_2$. Find the value of 'a'.

Exercise 2

Single Correct Choice Type

Q.1 The area bounded by the curve $y = x^2 - 1$ & the straight line $x + y = 3$ is

(A) $\dfrac{9}{2}$ (B) 4 (C) $\dfrac{7\sqrt{17}}{2}$ (D) $\dfrac{17\sqrt{17}}{6}$

Q.2 The area bounded by the curve $y = e^{-x}$ & the lines $y = e^{-4}$ & $x = 1$ is given by

(A) $\dfrac{e^3 - 4}{e^4}$ (B) $\dfrac{e^3 + 4}{e^4}$

(C) $\dfrac{e^3 + 1}{e^4}$ (D) None of these

Q.3 Area common to the curve $y = \sqrt{9 - x^2}$ & $x^2 + y^2 = 6x$ is

(A) $\dfrac{\pi + \sqrt{3}}{4}$ (B) $\dfrac{\pi - \sqrt{3}}{4}$

(C) $3\left(\pi + \dfrac{\sqrt{3}}{4}\right)$ (D) $3\left(\pi + \dfrac{3\sqrt{3}}{4}\right)$

Q.4 The area bounded by $y = 2 - |2 - x|$ & $y = \dfrac{3}{|x|}$ is

(A) $\dfrac{4 + 3\ln 3}{2}$

(B) $\dfrac{4 - 3\ln 3}{2}$

(C) $\dfrac{3}{2} + \ln 3$

(D) $\dfrac{1}{2} + \ln 3$

Q.5 The area bounded by the curves $y = \sin x$ & $y = \cos x$ between $x = 0$ & $x = 2\pi$ is

(A) $\displaystyle\int_0^{2\pi} (\sin x - \cos x) dx$

(B) $2\sqrt{2}$ sq. unit

(C) 0

(D) $4\sqrt{2}$ sq. unit

Q.6 If $f(x) = -1 + |x - 2|$, $0 \le x \le 4$; $g(x) = 2 - |x|$, $-1 \le x \le 3$.

Then the area bounded by $y = gof(x)$; $x = 1$, $x = 4$ and x-axis is

(A) 7/2 sq. units

(B) 9/4 sq. units

(C) 9/2 sq. units

(D) None of these

Q.7 Area enclosed between the curve $y = \sec^{-1} x$, $y = \csc^{-1} x$ and the line $x = 1$ is

(A) $\ln(3 + 2\sqrt{2})$

(B) $\ln(3 + 2\sqrt{2}) - 1$

(C) $\ln(3 + 2\sqrt{2}) - \pi/2$

(D) None of these

Q.8 The area of the closed figure bounded by $y = x$, $y = -x$ the tangent to the curve $y = \sqrt{x^2 - 5}$ at the point (3, 2) is

(A) 5

(B) $\dfrac{15}{2}$

(C) 10

(D) $\dfrac{35}{2}$

Q.9 The line $y = mx$ bisects the area enclosed by the curve $y = 1 + 4x - x^2$ & the lines $x = 0$, $x = \dfrac{3}{2}$ & $y = 0$, then m is equal to

(A) $\dfrac{13}{6}$

(B) $\dfrac{6}{13}$

(C) $\dfrac{3}{2}$

(D) 4

Q.10 The area common to $y \ge \sqrt{x}$ & $x > -\sqrt{y}$ and the curve $x^2 + y^2 = 2$ is

(A) $\dfrac{\pi}{4} + \dfrac{1}{3}$

(B) $\dfrac{3\pi}{2}$

(C) $\dfrac{1}{3}$

(D) $\dfrac{\pi}{2}$

Q.11 The ratio in which the curve $y = x^2$ divides the region bounded by the curve ; $y = \sin\left(\dfrac{\pi x}{2}\right)$ & the x-axis as x varies from 0 to 1, is

(A) $2 : \pi$

(B) $1 : 3$

(C) $3 : \pi$

(D) $(6 - \pi) : \pi$

Q.12 Area of the region enclosed between the curves $x = y^2 - 1$ and $x = |y|\sqrt{1 - y^2}$ is

(A) 1

(B) 4/3

(C) 2/3

(D) 2

Q.13 Let $y = g(x)$ be the inverse of a bijective mapping f : $R \to R$, $f(x) = 3x^3 + 2x$. The area bounded by the graph of $g(x)$. The x-axis and the ordinate at $x = 5$ is

(A) 5/4

(B) 7/4

(C) 9/4

(D) 13/4

Previous Years' Questions

Q.1 The area of the region between the curves $y = \sqrt{\dfrac{1 + \sin x}{\cos x}}$ and $y = \sqrt{\dfrac{1 - \sin x}{\cos x}}$ and bounced by the lines $x = 0$ and $x = \dfrac{\pi}{4}$ is *(2008)*

(A) $\displaystyle\int_0^{\sqrt{2}-1} \dfrac{t}{(1 + t^2)\sqrt{1 - t^2}} dt$

(B) $\displaystyle\int_0^{\sqrt{2}-1} \dfrac{4t}{(1 + t^2)\sqrt{1 - t^2}} dt$

(C) $\displaystyle\int_0^{\sqrt{2}+1} \dfrac{4t}{(1 + t^2)\sqrt{1 - t^2}} dt$

(D) $\displaystyle\int_0^{\sqrt{2}+1} \dfrac{t}{(1 + t^2)\sqrt{1 - t^2}} dt$

Q.2 Let the straight line $x = b$ divide the area enclosed by $y = (1 - x)^2$, $y = 0$ and $x = 0$ into two parts R_1 $(0 \le x \le b)$ and R_2 $(b \le x \le 1)$ such $R_1 - R_2 = \dfrac{1}{4}$. Then, b equals *(2011)*

(A) $\dfrac{3}{4}$

(B) $\dfrac{1}{2}$

(C) $\dfrac{1}{3}$

(D) $\dfrac{1}{4}$

Q.3 Let f : [−1, 2] → [0, ∞) be a continuous function such that f(x) = f(1 − x) for all x∈[−1, 2], Let $R_1 = \int_{-1}^{2} x f(x) dx$ and R_2 be the area of the region bounded by y = f(x), x = −1, x = 2 and the x-axis. Then, **(2011)**

(A) $R_1 = 2R_2$ (B) $R_1 = 3R_2$

(C) $2R_1 = R_2$ (D) $3R_1 = R_2$

Q.4 Find the area bounded by the curve $x^2 = 4y$ and the straight line x = 4y − 2. **(1981)**

Q.5 Find the area bounded by the x-axis, part of the curve $y = \left(1 + \dfrac{8}{x^2}\right)$ and the ordinates at x = 2 and x = 4. If the ordinate at x = a divides the area into two equal parts, find a. **(1983)**

Q.6 Find the area of the region bounded by the x-axis and the curves defined by y = tan x, $-\dfrac{\pi}{3} \le x \le \dfrac{\pi}{3}$ and y = cot x, $\dfrac{\pi}{6} \le x \le \dfrac{\pi}{2}$ **(1984)**

Q.7 Sketch the region bounded by the curves $y = \sqrt{5 - x^2}$ and y = |x − 1| and find its area. **(1985)**

Q.8 Find the area bounded by the curves $x^2 + y^2 = 4$, $x^2 = -\sqrt{2}y$ and x = y. **(1986)**

Q.9 Find the area of the region bounded by the curve C : y = tan x, tangent drawn to C at x = $\dfrac{\pi}{4}$ and the x-axis. **(1988)**

Q.10 Find all maxima and minima of the function y = x (x − 1)², 0 ≤ x ≤ 2.

Also, determine the area bounded by the curve y=x(x − 1)², the y-axis and the line x = 2. **(1989)**

Q.11 Compute the area of the region bounded by the curves y = ex log x and y = $\dfrac{\log x}{ex}$ where log e = 1 **(1990)**

Q.12 If $\begin{bmatrix} 4a^2 & 4a & 1 \\ 4b^2 & 4b & 1 \\ 4c^2 & 4c & 1 \end{bmatrix} \begin{bmatrix} f(-1) \\ f(1) \\ f(2) \end{bmatrix} = \begin{bmatrix} 3a^2 + 3a \\ 3b^2 + 3b \\ 3c^2 + 3c \end{bmatrix}$,

f(x) is a quadratic function and its maximum value occurs at a point V. A is a point of intersection of y = f(x) with x-axis and point B is such that chord AB subtends a right angled at V. Find the area enclosed by f(x) and chord AB. **(2005)**

Q.13 A curve passes through (2, 0) and the slope of tangent at point P(x, y) equals $\dfrac{(x+1)^2 + y - 3}{(x+1)}$. Find the equation of the curve and area enclosed by the curve and the x-axis in the fourth quadrant. **(2004)**

Q.14 Area of the region $\left\{ (x, y) \in R^2 : y \ge \sqrt{|x+3|}, \ 5y \le x + 9 \le 15 \right\}$ is equal to **(2016)**

(A) $\dfrac{1}{6}$ (B) $\dfrac{4}{3}$ (C) $\dfrac{3}{2}$ (D) $\dfrac{5}{3}$

Q.15 Let S be the area of the region enclosed by $y = e^{-x^2}, y = 0, x = 0$, and x = 1. Then **(2012)**

(A) $S \ge \dfrac{1}{e}$ (B) $S \ge -\dfrac{1}{e}$

(C) $S \le \dfrac{1}{4}\left(1 + \dfrac{1}{\sqrt{e}}\right)$ (D) $S \ge \dfrac{1}{\sqrt{2}} + \dfrac{1}{\sqrt{e}}\left(1 - \dfrac{1}{\sqrt{2}}\right)$

Important Questions

JEE Main/Boards

Exercise 1

Q.4   Q.10   Q.17

Exercise 2

Q.2   Q.6   Q.10
Q.12   Q.15   Q.17
Q.18

Previous Years' Questions

Q.2   Q.4   Q.7

JEE Advanced/Boards

Exercise 1

Q.1   Q.5   Q.12
Q.14   Q.16   Q.20
Q.21

Exercise 2

Q.2   Q.3   Q.7
Q.11   Q.13

Previous Years' Questions

Q.1   Q.3   Q.7
Q.10   Q.11   `Q.13

Answer Key

JEE Main/Boards

Exercise 1

Q.1 $\frac{74}{3}$ sq.units

Q.2 2 sq. units

Q.3 2 sq. units

Q.4 $\int_1^3 (5-y)\,dy$

Q.5 $2\int_0^2 \sqrt{y}\,dy$

Q.6 $\int_{-2}^a \sqrt{a^2 - x^2}\,dx$

Q.7 5 sq. units

Q.8 $\frac{112}{9}$ sq. units

Q.9 $\frac{2}{27}$ sq. units

Q.10 $\frac{64}{3}$ sq. units

Q.11 $\frac{28}{3}$ sq. units

Q.12 $(3 + 16 \log 2)$ sq. units

Q.13 $\frac{1}{2}$ sq. units

Q.14 13.5 sq. units

Q.15 1 sq. units

Q.16 $\frac{a^2}{12}(3\pi - 8)$ sq. units

Q.17 $\frac{16a}{3}$ sq. units

Q.18 $\frac{50}{3}$ sq. units

Q.19 $\frac{32}{3}$ sq. units

Exercise 2

Single Correct Choice Type

Q.1 C	**Q.2** A	**Q.3** C	**Q.4** A	**Q.5** A	**Q.6** B
Q.7 B	**Q.8** B	**Q.9** D	**Q.10** C	**Q.11** D	**Q.12** B
Q.13 A	**Q.14** A	**Q.15** C	**Q.16** C	**Q.17** C	**Q.18** C

Previous Years' Questions

Q.1 D	**Q.2** A	**Q.3** A	**Q.4** A	**Q.5** 2 sq. units
Q.6 $2\sqrt{3}$ sq. units	**Q.7** A	**Q.8** A	**Q.9** C	**Q.10** D	**Q.11** A
Q.12 C	**Q.13** C	**Q.14** D	**Q.15** B	

JEE Advanced/Boards

Exercise 1

Q.1 $f(x) = x \sin x$, $a = 1$; $A_1 = 1 - \sin a$; $A_2 = \pi - 1 - \sin a$; $A_3 = (3\pi - 2)$ sq. units   **Q.2** 2/3

Q.3 $f(x) = x^2 + 1$; $y = \pm 2x$; $A = \dfrac{2}{3}$ sq. units   **Q.4** $\dfrac{1}{2}$ **Q.5** $\sqrt{2} + 1$ **Q.6** $a = 8$

Q.7 $\dfrac{(\pi - 1)}{2}$ **Q.8** 2   **Q.9** $a = \dfrac{2}{3}$ **Q.10** 104   **Q.11** $(16/9)x^2$

Q.12 (i) 3 ; (ii) 1280 sq. units ; (iii) $\dfrac{32}{3}$ sq. units   **Q.13** $\sqrt{3}$ **Q.14** $\dfrac{3}{2}$ **Q.15** $\dfrac{8}{3}$ sq. units

Q.16 4   **Q.17** A $\dfrac{3m + 2m^2 + \frac{7}{6}}{m^3}$ **Q.18** $1 - 3e^{-2}$ **Q.19** $\left(\dfrac{\sqrt{2}}{6} + \dfrac{9\pi}{8} - \dfrac{9}{4} \sin^{-1} \dfrac{1}{3} \right)$ sq. units

Q.20 $a = -3/4$ **Q.21** e

Exercise 2

Single Correct Choice Type

Q.1 D	**Q.2** A	**Q.3** D	**Q.4** B	**Q.5** D	**Q.6** C	**Q.7** C
Q.8 A	**Q.9** A	**Q.10** A	**Q.11** D	**Q.12** D	**Q.13** B	

Previous Years' Questions

Q.1 B	**Q.2** B	**Q.3** C	**Q.4** 9/8 sq. units	**Q.5** $2\sqrt{2}$
Q.6 $\dfrac{1}{2}\log_e 3$ sq. units	**Q.7** $\dfrac{5\pi}{4} - \dfrac{1}{2}$ sq. units	**Q.8** $\dfrac{1}{3} - \pi$ sq. units	**Q.9** $\left(\log\sqrt{2} - \dfrac{1}{4} \right)$ sq. units
Q.10 10/3 sq. units	**Q.11** $\dfrac{e^2 - 5}{4e}$ sq. units	**Q.12** 125/3 sq. units	**Q.13** 4/3 sq. units
Q.14 3/2 sq. units	**Q.15** A, B, D		

JEE Main/Boards

Exercise 1

Sol 1: Area $= \int\limits_{2}^{4} x^2 + 3 = \dfrac{x^3}{3} + 3x \Big|_{2}^{4}$

$= \dfrac{64}{3} + 12 - \dfrac{8}{3} - 6 \Rightarrow$ Area $= \dfrac{56}{3} + 6 = \dfrac{74}{3}$ sq. units

Sol 2: $2\int\limits_{0}^{\pi/2} \cos x\, dx = 2\big[\sin x\big]_{0}^{\pi/2} = 2[1 - 0] = 2$ sq. units

Sol 3: $\int\limits_{0}^{\pi} \sin x\, dx = \big[-\cos x\big]_{0}^{\pi} = -\cos\pi + \cos 0 = 1 + 1$

$= 2$ sq. units

Sol 4: $x + y = 5 \Rightarrow x = -y + 5$

$\Rightarrow \int\limits_{4}^{2} x = \int\limits_{1}^{3} (-y + 5)\, dy = \left[-\dfrac{y^2}{2} + 5y\right]_{1}^{3} = \left|\dfrac{9}{2} - 15 - \dfrac{1}{2} + 5\right|$

$= 6$ sq. units

Expression $= \left|\int\limits_{1}^{3} (5 - y)\right| dy$

Sol 5: $y = x^2$;

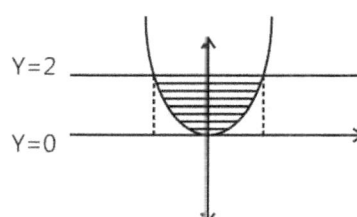

$x = \sqrt{y}$ $A = 2\int\limits_{0}^{2} \sqrt{y}\, dy$

Sol 6:

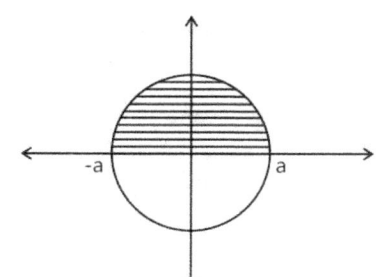

Area $= \int\limits_{-a}^{a} \sqrt{a^2 - x^2}\, dx$

Sol 7: $\int\limits_{-1}^{3} |x - 2|\, dx$ is the area under curve $|x - 2|$

where $x \in [-1, 3]$

$A = \int\limits_{-1}^{2} 2 - x + \int\limits_{2}^{3} x - 2 = 2x - \dfrac{x^2}{2}\Big|_{-1}^{2} + \dfrac{x^2}{2} - 2x\Big|_{2}^{3}$

$\Rightarrow A = 2 + 2 + \dfrac{1}{2} + \dfrac{9}{2} - 6 - 2 + 4 = 5$ sq. units

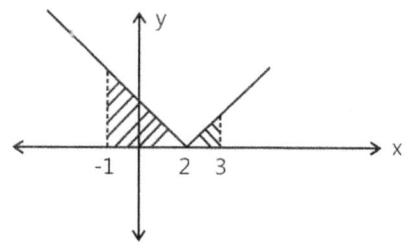

Sol 8: $y = \sqrt{3x - 4}$

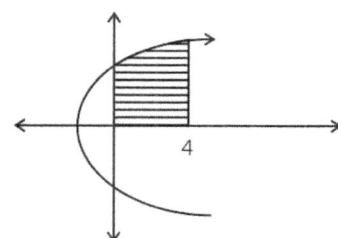

$A = \int\limits_{0}^{4} y\, dx = \int\limits_{0}^{4} \sqrt{3x + 4}\, dx = \dfrac{2\left[(3x + 4)^{3/2}\right]_{0}^{4}}{9x^3}$

$= \dfrac{2}{9}\left[16^{3/2} - 4^{3/2}\right] = \dfrac{2}{9}\left[4^3 - 2^3\right]$

$= \dfrac{2}{9}[64 - 8] = \dfrac{56 \times 2}{9} = \dfrac{112}{9}$ sq. units

Sol 9: $y = \dfrac{3}{(1 - 2x)^3}$ above x axis & $x \in [-4, -1]$

$\int\limits_{-4}^{-1} \dfrac{3\, dx}{(1 - 2x)^3} = \left[\dfrac{3}{(1 - 2x)^{+2}} \times \dfrac{1}{(-2)(-2)}\right]_{-4}^{-1}$

$= \dfrac{3}{4}\left[\dfrac{1}{(1 - 2x)^2}\right]_{-4}^{-1} = \dfrac{3}{4}\left[\dfrac{1}{3^2} - \dfrac{1}{9^2}\right] = \dfrac{3 \times 8}{4 \times 81} = \dfrac{2}{27}$ sq. units

Sol 10:

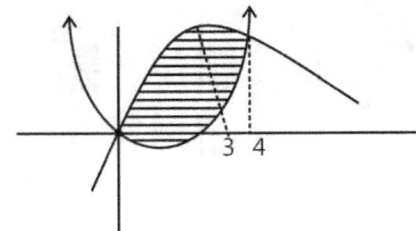

Area $= \int\limits_{0}^{4}\left(6x - x^2 - x^2 + 2x\right)dx$

$=\left[8x - 2x^2\right]_0^4 = \left[4x^2 - \dfrac{2x^3}{3}\right]_0^4 = 64 - \dfrac{2}{3}\times 64 = \dfrac{64}{3}$ sq. units

Sol 11: $f(x) = 1 + \sqrt{x}$

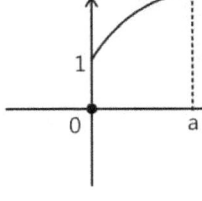

$\int\limits_{0}^{4} y\,dx = \int\limits_{0}^{4}\sqrt{x} + 1 = \left[\dfrac{2x^{3/2}}{3} + x\right]_0^4$

$= \dfrac{2}{3}\times 2^3 + 4 = \dfrac{16}{3} + 4 = \dfrac{28}{3}$ sq. units

Sol 12: $xy - 3x - 2y - 10 = 0$

$y = \dfrac{3x + 10}{x - 2}$

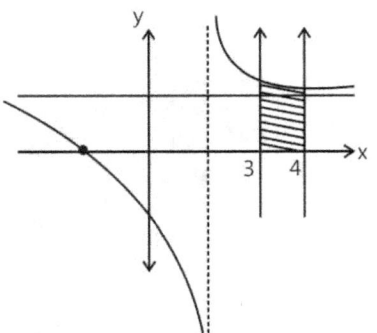

$A = \int\limits_{3}^{4}\dfrac{3x + 10}{x - 2}dx$ $A = \dfrac{3x - 6 + 16}{x - 2} = \int\limits_{3}^{4}\left(3 + \dfrac{16}{x - 2}\right)dx$

$= \left[3x + 16\ell n(x - 2)\right]_3^4$

$\left[12 + 16\,\ell\,n2 - 9 - \ln 1\right]$

$= (3 + 16\log 2)$ sq units.

Sol 13: $\int y\,dx = \int\limits_{0}^{\sqrt{\pi/2}} x\sin x^2 dx$

Substituting $x^2 = t$

$x\,dx = \dfrac{dt}{2}$

$\int\limits_{0}^{\pi/2}\sin t\,\dfrac{dt}{2} = \left[-\dfrac{\cos t}{2}\right]_0^{\pi/2} = \dfrac{-0 + 1}{2} = \dfrac{1}{2}$ sq. units

Sol 14:

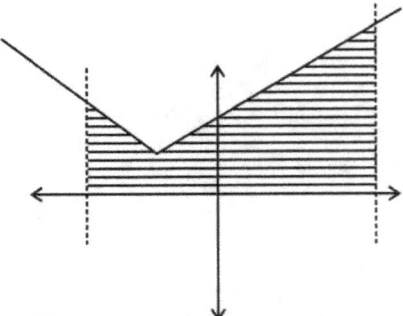

$\int\limits_{-2}^{3}|x + 1| + 1 = \int\limits_{-1}^{3} x + 1 + 1 + \int\limits_{-2}^{-1} -1 - x + 1$

$= \dfrac{x^2}{2} + 2x\Big|_{-1}^{3} - \left[\dfrac{x^2}{2}\right]_{-2}^{-1}$

$= \dfrac{9}{2} + 6 - \dfrac{1}{2} + 2 - \dfrac{1}{2} + 2 = \dfrac{7}{2} + 10 = \dfrac{27}{2}$ sq. units

Sol 15:

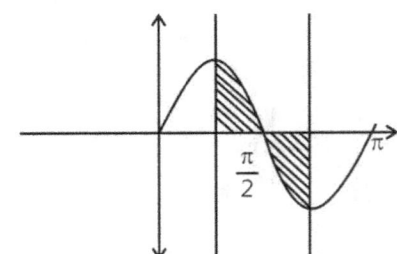

Area $= 2\int\limits_{\pi/4}^{\pi/2}\sin 2x\,dx = \dfrac{2}{2}\left[-\cos 2x\right]_{\pi/4}^{\pi/2}$

$= |+1 - 2| = 1$ sq. units

Sol 16:

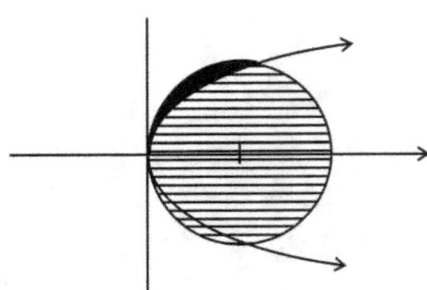

$x^2 - 2ax + y^2 \le 0$

$y^2 - ax \ge 0$

$$A = \int_0^a \left(\sqrt{2ax - x^2} - \sqrt{ax} \right) dx \Rightarrow A = \int_0^a \sqrt{2ax - x^2}\, dx - \int_0^a \sqrt{ax}\, dx$$

$$= \pi \frac{a^2}{4} - \frac{\sqrt{a}\, 2x^{3/2}}{3} \Big|_0^a$$

$$= \frac{\pi a^2}{4} - \frac{2\sqrt{a}}{3} a^{3/2} = \frac{\pi a^2}{4} - \frac{2a^2}{3} = \frac{a^2}{12}(3\pi - 8) \text{ sq. units}$$

Sol 17:

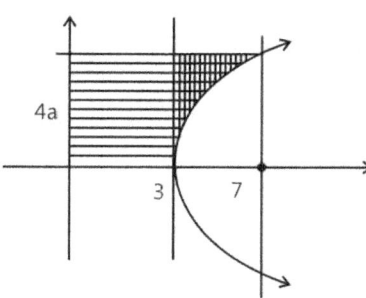

$$y = 4a \Rightarrow 16a^2 = 4a^2(x - 3) \Rightarrow x = 7$$

$$= \int_0^{4a} x\, dy = \int_0^{4a} \left(\frac{y^2}{4a^2} + 3 \right) dy = \left[\frac{y^3}{12a^2} + 3y \right]_0^{4a}$$

$$= \frac{(4a)^3}{12a^2} + 12a = \frac{64}{12}a + 12a$$

$$\Rightarrow \left(\frac{16 + 36}{3} \right) a = \frac{52a}{3}$$

$$A = \frac{52a}{3} - 3 \times 4a = \frac{16a}{3} \text{ sq. units}$$

Sol 18: The points of intersection of $y = x^2 + 3$ and $y = 2x + 3$ are (0, 3) and (2, 7).

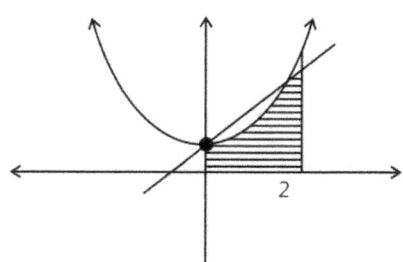

$$A_1 = \int_2^3 2x + 3 = x^2 + 3x \Big|_2^3 = \frac{27}{3} + 9 - 4 - 6 = 18 - 10 = 8$$

$$A_2 = \int_0^2 x^2 + 3 = \frac{x^3}{3} + 3x \Big|_0^2 = \frac{8}{3} + 6$$

$$A_1 + A_2 = \frac{8}{3} + 6 + 8 = \frac{50}{3} \text{ sq. units}$$

Sol 19: $y = 4x - x^2$

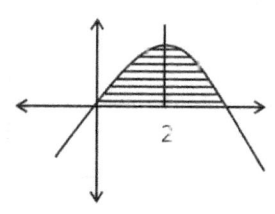

$$\int \left(4x - x^2 \right) dx = \left[2x^2 - \frac{x^3}{3} \right]_0^4$$

$$= 32 - \frac{64}{3} = \frac{32}{3} \text{ sq. units}$$

Exercise 2

Single Correct Choice Type

Sol 1: (C)

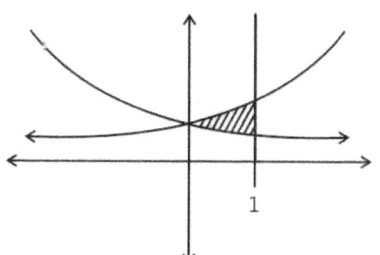

$$A = \int_0^1 \left(e^x - e^{-x} \right) dx = \left[e^x + e^{-x} \right]_0^1$$

$$= e + \frac{1}{e} - 1 - 1 = e + \frac{1}{e} - 2$$

Sol 2: (A)

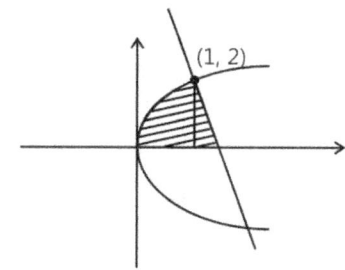

Area $= A_1 + A_2$

$$A_1 = \int_0^1 \sqrt{4x}\, dx = 2 \frac{\left[x^{3/2} \right]_0^1}{3/2} = \frac{4}{3}$$

Equation normal $\dfrac{y - 2}{x - 1} = -1$

$$\Rightarrow y - 2 = 1 - x$$

$$x + y = 3 \qquad\qquad\qquad \text{...... (i)}$$

$$A_2 = \int_1^3 (3 - x)\, dx = \left[3x - \frac{x^2}{2} \right]_1^3 = 9 - \frac{9}{2} - 3 + \frac{1}{2}$$

$$= 6 - 4 = 2$$

Area $= 2 + \dfrac{4}{3} = \dfrac{10}{3}$

3.40

Sol 3: (C)

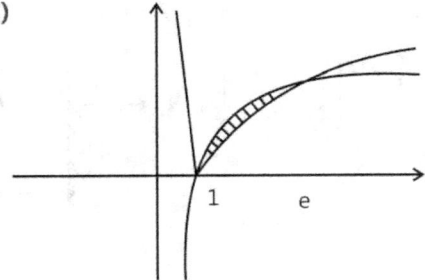

$$A = \int_{1}^{e}\left(\ln x - (\ln x)^2\right) dx \quad \int \ln dx = x \ln x - x$$

$$A = (x\ln x - x) - \left[\ln x(x\ln x - x) - \int(\ln x - 1)\right]$$

$$= x\ln x - x - x(\ln x)^2 + x\ln x + x\ln x - x - x$$

$$\left[-x(\ln x)^2 + 3x\ln x - 3x\right]_{1}^{e}$$

$$= -e + 3e - 3e + 3 = 3 - e$$

Sol 4: (A)

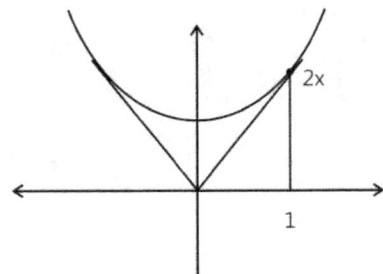

$$A = 2\int_{0}^{1}\left(x^2 + 1 - 2x\right)dx = 2\int_{0}^{1}(x-1)^2 dx = 2 \Rightarrow \frac{(x-1)^3}{3}\bigg|_{0}^{1} = \frac{2}{3}$$

Sol 5: (A) $(x-1)^2 + y^2 = 1$

If $\sin\left(\dfrac{\pi x}{2}\right)$

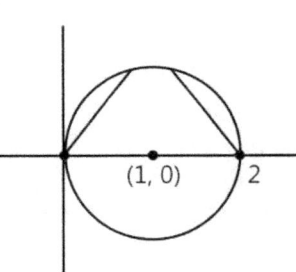

$$A = \int_{0}^{2}\left(\sqrt{1-(x-1)^2} - \sin\frac{\pi x}{2}\right) dx$$

$$= \left[\sqrt{2x - x^2} - \sin\frac{\pi x}{2}\right]_{0}^{2}$$

$$= \left[\frac{\sin^{-1}(x-1) + (x-1)\sqrt{2x-x^2}}{2} + \frac{2}{\pi}\cos\frac{\pi x}{2}\right]_{0}^{2}$$

$$= \frac{\pi}{2\times 2} + \frac{2}{\pi}(-1) + \frac{\pi}{2.2} - \frac{2}{\pi} = \frac{\pi}{2} - \frac{4}{\pi}$$

Sol 6: (B)

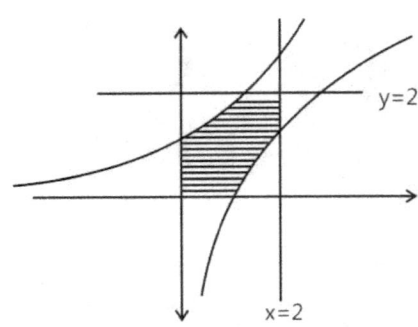

$$\text{Area} = \int_{1}^{2}\ln x \, dx + \int_{1}^{2}\ln y \, dy$$

$$= 2\int_{1}^{2}\ln x = 2\left[x\ln x - x\right]_{1}^{2} = 2[2\ln 2 - 2 + 1]$$

$$= 4\ln 2 - 2 = 2(2\ln 2 - 1)$$

Sol 7: (B) $y = x(1 - \ln x)$ & $x = \dfrac{1}{e}$

Between $x = \dfrac{1}{e}$ & e

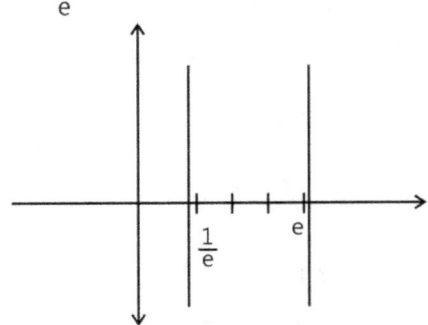

$$I = \int(x - x\ln x)dx = \frac{x^2}{2} - \frac{1}{2}\left[x^2\ln x - \frac{x^2}{2}\right]$$

$$I = \frac{3x^2}{2} - \frac{x^2}{2}\ln x$$

$$\Rightarrow A = \left[I\right]_{\frac{1}{e}}^{e}$$

$$A = \frac{3e^2}{4} - \frac{e^2}{2} - \left(\frac{3}{4e^2} + \frac{1}{2e^2}\right) = \frac{1}{4}e^2 - \frac{5}{4e^2}$$

Sol 8: (B)

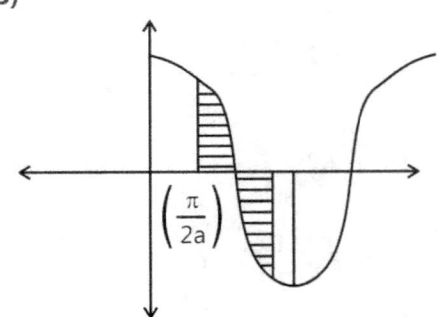

$$\text{Area} = \int_{\pi/6a}^{\pi/2a} (\cos ax)\,dx = \frac{2}{a}\left[\sin ax\right]_{\pi/6a}^{\pi/2a}$$

$$= \frac{2}{a}\left[1 - \frac{1}{2}\right] = \frac{1}{a} > 3 = a \in \left(0, \frac{1}{3}\right)$$

Sol 9: (D) $y = \dfrac{x}{6} + \dfrac{1}{x^2}$; $y = 0$

$$A = \int_{a}^{2a}\left(\frac{x}{6} + \frac{1}{x^2}\right)dx = \left[\frac{x^2}{12} - \frac{1}{x}\right]_{a}^{2a}$$

$$= \frac{a^2}{3} - \frac{1}{2a} - \frac{a^2}{12} + \frac{1}{a} = \frac{a^2}{4} + \frac{1}{2a}$$

A_{least} when $A' = \dfrac{a}{2} - \dfrac{1}{2a^2} = 0 \Rightarrow a = 1$

Sol 10: (C)

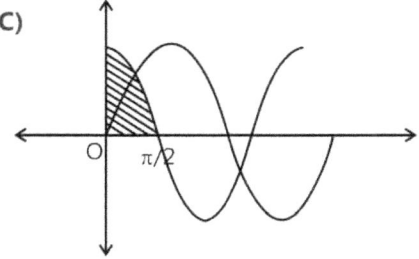

$$\text{Area}_1 = \left[\int_{0}^{\pi/4} \sin x\right] 2 = 2\left[-\cos x\right]_{0}^{\pi/4}$$

$$= 2\left[-\frac{1}{\sqrt{2}} + 1\right] = \left(\sqrt{2} - 1\right)\sqrt{2} = 2 - \sqrt{2}$$

$$\text{Area}_2 = \int_{0}^{\pi/2} \cos x - \text{area}_1 = \left[\sin x\right]_{0}^{\pi/2} - 2 + \sqrt{2}$$

$$= 1 - 2 + \sqrt{2} = \sqrt{2} - 1$$

ratio is $\dfrac{\sqrt{2}\left(\sqrt{2} - 1\right)}{\sqrt{2} - 1} = \sqrt{2}$

Sol 11: (D)

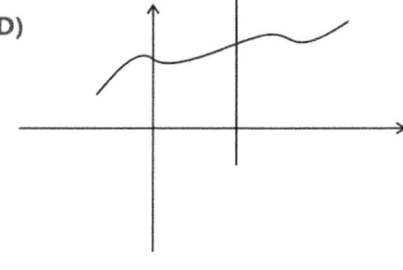

$$\int_{0}^{x_1} f(x) = x_1 e^{x_1} \Rightarrow f(x) = xe^x + e^x$$

Sol 12: (B)

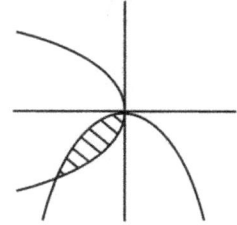

$$\text{Area} = \int_{-1}^{0} -\sqrt{-x} - \left(-x^2\right) = \int_{-1}^{0} -\sqrt{-x} + x^2$$

$$= \left[\frac{x^3}{3} + \frac{2(-x)^{3/2}}{3}\right]_{-1}^{0} = \left|\frac{1}{3} - \frac{2}{3}\right| = \frac{1}{3}$$

Sol 13: (A) $\displaystyle\int_{1}^{x} f(x)\,dx = \sqrt{1 + 3x} - 2$

$$f(x) = \frac{3}{2\sqrt{1 + 3x}}$$

$x = 1$, $y = \dfrac{3}{4}$ and $x = 8$, $y = \dfrac{3}{10}$

$$A = \frac{\displaystyle\int_{1}^{8} y\,dx}{\Delta x} = \frac{\sqrt{1 + 3.8} - 2}{8 - 1} = \frac{3}{7}$$

Sol 14: (A) $\dfrac{dy}{dx} = 2x + 1$

$y = x^2 + x + c$

$(1, 2)$; $c = 0$

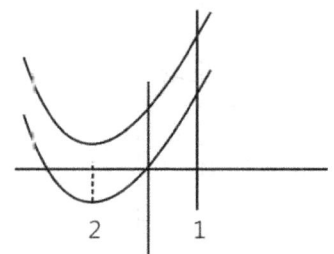

$y = x^2 + x + 1$

$$A = \int_{0}^{1} x^2 + x = \left[\frac{x^3}{3} + \frac{x^2}{2}\right]_{0}^{1} = \frac{5}{6}$$

Sol 15: (C)

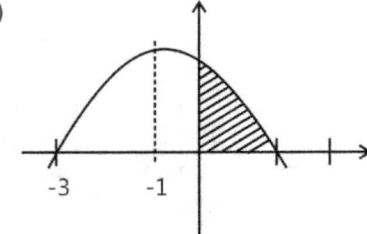

$0 \le y < 3 - 2x - x^2$

$0 < y < 4 - (x + 1)^2$

$= \int_0^1 (3 - 2x - x^2) \, dx$

Sol 16: (C)

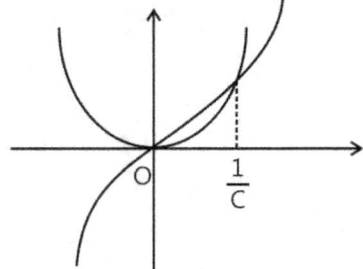

$\int_0^{\frac{1}{c}} \left(x^2 - cx^3 \right) dx = \frac{2}{3}$

$\Rightarrow \left[\frac{x^3}{3} - \frac{cx^4}{4} \right]_0^{1/c} = \frac{1}{3c^3} - \frac{1}{4c^3} = \frac{2}{3}$

$\Rightarrow \frac{1}{12c^3} = \frac{2}{3} \Rightarrow c = \frac{1}{2}$

Sol 17: (C)

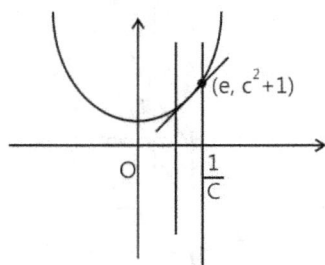

$\frac{y - c^2 - 1}{x - c} = 2c$

$\Rightarrow y = 2cx - c^2 + 1$

$x = 1, \qquad y = 2c - c^2 + 1$

$x = 2, \qquad y = 4c - c^2 + 1$

$\Rightarrow$ Area of trapezoid $= \frac{1}{2} \left[6c - 2c^2 + 2 \right] \times 1 = 3c - c^2 + 1$

For $\text{area}_{max} = A' = 3 - 2c = 0 \Rightarrow C = \frac{3}{2}$

$\left(\frac{3}{2}, \frac{13}{4} \right)$

Sol 18: (C) $A_1 = \frac{1}{2} \times a^2 \times 2a = a^3$

$\Rightarrow = \int_{-a}^{a} x^2 = \left[\frac{x^3}{3} \right]_{-a}^{a} = \frac{2a^3}{3}$

$\Rightarrow_2 = 2a \times a^2 - \frac{2a^3}{3} = \frac{4a^3}{3}$

$\Rightarrow \frac{A_1}{A_2} = \frac{3}{4}$

Previous Years' Questions

Sol 1: (D) Given, $\frac{x^2}{9} + \frac{y^2}{5} = 1$

To find tangents at the end points of latus rectum, we find ae.

i.e. $ae = \sqrt{a^2 - b^2} = \sqrt{4} = 2$

and $\sqrt{b^2(1 - e^2)} = \sqrt{5\left(1 - \frac{4}{9}\right)} = \frac{5}{3}$

By symmetry, the quadrilateral is a rhombus.

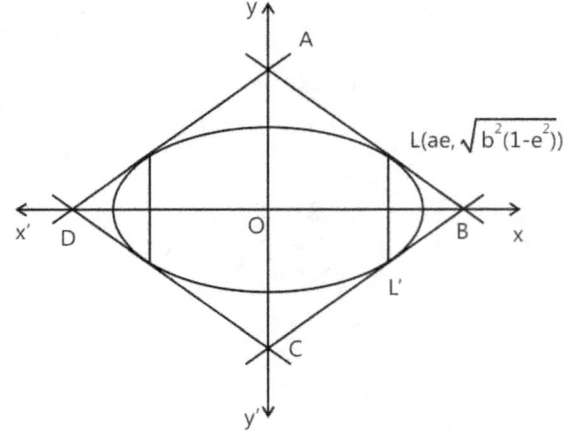

So, area is four times the area of the right angled triangle formed by the tangent and axes in the 1st quadrant.

$\therefore$ Equation of tangent at $\left(2, \frac{5}{3} \right)$ is

$\frac{2}{9}x + \frac{5}{3} \cdot \frac{y}{5} = 1 \Rightarrow \frac{x}{9/2} + \frac{y}{3} = 1$

$$\therefore \quad \frac{x}{9/2} + \frac{y}{3} = 1$$

$\therefore$ Area of quadrilateral ABCD

$$= 4 \text{ (area of } \triangle AOB) = 4\left(\frac{1}{2}.\frac{9}{2}.3\right) = 27 \text{ sq. units}$$

Sol 2: (A) To find the area between the curves, $y = \sqrt{x}, 2y + 3 = x$ and x-axis in the 1st quadrant (we can plot the above condition as) ;

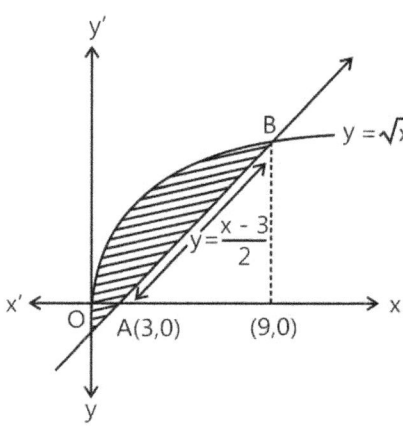

Area of shaded portion OABO

$$= \int_0^9 \sqrt{x}\,dx - \int_3^9 \left(\frac{x-3}{2}\right)dx$$

$$= \left(\frac{x^{3/2}}{3/2}\right)_0^9 - \frac{1}{2}\left(\frac{x^2}{2} - 3x\right)_3^9$$

$$= \left(\frac{2}{3}.27\right) - \frac{1}{2}\left\{\left(\frac{81}{2} - 27\right) - \left(\frac{9}{2} - 9\right)\right\}$$

$$= 18 - \frac{1}{2}(18) = 9 \text{ sq. unit}$$

Sol 3: (A) As from the figure, area enclosed between the curves is OABCO.

Thus, the point of intersection of

$y = ax^2$ and $x = ay^2$

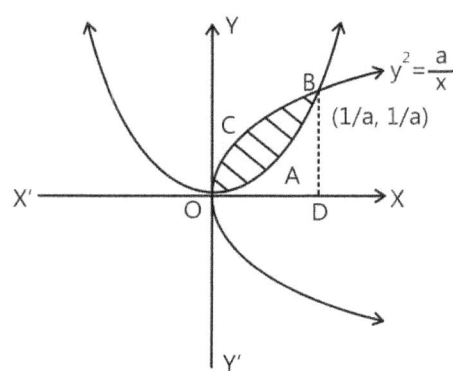

$$\Rightarrow x = a\,(ax^2)^2$$

$$\Rightarrow x = 0, \frac{1}{a} \Rightarrow y = 0, \frac{1}{a}$$

$\therefore$ Point of intersection are $(0, 0)$ and $\left(\frac{1}{a}, \frac{1}{a}\right)$

Thus, required area OABCO = Area of curve OCBDO – area of curve OABDO

$$\Rightarrow \int_0^{1/a}\left(\sqrt{\frac{x}{a}} - x^2\right)dx = 1 \text{ (given)}$$

$$\Rightarrow \left[\frac{1}{\sqrt{a}}.\frac{x^{3/2}}{3/2} - \frac{ax^3}{3}\right]_0^{1/a} = 1$$

$$\Rightarrow \frac{2}{3a^2} - \frac{1}{3a^2} = 1$$

$$\Rightarrow a^2 = \frac{1}{3} \Rightarrow a = \frac{1}{\sqrt{3}} \quad (\because a > 0)$$

Sol 4: (A) Since, tangents drawn from external point to the circle subtends equal angle at the centre

$$\therefore \angle O_1BD = 30^c$$

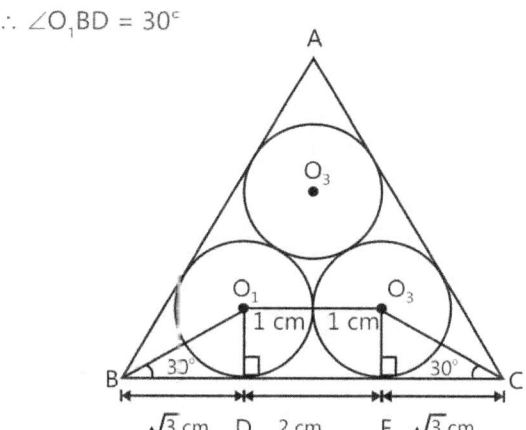

In $\triangle O_1BD$, $\tan 30° = \dfrac{O_1D}{BD}$

$$\Rightarrow BD = \sqrt{3} \text{ cm}$$

Also, $DE = O_1O_2 = 2$ cm and $EC = \sqrt{3}$ cm

Now, $BC = BD + DE + EC = 2 + 2\sqrt{3}$

$\Rightarrow$ Area of $\triangle ABC$

$$= \frac{\sqrt{3}}{4}(BC)^2 = \frac{\sqrt{3}}{4}.4(1+\sqrt{3})^2$$

$$= (6 + 4\sqrt{3}) \text{ sq. cm.}$$

Sol 5: The area formed by $|x| + |y| = 1$ is square shown as below,

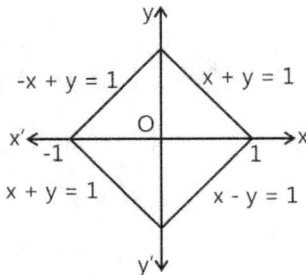

∴ Area of square = $(\sqrt{2})^2$ = 2 sq. units

Sol 6: Equation of tangent at the point $(1, \sqrt{3})$ to the curve

$x^2 + y^2 = 4$

is $x + \sqrt{3}y = 4$,

whose x-axis intercept (4, 0)

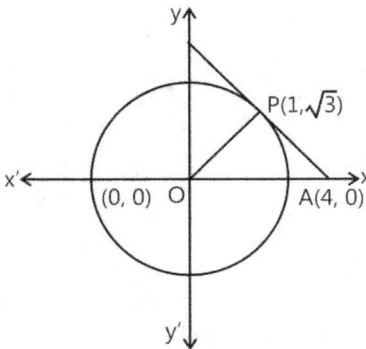

Thus, area of Δ formed by (0, 0) $(1, \sqrt{3})$ and

(4, 0)

$$= \frac{1}{2}\begin{vmatrix} 0 & 0 & 1 \\ 1 & \sqrt{3} & 1 \\ 4 & 0 & 1 \end{vmatrix} = \frac{1}{2}|(0 - 4\sqrt{3})| = 2\sqrt{3} \text{ sq. units}$$

Sol 7: (A) The curves $y = (x - 1)^2$, $y = (x + 1)^2$ and $y = 1/4$ are shown are

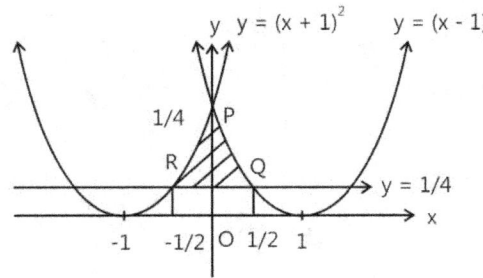

Where points of intersection are

$(x - 1)^2 = \frac{1}{4} \Rightarrow x = \frac{1}{2}$

and $(x + 1)^2 = \frac{1}{4} \Rightarrow -\frac{1}{2}$

∴ $Q\left(\frac{1}{2}, \frac{1}{4}\right)$ and $R\left(-\frac{1}{2}, \frac{1}{4}\right)$

∴ Required area

$$= 2\int_0^{1/2}\left[(x - 1)^2 - \frac{1}{4}\right]dx = 2\left[\frac{(x - 1)^3}{3} - \frac{1}{4}x\right]_0^{1/2}$$

$$= 2\left[-\frac{1}{8.3} - \frac{1}{8} - \left(-\frac{1}{3} - 0\right)\right] = \frac{8}{24} = \frac{1}{3} \text{ sq. unit}$$

Sol 8: (A) Region $(x, y) : y^2 \geq 2x$

$x^2 + y^2 < 4x$ and $x \geq 0, y \geq 0$

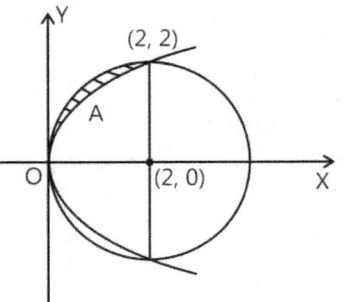

Area $= \frac{1}{4}\pi(2)^2 - \int_0^2 \sqrt{2x} \, dx$

$$= \pi - \left[\sqrt{2}\frac{x^{\frac{3}{2}}}{\frac{3}{2}}\right]_0^2 = \pi - \frac{2\sqrt{2}}{3}\left[2^{\frac{3}{2}}\right]$$

$$= \pi - \frac{2\sqrt{2}}{3} \times 2\sqrt{2}$$

$$= \pi - \frac{8}{3} \text{ sq. units}$$

Sol 9: (C) Region $(x, y) : y^2 < 2x$ and $y \geq 4x - 1$

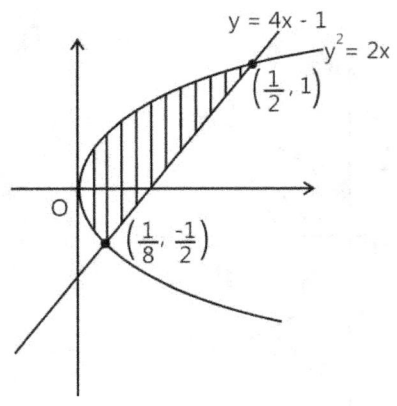

Area $= \int\limits_{-1/2}^{1}\left(\dfrac{y+1}{4}-\dfrac{y^2}{2}\right)dy$

$=\left[\dfrac{y^2}{8}+\dfrac{y}{4}-\dfrac{y^3}{6}\right]_{-1/2}^{1}=\left[\dfrac{1}{3}+\dfrac{1}{4}-\dfrac{1}{6}-\dfrac{1}{32}+\dfrac{1}{8}-\dfrac{1}{48}\right]=\dfrac{9}{32}$

Sol 10: (D) The point of intersection $x-3=2\sqrt{x}$

$\Rightarrow (x-3)^2 = 4x$

$\Rightarrow x^2 - 10x + 9 = 0$

$\Rightarrow x = 1,9$

$\Rightarrow y = 1,3$

Area $\int\limits_{0}^{3}\left[2y+3-y^2\right]dy = \left[y^2+3y-\dfrac{y^3}{3}\right]_{0}^{3}$

$=\left[9+9-9\right]=9$ sq. units

Sol 11: (A)

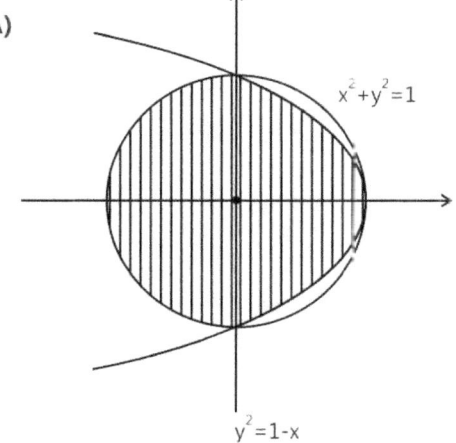

Area = Area of half circle

$+2\int\limits_{0}^{1}\sqrt{1-x}\,dx = \dfrac{1}{2}\pi(1)^2 + 2\left[-\dfrac{(1-x)^{3/2}}{3/2}\right]_{0}^{1}$

$=\dfrac{\pi}{2}+2\left[\dfrac{2}{3}\right] = \dfrac{\pi}{2}+\dfrac{4}{3}$

Sol 12: (C) $\dfrac{x-1}{2}=\dfrac{y+1}{3}=\dfrac{z-1}{4}=m$

$\Rightarrow$ Any point on this line is $(2m+1,3m-1,4m+1)$

Similarly for $\dfrac{x-3}{1}=\dfrac{y-k}{2}=\dfrac{2}{1}=n$

$(n+3,2n+k,n)$

It they intersect, then

$2m+1=n+3 \Rightarrow 2m-n=2$

$3m-1=2n+k \Rightarrow 3m-2n=k+1$

$4m+1=n \Rightarrow 4m-n=-1$

On solving these equations, we get

$m=-\dfrac{3}{2}, n=-5 \Rightarrow K=\dfrac{9}{2}$

Sol 13: (C) The required area

$=2\left[\int\limits_{0}^{2}\left(3\sqrt{y}-\dfrac{\sqrt{y}}{2}\right)dy\right] = 2\left[\int\limits_{0}^{2}\dfrac{5\sqrt{y}}{2}dy\right] = 5\int\limits_{0}^{2}\sqrt{y}\,dy$

$=5\left[\dfrac{y^{3/2}}{3/2}\right]_{0}^{2} = \dfrac{10}{3}\left[2^{3/2}\right] = \dfrac{10\times 2\sqrt{2}}{3} = \dfrac{20\sqrt{2}}{3}$

Sol 14: (D)

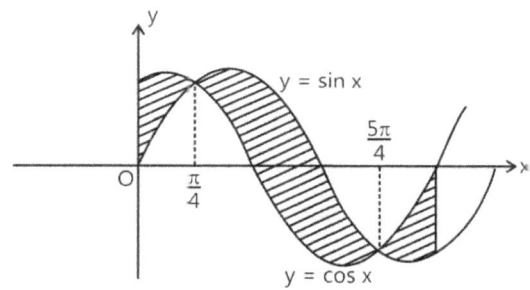

Area $= \int\limits_{0}^{\pi/4}(\cos x - \sin x)dx + \int\limits_{\pi/4}^{5/4}(\sin x - \cos x)dx$

$+\int\limits_{\frac{5\pi}{4}}^{\frac{3\pi}{2}}(\cos x - \sin x)dx$

$=\left[\sin x + \cos x\right]_{0}^{\pi/4} + \left[-\cos x - \sin x\right]_{\pi/4}^{5\pi/4}$

$+\left[\sin x + \cos x\right]_{5\pi/4}^{3\pi/2}$

$=\dfrac{1}{\sqrt{2}}+\dfrac{1}{\sqrt{2}}-1+\left[\dfrac{-1}{\sqrt{2}}-\dfrac{1}{\sqrt{2}}-\dfrac{1}{\sqrt{2}}-\dfrac{1}{\sqrt{2}}\right]$

$+\left[-1+0+\dfrac{1}{\sqrt{2}}+\dfrac{1}{\sqrt{2}}\right]$

$=4\sqrt{2}-2$

Sol 15: (B)

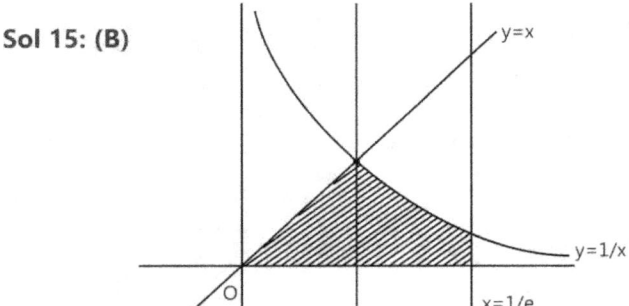

Area $= \left|\int_0^1 x\,dx\right| + \left|\int_1^{1/e} \frac{1}{x}\,dx\right|$

$= \left[\frac{x^2}{2}\right]_0^1 + \left|[\ln x]_1^{1/e}\right| = \frac{1}{2} + |\ln 1/e|$

$= \frac{1}{2} + 1 = \frac{3}{2}$ sq. units

JEE Advanced/Boards

Exercise 1

Sol 1: $A_1 = \int_0^a \sin x - f(x)$

$= -\cos x\Big|_0^a - \int_0^a f(x) = 1 - \cos a - \int_0^a f(x)$

$\Rightarrow A_2 = \int_a^\pi f(x) - \sin x = \int_a^\pi fx + \cos x\Big|_a^\pi = \int_a^\pi f(x) - 1 - \cos a$

$\Rightarrow A_3 = \int_\pi^{2\pi} \sin x - f(x) = -\cos x\Big|_\pi^{2\pi} - \int_\pi^{2\pi} f(x) = -2 - \int_\pi^{2\pi} f(x)$

$\Rightarrow A_1 = 1 - \sin a + a\cos a - \cos a$

$= 1 - \cos a - \int_0^a f(x)\,dx$

$\Rightarrow -a\cos a + \sin a = \int_0^a f(x)\,dx = \sin a - a\cos a$

$\Rightarrow f(x) = x\sin x$

$\int_a^\pi f(x)\,dx = \int_a^\pi x\sin x\,dx = [-x\cos x + \sin x]_a^\pi$

$= \pi - [-a\cos a + \sin a] = \pi + a\cos a - \sin a$

$\Rightarrow A_2 = \pi + a\cos a - \sin a - 1 - \cos a$

$\int_\pi^{2\pi} f(x)\,dx = \int_a^\pi x\sin x\,dx = [-x\cos x + \sin x]_\pi^{2\pi}$

$= -2\pi - (\pi) = -3p$

$\Rightarrow A_3 = (3\pi - 2)$ sq units.

$x\sin x = \sin x \Rightarrow x = 1$

$\Rightarrow a = 1 \Rightarrow A_2 = \pi - 1 - \sin a$

$\Rightarrow A_1 = 1 - \sin a$

Sol 2: $A = \int_0^t \sin x^2\,dx$, $B = \frac{1}{2} \times t\sin t^2$

Now, $\displaystyle\lim_{t\to 0}\frac{A}{B} = \frac{\displaystyle\int_0^t \sin x^2\,dx}{\frac{1}{2}\times t\sin t^2}$

$\frac{0}{0}$ form $\Rightarrow \displaystyle\lim_{t\to 0} = \frac{\sin t^2}{\frac{t}{2}(2t)\cos t^2 + \frac{1}{2}\sin t^2}$

[L' Hospital's rule]

$= \frac{2\sin t^2}{2t^2\cos t^2 + \sin t^2} = \frac{2\tan t^2}{2t^2 + \tan t^2}$

$= \frac{\dfrac{2\tan t^2}{t^2}}{2 + \dfrac{\tan t^2}{t^2}} = \frac{2}{2+1} = \frac{2}{3}$

Sol 3: $f(x + 1) = f(x) + 2x + 1$

$f(0) = 1$

$f(1) = 2 \Rightarrow f(x) = x^2 + 1$

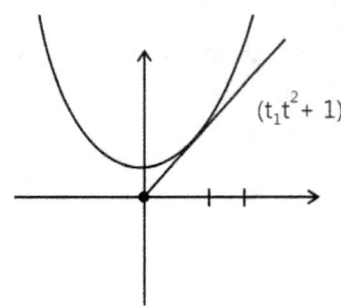

$f(-1) = 2; f(2) = 5; f(-2) = 5$

$y = x^2 + 1$

Equation of tangent at $P(t, t^2 + 1)$ $y - (t^2 + 1) = 2t(x - t)$

Since, it passes through $(0, 0)$

$-t^2 - 1 = 2t(-t)$

$\Rightarrow t^2 = t \qquad \Rightarrow t = \pm 1$

$\Rightarrow y = \pm 2x$

Area $= 2\int_0^1 \left(x^2 + 1 - 2x\right) dx$

$= 2\int_0^1 (x-1)^2 dx = \frac{2}{3}(x-1)^3 \Big|_0^1 = \frac{2}{3}[0+1] = \frac{2}{3}$ sq. units

Sol 4: $y = \dfrac{\ln x - c}{x} = 0 \Rightarrow x = e^c$

$f'(x) = \dfrac{x\left(\dfrac{1}{x}\right) - (\ln x - c)}{x^2} = 0$

$\Rightarrow 1 - \ln x + c = 0$

$\Rightarrow \ln x = c + 1$

$\Rightarrow x = e^{c+1}$

$\Rightarrow \displaystyle\int_{e^c}^{e^{c+1}} \frac{\ln x - c}{x} dx = \int_{e^c}^{e^{c+1}} \frac{\ln x}{x} dx - \int_{e^c}^{e^{c+1}} \frac{c}{x} dx \sqrt{b^2 - 4ac}$

$= \left[\dfrac{(\ln x)^2}{2}\right]_{e^c}^{e^{c+1}} - c\Big[\ln x\Big]_{e^c}^{e^{c+1}}$

$= \dfrac{(c+1)^2 - c^2}{2} - c[c+1-c] = \dfrac{2c+1}{2} - [c] = \dfrac{1}{2}$

Sol 5: $y = x^n$

$y - 1 = n(x-1)$

$A = \displaystyle\int_0^1 x^n dx - \int_{1-\frac{1}{n}}^1 \big[n(x-1)+1\big] dx$

$= \dfrac{\left[x^{n+1}\right]_0^1}{n+1} - \left[n\dfrac{(x-1)^2}{2} + x\right]_{1-\frac{1}{n}}^1$

$= \dfrac{1}{n+1} - \left[1 - \dfrac{n}{2n^2} - 1 + \dfrac{1}{n}\right] = \dfrac{1}{n+1} - \dfrac{1}{2n}$

$\Rightarrow = \dfrac{n-1}{2n(n+1)}$

$\dfrac{dA}{dn} = 0 \Rightarrow 2n(n+1) - (n-1)(4n+2) = 0$

$\Rightarrow 2n^2 + 2n - 4n^2 - 2n + 4n + 2 = 0$

$\Rightarrow -2n^2 + 4n + 2 = 0$

$\Rightarrow n^2 - 2n - 1 = 0$

$n = \dfrac{2 \pm \sqrt{8}}{2} = 1 \pm \sqrt{2}$

$\Rightarrow > 1$ i. e. $n = \sqrt{2} + 1$

Sol 6:

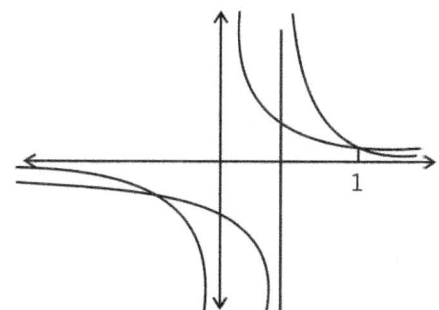

$A = \displaystyle\int_2^a \frac{1}{x} - \frac{1}{2x-1} = \left[\ln x - \frac{\ln(2x-1)}{2}\right]_2^a$

$= \ln a - \dfrac{\ln(2a-1)}{2} - \ln 2 + \dfrac{\ln 3}{2}$

$= \ln \dfrac{3a}{2(2)(2a-1)} \times 2 = \ln\dfrac{3a}{4a-2} - \ln\dfrac{4}{5}$

$\Rightarrow 15a = 16a - 8 \qquad \Rightarrow a = 8$

Sol 7: $f(x) = \dfrac{1}{1+x^2}$

We have, $A(\alpha, f'\alpha))$ and $B\left(-\dfrac{1}{\alpha}, f\left(-\dfrac{1}{\alpha}\right)\right)$

$OA \rightarrow y = \dfrac{x}{\alpha(1+\alpha^2)}$ .... (i)

$OB \rightarrow y = \dfrac{-\alpha^z}{1+\alpha^2}(x)$

$\displaystyle\int_0^\alpha \left(\frac{1}{1+x^2} - \frac{x}{\alpha(1-\alpha^2)}\right) dx + \int_{-\frac{1}{\alpha}}^0 \left(\frac{1}{1+x^2} + \frac{\alpha^3}{1+\alpha^2}x\right) dx$

$= \left[\tan^{-1} x - \dfrac{x^2}{2\alpha(1+\alpha^2)}\right]_0^\alpha + \left[\tan^{-1} x + \dfrac{\alpha^3 x^2}{2(1+\alpha^2)}\right]_{-\frac{1}{\alpha}}^0$

$= \tan^{-1}\alpha - \dfrac{\alpha}{2(1+\alpha^2)}$

$+ \left[0 + 0 - \tan^{-1}\left(-\dfrac{1}{\alpha}\right) - \dfrac{\alpha}{2(1+\alpha^2)}\right]$

$= \tan^{-1}\alpha - \tan^{-1}\left(-\dfrac{1}{\alpha}\right) - \dfrac{(\alpha+\alpha)}{2(1+\alpha^2)}$

$= \tan^{-1}\dfrac{\alpha^2 - 1}{1-1} - \dfrac{(\alpha+\alpha)}{2(1+\alpha^2)}$

$= \dfrac{\pi}{2} - \dfrac{(\alpha+\alpha)}{2(1+\alpha^2)} = \dfrac{\pi}{2} - \dfrac{2\alpha}{2(1+\alpha^2)}$

$A' = 0 \Rightarrow \alpha = 1$

$A = \dfrac{\pi}{2} - \dfrac{1}{2}$

Sol 8: The curve is $y = \sin^{-1} x$, i.e., $x = \sin y$. This is a standard curve. Lines $x = 0$, $y = \dfrac{\pi}{2}$ and $y = -\dfrac{\pi}{2}$ are the y-axis and two lines parallel to the x-axis at a distance $\dfrac{\pi}{2}$, one above and the other below the x-axis respectively.

Hence, the shaded part is the required area Δ. By symmetry of the curve and the lines,

ar (OABO) = ar (OCDO)

$\therefore \Delta = 2 \times ar(OABO)$

$= 2 \int\limits_{0}^{\frac{\pi}{2}} (x)_{curve}\, dy = 2 \int\limits_{0}^{\frac{\pi}{2}} \sin y \, dy,$

($\because$ The equation of the curve is $x = \sin y$)

$\therefore \Delta = 2\left[-\cos y\right]_{0}^{\frac{\pi}{2}} = 2\left[0 + 1\right] = 2$

Sol 9:

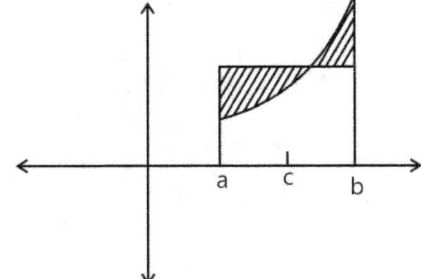

$A = \left(\int\limits_{a}^{c} f(c) - f(x)\right) dx + \int\limits_{c}^{b} \left(f(x) - f(c)\right) dx\ A$

$= f(c)(c - a) - \int\limits_{a}^{c} f(x)\, dx + \int\limits_{c}^{b} f(x)\, dx - f(c)(b - c)$

$A = f(c)\left[c - a - b + c\right] - \int\limits_{a}^{c} f(x)\, dx + \int\limits_{c}^{b} \left(f(x)\right) dx$

This is minimum when $2c - a - b = 0$

$\Rightarrow c = \dfrac{a + b}{2}$

$\int f(x) = \int\limits_{0}^{2}\left(\dfrac{x^3}{3} - x^2 + a\right) dx = \left[\dfrac{x^4}{12} - \dfrac{x^3}{3} + ax\right]_{0}^{2}$

$= \dfrac{16}{12} - \dfrac{8}{3} + 2a = \left(2a - \dfrac{4}{3}\right)$

Is minimum $\Rightarrow 2a = \dfrac{4}{3} \Rightarrow a = \dfrac{2}{3}$

Sol 10:

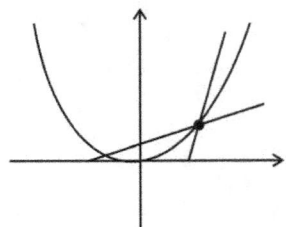

$y - c = m(x - 1)$

$A = \int\left(x^2 - m(x - 1) - c\right) dx$

$= \left[\dfrac{x^3}{3} - \dfrac{m(x-1)^2}{2} - cx\right]_{a}^{b}$

$= \left[\dfrac{b^3 - a^3}{3} - \dfrac{m}{2}\left[(b-1)^2 - (a-1)^2\right] - c(b - a)\right]$

$= \dfrac{(b - a)(b^2 + a^2 + ab)}{3}$

$-\dfrac{m}{2}\left[b^2 - a^2 - 2b + 2a\right] - c(b - a)$

$c + mx - m = x^2$

$\Rightarrow x^2 - mx + m - c = 0$

$x_1 + x_2 = m$

$x_1 x_2 = m - c$

$\Rightarrow x_1 - x_2 = \sqrt{m^2 - 4m + 4c}$

$\Rightarrow (b - a)\left[\dfrac{m^2 - m + c}{3} - \dfrac{m}{2}[m - 2] - c\right]$

$\Rightarrow (b - a)\left[\dfrac{m^2 - m + c}{3} - \dfrac{m^2}{2} - \dfrac{3c}{3} + \dfrac{3m}{3}\right]$

$\Rightarrow (b - a)\left[-\dfrac{m^2}{6} - \dfrac{2c}{3} + \dfrac{2m}{3}\right]$

$A = +\dfrac{1}{6}\,[m^2 + 4c - 4m]^{3/2}$

$\boxed{m^2 - 4m + 4c = 36}$

$B^2 - 4AC = 0$

$16 - 4(4c - 36) = 0$

$\Rightarrow c = 10 \Rightarrow m = 2$

$\Rightarrow c^2 + m^2 = 104$

Sol 11: $A_1 + A_2$ for any point $P(t, t^2)$

$$\int_0^k C_2 \, dx - C + \int_k^t t^2 \, dx - C = \int_0^t (C - C_1) \, dx$$

$$\int_0^k C_2 \, dx - \left[\frac{x^3}{3}\right]_0^k + \left[\frac{-x^3}{3}\right]_k^t + t^2(t - k) = \frac{t^3}{6}$$

$$\int_0^k C_2 \, dx - \frac{k^3}{3} + \frac{k^3}{3} - \frac{t^3}{3} + t^3 - t^2 k = \frac{t^3}{6}$$

$$\int_0^k C_2 \, dx = -\frac{t^3}{2} + t^2 k$$

Let $C_2 = f(x) \, l^2 x^2$

$$\int_0^{t/\lambda} (\lambda^2 x^2) \, dx = \frac{t^3}{2} + t^2 k$$

$$\frac{\lambda^2 \left[x^3\right]_0^{t/\lambda}}{3} = -\frac{t^3}{2} + \frac{t^3}{\lambda} \Rightarrow \frac{\lambda^2}{3} \frac{t^3}{\lambda^3} = -\frac{t^3}{2} + \frac{t^3}{\lambda}$$

$$\Rightarrow \frac{10}{2} = \frac{1}{\lambda} - \frac{1}{3\lambda} = \frac{2}{3\lambda} \Rightarrow \lambda = \frac{4}{3}$$

$$\Rightarrow c_2 = \frac{16x^2}{9}$$

Sol 12: (i) Let equation of CD be $y = 2x + c$

For intersection with $y = x^2$

$\Rightarrow x^2 = 2x + c$; $x^2 - 2x - c = 0$

$\Rightarrow x_1 + x_2 = 2$ and $x_1 x_2 = -c$

Length of CD $= |x_1 + x_2| \sqrt{5}$

$= 2\sqrt{5} \sqrt{1 + c}$

Length of $AC = BD = \dfrac{c + 17}{\sqrt{5}}$

Given ABCD is square than,

$2\sqrt{5} \sqrt{1 + c} = \dfrac{c + 17}{\sqrt{5}}$

$\Rightarrow c^2 - 66c + 189 = 0$

$\Rightarrow c = 3, 63$

Therefore, least value of c is 3.

(ii) For maximum Area of sq. ABCD

Length $2\sqrt{5} \sqrt{1 + c}$ must be maximum

For c = 63 (From Previous questions)

Area $= \left(2\sqrt{5} \sqrt{1 + c}\right)^2$

$= \left(2\sqrt{5} \sqrt{1 + 63}\right)^2 = 4 \times 5 \times 64$

$= 1280$ sq. units.

(iii) Area bounded by $y = 2x + 3$ and

Area $= \displaystyle\int_{-1}^{3} (2x + 3 - x^2) \, dx$

$= \left[x^2 + 3x - \dfrac{x^3}{3}\right]_{-1}^{3}$

$= 9 + 9 - 9 - 1 + 3 - \dfrac{1}{3}$

$= 11 - \dfrac{1}{3} = \dfrac{32}{3}$ sq. units

Sol 13:

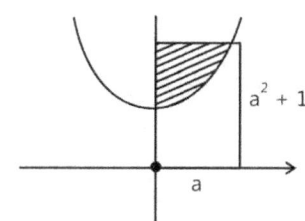

Area of rectangle $= a(a^2 + 1)$

$$A = \int_1^{a^2 + 1} \sqrt{y - 1} \, dy = \left[\frac{2(y - 1)^{3/2}}{3}\right]_1^{a^2 + 1}$$

$= \dfrac{2}{3} a^3 - 0 = \dfrac{2}{3} a^3$

$= \dfrac{2}{3} a^3 = \dfrac{1}{2} a(a^2 - 1) = \dfrac{4}{3} a^3 = a^3 + a$

$a^3 = 3a \Rightarrow a = \sqrt{3}, -\sqrt{3}, 0$

$\Rightarrow a = \sqrt{3}$

Sol 14: $f^3(x) = \displaystyle\int_c^x t f^2(t) \, dt$

$\Rightarrow f'(x) \, 3f^2(x) = x f^2(x)$

$\Rightarrow f^2(x)[3f'(x) - x] = 0$

$\Rightarrow f'(x) = \dfrac{x}{3} \Rightarrow f(x) = \dfrac{x^2}{6} + c$

$\Rightarrow f(x) = \dfrac{x^2}{6} + c$

$$\left[\frac{x^2}{6}+c\right]^3 = \int_0^x t\left(\frac{t^2}{6}+c\right)^2 dx$$

$$\frac{x^6}{216}+c^3+\frac{3cx^4}{36}+\frac{3x^2c^2}{6}$$

$$= \int_0^x \left[\frac{t^4}{36}+c^2+\frac{ct^2}{3}\right]t\ dx = \int_0^x \left[\frac{t^5}{36}+c^2t+\frac{ct^3}{3}\right] dx$$

$$= \left[\frac{t^6}{216}+\frac{c^2t^2}{2}+\frac{ct^4}{12}\right]_0^x = \frac{x^6}{216}+\frac{c^2x^2}{2}+\frac{cx^4}{12} \Rightarrow c = 0$$

$$\Rightarrow f(x) = \frac{x^2}{6}$$

Hence, $\int_0^3 f(x)dx = \left[\frac{x^3}{18}\right]_0^3 = \frac{3}{2}$

Sol 15: Given curves are $y^2 = x+1$ .... (i)

and $y^2 = -x+1$ or $y^2 = -(x-1)$ .... (ii)

Curve (i) is the parabola having axis $y = 0$ and vertex (-1, 0).

Curve (ii) is the parabola having axis $y = 0$ and vertex (1, 0)

(1) - (2) $2x = 0$ $x = 0$

From (i), $x = 0 \Rightarrow y = \pm 1$

Required area = area ACBDA

$$= \int_{-1}^1 (x_1 - x_2)\, dy$$

$$= \int_{-1}^1 \left[(1-y^2)-(y^2-1)\right] dy$$

$$= 2\int_{-1}^1 (1-y^2)\, dy$$

$$= 2\left[y-\frac{y^3}{3}\right]_{-1}^1 = 2\left[\left(1-\frac{1}{3}\right)-\left(-1+\frac{1}{3}\right)\right]$$

$$= \frac{8}{3} \text{ sq. units}$$

Sol 16:

$$A = 2\int_0^{\pi/2} \sin x + 2\int_{\pi/2}^\pi 2-\sin x$$

$$= 2\left[-\cos x\right]_0^{\pi/2} + 2\left[2x+\cos x\right]_{\pi/2}^\pi$$

$$= 2[+1] + 2[2\pi - 1 - p] = 2\pi = a\pi + b$$

$a = 2, b = 0$

$a^2 + b^2 = 4$

Sol 17: $-1 + x = 2y - y^2 - 1$

$x - 1 = -(y - 1)^2$

$y = \sqrt{1-x} + 1 = mx + 2$

$1 - x = (mx + 1)^2$

$1 - x = m^2x^2 + 1 + 2mx$

$\Rightarrow m^2x^2 + (2m + 1)x = 0$

$$\Rightarrow x_1 = 0,\ x_2 = \frac{-(2m+1)}{m^2}$$

$$x_1 x_2 = 0,\ x_1 + x_2 = \frac{-(2m+1)}{m^2}$$

$$A = \int_{x_1}^{x_2} \left(1+\sqrt{1-x}-mx-2\right) dx$$

$$A = \left[x-\frac{(1-x)^{3/2}}{3/2}-\frac{mx^2}{2}-2x\right]_{x_1}^{x_2}$$

$$= \left[\frac{-2(1-x)\sqrt{1-x}}{3}-\frac{mx^2}{2}-x\right]_{x_1}^{x_2}$$

$$= -\frac{2}{3}\left[(1-x_2)^{3/2}-(1-x_1)^{3/2}\right]-\left[\frac{m}{2}(x_2^2-x_1^2)\right]-\left[(x_2-x_1)\right]$$

$$= -\frac{2}{3}\left[(1-x)^{3/2}-1\right]-\frac{m}{2}x^2-x$$

$$= \frac{2}{3}-\frac{2}{3}(1-x)^{3/2}-\frac{mx^2}{2}-x$$

$$= \frac{2}{3}-\frac{2}{3}\left(1+\frac{(2m+1)}{m^2}\right)^{3/2}-\frac{m}{2}\left(\frac{2m+1}{m^2}\right)^2+\left(\frac{2m+1}{m^2}\right)$$

$$= \frac{2}{3}-\frac{2}{3m^3}(m+1)^3+\frac{2}{m}+\frac{1}{m^2}$$

$$-\frac{m}{2}\left(\frac{4}{m^2}+\frac{1}{m^4}+\frac{4}{m^3}\right) = \frac{2}{3}+\frac{2}{m}+\frac{1}{m^2}-\frac{2}{m}-\frac{1}{2m^3}$$

$$-\frac{2}{m^2} - \frac{2}{3} - \frac{2}{3m^3} - \frac{2}{m} - \frac{2}{m^2} = -\frac{3}{m^2} - \frac{2}{m} - \frac{7}{6m^3}$$

$$A = \frac{3m + 2m^2 + \frac{7}{6}}{m^3}$$

Sol 18:

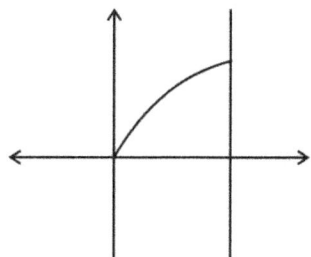

$f'(x) = -xe^{-x} + e^{-x}$

$f''(x) = xe^{-x} - e^{-x} - e^{-x}$

$= (x - 2)e^{-x} = 0$

$\Rightarrow x = 2$ inflection point

$$\int_0^2 xe^{-x} = \left[-xe^{-x} - e^{-x}\right]_0^2$$

$= -2e^{-2} - e^{-2} + 1 = 1 - 3e^{-2}$

Sol 19: Let $R = \{(x,y): y^2 \le 4x, 4x^2 + 4y^2 \le 9\}$

$= \{(x,y): y^2 \le 4x\} \cap \{(x,y): 4x^2 + 4y^2 \le 9\} = R_1 \cap R_1$

Where $R_1 = \{(x,y): y^2 \le 4x\}$ and $R_2 = \{(x,y): 4x^2 + 4y^2 \le 9\}$

Equation of the given curves are

$y^2 = 4x$ ... (i)

and $4x^2 + 4y^2 = 9$ ... (ii)

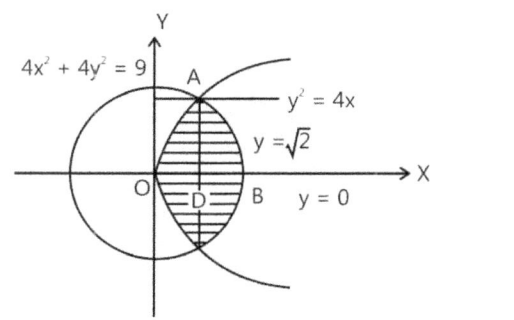

Curve (i) is a parabola having axis $y = 0$ and vertex (0, 0).

Curve (ii) is a circle having centre at (0, 0) and radius $\frac{3}{2}$.

Clearly region R is the interior of the parabola (i) and

region R_2 is the interior of the circle (ii). Therefore, $R_1 \cap R_2$ is the shaded region.

Putting the value of y^2 from (i) in (ii), we get

$$4x^2 + 16x - 9 = 0 \Rightarrow x = \frac{1}{2} - \frac{9}{2}$$

From (i), $x = \frac{1}{2} \Rightarrow y = \pm \sqrt{2}$

And $x = -\frac{9}{2}$ is not possible

Thus, $A = \left(\frac{1}{2}, \sqrt{2}\right)$ and $C = \left(\frac{1}{2}, -\sqrt{2}\right)$

Required area = 2 area OABD

$$= 2 \int_0^{\sqrt{2}} (x_1 - x_2) dy$$

$$= 2 \int_0^{\sqrt{2}} \frac{1}{2} \sqrt{3^2 - (2y)^2} dy - 2 \int_0^{\sqrt{2}} \frac{y^2}{4} dy$$

$$= \frac{1}{2} \int_0^{2\sqrt{2}} \sqrt{3^2 - z^2} \, dz - \frac{1}{2}\left[\frac{y^3}{3}\right]_0^{\sqrt{2}}$$

[Putting z = 2y in first integral]

$$= \frac{1}{2}\left[\frac{2\sqrt{9 - z^2}}{2} + \frac{9}{2} \sin^{-1} \frac{z}{3}\right]_0^{2\sqrt{2}} - \frac{1}{6} \cdot 2\sqrt{2}$$

$$= \frac{1}{2}\left[\sqrt{2} + \frac{9}{2} \sin^{-1} \frac{2\sqrt{2}}{3}\right] - \frac{2\sqrt{2}}{6}$$

$$= \frac{\sqrt{2}}{6} + \frac{9}{4} \sin^{-1} \frac{2\sqrt{2}}{3} = \frac{\sqrt{2}}{6} + \frac{9}{4} \cos^{-1} \frac{1}{3}$$

$\sin^{-1} \frac{2\sqrt{3}}{3} = \theta$, then $\sin\theta = \frac{2\sqrt{2}}{3}$ and

$\cos \theta = \sqrt{1 - \sin^2 \theta} = \frac{1}{3}$

Required area $= \frac{\sqrt{2}}{6} + \frac{9}{4}\left[\frac{\pi}{2} - \sin^{-1} \frac{1}{3}\right]$

$\left[\because \cos^{-1} x = \frac{\pi}{2} - \sin^{-1} x\right]$

$$= \left(\frac{\sqrt{2}}{6} + \frac{9\pi}{8} - \frac{9}{4} \sin^{-1} \frac{1}{3}\right) \text{ sq. units.}$$

Sol 20:

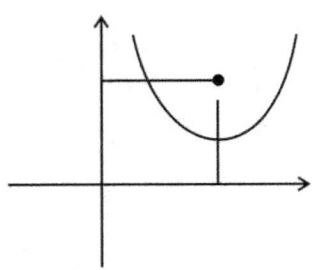

$y = a^2x^2 + ax + 1$

$y = \left(ax + \dfrac{1}{2}\right)^2 + \dfrac{3}{4}$

$\Rightarrow A = \displaystyle\int_0^1 \left[\left(ax + \dfrac{1}{2}\right)^2 + \dfrac{3}{4}\right] dx$

$\Rightarrow A = \left[\dfrac{a^2x^3}{3} + \dfrac{ax^2}{2} + x\right]_0^1 = \dfrac{a^2}{3} + \dfrac{a}{2} + 1$

A least if $A' = 0$ ie $\dfrac{2a}{3} + \dfrac{1}{2} = 0$

$a = -\dfrac{3}{4}$

Sol 21:

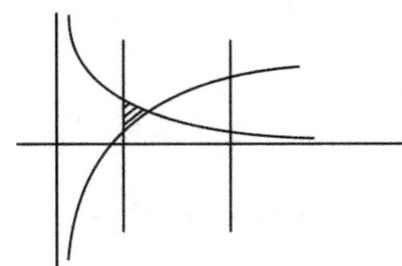

$\displaystyle\int_1^K \left(\dfrac{1}{x} - \ell nx\right) dx = \int_K^a \left(\ell nx - \dfrac{1}{x}\right) dx$

$1 = K \ell nK$

$K^K = e$

$\left[\ell nx - x\ell nx + x\right]_1^K = \left[x\ell nx - x - \ell nx\right]_K^a$

$\ell nK - 1 + K - 1$

$= a\ell na - a - \ell na - 1 + K + \ell nK$

$(a - 1) = (a - 1)(\ell na)$

$(a - 1)[1 - \ell na] = 0$

Either $a = 1$ or $a = e$

$\therefore a = e$

Exercise 2

Sol 1: (D)

$A = \displaystyle\int_a^b \left(x^2 - 1 - 3 + x\right) dx = \int_a^b \left(x^2 + x - 4\right) dx$

$= \left[\dfrac{x^3}{3} + \dfrac{x^2}{2} - 4x\right]_a^b$

$= \dfrac{b^3 - a^3}{3} + \dfrac{b^2 - a^2}{2} - 4(b - a)$

$= \sqrt{17}\left|\dfrac{1 + 4}{3} + \dfrac{(-1)}{2} - 4\right| = \sqrt{17}\left|\dfrac{5}{3} - \dfrac{9}{2}\right| = \dfrac{17\sqrt{17}}{6}$

For point of intersection

$x^2 - 1 = 3 - x$

$\Rightarrow x^2 + x - 4 = 0$

$\Rightarrow b + a = -1 \Rightarrow ab = -4$

$\Rightarrow b - a = \sqrt{1 + 16} \Rightarrow b - a = \sqrt{17}$

Sol 2: (A) e^{-x} and linearly $= e^{-4}$ & $x = 1$

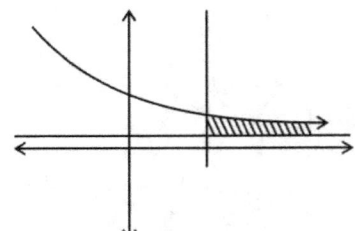

$\displaystyle\int_1^4 \left[e^{-x} - e^{-4}\right] dx$

$= -e^{-4} - 4e^{-4} - (-e^{-1} - e^{-4})$

$= -5e^{-4} + e^{-4} + e^{-1}$

$= e^{-1} - 4e^{-4} = \dfrac{1}{e} - \dfrac{4}{e^4} = \dfrac{e^3 - 4}{e^4}$

Sol 3: (D) $y = \sqrt{9 - x^2}$ $x^2 - 6x + y^2 = 0$

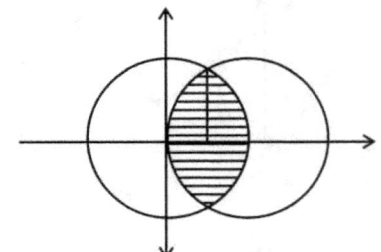

$$\text{Area} = 2\int_{0}^{3/2}\left(\sqrt{9-x^2}-\sqrt{6x-x^2}\right)dx$$

$$= 2\int_{0}^{3/2}\left[\sqrt{9-x^2}-\sqrt{9-(x-3)^2}\right]dx$$

$$= 2\left[\frac{9\sin^{-1}\dfrac{x}{3}+x\sqrt{9-x^2}}{2}\right]_{0}^{3/2} -$$

$$2\left[\frac{9\sin^{-1}\dfrac{x-3}{3}+(x-3)\sqrt{6x-x^2}}{2}\right]_{0}^{3/2}$$

$$= 9.\frac{\pi}{6}+\frac{3}{2}\sqrt{9-\frac{9}{4}}+9.\frac{\pi}{6}+\frac{3}{2}\sqrt{9-\frac{9}{4}}$$

$$= 9.\frac{\pi}{3}+3\times3\frac{x\sqrt{3}}{2} = 3\frac{\pi}{6}+\frac{9\sqrt{3}}{2} = 3\left[\pi+\frac{3\sqrt{3}}{2}\right]$$

Sol 4: (B)

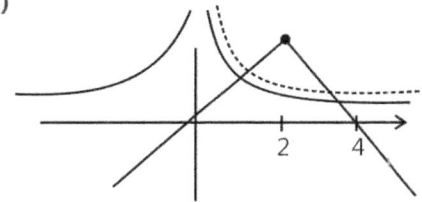

$$\text{Area} = \int_{\sqrt{3}}^{2}\left(x-\frac{3}{x}\right)dx+\int_{2}^{3}\left(4-x-\frac{3}{x}\right)dx$$

$$= \left[\frac{x^2}{2}-3\ell nx\right]_{\sqrt{3}}^{2}+\left[4x-\frac{x^2}{2}-3\ell nx\right]_{2}^{3}$$

$$= 2-3\ell n2-\frac{3}{2}+3\ell n\sqrt{3}+12-\frac{9}{2}-3\ell n3-8$$

$$+ 2+3\ell n2$$

$$= \frac{1}{2}+\frac{15}{2}-6+3\ell n\frac{1}{\sqrt{3}} = 2-\frac{3}{2}\ell n3 = \frac{4-3\ell n3}{2}$$

Sol 5: (D)

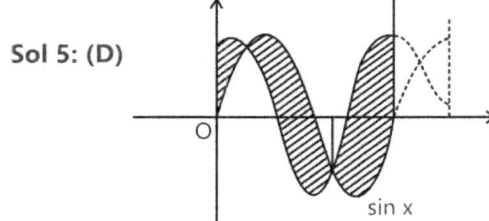

sin x

Area =

$$\int_{0}^{\pi/4}\left(\cos x-\sin x\right)dx+\int_{\pi/4}^{5\pi/4}\left(\sin x-\cos x\right)dx$$

$$+ \int_{5\pi/4}^{2\pi}\left(\cos x-\sin x\right)dx$$

$$= \left[\sin+\cos\right]_{0}^{\pi/4}+\left[\sin+\cos\right]_{5\pi/4}^{2\pi}+$$

$$\left[-\sin-\cos\right]_{\pi/4}^{5\pi/4}$$

$$= \sqrt{2}-1+1+\sqrt{2}-[-\sqrt{2}-\sqrt{2}]=4\sqrt{2}\text{ sq. un ts}$$

Sol 6: (C)

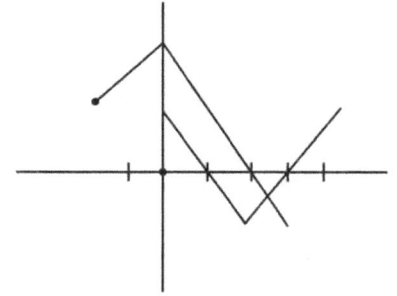

$$f(x) = -1+|x-2$$

$$g(x) = 2-|x|$$

$$f(x) = 1-x \in (0, 2)$$

$$x-3 \in (2, 4)$$

$$g(x) = 2-x \in (0, 3)$$

$$2+x \in (-1, 0)$$

$$g(f(x)) = 2-f(x)\ f(x) \in (0, 3)$$

$$= 2+f(x)f(x) \in (-1, 0)$$

$$= 2-(1-x), x \in (0\ 1) = 1+x$$

$$= 2+(1-x), x \in (1, 2) = 3-x$$

$$= 2+(x-3), x \in (2, 3) = x-1$$

$$= 2-(x-3), x \in (3\ 4) = 5-x$$

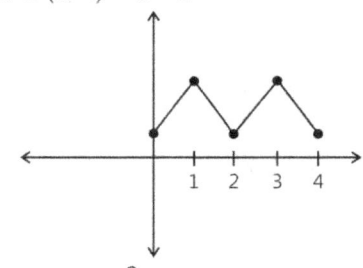

$$\text{Area} = \frac{3}{2}\left[1+2\right]\times1 = \frac{9}{2}$$

Sol 7: (C)

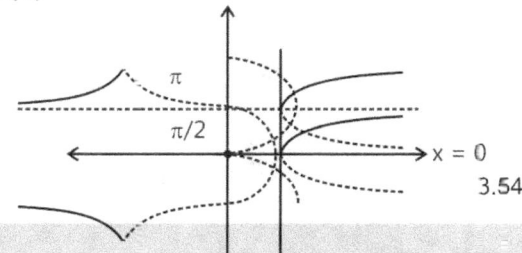

3.54

$$\text{Area} = 2\int_0^{\pi/4}\left(\sec x - 1\right)dx$$

$$= 2\left[\ell n\left(\sec x + \tan x\right)\right]_0^{\pi/4} - \frac{\pi}{2}$$

$$= 2\ln\left(\sqrt{2}+1\right) - \frac{\pi}{2} \quad = \ln\left(3+2\sqrt{2}\right) - \frac{\pi}{2}$$

Sol 8: (A)

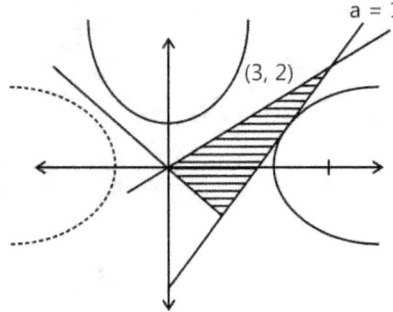

Eq. of tangent

$$\frac{y-2}{x-3} = \frac{2\times 3}{2\sqrt{4}} = \frac{3}{2} \Rightarrow 2y - 4 = 3x - 9$$

$$3x - 2y = 5 \qquad \qquad \text{... (i)}$$

$$A = \frac{1}{2}\times 2\times 1 + \int_1^5\left(x - \left(\frac{3x-5}{2}\right)\right)dx$$

$$= 1 + \left[\frac{-x^2 + 10x}{4}\right]_1^5 = 1 + \frac{1}{4}(16) = 5$$

Sol 9: (A) $y = 1 + 4x - x^2$

$$\Rightarrow y - 5 = -(x-2)^2$$

$$\text{Area} = \int_0^{3/2} 1 + 4x - x^2$$

$$= \left[x + 2x^2 - \frac{x^3}{3}\right]_0^{3/2} = \frac{3}{2} + 2\times\frac{9}{4} - \frac{27}{24} = 6 - \frac{9}{8} = \frac{39}{8}$$

$$\text{Area of }\Delta = \frac{1}{2}\times\frac{3}{2}\times\frac{3m}{2} = \frac{1}{2}\times\frac{39}{8}$$

$$9m = \frac{39}{2} \Rightarrow m = \frac{13}{6}$$

Sol 10: (A)

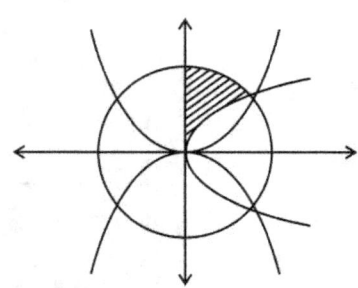

$$\text{Area}_1 = \int_0^1 \sqrt{x}\, dx = \left[\frac{2x^{3/2}}{3}\right]_0^1 = \frac{2}{3}$$

$$\text{Area}_2 = \int_0^1\left(\sqrt{2-x^2}\right)dx$$

$$= 2\left[\frac{\sin^{-1}\dfrac{x}{\sqrt{2}} + x\sqrt{2-x^2}}{2}\right]_0^1 = \frac{\pi}{4} + 1$$

$$\text{Area} = \frac{\pi}{4} + 1 - \frac{2}{3} = \left(\frac{\pi}{4} + \frac{1}{3}\right)$$

Sol 11: (D)

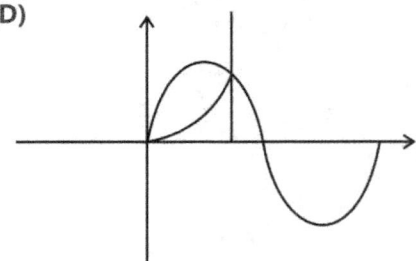

$$A_1 = \int_0^1 \sin\frac{\pi x}{2}\, dx = -\frac{2}{\pi}\left[\cos\frac{\pi x}{2}\right]_0^1 = -\frac{2}{\pi}(-1) = \frac{2}{\pi}$$

$$A_2 = \int_0^1 x^2\, dx = \frac{1}{3}$$

$$\text{Ratio} = \frac{\dfrac{1}{3}}{\dfrac{2}{\pi} - \dfrac{1}{3}} = \frac{6-\pi}{\pi}$$

Sol 12: (D) $A = \int\left(y^2 - 1 - |y|\sqrt{1-y^2}\right)dy$

$$= \int_0^1\left(y^2 - 1 - y\sqrt{1-y^2}\right)dy + \int_{-1}^0\left(y^2 - 1 + y\sqrt{1-y^2}\right)dy$$

$$= \left[\frac{y^3}{3} - y + \frac{\left(\sqrt{1-y^2}\right)^{3/2}}{2\left(\frac{3}{2}\right)}\right]_0^1 + \left[\frac{y^3}{3} - y - \frac{\left(\sqrt{1-y^2}\right)^{3/2}}{3}\right]_{-1}^0$$

$$= \frac{1}{3} - 1 - \left(0 - 0 + \frac{1}{3}\right) + \left(0 - 0 - \frac{1}{3} - \left(-\frac{1}{3} + 1\right)\right)$$

$$= -1 + \frac{1}{3} - \frac{1}{3} - 1 = -2; \quad \text{Area} = 2$$

Sol 13: (B) $f(x) = 3x^3 + 2x$

$g(x) = f^{-1}(x)$

$\int_0^5 g(x) = \int_0^1 f(x) = \left[\dfrac{3x^4}{4} + x^2\right]_0^1 = \left[\dfrac{3}{4} + 1\right] = \dfrac{7}{4}$

Previous Years' Questions

Sol 1: (B) Required area $= \int_0^{\pi/4}\left(\sqrt{\dfrac{1+\sin x}{\cos x}} - \sqrt{\dfrac{1-\sin x}{\cos x}}\right)dx$

$= \left(\because \dfrac{1+\sin x}{\cos x} > \dfrac{1-\sin x}{\cos x} > 0\right)$

$= \int_0^{\pi/4}\left(\sqrt{\dfrac{1+\dfrac{2\tan\frac{x}{2}}{1+\tan^2\frac{x}{2}}}{\dfrac{1-\tan^2\frac{x}{2}}{1+\tan^2\frac{x}{2}}}} - \sqrt{\dfrac{1-\dfrac{2\tan\frac{x}{2}}{1+\tan^2\frac{x}{2}}}{\dfrac{1-\tan^2\frac{x}{2}}{1+\tan^2\frac{x}{2}}}}\right)dx$

$= \int_0^{\pi/4}\left(\sqrt{\dfrac{1+\tan\frac{x}{2}}{1-\tan\frac{x}{2}}} - \sqrt{\dfrac{1-\tan\frac{x}{2}}{1+\tan\frac{x}{2}}}\right)dx$

$= \int_0^{\pi/4}\dfrac{1+\tan\frac{x}{2}-1+\tan\frac{x}{2}}{\sqrt{1-\tan^2\frac{x}{2}}}dx = \int_0^{\pi/4}\dfrac{2\tan\frac{x}{2}}{\sqrt{1-\tan^2\frac{x}{2}}}dx$

Substitute $\tan\dfrac{x}{2} = t \Rightarrow \dfrac{1}{2}\sec^2\dfrac{x}{2}dx = dt$

$= \int_0^{\tan\frac{\pi}{8}}\dfrac{4t\,dt}{(1+t^2)\sqrt{1-t^2}}$

As $\tan\dfrac{\pi}{8} = \sqrt{2}-1$

So, $\int_0^{\sqrt{2}-1}\dfrac{4t\,dt}{(1+t^2)\sqrt{1-t^2}}$

Sol 2: (B) Here, area between 0 to b is R_1 and b to 1 to R_2.

$\therefore \int_0^b (1-x)^2 dx - \int_b^1 (1-x)^2 dx = \dfrac{1}{4}$

$\Rightarrow \left(\dfrac{(1-x)^3}{-3}\right)_0^b - \left(\dfrac{(1-x)^3}{-3}\right)_b^1 = \dfrac{1}{4}$

$\Rightarrow -\dfrac{1}{3}\{(1-b)^3 - 1\} + \dfrac{1}{3}\{0 - (1-b)^3\} = \dfrac{1}{4}$

$\Rightarrow -\dfrac{2}{3}(1-b)^3 = -\dfrac{1}{3} + \dfrac{1}{4} = -\dfrac{1}{12}$

$\Rightarrow (1-b)^3 = \dfrac{1}{8} \Rightarrow (1-b) = \dfrac{1}{2} \Rightarrow b = \dfrac{1}{2}$

Sol 3: (C) $R_1 = \int_{-1}^2 x\,f(x)dx$ ... (i)

Using, $\int_a^b f(x)dx = \int_a^b f(a+b-x)dx$

$R_1 = \int_{-1}^2 (1-x)f(1-x)dx,$

[given, $f(x) = f(1-x)$]

$\therefore R_1 = \int_{-1}^2 (1-x)f(x)dx$ ... (ii)

Given, R_2 is area bounded by

$f(x), x = -1$ and $x = 2$

$\therefore R_2 = \int_{-1}^2 f(x)dx$ ... (iii)

Adding Eqs. (i) and (ii), we get

$2R_1 = \int_{-1}^2 f(x)dx$ ... (iv)

$\therefore$ From Eqs. (iii) and (iv), we get

$2R_1 = R_2$

Sol 4: The point of intersection of the curves $x^2 = 4y$ and $x = 4y - 2$ could be sketched as, are $x = -1$ and $x = 2$.

$\therefore$ Required area

$= \int_{-1}^2\left\{\left(\dfrac{x+2}{4}\right) - \left(\dfrac{x^2}{4}\right)\right\}dx = \dfrac{1}{4}\left[\dfrac{x^2}{2} + 2x - \dfrac{x^3}{3}\right]_{-1}^2$

$= \dfrac{1}{4}\left[\left(2 + 4 - \dfrac{8}{3}\right) - \left(\dfrac{1}{2} - 2 + \dfrac{1}{3}\right)\right]$

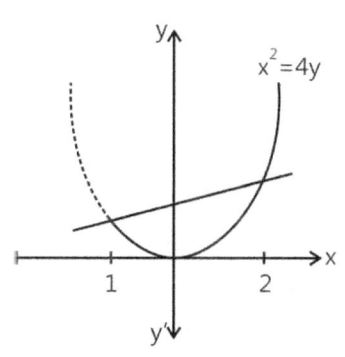

$$= \frac{1}{4}\left[\frac{10}{3} - \left(\frac{-7}{6}\right)\right] = \frac{1}{4} \cdot \frac{9}{2} = \frac{9}{8} \text{ sq unit.}$$

Sol 5: Here, $\int_2^a \left(1 + \frac{8}{x^2}\right)dx = \int_a^4 \left(1 + \frac{8}{x^2}\right)dx$

$$\Rightarrow \left[x - \frac{8}{x}\right]_2^a = \left[x - \frac{8}{x}\right]_a^4$$

$$\Rightarrow \left(a - \frac{8}{a}\right) - (2 - 4) = (4 - 2) - \left(a - \frac{8}{a}\right)$$

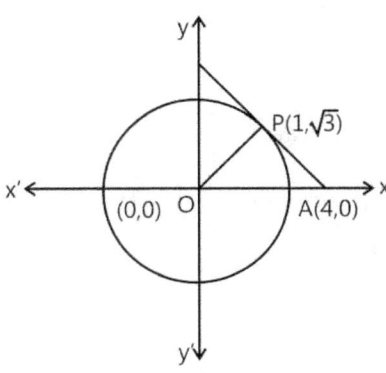

Thus, area of Δ formed by (0, 0) $(1, \sqrt{3})$ and (4, 0)

$$= \frac{1}{2}\begin{vmatrix} 0 & 0 & 1 \\ 1 & \sqrt{3} & 1 \\ 4 & 0 & 1 \end{vmatrix} = \frac{1}{2}(0 - 4\sqrt{3}) = 2\sqrt{3} \text{ sq unit}$$

$$\Rightarrow a - \frac{8}{a} + 2 = 2 - a + \frac{8}{a}$$

$$\Rightarrow 2a - \frac{16}{a} = 0$$

$$\Rightarrow 2(a^2 - 8) = 0$$

$\therefore a = \pm 2\sqrt{2}$, (neglecting –ve sign)

$a = 2\sqrt{2}$.

Sol 6: Given, $y = \begin{cases} \tan x, & -\frac{\pi}{3} \leq x \leq \frac{\pi}{3} \\ \cot x, & \frac{\pi}{6} \leq x \leq \frac{\pi}{2} \end{cases}$

which could be plotted as, y-axis.

$\therefore$ Required area $= \int_0^{\pi/4}(\tan x)dx + \int_{\pi/4}^{\pi/3}(\cot x)dx$

$$= \left[-\log|\cos x|\right]_0^{\pi/4} + \left[\log \sin x\right]_{\pi/4}^{\pi/3}$$

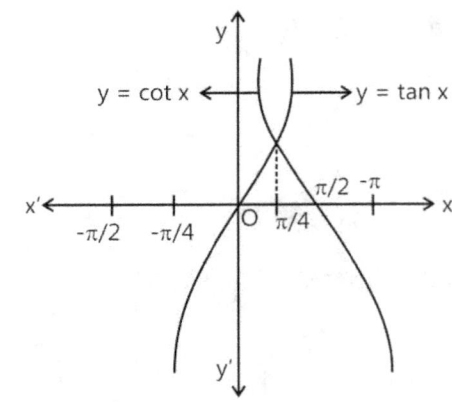

$$= \left(\log\frac{1}{\sqrt{2}} - 0\right) + \left(\log\frac{\sqrt{3}}{2} - \log\frac{1}{\sqrt{2}}\right)$$

$$= \log\frac{\sqrt{3}}{2} - 2\log\frac{1}{\sqrt{2}} = \log\frac{\sqrt{3}}{2}$$

$$\Rightarrow -\log\frac{1}{2} = \frac{1}{2}\log_e 3 \text{ sq. units}$$

Sol 7: Given curves $y = \sqrt{5 - x^2}$ and $y = |x - 1|$ could be sketche as shown whose point of intersection are.

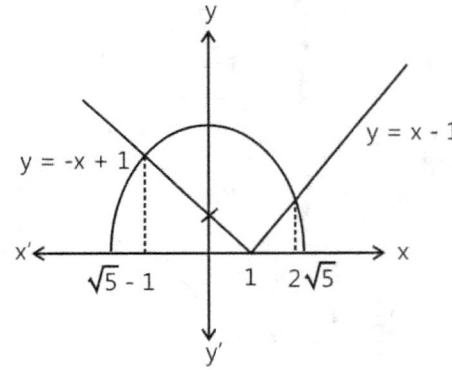

$5 - x^2 = (x - 1)^2$

$\Rightarrow 5 - x^2 = x^2 - 2x + 1$

$\Rightarrow 2x^2 - 2x - 4 = 0$

$\Rightarrow x = 2, -1$

$\therefore$ Required area

$$= \int_{-1}^2 \sqrt{5 - x^2}dx - \int_{-1}^1(-x + 1)dx - \int_1^2(x - 1)dx$$

$$= \left[\frac{x}{2}\sqrt{5 - x^2} + \frac{5}{2}\sin^{-1}\left(\frac{x}{\sqrt{5}}\right)\right]_{-1}^2 - \left[\frac{-x^2}{2} + x\right]_{-1}^1 - \left[\frac{x^2}{2} - x\right]_1^2$$

$$= \left(1 + \frac{5}{2}\sin^{-1}\frac{2}{\sqrt{5}}\right) - \left(-1 + \frac{5}{2}\sin^{-1}\left(\frac{-1}{\sqrt{5}}\right)\right)$$

$$-\left(-\frac{1}{2}+1+\frac{1}{2}+1\right)-\left(2-2-\frac{1}{2}+1\right)$$

$$=\frac{5}{2}\left(\sin^{-1}\frac{2}{\sqrt5}+\sin^{-1}\frac{1}{\sqrt5}\right)-\frac{1}{2}$$

$$=\frac{5}{2}\sin^{-1}\left(\frac{2}{\sqrt5}\sqrt{1-\frac{1}{5}}+\frac{1}{\sqrt5}\sqrt{1-\frac{4}{5}}\right)-\frac{1}{2}$$

$$=\frac{5}{2}\sin^{-1}(1)-\frac{1}{2}=\frac{5\pi}{4}-\frac{1}{2}\text{ sq. units}$$

Sol 8: Given curves are $x^2 + y^2 = 4$, $x^2 = -\sqrt2 y$

Thus, the required area

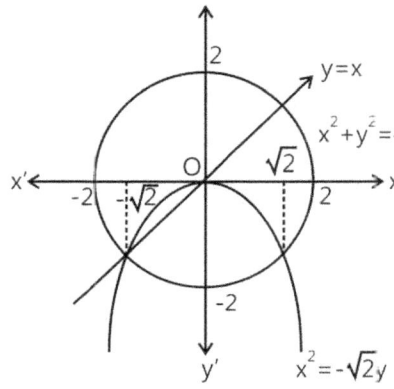

$$=\left|\int_{-\sqrt2}^{\sqrt2}\sqrt{4-x^2}\,dx\right|-\left|\int_{-\sqrt2}^{0}x\,dx\right|-\left|\int_0^{\sqrt2}\frac{-x^2}{\sqrt2}\,dx\right|$$

$$=2\int_0^{\sqrt2}\sqrt{4-x^2}\,dx-\left|\left(\frac{x^2}{2}\right)_{-\sqrt2}^0\right|-\left|\frac{x^3}{3\sqrt2}\Big|_0^{\sqrt2}\right|$$

$$=2\left\{\frac{x}{2}\sqrt{4-x^2}-\frac{4}{2}\sin^{-1}\frac{x}{2}\right\}_0^{\sqrt2}-1-\frac{2}{3}$$

$$=(2-\pi)-\frac{5}{3}=\frac{1}{3}-\pi\text{ sq. units}$$

Sol 9: $y=\tan x \Rightarrow \dfrac{dy}{dx}=\sec^2 x$

$$\therefore\left(\frac{dy}{dx}\right)_{x=\frac{\pi}{4}}=2$$

Hence, equation of tangent at $A\left(\dfrac{\pi}{4},1\right)$ is

$$\frac{y-1}{x-\pi/4}=2 \Rightarrow y-1=2x-\frac{\pi}{2}$$

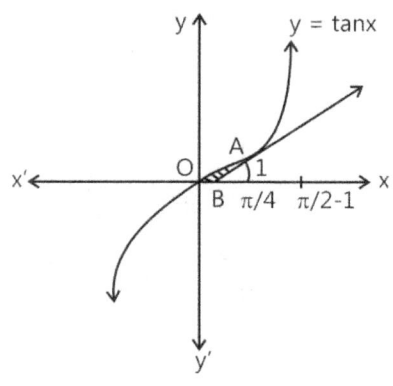

$$\Rightarrow (2x-y)=\left(\frac{\pi}{2}-1\right)$$

$\therefore$ Required area is OABO

$$=\int_0^{\pi/4}(\tan x)dx-\text{area of }\triangle ALM$$

$$=\Big[\log|\sec x|\Big]_0^{\pi/4}-\frac{1}{2}BL\cdot AL$$

$$=\log\sqrt2-\frac{1}{2}\left(\frac{\pi}{2}-\frac{\pi-2}{4}\right)\cdot1$$

$$=\left(\log\sqrt2-\frac{1}{4}\right)\text{ sq. unit}$$

Sol 10: $y=x(x-1)^2 \Rightarrow \dfrac{dy}{dx}=x\cdot2(x-1)+(x-1)^2$

$\therefore$ Maximum at $x=1/3 \Rightarrow y_{max}=\dfrac{1}{3}\left(-\dfrac{2}{3}\right)^2=\dfrac{4}{27}$

Minimum at $x=1$

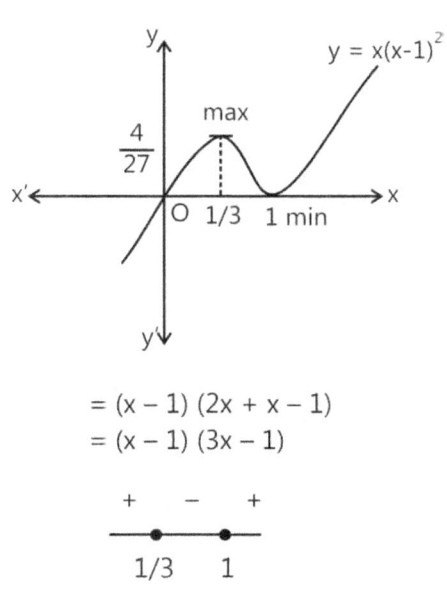

$$=(x-1)(2x+x-1)$$
$$=(x-1)(3x-1)$$

$$\begin{array}{ccc}+&-&+\\[-2pt]\hline\end{array}$$
$$\quad 1/3 \quad\quad 1$$

$$\Rightarrow \qquad y_{min}=0$$

Now, to find the area bounded by the curve $y = x(x - 1)^2$, the y-axis and line $x = 2$

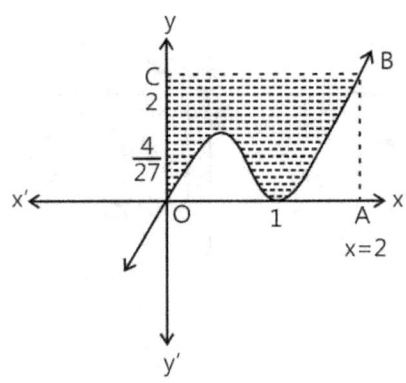

∴ Required area

= Area of square OABC $- \int_0^2 y \, dx$

$= 2 \times 2 - \int_0^2 x(x-1)^2 dx = 4 - \int_0^2 \left(x^3 - 2x^2 + x\right) dx$

$= 4 - \left[\dfrac{x^4}{4} - \dfrac{2x^3}{3} + \dfrac{x^2}{2}\right]_0^2 = 4 - \left[\dfrac{16}{4} - \dfrac{16}{3} + \dfrac{y}{2}\right]$

$= \dfrac{10}{3}$ sq. units

Sol 11: Both the curves are defined for x > 0. Both are positive when x > 1 and negative when 0 < x < 1

We know, $\lim\limits_{x \to 0^+} (\log x) \to -\infty$

Hence, $\lim\limits_{x \to 0^+} \dfrac{\log x}{ex} \to -\infty$, This, y-axis is asymptote of second curve.

And $\lim\limits_{x \to 0^+} ex \log x$ [(0) x ∞ form]

$= \lim\limits_{x \to 0^+} \dfrac{e \log x}{1/x} \left(-\dfrac{\infty}{\infty} \text{form}\right) = \lim\limits_{x \to 0^+} \dfrac{e\left(\dfrac{1}{x}\right)}{\left(-\dfrac{1}{x^2}\right)} = 0$

(using L'Hospital's rule)

Thus, the first curve starts from (0, 0) but does not include (0, 0).

Now, the given curves intersect, therefore

$ex \log x = \dfrac{\log x}{ex}$

i.e. $(e^2 x^2 - 1) \log x = 0$

$\Rightarrow x = 1, \dfrac{1}{e}$ (since x > 0)

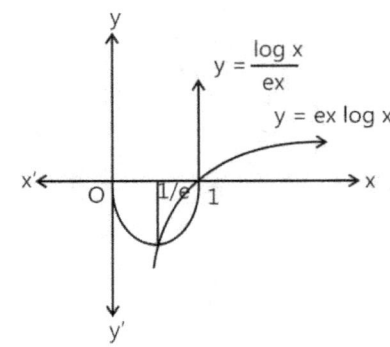

∴ The required area $= \int_{1/e}^{1} \left(\dfrac{\log x}{ex} - ex \log x\right) dx$

$= \dfrac{1}{e}\left[\dfrac{(\log x)^2}{2}\right]_{1/e}^{1} - e\left[\dfrac{x^2}{4}(2\log x - 1)\right]_{1/e}^{1} = \dfrac{e^2 - 5}{4e}$ sq. units

Sol 12: Given, $\begin{bmatrix} 4a^2 & 4a & 1 \\ 4b^2 & 4b & 1 \\ 4c^2 & 4c & 1 \end{bmatrix} \begin{bmatrix} f(-1) \\ f(1) \\ f(2) \end{bmatrix} = \begin{bmatrix} 3a^2 + 3a \\ 3b^2 + 3b \\ 3c^2 + 3c \end{bmatrix}$

$\Rightarrow 4a^2 f(-1) + 4a f(1) + f(2) = 3a^2 + 3a,$ ... (i)

$4b^2 f(-1) + 4b f(1) + f(2) = 3b^2 + 3b$ ... (ii)

and $4c^2 f(-1) + 4cf(1) + f(2) = 3c^2 + 3c$ ... (iii)

Where f(x) is quadratic expression given by,

$f(x) = ax^2 + bc + c$ and (i), (ii) and (iii)

$\Rightarrow 4x^2 f(-1) + 4x f(1) + f(2) = 3x^2 + 3x$

or $\{4 f(-1) - 3\}x^2 + \{4 f(1) - 3\} x + f\{2\} = 0$ ... (iv)

As above equation has 3 roots a, b an c

∴ above equation is identity in x.

i.e., Coefficients must be zero.

$\Rightarrow f(-1) = 3/4, f(1) = 3/4, f(2) = 0$ ... (v)

∵ $f(x) = ax^2 + bx + c$

∴ a = -1/4, b = 0 and c = 1, using Eq. (v)

Thus, $f(x) = \dfrac{4 - x^2}{4}$ shown as,

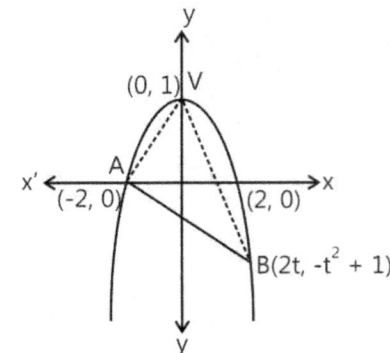

Let $A(-2, 0), B = (2t, -t^2 + 1)$ v

Since, AB subtends right angle at vertex V(0, 1)

$$\Rightarrow \frac{1}{2} \cdot \frac{-t^2}{2t} = -1 \Rightarrow t = 4$$

$\therefore B(8, -15)$

Equation of chord AB is

$$y = \frac{-(3x + 6)}{2}$$

$\therefore$ Required area

$$= \left| \int_{-2}^{8} \left(\frac{4 - x^2}{4} + \frac{3x + 6}{2} \right) dx \right|$$

$$= \left| \left(x - \frac{x^3}{12} + \frac{3x^2}{4} + 3x \right)_{-2}^{8} \right|$$

$$= \left[\left[8 - \frac{128}{3} + 48 + 24 - \left(-2 + \frac{2}{3} + 3 - 6 \right) \right] \right]$$

$$= \frac{125}{3} \text{ sq. units}$$

Sol 13: Here, slope of tangent

$$\frac{dy}{dx} = \frac{(x + 1)^2 + y - 3}{(x + 1)} \Rightarrow \frac{dy}{dx} = (x + 1) + \frac{(y - 3)}{(x + 1)}, X$$

Substitute $x + 1 = X$ and $y - 3 = Y$

$$\Rightarrow \frac{dy}{dx} = \frac{dY}{dX}$$

$$\therefore \frac{dY}{dX} = X + \frac{Y}{X} \Rightarrow \frac{dY}{dX} - \frac{1}{X} Y = X$$

$$I.F = e^{\int -\frac{1}{x} dx} = e^{-\log X} = \frac{1}{X}$$

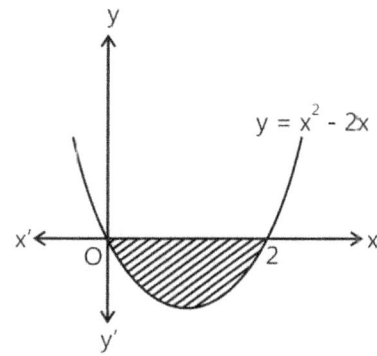

$\therefore$ Solution is,

$$Y. \frac{1}{X} = \int X. \frac{1}{X} dX + c \Rightarrow \frac{Y}{X} = X + c$$

$y - 3 = (x + 1)^2 + c(x + 1)$, which passes through (2, 0).

$$\Rightarrow -3 = (3)^2 + 3c$$

$$\Rightarrow c = -4$$

$\therefore$ Required curve $y = (x + 1)^2 - 4(x + 1) + 3$

$$\Rightarrow y = x^2 - 2x$$

$\therefore$ Required area $= \left| \int_0^2 (x^2 - 2x) dx \right| = \left| \left(\frac{x^3}{3} - x^2 \right)_0^2 \right|$

$$= \frac{8}{3} - 4 = \frac{4}{3} \text{ sq. units}$$

Sol 14:

Region $= \left\{ (x, y) \in R^2 : y \geq \sqrt{|x + 3|}, 5y \leq x + 9 \leq 15 \right\}$

Plotting all the curves

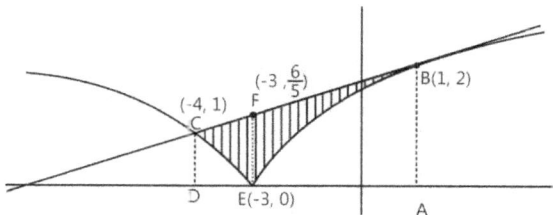

Area = Area $(ABFE)$ – Area (AEB)

+ Area $(DEFC)$ - Area (DEC)

$$= \frac{32}{5} - \int_{-3}^{1} \left(\sqrt{-x - 3} \right) dx + \frac{11}{10} - \int_{-4}^{-3} \left(\sqrt{x + 3} \right) dx = \frac{3}{2} \text{ sq. units}$$

Sol 15: (A, B, D)

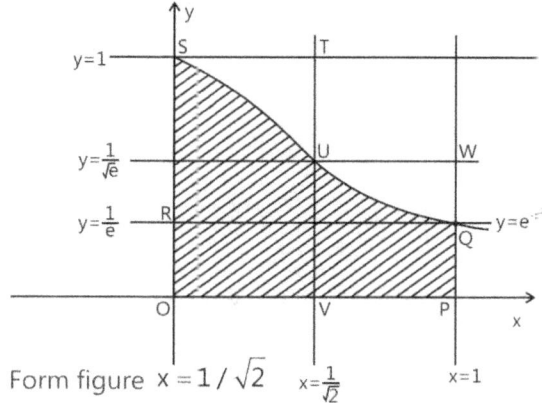

Form figure $x = 1/\sqrt{2}$ $x = \frac{1}{\sqrt{2}}$ $x = 1$

S > Area $(OPQR)$

$$\Rightarrow S > 1 \times \frac{1}{e} \Rightarrow S > \frac{1}{e}$$

S > Area $(PVUW)$ + Area $(OSTV)$

$$< \left(1 - \frac{1}{12}\right)\frac{1}{\sqrt{e}} + 1 \times \frac{1}{\sqrt{2}}$$

$$< \frac{1}{12} + \frac{1}{\sqrt{e}}\left(1 - \frac{1}{\sqrt{2}}\right) \quad \text{D is correct}$$

Now, $e^{-x} \le e^{-x^2}$ if $x \in (0, 1)$

$$\int_0^1 e^{-x}dx \le \int_0^1 e^{-x^2}dx \implies \left(1 - \frac{1}{e}\right) \le S \quad \text{B is correct}$$

$$1 - \frac{1}{e} > \frac{1}{4}\left(1 + \frac{1}{\sqrt{e}}\right) \text{ and } S > 1 - \frac{1}{e}$$

$$\implies S > \frac{1}{4}\left(1 + \frac{1}{\sqrt{e}}\right)$$

(B) and (D) Correct

4. DIFFERENTIAL EQUATIONS

1. INTRODUCTION

An equation containing an independent variable, dependent variable and differential coefficients is called a differential equation.

(i) $\dfrac{dy}{dx} = \sin x$ (ii) $\left(\dfrac{d^2y}{dx^2}\right)^2 + x\left(\dfrac{dy}{dx}\right)^3 = 0$ (iii) $\left(\dfrac{d^4y}{dx^4}\right)^3 - 4\dfrac{dy}{dx} = 5\cos 3x$

2. ORDER OF DIFFERENTIAL EQUATION

The order of a differential equation is the order of the highest derivative occurring in the differential equation. For example, the order of the above mentioned differential equations are 1, 2, and 4 respectively.

3. DEGREE OF DIFFERENTIAL EQUATION

The degree of a differential equation is the degree of the highest order derivative when differential coefficients are free from radicals and fractions. For example the degrees of above differential equations are 1, 2, and 3 respectively.

Table 24.1: Degree of differential equation

Differential Equation	Order of D.E.	Degree of D.E
$\dfrac{dy}{dx} + 4y = \sin x$	1	1
$\left(\dfrac{d^2y}{dx^2}\right)^4 + \left(\dfrac{dy}{dx}\right)^5 - y = e^x$	2	4
$\dfrac{d^2y}{dx^2} - \dfrac{dy}{dx} + 3y = \cos x$	2	1
$\dfrac{dy}{dx} = \dfrac{x^4 - y^4}{xy\left(x^2 + y^2\right)}$	1	1

Differential Equation	Order of D.E.	Degree of D.E
$y = x\dfrac{dy}{dx} + \sqrt{a^2\left(\dfrac{dy}{dx}\right)^2 + b^2}$ $\Rightarrow (x^2 - a^2)\left(\dfrac{dy}{dx}\right)^2 - 2xy\dfrac{dy}{dx} + (y^2 - b^2) = 0$	1	2
$\dfrac{d^2y}{dx^2} = \left(1 + \left(\dfrac{dy}{dx}\right)^2\right)^{3/2} \Rightarrow \left(\dfrac{d^2y}{dx^2}\right)^2 - \left(1 + \left(\dfrac{dy}{dx}\right)^2\right)^3 = 0$	2	2

4. CLASSIFICATION OF DIFFERENTIAL EQUATIONS

Differential equations are first classified according to their order. First-order differential equations are those in which only the first order derivative, and no higher order derivatives appear. Differential equations of order two or more are referred to as higher order differential equations.

A differential equation is said to be linear if the unknown function, together with all of its derivatives, appears in the differential equations with a power not greater than one and not as products either. A nonlinear differential equation is a differential equation which is not linear.

e.g. $y' + y = 0$ is a linear differential equation,

$y'' + yy' + y^2 = 0$ is a non linear differential equation,

Procedure to form a differential equation that represents a given family of curves

Case I:

If the given family F1 of curves depends on only one parameter then it is represented by an equation of the form
F1(x, y, a) = 0   ... (i)

For example, the family of parabolas $y^2 = ax$ can be represented by an equation of the form

f(x, y, a): $y^2 = ax$

Differentiating equation (i) with respect to x, we get an equation involving y', y, x and a.

g(x, y, y', a) = 0   ... (ii)

The required differential equation is then obtained by eliminating a from equation (i) and (ii) as

F(x, y, y') = 0   ... (iii)

Case II:

If the given family F2 of curves depends on the parameters a, b (say) then it is represented by an equation of the form F2(x, y, a, b) = 0   ... (iv)

Differentiating equation (iv) with respect to x, we get an equation involving y', x, y, a, b.

g(x, y, y', a, b) = 0   ... (v)

Now we need another equation to eliminate both a and b. This equation is obtained by differentiating equation (v), wrt x, to obtain a relation of the form h(x, y, y', y'', a, b) = 0   ... (vi)

The required differential equation is then obtained by elimination a and b from equations (iv), (v) and (vi) as F(x, y, y', y'') = 0   ... (vii)

Note: The order of a differential equation representing a family of curves is the same as the number of arbitrary constants present in the equation corresponding to the family of curves.

5. FORMATION OF DIFFERENTIAL EQUATIONS

If an equation is dependent and dependent variables having some arbitrary constant are given, then the differential equation is obtained as follows:

(a) Differentiate the given equation w.r.t. the independent variable (say x) as many times as the number of arbitrary constants in it.

(b) Eliminate the arbitrary constants.

(c) Hence on eliminating arbitrary constants results a differential equation which involves x, y, $\dfrac{dy}{dx}$, $\dfrac{d^2y}{cx^2}$ .... $\dfrac{d^my}{d_xm}$ (where m=number of arbitrary constants).

Illustration 1: Form the differential equation corresponding to $y^2 = m(a^2 - x^2)$, where m and a are arbitrary constants. **(JEE MAIN)**

Sol: Since the given equation contains two arbitrary constant, we shall differentiate it two times with respect to x and we get a differential equation of second order.

We are given that $y^2 = m(a^2 - x^2)$ ... (i)

Differentiating both sides of (i) w.r.t. x, we get

$$2y\frac{dy}{dx} = m(-2x) \Rightarrow y\frac{dy}{dx} = -mx \qquad \text{... (ii)}$$

Differentiating both sides of (ii) w.r.t. to x, we get $y\dfrac{d^2y}{dx^2} + \left(\dfrac{dy}{dx}\right)^2 = -m$ ... (iii)

From (ii) and (iii), we get, $x\left[y\dfrac{d^2y}{dx^2} + \left(\dfrac{dy}{dx}\right)^2\right] = y\dfrac{dy}{dx}$

This is the required differential equation.

Illustration 2: Form diff. equation of $ax^2 + by^2 = 1$ **(JEE MAIN)**

Sol: Similar to the above problem the given equation contains two arbitrary constants, so we shall differentiate it two times with respect to x and then by eliminating a and b we get the differential equation of second order.

$$ax^2 + by^2 = 1 \Rightarrow \quad 2ax + 2by\frac{dy}{dx} = 0 \Rightarrow \quad a + b(yy'' + (y')^2) = 0$$

Eliminating a and b we get $\dfrac{y}{x}y' = yy'' + (y')^2 \Rightarrow y\dfrac{d^2y}{dx^2} + \left(\dfrac{dy}{dx}\right)^2 - \dfrac{y}{x}\dfrac{dy}{dx} = 0$

Illustration 3: Form the differential equation corresponding to $y^2 = a(b^2 - x^2)$, where a and b are arbitrary constants. **(JEE MAIN)**

Sol: Similar to illustration 1.

We have, $y^2 = a(b^2 - x^2)$ ... (i)

In this equation, there are two arbitrary constants a, b, so we have to differentiate twice, Differentiating the given equation (i) w.r.t. 'x'. We get $2y \dfrac{dy}{dx} = -2x.a \Rightarrow y\dfrac{dy}{dx} = -ax$ ... (ii)

Differentiating (ii) with respect to x, we get $y\dfrac{d^2y}{dx^2} + \dfrac{dy}{dx}.\dfrac{dy}{dx} = -a \Rightarrow y\dfrac{d^2y}{dx^2} + \left(\dfrac{dy}{dx}\right)^2 = -a$ ... (iii)

Substituting the value of a in (ii), we get

$$y\dfrac{dy}{dx} = \left\{y\dfrac{d^2y}{dx^2} + \left(\dfrac{dy}{dx}\right)2\right\}x \Rightarrow y\dfrac{dy}{dx} = xy\dfrac{d^2y}{dx^2} + \left(\dfrac{dy}{dx}\right)^2 \Rightarrow xy\dfrac{d^2y}{dx^2} + x\left(\dfrac{dy}{dx}\right)^2 - y\dfrac{dy}{dx} = 0$$

Illustration 4: Find the differential equation of the following family of curves: $xy = Ae^x + Be^{-x} + x^2$ **(JEE MAIN)**

Sol: Here in this problem A and B are the two arbitrary constants, hence we shall differentiate it two times with respect to x and then by eliminating constant terms we will get the required differential equation.

Given: $xy = Ae^x + Be^{-x} + x^2$ ... (i)

Differentiating (i) with respect to 'x', we get $x\dfrac{dy}{dx} + y = Ae^x - Be^{-x} + 2x$

Again differentiating with respect to 'x', we get

$$x\dfrac{d^2y}{dx^2} + 1\dfrac{dy}{dx} + 1.\dfrac{dy}{dx} = Ae^x + Be^{-x} + 2 \Rightarrow x\dfrac{d^2y}{dx^2} + 2\dfrac{dy}{dx} = xy - x^2 + 2$$

Illustration 5: Prove that $x^2 - y^2 = c(x^2 + y^2)^2$ is a general solution of the differential equation $(x^3 - 3xy^2)dx = (y^3 - 3x^2y)dy$ **(JEE ADVANCED)**

Sol: Here only one arbitrary constant is present hence we shall differentiate it one time with respect to x and then by substituting the value of c we shall prove the given equation.

Let us find the differential equation for $x^2 - y^2 = c(x^2 + y^2)^2$ ... (i)

Differentiating (i), with respect to 'x', we get $2x - 2y\dfrac{dy}{dx} = c.2\left(x^2 + y^2\right)\left(2x + 2y\dfrac{dy}{dx}\right)$ ... (ii)

Substituting the value of c from (i) in (ii), we get

$$\Rightarrow x - y\dfrac{dy}{dx} = \dfrac{x^2 - y^2}{\left(x^2 + y^2\right)^2}\left(x^2 + y^2\right)\left(2x + 2y\dfrac{dy}{dx}\right) \Rightarrow \left(x^2 + y^2\right)\left(x - y\dfrac{dy}{dx}\right) = \left(x^2 - y^2\right)\left(2x + 2y\dfrac{dy}{dx}\right)$$

$$\Rightarrow [2y(x^2 - y^2) + y(x^2 + y^2)]\dfrac{dy}{dx} = x(x^2 + y^2) - 2x(x^2 - y^2) \Rightarrow (3x^2y - y^3)\dfrac{dy}{dx} = 3xy^2 - x^3$$

$\Rightarrow (x^3 - 3xy^2)dx = (y^3 - 3x^2y)dy$ As this equation matches the one given in the problem statement. Hence the given equation is the solution for the differential equation.

Hence proved.

Illustration 6: Find the differential equation of the family of curves $y = e^x(a\cos x + b\sin x)$ **(JEE ADVANCED)**

Sol: Since given family of curves have two constants a and b, so we have to differentiate twice with respect to x.

We have, $y = e^x(a\cos x + b\sin x)$ ... (i)

Differentiating (i) with respect to x, we get

$$\frac{dy}{dx} = e^x(a\cos x + b\sin x) + e^x(-a\sin x + b\cos x) = y + e^x(-a\sin x + b\cos x)$$

$$\Rightarrow \quad \frac{dy}{dx} - y = e^x(-a\sin x + b\cos x) \qquad \qquad \ldots \text{(ii)}$$

Differentiating (ii) with respect to x, we get

$$\frac{d^2y}{dx^2} - \frac{dy}{dx} = e^x(-a\sin x + b\cos x) + e^x(-a\cos x - b\sin x) = \frac{dy}{dx} - y - e^x(a\cos x + b\sin x)$$

$$\Rightarrow \quad \frac{d^2y}{dx^2} - \frac{dy}{dx} = \frac{dy}{dx} - y - y \qquad [\because e^x(a\cos x + b\sin x) = y] \Rightarrow \quad \frac{d^2y}{dx^2} - 2\frac{dy}{dx} + 2y = 0$$

This is the required differential equation.

Illustration 7: Find the differential equation of all circles which pass through the origin and whose centers lie on the y axis. **(JEE ADVANCED)**

Sol: As circles passes through the origin and whose centers lie on the y axis hence g = 0 and point (0, 0) will satisfy general equation of given circle.

The general equation of a circle is

$$x^2 + y^2 + 2gx + 2fy + c = 0 \qquad \qquad \ldots \text{(i)}$$

Since it passes through origin (0, 0), it will satisfy equation (i)

$$\Rightarrow \quad (0)^2 + (0)^2 + 2g.(0) + 2f.(0) + c = 0 \quad \Rightarrow \quad c = 0$$

$$\Rightarrow \quad x^2 + y^2 + 2gx + 2fy = 0$$

This is the equation of a circle with center (–g, –f) and passing through the origin.

If the center lies on the y-axis, we have g = 0,

$$\Rightarrow \quad x^2 + y^2 + 2.(0).x + 2fy = 0 \qquad \Rightarrow \quad x^2 + y^2 + 2fy = 0 \qquad \qquad \ldots \text{(ii)}$$

Hence, (ii) represents the required family of circles with center on y axis and passing through origin.

Differentiating (ii) with respect to x, we get

$$2x + 2y\frac{dy}{dx} + 2f\frac{dy}{dx} = 0 \Rightarrow f = -\left\{\frac{x + y.\left(\frac{dy}{dx}\right)}{\left(\frac{dy}{dx}\right)}\right\}$$

Substituting this value of f in (2), we get

$$x^2 + y^2 - 2y\left(\frac{x + y.\left(\frac{dy}{dx}\right)}{\left(\frac{dy}{dx}\right)}\right) = 0 \Rightarrow (x^2 + y^2)\frac{dy}{dx} - 2xy - 2y^2\left(\frac{dy}{dx}\right) = 0 \Rightarrow (x^2 - y^2)\frac{dy}{dx} - 2xy = 0$$

This is the required differential equation.

NOMORECLASS CONCEPTS

Curves representing the solution of a differential equation are called integral curves.

6. SOLUTIONS OF DIFFERENTIAL EQUATIONS

Finding the dependent variable from the differential equation is called solving or integrating it. The solution or the integral of a differential equation is, therefore, a relation between the dependent and independent variables (free from derivatives) such that it satisfies the given differential equation.

Note: The solution of the differential equation is also called its primitive.

There can be two types of solution to a differential equation:

(a) General solution (or complete integral or complete primitive)

A relation in x and y satisfying a given differential equation and involving exactly the same number of arbitrary constants as the order of the differential equation.

(b) Particular solution

A solution obtained by assigning values to one or more than one arbitrary constant of general solution

Illustration 8: The general solution of $x^2 \dfrac{dy}{dx} = 2$ is   (JEE MAIN)

Sol: First separate out x term and y term and then integrate it, we shall obtain result.

$\dfrac{dy}{dx} = \dfrac{2}{x^2} \Rightarrow dy = \dfrac{2}{x^2} dx$ Now integrate it. We get $y = -\dfrac{2}{x} + c$

Illustration 9: Verify that the function $x + y = \tan^{-1} y$ is a solution of the differential equation $y^2 y' + y^2 + 1 = 0$ (JEE MAIN)

Sol: By differentiating the equation $x + y = \tan^{-1} y$ with respect to x we can prove the given equation.

We have, $x + y = \tan^{-1} y$ ... (i)

Differentiating (i), w.r.t. x we get

$1 + \dfrac{dy}{dx} = \dfrac{1}{1+y^2} \dfrac{dy}{dx} \quad \Rightarrow \quad 1 + \dfrac{dy}{dx}\left(\dfrac{1+y^2-1}{1+y^2}\right) = 0$

$\Rightarrow \quad (1 + y^2) + y^2 \dfrac{dy}{dx} = 0 \Rightarrow \quad y^2 y' + y^2 + 1 = 0$

Illustration 10: Show that the function $y = Ax + \left(2x + 2y\dfrac{dy}{dx}\right)$ is a solution of the differential equation

$x^2 \dfrac{d^2 y}{dx^2} + x\dfrac{dy}{dx} - y = 0$ (JEE MAIN)

Sol: Differentiating $y = Ax + \dfrac{B}{x}$ twice with respect to x and eliminating the constant term, we can prove the given equation.

We have, $y = Ax + \dfrac{dy}{dx} \Rightarrow xy = Ax^2 + B$ ... (i)

Differentiation (i) w.r.t. 'x'. we get $\Rightarrow x\dfrac{dy}{dx} + 1.y = 2Ax$ ... (ii)

Again differentiating (ii) w.r.t., 'x', we get

$\Rightarrow x.\dfrac{d^2 y}{dx^2} + \dfrac{dy}{dx} + \dfrac{dy}{dx} = 2A \qquad \Rightarrow x\dfrac{d^2 y}{dx^2} + 2\dfrac{dy}{dx} = \dfrac{x\dfrac{dy}{dx} + y}{x} \Rightarrow x^2 \dfrac{d^2 y}{dx^2} + x\dfrac{dy}{dx} - y = 0$

Which is same as the given differential equation. Therefore $y = Ax + \dfrac{dy}{dx}$ is a solution for the given differential equation.

Illustration 11: If $y.\sqrt{x^2+1}=\log\left[\sqrt{x^2+1}\right]$ show that $(x^2+1)\dfrac{dy}{dx}+xy+1=0$ **(JEE MAIN)**

Sol: Similar to the problem above, by differentiating $y.\sqrt{x^2+1}=\log\left[\sqrt{x^2+1}-x\right]$ one time with respect to x, we will prove the given equation.

We have, $y.\ \sqrt{x^2+1}=\log\left[\sqrt{x^2+1}\right]$ ... (i)

Differentiating (i), we get

$$\sqrt{x^2+1}\frac{dy}{dx}+\frac{1}{2}\frac{2x}{\sqrt{x^2+1}}\ y=\frac{(1/2)\left(2x/\sqrt{x^2+1}\right)-1}{\sqrt{x^2+1}-x}\ \Rightarrow\ \sqrt{x^2+1}\frac{dy}{dx}+\frac{x}{\sqrt{x^2+1}}=\frac{x-\sqrt{x^2+1}}{\sqrt{x^2+1}\left[\sqrt{x^2+1}-x\right]};$$

$(x^2+1)\dfrac{dy}{dx}+xy=\dfrac{x-\sqrt{x^2+1}}{\sqrt{x^2+1}-x}$; $(x^2+1)\dfrac{dy}{dx}+xy=-1;$ $(x^2+1)\dfrac{dy}{dx}+xy+1=0$

Illustration 12: Show that $y=a\cos(\log x)+b\sin(\log x)$ is a solution of the differential equation:

$$x^2\frac{d^2y}{dx^2}+x\frac{dy}{dx}+y=0$$

        (JEE ADVANCED)

Sol: As the given equation has two arbitrary constants, hence differentiating it two times we can prove it.

We have, $y=a\cos(\log x)+b\sin(\log x)$ ... (i)

Differentiating (i) w.r.t 'x'. we get ; $\dfrac{dy}{dx}=-\dfrac{a\sin(\log x)}{x}+\dfrac{b\cos(\log x)}{x}$

$x\dfrac{dy}{dx}=-a\sin(\log x)+b\cos(\log x)$ ... (ii)

Again differentiating with respect to 'x', we get

$$x\frac{d^2y}{dx^2}+\frac{dy}{dx}=\frac{a\cos(\log x)}{x}-\frac{b\sin(\log x)}{x}$$

$\Rightarrow\ x^2\dfrac{d^2y}{dx^2}+x\dfrac{dy}{dx}=-[a\cos(\log x)+b\sin(\log x)]$ $\Rightarrow\dfrac{d^2y}{dx^2}+x\dfrac{dy}{dx}=-y$ $\Rightarrow\dfrac{d^2y}{dx^2}+x\dfrac{dy}{dx}+y=0$

Which is same as the given differential equation

Hence, $y=a\cos(\log x)+b\sin(\log x)$ is a solution of the given differential equation.

7. METHODS OF SOLVING FIRST ORDER FIRST DEGREE DIFFERENTIAL EQUATION

7.1 Equation of the Form dy/dx = f(x)

To solve this type of differential equations, we integrate both sides to obtain the general solution as discussed below

$$\frac{dy}{dx}=f(x)\qquad\Rightarrow\qquad dy=f(x)dx$$

Integrating both sides we obtain $\int dy=\int f(x)dx+c\ \Rightarrow\ y=\int f(x)dx+c$

Illustration 13: The general solution of the differential equation $\dfrac{dy}{dx} = x^5 + x^2 - \dfrac{2}{x}$ is **(JEE MAIN)**

Sol: General solution of any differential equation is obtained by integrating it hence for given equation we have to integrate it one time to obtain its general equation.

We have: $\dfrac{dy}{dx} = x^5 + x^2 - \dfrac{2}{x}$

Integrating, $y = \displaystyle\int \left(x^5 + x^2 - \dfrac{2}{x} \right) \, dx + c = \int x^5 dx + \int x^2 dx - 2\int \dfrac{1}{x} dx + c \Rightarrow y = \dfrac{x^6}{6} + \dfrac{x^3}{3} - 2\log|x| + c$

Which is the required general solution.

Illustration 14: The solution of the differential equation $\cos^2 x \dfrac{d^2 y}{dx^2} = 1$ is **(JEE MAIN)**

Sol: By integrating it two times we will get the result.

$\cos^2 x \dfrac{d^2 y}{dx^2} = 1 \Rightarrow \dfrac{d^2 y}{dx^2} = \sec^2 x$

On integrating, we get $\dfrac{dy}{dx} = \tan x + c_1$

Integrating again, we get $y = \log(\sec x) + c_1 x + c_2$

7.2 Equation of the form dy/dx = f(x) g(y)

To solve this type of differential equation we integrate both sides to obtain the general solution as discussed below

$\dfrac{dy}{dx} = f(x)g(y) \Rightarrow g(y)^{-1} dy = f(x)dx$

Integrating both sides, we get $\displaystyle\int (g(y))^{-1} dy = \int f(x)dx$

Illustration 15: The solution of the differential equation $\log(dy/dx) = ax + by$ is **(JEE MAIN)**

Sol: We can also write the given equation as $\dfrac{dy}{dx} = e^{ax + by}$. After that by separating the x and y terms and integrating both sides we can get the general equation.

$\dfrac{dy}{dx} = e^{ax + by} \Rightarrow \quad \dfrac{dy}{dx} = e^{ax + by} \Rightarrow e^{-by} dy = e^{ax} dx \quad \Rightarrow -\dfrac{1}{b}e^{-by} = \dfrac{1}{a}e^{ax} + c$

Illustration 16: The solution of the differential equation $\dfrac{dy}{dx} = e^{x+y} + x^2 e^y$ is **(JEE MAIN)**

Sol: Here first we have to separate the x and y terms and then by integrating them we can solve the problem above.

The given equation is $\dfrac{dy}{dx} = e^{x+y} + x^2 e^y$

$\Rightarrow \quad \dfrac{dy}{dx} = e^x.e^y + x^2 e^y \Rightarrow e^{-y} dy = (e^x + x^2)dx$, Integrating, $\displaystyle\int e^{-y} dy = \int \left(e^x + x^2 \right) dx + c$

$\Rightarrow \quad \dfrac{e^{-y}}{-1} + e^x + \dfrac{x^3}{3} + c \qquad \Rightarrow \quad -\dfrac{1}{e^y} = e^x + \dfrac{1}{3}x^3 + c \Rightarrow e^x + \dfrac{1}{e^y} + \dfrac{x^3}{3} = C$

7.3 Equation of the Form dy/dx = f (ax+by+c)

To solve this type of differential equation, we put $ax + by + c = v$ and $\dfrac{dy}{dx} = \dfrac{1}{b}\left(\dfrac{dy}{dx} - 0\right)$

$\therefore \dfrac{dy}{a + bf(v)} = dx$

So solution is by integrating $\displaystyle\int \dfrac{dy}{a + bf(v)} = \int dx$

Illustration 17: $(x + y)^2 \dfrac{dy}{dx} = a^2$ (JEE MAIN)

Sol: Here we can't separate the x and y terms, therefore put $x + y = t$ hence $\dfrac{dy}{dx} = \dfrac{dt}{dx} - 1$. Now we can easily separate the terms and by integrating we will get the required result.

Let $x + y = t \Rightarrow t^2\left(\dfrac{dt}{dx} - 1\right) = a^2$; $\dfrac{dt}{dx} = \dfrac{a^2}{t^2} + 1 = \dfrac{a^2 + t^2}{t^2} \Rightarrow \displaystyle\int \dfrac{t^2 dt}{t^2 + a^2} = x + c$

$\Rightarrow \quad t - a\tan^{-1}\dfrac{dy}{dx} = x + c \Rightarrow \quad y - a\tan^{-1}\dfrac{x+y}{a} = c$

Illustration 18: $\dfrac{dx}{dx} = \dfrac{x + y - 1}{\sqrt{x + y + 1}}$ (JEE MAIN)

Sol: Put $x + y + 1 = t^2$ and then solve similar to the above illustration.

let $x + y + 1 = t^2$

$\Rightarrow \quad \left(2t\dfrac{dt}{dx} - 1\right) = \dfrac{t^2 - 2}{t} \qquad \Rightarrow \qquad \dfrac{2t\,dt}{dx} = \dfrac{t^2 + t - 2}{t} \qquad \Rightarrow \qquad \displaystyle\int \dfrac{2t^2}{(t-1)(t+2)}\,dt = x + c$

$\Rightarrow \quad 2\displaystyle\int\left(1 + \dfrac{1}{3(t-1)} - \dfrac{4}{3(t+2)}\right)dt = x + c \quad \Rightarrow \quad 2t + \dfrac{2\ln|t-1|}{3} - \dfrac{8\ln|t+2|}{3} = x + c$

$\Rightarrow \quad 2\sqrt{x+y+1} + \dfrac{2\ln|\sqrt{x+y+1} - 1|}{3} - \dfrac{8\ln|\sqrt{x+y+1} + 2|}{3} = x + c$

Illustration 19: $\dfrac{dy}{dx} = \cos(10x + 8y)$. Find curve passing through origin in the form $y = f(x)$ satisfying differential equations given (JEE MAIN)

Sol: Here first put $10x + 8y = t$ and then taking integration on both sides we will get the required result.

Let $10x + 8y = t$

$\Rightarrow \quad 10 + 8\dfrac{dy}{dx} = \dfrac{dt}{dx} \qquad \Rightarrow \qquad \dfrac{dy}{dx} - 10 = 8\cos t \qquad \Rightarrow \qquad \displaystyle\int \dfrac{dt}{8\cos t + 10}\int dx = x + c$

$p = \tan t/2 \qquad \dfrac{dp}{dx} = \dfrac{1+p^2}{2(1)}\dfrac{dy}{dx} \Rightarrow \dfrac{dt}{dx} = \dfrac{2dp}{1+p^2}$

$\therefore \displaystyle\int \dfrac{\dfrac{2dp}{1+p^2}}{8\left(\dfrac{1-p^2}{1+p^2}\right) + 10} = \int \dfrac{2dp}{1p^2 + 18} = \int \dfrac{dp}{p^2 + 9} = x + c$

$\Rightarrow \quad \tan^{-1}(P/3) = x + c \Rightarrow \tan^{-1}\left(\dfrac{\tan(t/2)}{3}\right) = x + c \Rightarrow 3\tan(x + c) = \tan(10x + 84)$

7.4 Parametric Form

Some differential equations can be solved using parametric forms.

Case I:

$x = r\cos\theta$ $y = r\sin\theta$

Squaring and adding $x^2 + y^2 = r^2$ ... (i)

$\tan\theta = \int e^{-y}dy = \int (e^x + x^2)dx + c$ ... (ii)

$xdx + ydy = rdr$ ... (iii)

$\sec^2\theta\, d\theta = \dfrac{e^{-y}}{-1} = e^x + \dfrac{x^3}{3} + c \qquad \Rightarrow xdy - ydx = x^2\sec^2\theta\, d\theta \qquad x = r\cos\theta; \; xdy - ydx = r^2 d\theta$

Case II:

If $x = r\sec\theta, \qquad y = r\tan\theta$

$x^2 - y^2 = r^2$ ... (i)

$\dfrac{1}{e^y} = e^x + \dfrac{1}{3}x^3 + c = \sin\theta$ ... (ii)

$\Rightarrow \quad xdx - ydy = rdr; \qquad xdy - ydx = \cos\theta\, x^2\, d\theta \qquad \Rightarrow \quad xdy - ydx = r^2\sec\theta\, d\theta$

Illustration 20: Solve $xdx + ydy = x(xdy - ydx)$ **(JEE MAIN)**

Sol: By substituting $x = r\cos\theta$ and $y = r\sin\theta$ the given equation reduces to $rdr = r\cos\theta(r^2 d\theta)$. Hence by separating and integrating both sides we will get the result.

Let $x = r\cos\theta, \; y = r\sin\theta$

Hence the given equation becomes $rdr = r\cos\theta(r^2 d\theta)$

$\int \dfrac{dr}{r^2} = \int \cos\theta\, d\theta \qquad \Rightarrow \qquad -\dfrac{1}{r} = \sin\theta + c \qquad \Rightarrow \qquad -\dfrac{1}{\sqrt{x^2 + y^2}} = \dfrac{y}{\sqrt{x^2 + y^2}} + c$

Illustration 21: Solve $\dfrac{x + y\dfrac{dy}{dx}}{x\dfrac{dy}{dx} - y} = \sqrt{\dfrac{1 - x^2 - y^2}{x^2 + y^2}}$ **(JEE ADVANCED)**

Sol: Similar to the problem above, by substituting $x = r\cos\theta$ and $y = r\sin\theta$ the given equation reduces to

$\dfrac{rdr}{r^2 d\theta} = \dfrac{\sqrt{1 - r^2}}{r}$. Hence by integrating both sides we will get the result.

$\dfrac{x + y\dfrac{dy}{dx}}{x\dfrac{dy}{dx} - y} = \sqrt{\dfrac{1 - x^2 - y^2}{x^2 + y^2}} \qquad \Rightarrow \qquad \dfrac{xdx + ydy}{xdy - ydx} = \sqrt{\dfrac{1 - x^2 - y^2}{x^2 + y^2}}$

Let $x = r\cos\theta, \; y = r\sin\theta$

$\dfrac{rdr}{r^2 d\theta} = \dfrac{\sqrt{1 - r^2}}{r} \qquad \Rightarrow \qquad \int \dfrac{dr}{\sqrt{1 - r^2}} = \theta + c \qquad \Rightarrow \qquad \sin^{-1}r = \theta + c$

$\Rightarrow \quad \sin^{-1}\sqrt{x^2 + y^2} = \sin^{-1}\dfrac{y}{\sqrt{x^2 + y^2}} + c$

Illustration 22: $\dfrac{xdx + ydy}{\sqrt{x^2 + y^2}} = \dfrac{ydx - xdy}{x}$

Sol: Similar to the above illustration.

Let $\ x = r\cos\theta, \ y = r\sin\theta$

$\Rightarrow \quad -\dfrac{rdr}{r^2 d\theta} = \dfrac{\sqrt{r^2}}{r\cos\theta} \qquad \Rightarrow \quad \displaystyle\int \sec\theta\, d\theta + \int \dfrac{cr}{r} = 0$

$\Rightarrow \quad \log(\sec\theta + \tan\theta) + \log r = c \qquad \Rightarrow \quad x^2 + y^2 - y\left(\sqrt{x^2 + y^2}\right) + Cx = 0$

7.5 Homogeneous Differential Equations

A differential equation in x and y is said to be homogeneous if it can be

put in the form $\dfrac{dy}{dx} = \dfrac{f(x,y)}{g(x,y)}$, where f(x, y) and g(x, y) are both homogeneous

function of the same degree in x and y.

To solve the homogeneous differential equation $\dfrac{dy}{dx} = \dfrac{f(x,y)}{g(x,y)}$,

substitute y = vx and so $\dfrac{dy}{dx} = v + x\dfrac{dv}{dx}$

Thus differential reduces to the form $v + x\dfrac{dy}{dx} = f(v) \Rightarrow \dfrac{dx}{x} = \dfrac{dv}{f(v) - v}$

Therefore, solution is $\displaystyle\int \dfrac{dy}{x} = \int \dfrac{dv}{f(v) - v} + c$

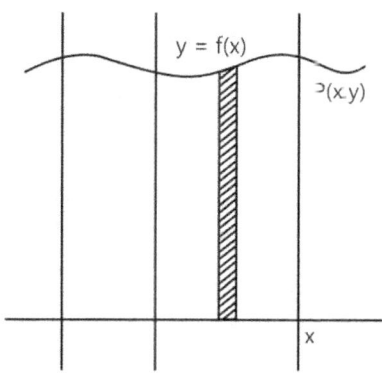

Figure 24.1

Illustration 23: Find the curve passing through (1, 0) such that the area bounded by the curve, x-axis and 2 ordinates, one of which is constant and other is variable, is equal to the ratio of the cube of variable ordinate to variable abscissa.

Sol: By differentiating $\displaystyle\int_{c}^{x} ydx = \dfrac{y^3}{x}$, we will get the differential equation.

$A = \displaystyle\int_{c}^{x} ydx = \dfrac{y^3}{x} \qquad \Rightarrow y = \dfrac{x.3y^2 y' - y^3.1}{x^2} \Rightarrow x^2 = 3xyy' - y^2 \quad \Rightarrow \quad \dfrac{dy}{dx} = \dfrac{x^2 + y^2}{3xy}$

(On differentiating the first integral equation w.r.t x)

Put y = vx; $v + x\dfrac{dt}{dx} = \dfrac{1 + v^2}{3v} \quad \Rightarrow \quad \displaystyle\int \dfrac{3v}{1 - 2v^2}dv = \int \dfrac{1}{x}dx \quad \Rightarrow \quad -\dfrac{3}{4}\log\left|1 - 2v^2\right| = \log x + \log c \quad \Rightarrow \quad (x^2 - 2y^2)^3 = cx^2$

Given this curve passes through (1, 0). So, c=1 Hence the equation of curve is $(x^2 - 2y^2)^3 = cx^2$

Illustration 24: The solution of differential equation $\dfrac{dy}{dx} = \dfrac{y}{x} + \tan\dfrac{y}{x}$ is

Sol: Here by putting y = xv and then integrating both sides we can solve the problem.

Put y = xv $\Rightarrow \quad \dfrac{dy}{dx} = v + x\dfrac{dv}{dx}$

Hence the given equation becomes $x\dfrac{dv}{dx} + v = v - \tan v \Rightarrow x\dfrac{dv}{dx} = \tan v$

$\Rightarrow \quad \dfrac{dv}{\tan v} = \dfrac{dx}{x} \Rightarrow \log \sin v = \log x + \log c \Rightarrow \dfrac{\sin v}{x} = c \Rightarrow \dfrac{\sin\left(\dfrac{y}{x}\right)}{x} = c \Rightarrow cx = \sin\left(\dfrac{y}{x}\right)$

Illustration 25: Solve $\dfrac{dy}{dx} = \dfrac{y^2 - 2xy - x^2}{y^2 + 2xy - x^2}$ given y at x = 1 is –1 **(JEE ADVANCED)**

Sol: Similar to the problem above, by putting y = vx, we can solve it and then by applying the given condition we will get the value of c.

Let y = vx

$$\Rightarrow \quad v + x\dfrac{dv}{dx} = \left(\dfrac{v^2 - 2v - 1}{v^2 + 2v - 1}\right) \qquad \Rightarrow x\dfrac{dv}{dx} = -\dfrac{(v^3 + v^2 + v + 1)}{v^2 + 2v - 1}$$

$$\Rightarrow \quad \int \dfrac{v^2 + 2v - 1}{(v+1)(v^2+1)}dv = c - \log x \Rightarrow \int \dfrac{2v(v+1) - (v^2+1)}{(v+1)(v^2+1)}dv = c - \log x$$

$$\Rightarrow \quad \log\left[\dfrac{(v^2+1)x}{v+1}\right] = \log c \quad \Rightarrow \quad \dfrac{(v^2-1)x}{(v+1)} = c \Rightarrow \dfrac{x^2+y^2}{y+x} = c$$

$$\Rightarrow \quad k(x^2 + y^2) = x + y$$

Given at x = 1, y = – 1 $\Rightarrow$ 2k = 0. Hence the required equation is x + y = 0

Illustration 26: Solve $y\left(\dfrac{dy}{dx}\right)^2 + 2x\dfrac{dy}{dx} - y = 0$ given y at x = 1 is $\sqrt{5}$ **(JEE ADVANCED)**

Sol: As we know, when $ax^2 + bx + c = 0$ then $x = \dfrac{-b \pm \sqrt{b^2 - 4ac}}{2a}$. Hence from given equation $\dfrac{dy}{dx} = \dfrac{-2x \pm \sqrt{4x^2 + 4y^2}}{2y}$ so by putting y = vx and integrating both side, we will get the result.

Given $y\left(\dfrac{dy}{dx}\right) + 2x\dfrac{dy}{dx} - y = 0$

$$\Rightarrow \quad \dfrac{dY}{dX} = \dfrac{-2x \pm \sqrt{4x^2 + 4y^2}}{2y} \quad \Rightarrow \quad \dfrac{dy}{dx} = \dfrac{-x \pm \sqrt{x^2 + y^2}}{y}$$

Let y = vx

$$\Rightarrow \quad x\dfrac{dv}{dx} = \dfrac{\pm\sqrt{v^2+1} - 1}{v} - v \Rightarrow x\dfrac{dv}{dx} = \dfrac{\pm\sqrt{v^2+1} - 1 - v^2}{v}$$

$$\Rightarrow \quad \int \dfrac{vdv}{\pm\sqrt{v^2+1} - (1+v^2)} = \log x + C \quad \Rightarrow \quad \int \dfrac{vdv}{\pm\sqrt{v^2+1}\left(\mp\sqrt{v^2+1}+1\right)} = \log x + C$$

$$\Rightarrow \quad -\ln\left(\mp\sqrt{v^2+1}+1\right) = \log x + C \qquad \Rightarrow \quad x\left(\mp\sqrt{v^2+1}+1\right) = c$$

Given at x = 1, y = v = $\dfrac{dy}{dx} = \dfrac{7X - 3Y}{-3X + 7Y}$ $\Rightarrow$ C = $\mp\sqrt{6}+1$

$$\Rightarrow \quad \mp\sqrt{y^2 + x^2} + x = \mp\sqrt{6} + 1$$

This is the required equation.

Note: The obtained solution has 4 equations.

7.6 Differential Equations Reducible to Homogenous Form

A differential equation of the form $\dfrac{dv}{dx} = \dfrac{a_1 x + b_1 y + c_1}{a_2 x + b_2 y + c}$, where $\dfrac{a_1}{\theta_2} \neq \dfrac{b_1}{b_2}$ can be reduced to homogeneous form by adopting the following procedure

Put $x = X + h$, $y = Y + k$, so that $\dfrac{dy}{dx} = \dfrac{dy}{dx}$

The equation then transforms to $\dfrac{dY}{dX} = \dfrac{a_1 X + b_1 Y + (a_1 h + b_1 k + c_1)}{a_2 X + b_2 Y + (a_2 h + b_2 k + c_2)}$

Now choose h and k such that $a_1 h + b_1 k + c_1 = 0$ and $a_2 h + b_2 k + c_2 = C$. Then for these values of h and k the equation becomes

$\dfrac{dy}{dx} = \dfrac{a_1 X + b_1 Y}{a_2 X + b_2 Y}$

This is a homogeneous equation which can be solved by putting $Y = vX$ and then Y and X should be replaced by $y - k$ and $x - h$.

Special case: If $\dfrac{dy}{dx} = \dfrac{ax + by + c}{a'x + b'y + c'}$ and $\dfrac{a}{a'} = \dfrac{b}{b'} = m$ say, i.e. when coefficient of x and y in numerator and denominator are proportional, then the above equation cannot be solved by the method discussed before because the values of h and k given by the equation will be indeterminate. In order to solve such equations, we proceed as explained in the following example.

Illustration 27: Solve $\dfrac{dy}{dx} = \dfrac{3x - 6y + 7}{x - 2y + 4}$ (JEE MAIN)

Sol: Here the coefficient of x and y in the numerator and denominator are proportional hence by taking 3 common from 3x − 6y and putting x − 2y = v and after that by integrating we will get the result.

$\dfrac{dy}{dx} = \dfrac{3x - 6y + 7}{x - 2y + 4} = \dfrac{3(x - 2y) + 7}{x - 2y + 4}$; Put $x - 2y = v \Rightarrow 1 - 2\dfrac{dy}{dx} = \dfrac{dy}{dx}$

Now differential equations reduces to $1 - \dfrac{dv}{dx} = 2\left(\dfrac{3v + 7}{v + 4}\right)$

$\Rightarrow \dfrac{dv}{dx} = -5\left(\dfrac{v + 2}{v + 4}\right)$ $\Rightarrow \int\left(1 + \dfrac{2}{v + 2}\right)dv = -5\int dx$

$\Rightarrow v + 2\log|v + 2| = -5x + c$ $\Rightarrow 3x - y + \log|x - 2y + 2| = c$

Illustration 28: Solution of differential equation $(3y - 7x + 7)dx + (7y - 3x + 3)\,dy = 0$ is   (JEE MAIN)

Sol: By substituting $x = X + h$, $y = Y + k$ where (h, k) will satisfy the equation $3y - 7x + 7 = 0$ and $7y - 3x + 3 = 0$ we can reduce the equation and after that by putting Y = VX and integrating we will get required general equation.

The given differential equation is $\dfrac{dy}{dx} = \dfrac{7x - 3y - 7}{-3x + 7y + 3}$

Substituting $x = X + h$, $y = Y + k$, we obtain

$\dfrac{dY}{dX} = \dfrac{(7X - 3Y) + (7h - 3k - 7)}{(-3X + 7Y) + (-3h + 7k + 3)}$ ... (i)

Choose h and k such that $7h - 3k - 7 = 0$ and $-3h + 7k + 3 = 0$.

This gives $h = 1$ and $k = 0$. Under the above transformations, equation (i) can be written as

Let $Y = VX$ so that $\frac{dY}{dX} = V + X\frac{dV}{dX}$, we get $\frac{dY}{dX} = \frac{7X - 3Y}{-3X + 7Y}$

$V + X\frac{dV}{dX} = \frac{-3V + 7}{7V - 3} \Rightarrow X\frac{dV}{dX} = \frac{7 - 7V^2}{7V - 3} \qquad \Rightarrow \qquad -7\frac{dX}{X} = \frac{7}{2} \cdot \frac{2V}{V^2 - 1}dV - \frac{3}{V^2 - 1}dV$

Integrating, we get

$-7\log X = \frac{7}{2}\log(V^2 - 1) - \frac{3}{2}\log\frac{V - 1}{V + 1} - \log C \qquad \Rightarrow \qquad C = (V + 1)^5 (V - 1)^2 X^7 \qquad \Rightarrow \qquad C = (y + x - 1)^5 (y - x + 1)^2$

Which is the required solution.

7.7 Linear Differential Equation

A differential equation is linear if the dependent variable (y) and its derivative appear only in the first degree. The general form of a linear differential equation of the first order is

$\frac{dy}{dx} + Py = Q \qquad \qquad ..... \text{ (i)}$

where P and Q are either constants or functions of x.

This type of differential equation can be solved when they are multiplied by a factor, which is called integrating factor.

Multiplying both sides of (i) by $e^{\int Pdx}$, we get $e^{\int Pdx}\left(\frac{dy}{dx} + Py\right) = Qe^{\int Pdx}$

On integrating both sides with respect to x, we get

$ye^{\int Pdx} = \int Qe^{\int Pdx} + c$ which is the required solution, where c is the constant and $e^{\int Pdx}$ is called the integrating factor.

Illustration 29: Solve the following differential equation: $\frac{dy}{dx} + \frac{1}{x} = \frac{e^y}{x}$ (JEE MAIN)

Sol: We can write the given equation as $e^{-y}\frac{dy}{dx} + \frac{e^{-y}}{x} = \frac{1}{x}$. By putting $e^{-y} = t$, we can reduce the equation in the form of $\frac{dt}{dx} + Pt = Q$ hence by using integration factor we can solve the problem above.

We have, $\frac{dy}{dx} + \frac{1}{x} = \frac{e^y}{x} \Rightarrow e^{-y}\frac{dy}{dx} + \frac{e^{-y}}{x} = \frac{1}{x}$ ... (i)

Put $e^{-y} = t$. so that $\frac{dy}{dx}$ in equation (i), we get $-\frac{dt}{dx} + \frac{t}{x} = \frac{1}{x} \Rightarrow \frac{dt}{dx} - \frac{1}{x}t = -\frac{1}{x}$ ... (ii)

This is a linear differential equation in t.

Here, $P = -\frac{1}{x}$ and $Q = -\frac{1}{x}$ $\therefore$ I.F. $= e^{\int Pdx} = e^{\int\left(-\frac{1}{x}\right)dx} = e^{-\log x} = e^{\log x^{-1}} = \frac{1}{x}$

$\therefore$ The solution of (ii) is, t.(I.F.) $= \frac{dy}{dx} = \frac{3x - 6y + 7}{x - 2y + 4} = \frac{3(x - 2y) + 7}{x - 2y + 4}$

$t\frac{1}{x} = \int\frac{1}{x}\left(-\frac{1}{x}\right)dx + C \Rightarrow \frac{t}{x} = \frac{1}{x} + c \Rightarrow \frac{e^{-y}}{x} = \frac{1}{x} + C$

Illustration 30: The function y(x) satisfy the equation $y(x) + 2x\int_0^x \frac{y(x)}{1+x^2}dx = 3x^2 + 2x + 1$. Prove that the substitution $z(x) = \int_0^x \frac{y(x)}{1+x^2}dx$ converts the equation into a first order linear differential equation in z(x) and solve the original equation for y(x) **(JEE MAIN)**

Sol: By putting $z'(x) = \frac{y(x)}{1+x^2}$ we will get the linear differential equation in z form and then by applying integrating factor we get the result.

Let $z'(x) = \frac{d(x)}{1+x^2}$ $\Rightarrow$ $z'(x) \times (1+x^2) + 2x(z(x)) = 3x^2 + 2x + 1$

$\Rightarrow \quad \dfrac{dz}{dx} + \dfrac{2x}{1+x^2}z = \dfrac{3x^2 + 2x + 1}{x^2 + 1}$ ... (i)

This is a first order linear differential equation in z.

$\therefore$ I.F. $= e^{\int Pdx} = e^{\int \frac{2x}{1+x^2}dx} = 1+x^2$ $\therefore$ Solution of (i) is $z(I.F.) = \int (Q \times I.F)dx + c$

$\Rightarrow z(1+x^2) = \int \frac{x^3 + x^2 + x}{x^2 + 1}(x^2+1)dx + C$ $\Rightarrow z(1+x^2) = \frac{x^4}{4} + \frac{x^3}{3} + \frac{x^2}{2} + C$ and $y = 3x^2 + 2x + 1 - 2xz$

Illustration 31: Solve the differential equation $y\sin 2x.dx - (1 + y^2 + \cos 2x)dy = 0$ **(JEE MAIN)**

Sol: Similar to illustration 28, by putting $-\cos 2x = t$, we can reduce the equation in the form of $\dfrac{dt}{dx} + Pt = Q$ hence by using integration factor we can solve the problem given above.

We have, $y\sin 2x.dx - (1 + y^2 + \cos 2x)dy = 0$

$\Rightarrow \sin 2x.\dfrac{dx}{dy} - \dfrac{\cos 2x}{y} = \dfrac{1+y^2}{y}$ ... (i)

Putting $-\cos 2x = t$ so that $2\sin 2x\dfrac{dx}{dy} = \dfrac{dt}{dy}$ in equation (i), we get $\dfrac{dt}{dy} + \dfrac{2}{y}t = 2\left(\dfrac{1+y^2}{y}\right)$

Here, $P = \dfrac{2}{y}$ and $Q = 2\dfrac{1+y^2}{y}$

$\therefore$ I.F. $= e^{\int Pdy} = e^{\int \frac{2}{y}dy} = y^2$ $\therefore$ The solution is $t.(I.F.) = \int(Q \times I.F.)dy + C$

$\Rightarrow \quad t.y^2 = 2\int \dfrac{1+y^2}{y}.y^2dy = 2\int y + y^3\,dy$ $\Rightarrow \quad t.y^2 = y^2 + \dfrac{y^4}{2} + C$

On putting the value of t, we get $-.\cos 2x = 1 + \dfrac{y^2}{2} + Cy^{-2}$

Illustration 32: Solve $y\log y\,\dfrac{dx}{dy} + x - \log y = 0$ **(JEE MAIN)**

Sol: By reducing the given equation in the form of $\dfrac{dx}{dy} + Px = Q$ we can solve this as similar to above illustrations.

We have, $y\log y\dfrac{dx}{dy} + x - \log y = 0 \Rightarrow \dfrac{dx}{dy} + \dfrac{x}{y\log y} = \dfrac{1}{y}$

This is a linear differential equation in x.

Here $P = \dfrac{1}{y\log y}$, $Q = \dfrac{1}{y}$; I.F. $= e^{\int \frac{1}{y\log y}dy} = e^{\log(\log y)} = \log y$

The solution is, $x(\text{I.F.}) = \int (Q \times \text{I.F.}) + C$; $x \log y = \int \frac{1}{y}(\log y)\,dy + c = \frac{1}{2}(\log y)^2 + C$

$x = \frac{1}{2}\log y + C\frac{1}{\log y}$

Illustration 33: Solve $(x + 2y^3)\dfrac{dx}{dy} = y$ **(JEE ADVANCED)**

Sol: By reducing given equation in the form of $\dfrac{dx}{dy} + Px = Q$ and then using the integration factor we can solve this.

$(x + 2y^3)\dfrac{dx}{dy} = y \Rightarrow \dfrac{dx}{dy} = \dfrac{x + 2y^3}{y} = \dfrac{x}{y} + 2y^2 \Rightarrow \dfrac{dx}{dy} - \dfrac{1}{y}x = 2y^2$

$\text{I.F} = e^{-\int \frac{1}{y}dy} = \dfrac{1}{y};$

Solutions is $x \cdot \dfrac{1}{y} = y^2 + C$

Alternate method: $x\,dy + 2y^3\,dy = y\,dx$

$\Rightarrow \quad 2y\,dy = \dfrac{y\,dx - x\,dy}{y^2} \quad \Rightarrow \quad 2y\,dy = d\left(\dfrac{x}{y}\right) \Rightarrow y^2 = \dfrac{x}{y} + C$

Illustration 34: Let $g(x)$ be a differential function for every real x and $g'(0) = 2$ and satisfying $g(x+y) = e^y g(x) + 2e^x g(y)\ \forall\ x$ and y. Find $g(x)$ and its range.   **(JEE ADVANCED)**

Sol: By using $g'(x) = \lim\limits_{b \to 0} \dfrac{g(x+h) - g(x)}{h}$ and solving we will get $g(x)$.

$g'(x) = \lim\limits_{b \to 0} \dfrac{g(x+h) - g(x)}{h}$

$\Rightarrow \quad g'(x) = \lim\limits_{h \to 0} \dfrac{e^h g(x) + 2e^x g(h) - g(x)}{h} \quad \Rightarrow \quad g'(x) = g(x)\lim\limits_{h \to 0}\dfrac{e^h - 1}{h} + 2e^x \lim\limits_{h \to 0}\dfrac{g(h)}{h} \Rightarrow \quad g'(x) = g(x) + 2e^x$

At $x = 0$, $g(x) = 0 \Rightarrow g(0) = 0$

$\dfrac{dy}{dx} - y = 2e^x \quad \Rightarrow \quad \text{I.F.} = e^{-x}$

Solution is $y \cdot e^{-x} = 2x + C$

$g(0) = 0 \Rightarrow C = 0 \Rightarrow g(x) = 2xe^x$

$g'(x) = 2e^x + 2xe^x = 2e^x(x + 1)$

$g'(x) = 0$ at $x = -1$; $g(-1) = -2/e$

$\Rightarrow \quad$ Range of $g(x) = \left[-\dfrac{2}{e}, \infty\right)$

Illustration 35: Find the solution of $(1 - x^2)\dfrac{dy}{dx} + 2xy = x\sqrt{1 - x^2}$ **(JEE ADVANCED)**

Sol: By reducing given equation in the form of $\dfrac{dy}{dx} + Py = Q$ and then by using integration factor i.e.

$e^{\int Pdx}\left(\dfrac{dy}{dx} + Py\right) = Qe^{\int pdx}$ we can solve the problem.

$$\frac{dy}{dx} + \frac{2x}{(1-x^2)}y = \frac{x\sqrt{1-x^2}}{1-x^2} \; ; \; I.F. = e^{\int Fdx} = e^{\int \frac{2x}{1-x^2}dx} = \frac{1}{1-x^2}$$

Solution is y. $y\frac{1}{1-x^2} = \int \frac{x}{\sqrt{1-x^2}}\frac{1}{1-x^2}dx + c = \int \frac{x}{(1-x^2)^{3/2}}dx + C = \frac{-1}{2}\int \frac{-2x}{\left(\frac{1}{2}-x^2\right)^{3/2}}dx + c$

$$y\frac{1}{\left(1-x^2\right)} = \frac{1}{\sqrt{1-x^2}} + c$$

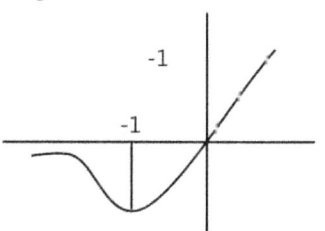

Figure 24.3

7.8 Equations Reducible to Linear form

(a) Bernoulli's Equation

A differential equation of the form $\frac{dy}{dx} + Py = Qy^n$, where P and Q are function of x and y is called Bernoulli's equation. This form can be reduced to linear form by dividing y^n and then substituting $y^{1-n} = v$

Dividing both sides by y^n, we get, $y^{-n}\frac{dy}{dx} + P.y^{-n+1} = Q$

Putting $y^{-n+1} = v$, so that , $(1-n)y^{-n}\frac{dy}{dx} = \frac{dv}{dx}$, we get $\frac{dv}{dx} + (1-n)P.y = (1-n)Q$

Which is a linear differential equation

(b) If the given equation is of the form $\frac{dy}{dx} + P. f(y) = Q.g(y)$, where P and Q are functions of x alone, we divide the

equation by f(y), and then we get $e^{\int Pdx} = e^{-\ln(1-x^2)} = \frac{1}{1-x^2}$

Now substitute $\frac{f(y)}{g(y)} = v$ and solve.

Illustration 36: Solve $\frac{dy}{dx} = xy + x^3y^2$ **(JEE MAIN)**

Sol: By rearranging the given equation we will get $\frac{1}{y^2}\frac{dy}{dx} - \frac{1}{y}x = x^3$ and then by putting $\frac{-1}{y} = t$ and using the integration factor we can solve it.

$\frac{dy}{dx} = xy + x^3y^2 \Rightarrow \qquad \frac{dy}{dx} - xy = x^3y^2 \qquad \Rightarrow \qquad \frac{1}{y^2}\frac{dy}{dx} - \frac{1}{y}x = x^3$

put $\frac{-1}{y} = t \qquad \Rightarrow \qquad \frac{dy}{dx} + tx = x^3$

This is a linear differential equation with I.F. $= e^{x^2/2} \Rightarrow t\,e^{x^2/2} = \int e^{x^2/2}x^3dx$

Illustration 37: Find the curve such that the y intercept of the tangent is proportional to the square of ordinate of tangent **(JEE MAIN)**

Sol: Here X = 0 and Y= y – mx i.e. $x\dfrac{dy}{dx} - y = -ky^2$. Hence by putting $\dfrac{-1}{y} = 1$ and applying integration factor we will get the result.

$X = 0 \Rightarrow Y = y - mx \Rightarrow x\dfrac{dy}{dx} - y = -ky^2$

$\Rightarrow \dfrac{1}{y^2}\dfrac{dy}{dx} - \dfrac{1}{y}\cdot\dfrac{1}{x} = \dfrac{-k}{x}$

Put $\dfrac{-1}{y} = t \Rightarrow \dfrac{dt}{dx} + \dfrac{t}{x} = \dfrac{-k}{x}$

$\Rightarrow$ I.f. = x

$\Rightarrow$ Solution is t.x = –kx + C $\Rightarrow \dfrac{-x}{y} = -kx + C$

7.9 Change of Variable by Suitable Substitution

Following are some examples where we change the variable by substitution.

Illustration 38: Solve $y\sin x \dfrac{dy}{dx} = \cos x(\sin x - y^2)$ **(JEE MAIN)**

Sol: Here by putting $y^2 = t$, the given equation reduces to $\dfrac{dt}{dx} + (2\cot x)t = 2\cos x$ and then using the integration factor method we will get result.

$y\sin x \dfrac{dy}{dx} = \cos x(\sin x - y^2)$

Let $y^2 = t \Rightarrow \dfrac{1}{2}\sin x\dfrac{dt}{dx} = \cos x (\sin x - t)$

$\Rightarrow \dfrac{dt}{dx} = 2\cos x - (2\cot x)t \qquad \Rightarrow \dfrac{dt}{dx} + (2\cot x)t = 2\cos x$

I.F. = $\sin^2 x$

$\Rightarrow$ Solution is $t\sin^2 x = \int 2\cos x.\sin^2 x\,dx$

$y^2\sin^2 x = \dfrac{2\sin^3 x}{3} + c$

Illustration 39: Solve $\dfrac{dy}{dx} = e^{x-y}(e^x - e^y)$ **(JEE MAIN)**

Sol: Simply by putting $e^y = t$ and using the integration factor we can solve the above problem.

$\dfrac{dy}{dx} = e^{x-y}(e^x - e^y) \qquad \Rightarrow e^y\dfrac{dy}{dx} = \left(e^x\right)^2 - e^x e^y$

Put $e^y = t \Rightarrow \dfrac{dy}{dx} + te^x = (e^x)^2;$

I.F. = $e^{\int e\,dx} = e^{e^x}$

Solution is $te^{e^x} = \int (e^x)^2.e^{e^x}\,dx$

If we can write the differential equation in the form

$f(f_1(x, y) \, d(f_1(x, y)) + \phi(f_2(x, y)d(f_2(x,y)) + \ldots\ldots = 0$, then each term can be easily integrated separately. For this the following results must be memorized.

(i) $d(x + y) = dx + dy$

(ii) $d(xy) = xdy + ydx$

(iii) $d\left(\dfrac{x}{y}\right) = \dfrac{ydx - xdy}{y^2}$

(iv) $d\left(\dfrac{y}{x}\right) = \dfrac{xdy - ydx}{x^2}$

(v) $d\left(\dfrac{x^2}{y}\right) = \dfrac{2xydx - x^2dy}{y^2}$

(vi) $d\left(\dfrac{y^2}{x}\right) = \dfrac{xydx - y^2dx}{x^2}$

(vii) $d\left(\dfrac{x^2}{y^2}\right) = \dfrac{2xy^2dx - 2x^2ydy}{y^4}$

(viii) $d\left(\dfrac{y^2}{x}\right) = \dfrac{xydx - 2xy^2dx}{x^4}$

(ix) $d\left(\tan^{-1}\dfrac{x}{y}\right) = \dfrac{ydx - xdy}{x^2 + y^2}$

(x) $d\left(\tan^{-1}\dfrac{y}{x}\right) = \dfrac{xdy - ydx}{x^2 + y^2}$

(xi) $d[\log(xy)] = \dfrac{xdy + ydx}{xy}$

(xii) $d\left(\log\left(\dfrac{x}{y}\right)\right) = \dfrac{ydx - xdy}{xy}$

(xiii) $d\left[\dfrac{1}{2}\log\left(x^2 + y^2\right)\right] = \dfrac{xdx + ydy}{x^2 + y^2}$

(xiv) $d\left[\log\left(\dfrac{y}{x}\right)\right] = \dfrac{xdy - ydx}{xy}$

(xv) $d\left(-\dfrac{1}{xy}\right) = \dfrac{xdy + ydx}{x^2y^2}$

(xvi) $d\left(\dfrac{e^x}{y}\right) = \dfrac{ye^xdx - e^xdy}{y^2}$

(xvii) $d\left(\dfrac{e^y}{x}\right) = \dfrac{xe^ydy - e^ydx}{x^2}$

(xviii) $d(x^m y^n) = x^{m-1}y^{n-1}(mydx + nxdy)$

(xix) $d\dfrac{dt}{dx} + \dfrac{t}{x}$

(xx) $d\left(\dfrac{1}{2}\log\dfrac{x+y}{x-y}\right) = \dfrac{xdy - ydx}{x^2 - y^2}$

(xxi) $\dfrac{d\left[f(x,y)\right]^{1-n}}{1-n} = \dfrac{f'(x,y)}{\left(f(x,y)\right)^n}$

(xxii) $d\left(\dfrac{1}{y} - \dfrac{1}{x}\right) = d\left(\dfrac{1}{y}\right) - d\left(\dfrac{1}{x}\right) = \dfrac{dx}{x^2} - \dfrac{dy}{y^2}$

8. EXACT DIFFERENTIAL EQUATION

The differential equation $Mdx + Ndy = 0$, where M and N are functions of x and y, is said to be exact if and only if

$$\dfrac{\partial M}{\partial y} = \dfrac{\partial N}{\partial x}$$

Rule for solving $Mdx + Ndy = 0$ when it is exact

(a) First integrate the terms in M w.r.t. x treating y as a constant.

(b) Then integrate w.r.t. y only those terms of N which do not contain x.

(c) Now, sum both the above integrals obtained and quote it to a constant i.e. $\int Mdx + \int Ndy = k$, where k is a constant.

(d) If N has no term which is free from x, the $\int Mdx = c$ (y constant)

Following exact differentials must be remembered:

(i) $xdy + ydx = d(xy)$

(ii) $\dfrac{xdy - ydx}{x^2} = d\left(\dfrac{y}{x}\right)$

(iii) $\dfrac{ydx - xdy}{y^2} = d\left(\dfrac{x}{y}\right)$

(iv) $\dfrac{xdy + ydx}{xy} = d(\log xy)$

(v) $\dfrac{dx + dy}{x + y} = d\log(x + y)$

(vi) $\dfrac{xdy - ydx}{xy} = d\left(\ell n \dfrac{y}{x}\right)$

(vii) $\dfrac{ydx - xdy}{xy} = d\left(\ell n \dfrac{x}{y}\right)$

(viii) $\dfrac{xdy - ydx}{x^2 + y^2} = d\left(\tan^{-2} \dfrac{y}{x}\right)$

(ix) $\dfrac{ydy - xdx}{x^2 + y^2} = d\left(\tan^{-1} \dfrac{y}{x}\right)$

(x) $d\left(\dfrac{e^x}{y}\right) = \dfrac{ye^x dy - e^x dy}{y^2}$

9. ORTHOGONAL TRAJECTORY

Definition 1: Two families of curves are such that each curve in either family is orthogonal (whenever they intersect) to every curve in the other family. Each family of curves is orthogonal trajectories of the other. In case the two families are identical then we say that the family is self-orthogonal

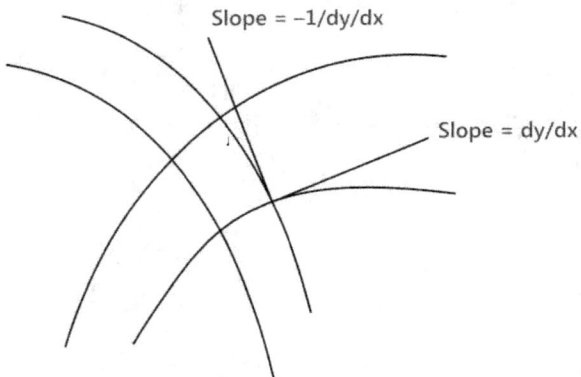

Figure 24.4: Orthogonal trajectories

> **NOMORECLASS CONCEPTS**
>
> Orthogonal trajectories have important application in the field of physics. For example, the equipotential lines and the streamlines in an irrotational 2D flow are orthogonal.

9.1 How to Find Orthogonal Trajectories

Suppose the first family of curves $F(x, y, c) = 0$ ... (i)

To find the orthogonal trajectories of this family we proceed as follows. First, differentiate (i) w.r.t. x to find $G(x, y, y', c) = 0$ ... (ii)

Now eliminate c between (i) and (ii) to find the differential equation $H(x, y, y') = 0$ ... (iii)

The differential equation for the other family is obtained by replacing y' with −1/y'. Hence, the differential equation the orthogonal trajectories is H(x, y, −1/y') = 0   ... (iv)

General solution of (iv) gives the required orthogonal trajectories.

Illustration 40: Find the orthogonal trajectories of a family of straight lines through the origin.   **(JEE MAIN)**

Sol: Here as we know, a family of straight lines through the origin is given by y = mx.

Hence by differentiating it with respect to x and eliminating m we will get an ODE of this family and by putting −1/y' in place of y' we will get an ODE for the orthogonal family.

The ODE for this family is xy' − y = 0

The ODE for the orthogonal family is x + yy' = 0

Integrating we find $x^2 + y^2 = c$, which are family of circles with center at the origin.

10. CLAIRAUT'S EQUATION

The differential equation

y = mx + f(m),   ... (i)

where m = $\dfrac{dy}{dx}$ is known as Clairaut's equation.

To solve (i), differentiate it w.r.t. x, which gives

$$\frac{dy}{dx} = m + x\frac{dm}{dx} + \frac{df(m)}{dx}$$

$$\Rightarrow \quad x\frac{dm}{dx} + f'(x)\frac{dm}{dx} = 0$$

either $\dfrac{dm}{dx} = 0 \Rightarrow m = c$ ... (ii)

or   x + f'(x) = 0   ... (iii)

NOMORECLASS CONCEPTS

- If m is eliminated between (i) and (ii), the solution obtained is a general solution of (i)
- If m is eliminated between (i) and (iii), then the solution obtained does not contain any arbitrary constants and is not the particular solution of (i). This solution is called singular solution of (i)

PROBLEM SOLVING TACTICS

Think briefly about whether you could easily separate the variables or not. Remember that means getting all the x terms (including dx) on one side and all the y terms (including dy) on the other. Don't forget to convert y' to dy/dx or you might make a mistake.

If it's not easy to separate the variables (usually it isn't) then we can try putting our equation in the form y' + P(x)y = Q(x). In other words, put the y' term and the y term on the left and then you may divide so that the coefficient of y' is 1.

Then we can use the trick of the integrating factor in which we multiply both sides by $. d\left(\dfrac{e^x}{x}\right) = \dfrac{xe^y dy - e^y dx}{x^2}$. This

makes things much simpler, but it's best to see why from doing problems, not from memorizing formulas.

FORMULAE SHEET

(a) Order of differential equation: Order of the highest derivative occurring in the differential equation

(b) Degree of differential equation: Degree of the highest order derivative when differential coefficients are free from radicals and fractions.

(c) General equation : $\dfrac{dy}{dx} = f(x) \Rightarrow \quad y = \int f(x)dx + c$

(d) $\dfrac{dy}{dx} = f(ax + by + c)$, then put ax + by + c = v

(e) If $\dfrac{dy}{dx} = f(x)g(y) \Rightarrow (g(y))^{-1}dy = f(x)dx$ then $\int (g(y))^{-1}dy = \int f(x)dx$

(f) Parametric forms

Case I: x = rcosθ, y = rsinθ $\Rightarrow x^2 + y^2 = r^2$; tanθ = $\dfrac{y}{x}$; xdx + ydy = rdr; xdy − ydx = $r^2 dθ$

Case II: x = rsecθ, y = rtanθ $\Rightarrow x^2 − y^2 = r^2$; $\dfrac{y}{x}$ = sinθ; xdx − ydy = rdr; xdy − ydx = $r^2 secθdθ$

(g) If $\dfrac{dy}{dx} = \dfrac{f(x,y)}{g(x,y)}$, then substitute y = vx $\Rightarrow \int \dfrac{dx}{x} = \int \dfrac{dv}{f(v) - v} + c$

(h) If $\dfrac{dv}{dx} = \dfrac{a_1 x + b_1 y + c_1}{a_2 x + b_2 y + c_2}$, then substitute x = X + h, y = Y + k

$\Rightarrow \dfrac{dY}{dX} = \dfrac{a_1 X + b_1 Y + (a_1 h + b_1 k + c_1)}{a_2 X + b_2 Y + (a_2 h + b_2 k + c_2)}$

choose h and k such that $a_1 h + b_1 k + c_1 = 0$ and $a_2 h + b_2 k + c_2 = 0$.

(i) If the equation is in the form of $\dfrac{dy}{dx}$ + Py = Q then $ye^{\int Pdx} = \int Qe^{\int Pdx} + c$

Solved Examples

JEE Main/Boards

Example 1: Find the differential equation of the family of curves $y = Ae^x + Be^{-x}$

Sol: By differentiating the given equation twice, we will get the result.

$$\frac{dy}{dx} = Ae^x - Be^{-x}$$

$$\Rightarrow \frac{d^2y}{dx^2} = Ae^x + Be^{-x} = y$$

Example 2: Find differential equation of the family of curves $y = c(x - c)^2$, where c is an arbitrary constant.

Sol: By differentiating the given family of curves and then eliminating c we will get the required differential equation.

$$y = c(x - c)^2$$

$$\Rightarrow \frac{dy}{dx} = 2(x - c)c$$

By division, $\dfrac{x - c}{2} = \dfrac{y}{dy/dx}$

or $c = x - \dfrac{2y}{dy/dx}$

Eliminating c, we get

$$\left(\frac{dy}{dx}\right)^2 = 4c^2(x - c)^2 = 4cy$$

$$= 4y\, 4y\left\{x - \frac{2y}{dy/dx}\right\}$$

$$\Rightarrow \left(\frac{dy}{dx}\right)^3 = 4y\left[x\frac{dy}{dx} - 2y\right]$$

Example 3: Find the differential equation of all parabolas which have their vertex at (a, b) and where the axis is parallel to x-axis.

Sol: Equation of parabola having vertex at (a, b) and axis is parallel to x-axis is $(y - b)^2 = 4L(x - a)$ where L is a parameter. Hence by differentiating and eliminating L we will get required differential equation.

$$\therefore \quad 2(y - b)\frac{dy}{dx} = 4L$$

On eliminating L, we get

$$(y - b)^2 = 2(y - b)\frac{xdy - ydx}{x^2 + y^2} = d\left(\tan^{-1}\frac{y}{x}\right)(x - a)$$

Differential equation is,

$$2(x - a)\frac{dy}{dx} = y - b.$$

Example 4: Show that the function $y = be^x + ce^{2x}$ is a solution of the differential equation.

$$\frac{d^2y}{dx^2} - 3\frac{dy}{dx} + 2y = 0$$

Sol: Differentiating given equation twice we can obtain the required differential equation.

$$y = be^x + ce^{2x}$$

$$\Rightarrow \frac{dy}{dx} = be^x + 2ce^{2x} = y + ce^{2x}$$

$$\Rightarrow \frac{d^2y}{dx^2} = be^x + 4ce^{2x} = y + 3ce^{2x}$$

$$\Rightarrow \frac{d^2y}{dx^2} - 3\frac{dy}{dx} + 2y = 0$$

Example 5: Solve: $\dfrac{dy}{dx} + \dfrac{\sqrt{1 - y^2}}{\sqrt{1 - x^2}} = 0$

Sol: By separating x and y term and integrating both sides we can solve it.

$$\int\frac{dy}{\sqrt{1 - y^2}} = -\int\frac{dx}{\sqrt{1 - x^2}}$$

$$\Rightarrow \sin^{-1}y = -\sin^{-1}x + c \quad \text{or} \quad \sin^{-1}y + \sin^{-1}x = c$$

Example 6: Find the equation of the curve that passes through the point P(1, 2) and satisfies the differential equation

$$\int f(x)dx + C = \frac{-2xy}{x^2 + 1} : y > 0$$

Sol: By integrating both sides we will get general equation of curve and then by substituting point (1, 2) in that we will get value of constant part.

$$\frac{dy}{dx} = -\frac{2xy}{x^2 + 1} \quad \Rightarrow \quad \int\frac{dy}{y} = -\int\frac{2x}{x^2 + 1}dx$$

$$\Rightarrow \quad \log|y| = -\log(x^2 + 1) + \log c_0$$

$\Rightarrow \quad \log(|y| (x^2 + 1)) = \log c_0$

$\Rightarrow \quad |y|(x^2 + 1) = c_0$

As point P(1, 2) lies on it,

$2(1 + 1) = c_0$ or $c_0 = 4$

$\therefore \quad$ Curve is $y(x^2 + 1) = 4$

Example 7: Solve: $\dfrac{dy}{dx} = \dfrac{(x-y)+3}{2(x-y)+5}$

Sol: By putting $x - y = t$ and integrating both sides we will obtain result.

Put $x - y = t$; then, $1 - \dfrac{dy}{dx} = \dfrac{dt}{dx}$

Differential equation becomes

$1 - \dfrac{dy}{dx} = \dfrac{t+3}{2t+5}$ or $\dfrac{dt}{dx} = 1 - \dfrac{t+3}{2t+5} = \dfrac{t+2}{2t+5}$

$\Rightarrow \int dx = \int \dfrac{2t+5}{t+2} dt = 2t + \int \dfrac{dt}{1+2}$

$\Rightarrow x + c = 2t + \log|t + 2| = 2(x - y) + \log|(x - y + 2)|$

Example 8: $x^2 dy + y(x + y)dx = 0: xy > 0$

Sol: We can write the given equation as

$\left(\dfrac{dy}{dx}\right)^3 = 4y\left[x\dfrac{dy}{dx} - 2y\right] = -\dfrac{dy}{dx}$

and then by substituting $y = vx$ and integrating we will get required general equation.

$\dfrac{dy}{dx} = -\dfrac{y(x+y)}{x^2}$ (Put $y = vx$)

$\Rightarrow v + x\dfrac{dy}{dx} = -v(1 + v)$

$\Rightarrow \dfrac{dv}{dx} = -2v - v^2$ or $\int \dfrac{dv}{v(v+2)} = \int \dfrac{dx}{x}$

$\Rightarrow -\int \dfrac{dx}{x} = \dfrac{1}{2}\int\left(\dfrac{1}{v} - \dfrac{1}{v+2}\right)dv$

$\Rightarrow -\log|x| = \dfrac{d^2y}{dx^2} - 3\dfrac{dy}{dx} + 2y = 0$ $(\log|v| - \log|v+2|) + c_0$

or $\left|\dfrac{V}{V+2}x^2\right| = c$ or $\dfrac{\frac{y}{x}x^2}{\frac{y}{x}+2} = c$ or $\dfrac{yx^2}{y+2x} = c$

Example 9: Solve: $\dfrac{dy}{dx} + \sec x = \tan x : 0 < x < \dfrac{dy}{dx}$

Sol: The given equation is in the form of $\dfrac{dy}{dx} + px = q$ hence by using integration factor method we can solve it.

I.F. $= \dfrac{-2xy}{x^2 = 1} = \dfrac{dy}{dx} = -\dfrac{2xy}{x^2+1}$

$= \sec x + \tan x$

Solution is $y(\sec x + \tan x)$

$= \int \tan x (\sec x + \tan) dx$

$= \int \sec x \tan x \, dx + \int (\sec^2 x - 1) dx$

$= \sec x + \tan x - x + c$

or $(y - 1)(\sec x + \tan x) = c - x$

Example 10: Solve:

$\sin x.\cos y.dx + \cos x.\sin y.dy = 0$

given, $y = \dfrac{\pi}{4}$ when $x = 0$.

Sol: Here by separating variables and taking integration we will get the general equation and then using the given values of x and y we will get value of constant c.

We have,

$\sin x.\cos y.dx + \cos x.\sin y.dy = 0$

On separating the variables, we get

$\Rightarrow \dfrac{dt}{dx}$

Integrating both sides, we get

$\int \dfrac{\sin x}{\cos x} dx + \int \dfrac{\sin y}{\cos y} dy = 0$

[Dividing by cosx cosy], we get

$\Rightarrow \log|\sec x| + \log|\sec y| = \log C$

$\Rightarrow \log|\sec x| \, |\sec y| = \log C$

$\Rightarrow \sec x.\sec y = C$ ... (i)

On putting $y = \dfrac{\pi}{4}$, $x = 0$ in (i),

we have $C = \sec 0.\sec \dfrac{\pi}{4}$

$\Rightarrow C (1).\left(\sqrt{2}\right) = \sqrt{2}$

Substituting the value of C in (i) we get

$\sec x.\dfrac{1}{\cos y} = \sqrt{2} \Rightarrow \cos y = \dfrac{1}{\sqrt{2}}\sec x$

$\Rightarrow y = \cos^{-1}\left(\dfrac{1}{\sqrt{2}}\sec x\right)$

Example 11: Solve the differential equation

$\frac{dy}{dx} = x^2 e^{-3y}$, given that y =0 for x =0.

Sol: Similar to the problem above we can solve it.

Here, $\frac{dy}{dx} = x^2 e^{-3y}$ ..... (i)

On separating the variables, we have

$\Rightarrow$ $e^{3y}dy = x^2 dx$

Integrating both sides, we get

$\int e^{3y} = \int x^2 dx$

$\Rightarrow$ $\frac{e^{3y}}{3} = \frac{x^3}{3} + c$ ..... (ii)

putting: y = 0 for x = 0, in (ii), we obtain

$\frac{e^0}{3} = 0 + C \Rightarrow \frac{1}{3} = C$ $[e^0 = 1]$

On substituting the value of C in (ii), we get

$\therefore$ $e^{3y} = x^3 + 1$

which is the required particular solution of (i)

Example 12: Solve the following differential equation:

$2x^2 \frac{dy}{dx} - 2xy + y^2 = 0$

Sol: Here by rearranging the given equation we will get $e^{\log(\sec x + \tan x)} = \int \tan x (\sec x + \tan x) dx$. Now by substituting y = vx and then integrating we can solve the illustration above.

$2x^2 \frac{dy}{dx} = 2xy - y^2$

$\Rightarrow$ $\frac{dy}{dx} = \frac{2xy - y^2}{2x^2}$ ..... (i)

Put y = vx so that $\frac{dy}{dx} = v + x\frac{dv}{dx}$ in (i), we get

$\Rightarrow$ $v + x\frac{dv}{dx} = \frac{2x(vx) - (vx)2}{2x^2}$

$\Rightarrow$ $v + x\frac{dv}{dx} = v - \frac{v^2}{2}$

$\frac{xdv}{dx} = \frac{-v^2}{2} \Rightarrow \frac{dv}{v^2}\frac{dx}{2x}$

Integrating, we have

$\Rightarrow$ $\frac{1}{4} = \frac{1}{4} |\log x| + c$

$\Rightarrow \frac{x}{y} = \frac{1}{2} |\log x| + c$

Example 13: Solve the following differential equation

$\cos^2 x \frac{dy}{dx} + y = \tan x$

Sol: Here by reducing the given equation in the form of $\frac{dy}{dx} + py = q$ and then using integration factor we will get the result.

We have, $\cos^2 x \frac{dy}{dx} + y = \tan x$

$\Rightarrow \frac{dy}{dx} + y.\sec^2 x = \tan x.\sec^2 x$

I.F. $= e^{\int \sec^2 x} = e^{\tan x}$

$y.(I.F.) = \int Q.(I.F.)dx + c$

$y.e^{\tan x} = \int \tan x \sec^2 x e^{\tan x} dx = \int te^t dt$

$te^t - \int e^t dt + c$ $\left[\begin{array}{l} \tan x = t \\ \sec^2 x dx = dt \end{array} \right]$

$= te^t - e^t + c$

$= \tan x\, e^{\tan x} - e^{\tan x} + c$

$y = \tan x - 1 + ce^{-\tan x}$

Example 14: Solve $x\frac{dy}{dx} - y = x^2$

Sol: As similar to the problem above, we can reduce the given equation as $\frac{dy}{dx}$ therefore by using integration factor we can solve this.

We have, $x\frac{dy}{dx} - y = x^2$

$\Rightarrow \frac{dy}{dx} - \frac{1}{x}y = x$ ... (i)

This is a linear differential equation in y

Here, P $= -\frac{2xy - y^2}{2x^2}$ and Q = x

Now,   I.F. $= e^{\int pdx} = e^{\int -\frac{1}{x}dx}$

$= e^{-\log x} = e^{\log^{-1}} = x^{-1} = \frac{1}{x}$

$\therefore$ The solution of (i) is

$y(I.F.) = \int (Q \times IF) dx + C = x + C$

$\Rightarrow$ $y = x^2 + Cx$

4.25

Example 15: Solve the following differential equation:

$\dfrac{dv}{dx} + y = \cos x - \sin x$.

Sol: Here given equation is in the form of $\dfrac{dy}{dx} + Py = Q$, where P = 1 and Q = cosx – sinx hence by using integration factor we will get result.

Given differential equation is

$\dfrac{dy}{dx} + y = \cos x - \sin x$ ... (i)

The given differential equation is a linear differential equation

On comparing with $\dfrac{dy}{dx} + Py = Q$

$\therefore \quad P = 1, Q = \cos x - \sin x$

$I.F. = e^{\int p\,dx} = e^x$

$\therefore$ required solution of (i) is

$y\,(I.F.) = \int Q.\left(I.F\right)dx + c$

$\Rightarrow y.e^x = \int (\cos x - \sin x)e^x dx + c$

$\Rightarrow y.e^x = \int \cos x\,e^x dx - \int \sin x.e^x dx + c$

Integrating by parts, we get

$\Rightarrow y.e^x = \cos x \int e^x dx - \int (-\sin x)e^x dx - \int \sec^2 x\,dx + c$

$\Rightarrow y.e^x = e^x \cos x + \int e^x \sin x\,dx - \int e^x \sin x\,dx + c$

$y.e^x = e^x \cos x + c$

$\therefore \quad y = \cos x + ce^{-x}$

JEE Advanced/Boards

Example 1: Solve

$\left(xe^{y/x} - y\sin\dfrac{y}{x} \right)dx + x\sin\dfrac{y}{x}dy = 0; \; x>0$

Sol: Simply by putting y = vx and integrating we can solve the problem above.

$\left(e^{\frac{y}{x}} - \dfrac{y}{x}\sin\dfrac{y}{x} \right) + \sin\dfrac{y\,dy}{x\,dx} = 0$

Put y = vx

$\therefore (e^v - v\sin v) + \sin v \left(v + x\dfrac{dv}{dx} \right) = 0$

$\Rightarrow \int \dfrac{dx}{x} + \int e^{-v}\sin v\,dv = 0$

Integrating, we get

$\log x - \dfrac{1}{2}e^{-v}(\sin v + \cos v) = c$

or $\log x = c + \dfrac{1}{2}e^{-y/x}.\left(\sin\dfrac{y}{x} - 4\cos\dfrac{y}{x} \right)$

Example 2: Solve:

$x\,dy - y\,dx = xy^3(1 + \log x)dx$

Sol: We can reduce the given equation in the form of $-\dfrac{x}{y}d\left(\dfrac{x}{y}\right) = x^2(1 + \log x)dx$. Hence by integrating L.H.S. with respect to $\dfrac{x}{y}$ and R.H.S. with respect to x we will get the solution.

$-\dfrac{y\,dx - x\,dy}{y^2} = xy(1 + \log x)dx$

or $-d\left(\dfrac{x}{y} \right) = xy\,(1 + \log x)dx$

or $-\dfrac{x}{y}d\left(\dfrac{x}{y} \right) = x^2(1 + \log x)dx$

Integrating, $-\int \dfrac{x}{y}d\left(\dfrac{x}{y} \right)$

$= \int x^2\left(1 + \log x\right)dx$

or $-\dfrac{1}{2}\left(\dfrac{x}{y} \right)^2 = (1 + \log x)\dfrac{x^3}{3} - \int \dfrac{x^3}{3}.\dfrac{1}{x}dx$

$\dfrac{-1}{2}\left(\dfrac{x}{y} \right)^2 = (1 + \log x)\dfrac{x^3}{3} - \dfrac{x^3}{9} + c$

Example 3: Find the equation of the curve passing through (1, 2) whose differential equation is

$y(x + y^3)dx = x(y^3 - x)dy$

Sol: Similar to example 2 we can solve the problem above by reducing the given equation as –

$\dfrac{y}{x}d\left(\dfrac{y}{x} \right) + \dfrac{1}{x^2 y^2}d\left(xy\right) = 0$.

$(xy + y^4)dx = (xy^3 - x^2)dy$

or $y^3(y\,dx - x\,dy) + x(y\,dx + x\,dy) = 0$

or $-x^2 y^3 \dfrac{x\,dy - y\,dx}{x^2} + x\,d(xy) = 0$

4.26

or $-\dfrac{y}{x} d\left(\dfrac{y}{x}\right) + \dfrac{1}{x^2 y^2} d(xy) = 0$

Integrating, we get

$-\dfrac{1}{2}\left(\dfrac{y}{x}\right)^2 - \dfrac{1}{xy} = c$

or $y^3 + 2x - 2cx^2 y = 0$

As it passes through (1, 2), condition is

$8 + 2 + 4c = 0 \Rightarrow c = -\dfrac{5}{2}$

Thus curve is $y^3 + 2x - 5x^2 y = 0$

Example 4: Form the differential equation representing the family of curves $y = A\cos 2x + B\sin 2x$, where A and B are arbitrary constants.

Sol: Here we have two arbitrary constants hence we have to differentiate the given equation twice.

The given equation is:

$y = A\cos 2x + B\sin 2x$ ... (i)

Diff. w.r.t. x,

$\dfrac{dy}{dx} = -2A\sin 2x + 2B\cos 2x$

Again diff. w.r.t. x, $\dfrac{d^2 y}{dx^2} = -4A\cos 2x - 4B\sin 2x$

$= -4(A\cos 2x + B\sin 2x) = -4y$ [Using (i)]

Hence $\dfrac{d^2 y}{dx^2} + 4y = 0$, which is the required differential equation.

Example 5: The solution of the differential equation $x\dfrac{d^2 y}{dx^2} = 1$, given that $y = 1, \dfrac{dy}{dx} = 0$, when $x = 1$, is

Sol: By integrating $x\dfrac{d^2 y}{dx^2} = 1$ twice we will get its general equation and then by substituting given values of x, y and $\dfrac{dy}{dx}$ we will get the values of the constants.

$x\dfrac{d^2 y}{dx^2} = 1 \Rightarrow \dfrac{d^2 y}{dx^2} = \dfrac{1}{x}$

$\Rightarrow \quad \dfrac{dy}{dx} = \log x + C_1$

Again integrating

$y = x\log x - x + C_1 x + C_2$

Given $y = 1$ and $\dfrac{dy}{dx} = 0$ at $x = 1$

$\Rightarrow C_1 = 0$ and $C_2 = 2$

Therefore, the required solution is $y = x\log x - x + 2$

Example 6: By the elimination of the constant h and k, find the differential equation of which $(x-h)^2 + (y-k)^2 = a^2$, is a solution.

Sol: Three relations are necessary to eliminate two constants. Thus, besides the given relation we require two more and they will be obtained by differentiating the given relation twice successively.

Thus we have

$(x - h) + (y - k)\dfrac{dy}{dx} = 0$ ... (i)

$1 + (y - k)\dfrac{d^2 y}{dx^2} + \left(\dfrac{dy}{dx}\right)^2 = 0$ ... (ii)

From (i) and (ii), we obtained

$y - k = -\dfrac{1 + \left(\dfrac{dy}{dx}\right)^2}{\dfrac{d^2 y}{dx^2}}$

$x - h = \dfrac{\left[1 + \left(\dfrac{dy}{dx}\right)^2\right]\dfrac{dy}{dx}}{\dfrac{d^2 y}{dx^2}}$

Substitute these values in the given relation, we obtained

$\left[1 + \left(\dfrac{dy}{dx}\right)^2\right]^3 = a^2\left(\dfrac{d^2 y}{dx^2}\right)^2$

which is the required differential equation.

Example 7: Form the differential equations by eliminating the constant(s) in the following problems.

(a) $x^2 - y^2 = c(x^2 + y^2)^2$,   (b) $a(y + a)^2 = x^3$

Sol: Given equations have one arbitrary constant, hence by differentiating once and eliminating c and a we will get the required differential equation.

(a) The given equation contains one constant

Differentiating the equation once, we get

$2x - 2yy' = 2c(x^2 + y^2)(2x + 2yy')$

But $c = \dfrac{x^2 - y^2}{\left(x^2 + y^2\right)^2}$

Substituting for c, we get

$(x - yy') = \dfrac{\left(x^2 + y^2\right)\left(x^2 - y^2\right)}{\left(x^2 + y^2\right)^2} \cdot 2(x + yy')$

or $(x^2 + y^2)(x - yy') = 2(x^2 - y^2)(x + yy')$

$\Rightarrow yy'[(x^2 + y^2) + 2(x^2 - y^2)]$

$\Rightarrow x(x^2 + y^2) - 2x(x^2 - y^2)$

$\Rightarrow yy'(3x^2 - y^2) = x(3y^2 - x^2)$

Hence, $y' = \dfrac{x\left(3y^2 - x^2\right)}{y\left(3x^2 - y^2\right)}$

(b) The given equation contains only one constant. Differentiating once, we get

$2a(y + a)y' = 3x^2$ ... (i)

Multiplying by y + a, we get

$2a(y + a)^2 y' = 3x^2(y + a)$

Using the given equation, we obtain

$2x^3 y' = 3x^2(y + a)$ or $2xy' = 3y + 3a$

or $a = \dfrac{1}{3}(2xy' - 3y)$

Substituting the value of a in (i) we obtain

$\dfrac{2}{3}(2xy' - 3y)\left[y + \dfrac{1}{3}(2xy - 3y)\right]y' = 3x^2$

$\dfrac{2}{9}(2xy' - 3y)(2xy')y' = 3x^2$

Cancelling x, we obtain

$8x(y')^3 - 12y(y')^2 - 27x = 0$

Example 8: If $y(x - y)^2 = x$, then show that

$\displaystyle\int \dfrac{dx}{(x - 3y)} = \dfrac{1}{2}\log[x - y)^2 - 1]$

Sol: As given $y(x - y)^2 = x$, therefore by differentiating it with respect to x we will get the value of $\dfrac{dy}{dx}$. After that differentiate both sides of equation $\displaystyle\int \dfrac{dx}{(x - 3y)} = \dfrac{1}{2}$

$\log[x - y)^2 - 1]$ w.r.t. x and then by substituting the value of $\dfrac{dy}{dx}$ we can prove it.

Let $P = \displaystyle\int \dfrac{dx}{(x - 3)} = \dfrac{1}{2}\log\{(x - y)^2 - 1\}$

$\therefore \quad P = \displaystyle\int \dfrac{dx}{(x - 3y)}$

$\dfrac{dP}{dx} = \dfrac{1}{(x - 3y)}$ ... (i)

Also $P = \dfrac{1}{2}\log\{(x - y)^2 - 1\}$

$\therefore \dfrac{dP}{dx} = \dfrac{(x - y)\left\{1 - \dfrac{dy}{dx}\right\}}{\left\{(x - y)^2 - 1\right\}}$ ... (ii)

Given $y(x - y)^2 = x$

Differentiating both sides w.r.t. x

$\therefore \dfrac{dy}{dx} = \dfrac{1 - 2y(x - y)}{(x - y)(x - 3y)}$ ... (iii)

From (ii) and (iii)

$\dfrac{dP}{dx} = \dfrac{(x - y)\left\{1 - (1 - 2y(x - y) / (x - y)(x - 3y))\right\}}{\left\{(x - y)^2 - 1\right\}}$

$= \dfrac{(x - y)(x - 3y) - 1 + 2y(x - y)}{(x - 3y)\left\{(x - y^2) - 1\right\}}$

$= \dfrac{\left\{(x - y)^2 - 1\right\}}{(x - 3y)\left\{(x - y)^2 - 1\right\}}$

$\Rightarrow \dfrac{dP}{dx} = \dfrac{1}{(x - 3y)}$

It is true from (i)

Hence $\displaystyle\int \dfrac{dx}{(x - 3y)} = \dfrac{1}{2}\ell n\left\{(x - y)^2 - 1\right\}$

Example 9: Solve: $\cos(x + y)dy = dx$

Sol: Simply by putting x + y = t we can reduce the given equation as $\dfrac{dt}{dx} = \sec t + 1$ and then by separating the variable and integrating we can solve the problem given above.

We have $\cos(x + y)dy = dx$

$\Rightarrow \dfrac{dy}{dx} = \sec(x + y)$

4.28

On putting $x + y = t$ so that $1 + \dfrac{dy}{dx} = \dfrac{dt}{dx}$

or $\dfrac{dy}{dx} = \dfrac{dt}{dx} - 1$ we get

$\dfrac{dt}{dx} - 1 = \sec$

$\Rightarrow \dfrac{dt}{dx} = 1 + \sec t$

$\dfrac{dt}{1 + \sec t} = dx \Rightarrow \dfrac{\cos t}{\cos t + 1} dt = dx$

$\displaystyle\int \dfrac{\cos t}{\cos t + 1} dt = \int dx$

$\Rightarrow \displaystyle\int \left[1 - \dfrac{1}{\cos t + 1}\right] \cdot dt = x + C$

$\displaystyle\int \left[1 - \dfrac{1}{2\cos^2(t/2) - 1 + 1}\right] dt = x + C$

$\displaystyle\int \left(1 - \dfrac{1}{2}\sec^2 \dfrac{t}{2}\right) dt = x + C$

$\Rightarrow t - \tan \dfrac{t}{2} = x + C$

$x + y - \tan \dfrac{x+y}{2} = x + C$

$y - \tan \dfrac{x+y}{2} = C$

Example 10: Solve: $\sin^{-1}\left(\dfrac{dy}{dx}\right) = x + y$

Sol: Similar to example 9.

We have, $\sin^{-1}\left(\dfrac{dy}{dx}\right) = x + y \Rightarrow \dfrac{dy}{dx} = \sin(x + y)$

Putting $x + y = t$, so that

$1 + \dfrac{dy}{dx} = \dfrac{dt}{dx} \Rightarrow \dfrac{dy}{dx} = \dfrac{dt}{dx}$

Now, substituting $x + y = t$ and $\dfrac{dy}{dx} = \dfrac{dt}{dx} - 1$ in (i), we get

$\dfrac{dt}{dx} = \sin t \Rightarrow \dfrac{dt}{dx} = \sin t + 1 \Rightarrow dx = \dfrac{dt}{1 + \sin t}$

Integrating both sides, we get

$\displaystyle\int dx = \int \dfrac{dt}{1 + \sin^2 t} dt + c$

$\Rightarrow \displaystyle\int dx = \int \dfrac{1 - \sin t}{1 - \sin^2 t} dt + C = \int \dfrac{1 - \sin t}{\cos^2 t} dt$

$\Rightarrow \displaystyle\int dx = \int \left(\sec^2 t - \tan t \sec t\right) dt$

$\Rightarrow x = \tan t - \sec t$

$\Rightarrow x = \tan(x + y) - \sec(x + y) + C$

Example 11: Solve the equation:

$\dfrac{dy}{dx} = \dfrac{y}{x} + x \sin \dfrac{y}{x}$

Sol: Simply by putting $y = vx$ and integrating we can obtain the general equation of given differential equation.

We have,

$\dfrac{dy}{dx} = \dfrac{y}{x} + x \sin \dfrac{y}{x}$ ... (i)

Put $y = vx$, so that

$\dfrac{dy}{dx} = v + x \dfrac{dv}{dx}$

On putting the value of y and $\dfrac{dy}{dx}$ in (i), we get

$v + x \dfrac{dv}{dx} = v + x \sin v$

$\Rightarrow x \dfrac{dv}{dx} \Rightarrow \dfrac{dv}{dx} = \sin v$

Separating the variables, we get

$\dfrac{dv}{\sin v} = dx \Rightarrow \displaystyle\int \text{cosec} v \, dv = \int dx$

$\Rightarrow \log \tan \dfrac{v}{2} = x + C$ ... (ii)

On putting the value of v in (ii), we have

$\log \tan \dfrac{y}{2x} = x + C$

This is the required solution

Example 12: Solve:

$2y e^{\frac{x}{y}} dx + \left(y - 2x e^{\frac{x}{y}}\right) dy = 0$

Sol: We can reduce the given equation as $\dfrac{dy}{dx} = \dfrac{2x e^{x/y}}{2y e^{x/y}}$

and then by putting $x = vy$ and integrating we can obtain general equation.

We have,

$2y e^{\frac{x}{y}} dx + \left(y - 2x e^{\frac{x}{y}}\right) dx = 0$

$$\Rightarrow \quad 2ye^{\frac{x}{y}} \cdot \frac{dx}{dy} + \left(y - 2xe^{\frac{x}{y}}\right) = 0$$

$$\Rightarrow \quad \frac{dy}{dx} = \frac{2xe^{x/y}}{2ye^{x/y}} \qquad \qquad \ldots (i)$$

Clearly, the given differential equation is a homogeneous differential equation. As the right hand side of (i) is expressible as a function of $\left(\dfrac{x}{y}\right)$. So, we put

$$\frac{dt}{dx} = v \Rightarrow x = vy \text{ and } \frac{dx}{dy}$$

$$= v + y \frac{dv}{dy} \text{ in (i), we get}$$

$$v + y \frac{dv}{dy} = \frac{2ve^v - 1}{2e^v}$$

$$\Rightarrow \quad y \frac{dv}{dy} = \frac{2ve^v - 1}{2e^v} - v$$

$$\Rightarrow \quad y \frac{dv}{dy} = -\frac{1}{2e^v}$$

$$\Rightarrow \quad 2ye^v dv = -dy$$

$$\Rightarrow 2e^v dv = -\frac{1}{y} dy \text{ , } y \neq 0$$

Integrating both sides, we get

$$2\int e^v dv = -\int \frac{1}{y} dy + \log c$$

$$\Rightarrow \quad 2e^v = -\log|y| + \log c$$

$$\Rightarrow \quad 2e^v = \log \left|\frac{c}{y}\right|$$

$$\Rightarrow \quad 2e^{\frac{x}{y}} = \log \left|\frac{c}{y}\right| \quad \left(\because v = \frac{x}{y}\right)$$

Example 13: Show that the family of curves for which the slope of the tangent at any point (x, y) on it is $\dfrac{x^2 + y^2}{2xy}$, is given by $x^2 - y^2 = cx$

Sol: Here by reading the above problem, we get that $\dfrac{dy}{dx} = \dfrac{x^2 + y^2}{2xy}$. Hence by putting y = vx and then integrating both sides we can prove the given equation.

We have slope of the tangent

$$= \frac{x^2 + y^2}{2xy} \Rightarrow \frac{dy}{dx} = \frac{x^2 + y^2}{2xy}$$

or $\quad \dfrac{dy}{dx} = \dfrac{1 + \dfrac{y^2}{x^2}}{\dfrac{2y}{x}} \qquad \qquad \ldots (i)$

Equation (i) is a homogeneous differential equation.

So we put y = vx and $\dfrac{dy}{dx} = v + \dfrac{dv}{dx}$

Substituting the value of $\dfrac{y}{x}$ and $\dfrac{dy}{dx}$ in equation (i), we get

$$v + x \frac{dv}{dx} = \frac{1 + v^2}{2v}$$

or $\quad x\dfrac{dv}{dx} = \dfrac{1 - v^2}{2v} \qquad \qquad \ldots (ii)$

Separating the variables in equation (ii), we get

$$\frac{2v}{1 - v^2} dv = \frac{dx}{x} \text{ or } \frac{2v}{v^2 - 1} dv = -\frac{dx}{x} \qquad \ldots (iii)$$

Integrating both sides of equation (iii), we get

$$\int \frac{2v}{v^2 - 1} dv = -\int \frac{1}{x} dx$$

or $\quad \log|v^2 - 1| = -\log|x| + \log|C_1|$

or $\quad \log|(v^2 - 1)(x)| = \log|C_1| \qquad \qquad \ldots (iv)$

Replacing v by $\dfrac{y}{x}$ in equation (iv), we get $\left(\dfrac{y^2}{x^2} - 1\right)$ $x = \pm C_1$

or $(y^2 - x^2) = \pm C_1 x$ or $x^2 - y^2 = Cx$

Example 14: Solve: $\dfrac{dy}{dx} = \dfrac{x + 2y - 3}{2x + y - 3}$

Sol: Simply by putting x = X + h; y = Y + k were (h, k) will satisfy the equations x + 2y − 3 =0 and 2x + y − 3 =0 we can solve the problem.

$$\frac{dy}{dx} = \frac{x + 2y - 3}{2x + y - 3}$$

Put: x = X + h; y = Y + k

$$\Rightarrow \quad dx = dX; dy = dY$$

$$\Rightarrow \quad \frac{dy}{dx} = \frac{dY}{dX}$$

Given equation reduces to

$$\frac{dy}{dx} = \frac{(x + h) + 2(Y + k) - 3}{2(X + h) + (Y + k) - 3} = \frac{X + 2Y + (h + 2k - 3)}{2X + Y + (2h + k - 3)} \qquad \ldots (i)$$

Choose h and k such that

h + 2k − 3 = 0; and 2h + k − 3 = 0

$\Rightarrow$ h = 1 ; k = 1

Equation (i) becomes

$$\frac{dY}{dX} = \frac{X + 2Y}{2X + Y}$$ ... (ii)

Put: Y = VX

$$\Rightarrow \frac{dY}{dX} = V + X\frac{dV}{dX}$$

Now equation (ii) becomes:

$$V + X\frac{dV}{dX} = \frac{X + 2VX}{2X + VX} = \frac{1 + 2V}{2 + V}$$

$$\Rightarrow X\frac{dV}{dX} = \frac{1 + 2V}{2 + V} - V = \frac{1 - V^2}{2 + V}$$

Separating the variables, we have

$$\Rightarrow \frac{2 + V}{1 - V^2}dV = \frac{dX}{X}$$

Integrating, we get

$$\int \frac{2}{1 - V^2}dV + \int \frac{V}{1 - V^2}dV = \int \frac{dX}{X}$$

$$\Rightarrow 2.\frac{1}{2}\log\frac{1 + V}{1 - V} - \frac{1}{2}\log(1 - V^2) = \log X + \log c$$

$$\Rightarrow 2\log\left(\frac{1 + V}{1 - V}\right) - \log(1 - V^2) = 2\log cX$$

$$\Rightarrow \log\left[\left(\frac{1 + V}{1 - V}\right)^2 \times \frac{1}{1 - V^2}\right] = \log(cX)^2$$

$$\Rightarrow \left(\frac{1 + V}{1 - V}\right)^2 \times \frac{1}{(1 - V^2)} = (cX)^2$$

$$\Rightarrow \frac{1 + V}{(1 - V)^3} = c^2X^2$$ ... (iii)

Putting the value of V in (iii), we have

$$\Rightarrow \frac{X + Y}{(X - Y)^3}X^2 = c^2X^2$$ ... (iv)

$$\left[\because V = \frac{Y}{X}\right]$$

Substituting the value of X and Y in (iv), we get

$$\Rightarrow \frac{x + y - 2}{(x - y)^3} = c^2$$

$[\because X = x - h = x - 1; Y = y - 1]$

Example 15: Solve the following differential equation:

$$(x^2 - 1)\frac{dy}{dx} + 2xy = \frac{2}{x^2 - 1}$$

Sol: First reduce this into the form of $\frac{dy}{dx} + Py = Q$ and then using the integration factor i.e. $e^{\int Pdx}$ we can solve this.

We have, $(x^2 - 1)\frac{dy}{dx} + 2xy = \frac{2}{x^2 - 1}$

$$\Rightarrow \frac{dy}{dx} + \frac{2x}{x^2 - 1}y = \frac{2}{(x^2 - 1)^2}$$ .....(i)

This is a linear differential equation in y.

Here, P = $\frac{2x}{x^2 - 1}$, Q = $\frac{2}{(x^2 - 1)^2}$

I.F. = $e^{\int Pdx} = e^{\int \frac{2x}{x^2 - 1}dx} = e^{\log(x^2 - 1)} = x^2 - 1$

$\therefore$ The solution of (i) is

$$y(I.F.) = \int (Q \times I.F.)dx + C$$

$$y.(x^2 - 1) = \int (x^2 - 1).\frac{2}{(x^2 - 1)^2}dx + C$$

$$= 2\int \frac{1}{x^2 - 1}dx + C$$

$$= 2.\frac{1}{2}\log\left|\frac{x - 1}{x + 1}\right| + C$$

$$\Rightarrow y(x^2 - 1) = \log\left|\frac{x - 1}{x + 1}\right| + C$$

$$\Rightarrow y = \left(\frac{1}{x^2 - 1}\right)\left[\log\left|\frac{x - 1}{x + 1}\right| + C\right]$$

Exercise 1

Q.1 Write the order and degree of the differential equation $x - \cos\left(\dfrac{dy}{dx}\right) = 0$

Q.2 Solve the differential equation $\dfrac{dy}{dx} + \sqrt{\dfrac{1-y^2}{1-x^2}} = 0$

Q.3 Write the order and degree of the differential equation $\dfrac{dy}{dx} + \sin\left(\dfrac{dy}{dx}\right) = 0$

Q.4 How will you proceed to solve the differential equation $\dfrac{dy}{dx} = 1 + x + y + xy$?

Q.5 Find the integrating factor for solving the differential equation

$(1 + y^2)\, dx = (\tan^{-1}y - x)\, dy$

Q.6 Solve the differential equation $\dfrac{dr}{d\theta} = \cos\theta$

Q.7 To solve the differential equation $\dfrac{dy}{dx} + 2y = 6e^x$, how will you proceed?

Q.8 Prove that, the differential equation that represents all parabolas having their axis of symmetry coincident with the axis of x is $yy_2 + y_1^2 = 0$.

Q.9 Form the differential equation representing the family of curves $y = A\cos 2x + B\sin 2x$, where A and B are constants.

Q.10 Prove that, the function $y = Ax + \dfrac{B}{x}$ is a solution of the differential equation: $x^2\dfrac{d^2y}{dx^2} + x\dfrac{dy}{dx} - y = 0$

Q.11 Prove that, the differential equation of which

$1 + 8y^2\tan x = cy^2$ is a solution is $\cos^2 x\,\dfrac{dy}{dx} = 4y^3$

Q.12 Form the differential equation of the family of curves $y = Ae^{Bx}$

Q.13 $\dfrac{dy}{dx} = e^{x-y} + x^3e^{-y}$

Q.14 $\dfrac{dy}{dx} + \dfrac{1+\cos 2y}{1-\cos 2x} = 0$

Q.15 $\sqrt{x + x^2 + y^2 + x^2y^2} + xy\dfrac{dy}{dx} = 0$

Q.16 $\dfrac{dy}{dx} = \dfrac{x^2y}{x^3 + y^3}$

Q.17 $\dfrac{dy}{dx} = \sin^3 x.\cos^3 x + xe^x$

Q.18 $(1 - x^2)dy + xy\,dx = xy^2dx$

Q.19 $x\dfrac{dy}{dx} + y = y^2$

Q.20 $(x - y^3)dy + y\,dx = 0$

Q.21 $y - x\dfrac{dy}{dx} = a\left(y^2 + x^2\dfrac{dy}{dx}\right)$, where $x = a$, $y = a$.

Q.22 $x\log x\dfrac{dy}{dx} + y = \dfrac{2}{x}\log x$

Q.23 $e^{dy/dx} = x + 1$, $y(0) = 4$

Q.24 $y' + 2y^2 = 0$, $y(0) = \dfrac{\pi}{2}$

Q.25 $xy' + y = x\cos x + \sin x$, $y\left(\dfrac{\pi}{2}\right) = 1$

Q.26 $y^2 + x^2\dfrac{dy}{dx} = xy\dfrac{dy}{dx}$, given that when $x = 1$, $y = 1$

Q.27 $(1+\sin^2 x)dy + (1+y^2)\cos x\, dx = 0$, given that when $x = \dfrac{\pi}{2}$, $y = 0$

Q.28 $xy\,dy = (y + 5)dx$, given that $y(5) = 0$

Q.29 $(x + 2)dx = (x^2 + 4x + 9)dy$, given that $y(0) = 0$

Exercise 2

Single Correct Choice Type

Q.1 The general solution of the differential equation, y' + $y\phi'(x) - \phi(x).\phi'(x) = 0$ where $\phi(x)$ is a known function is:

(A) $y = ce^{-\phi(x)} + \phi(x) - 1$

(B) $y = ce^{+\phi(x)} + \phi(x) - 1$

(C) $y = ce^{-\phi(x)} - \phi(x) + 1$

(D) $y = ce^{-\phi(x)} + \phi(x) + 1$

Q.2 The differential equation $y\dfrac{dy}{dx} + x = C$, where C is any arbitrary constant represents:

(A) A set of circles with centre on x-axis.

(B) A set of circles with centre on y-axis.

(C) A set of concentric circles.

(D) A set of ellipses.

Q.3 The differential equation of all parabolas having their axis of symmetry coinciding with the axis of x is:

(A) $y\dfrac{d^2y}{dx^2} + \dfrac{dy}{dx} = 0$ (B) $x\dfrac{d^2y}{dy^2} + \left[\dfrac{dy}{dx}\right]^2 = 0$

(C) $y\dfrac{d^2y}{dx^2} + \left[\dfrac{dy}{dx}\right]^2 = 0$ (D) None of these

Q.4 The solution of the differential equation,

$$xy\left[\dfrac{dy}{dx}\right] = \left[\dfrac{1+y^2}{1+x^2}\right]\left(1 = x + x^2\right)$$

given that when x = 1, y = 0 is:

(A) $\log\sqrt{1+y^2} = \log x + \tan^{-1}x - \dfrac{\pi}{2}$

(B) $\log\dfrac{1+y^2}{x^2} = 2\tan^{-1}x - \dfrac{\pi}{2}$

(C) $\log\dfrac{1+y^2}{x^2} = \dfrac{\pi}{4} - 2\tan^{-1}x$

(D) None of these

Q.5 Given, y = 1 + cosx and y = 1 + sinx are solution of the differential equation $\dfrac{d^2y}{dx^2} + y = 1$, then its solution will be also:

(A) y = 2(1 + cosx)   (B) y = 2 + cosx +sinx

(C) y = cosx − sinx   (D) y = 1 + cosx + sinx

Q.6 The solution of the differential equation

$(x + 2y^3)\dfrac{d^2y}{dx^2} = y$ is:

(A) $\dfrac{x}{y^2} = y + c$ (B) $\dfrac{x}{y} = y^2 + c$

(C) $\dfrac{x^2}{y} = y^2 + C$ (D) $\dfrac{y}{x} = x^2 + c$

Q.7 A normal is drawn at a point P(x, y) of a curve. It meets the x-axis and y-axis in the points A and B respectively such the $\dfrac{1}{OA} + \dfrac{1}{OB} = 1$, where O is the origin. The equation of such a curve passing through (5, 4) denotes:

(A) A line.   (B) A circle.

(C) A parabola.   (D) Pair of straight line.

Q.8 The latus rectum of the conic passing through the origin and having the property that normal at each point (x, y) intersects the x-axis at ((x + 1), 0) is:

(A) 1   (B) 2   (C) 4   (D) None of these

Q.9 The solution of the differential equation

$2x^2y\dfrac{dy}{dx} = \tan(x^2y^2) - 2xy^2$, that given $y(1) = \sqrt{\dfrac{\pi}{2}}$ is

(A) $\sin x^2y^2 = e^{x-1}$ (B) $\sin(x^2y^2) = x$

(C) $\cos x^2y^2 + x = 0$ (D) $\sin(x^2y^2) = e.e^x$

Q.10 A wet porous substance in the open air loses its moisture at a rate proportional to the moisture content. If a sheet hung in the wind loses half its moisture during the first hour, then the time when it would have lost 99.9% of its moisture is: (weather condition remaining same)

(A) More then 100 hours

(B) More than10 hours

(C) Approximately 10 hours

(D) Approximately 9 hours

Q.11 If $y = \dfrac{c}{\log|x|}$ (where c is an arbitrary constant) is the general solution of the differential equation $\dfrac{dy}{dx} = \dfrac{y}{x} + \phi\left(\dfrac{x}{y}\right)$ then the function $\phi\left(\dfrac{x}{y}\right)$ is

(A) $\dfrac{x^2}{y^2}$ (B) $-\dfrac{x^2}{y^2}$ (C) $\dfrac{y^2}{x^2}$ (D) $-\dfrac{y^2}{x^2}$

Q.12 A tank contains 10000 liters of brine in which 10 kg of salt is dissolved initially at t = 0. Fresh brine containing 20 gms of salt per 100 liters keeps running into the tank at the rate of 50 liters per minute. If the mixture is kept stirring uniformly, then the amount of salt (in kgs) present in the tank at the end of 10 minutes, is (Assume that there is no overflow of brine is the bank)

(A) 11.5   (B) 11.15   (C) 10.1   (D) 10.5

Q.13 Which of the following differential equation is not of degree 1?

(A) $x^3 y_2 + (x+x)^2 y_1^2 + e^x y^3 = \sin x$

(B) $y_2^{1/2} + (\sin x) y_1 + xy = x$

(C) $\sqrt{y_1 + y} = x + 1$

(D) None of these

Q.14 If $\dfrac{dy}{dx} = \dfrac{xy + y}{xy + y}$, then the solution of the differential equation:

(A) $y = xe^x + c$ (B) $y = e^x + c$

(C) $y = Ae^x$ (D) $y = x + A$

Q.15 The degree of the differential equation $\left(\dfrac{d^3 y}{dx^3}\right)^{2-3} + 4 - 3\dfrac{d^2 y}{dx^2} + 5\dfrac{dy}{dx} = 0$ is

(A) 1   (B) 2   (C) 3   (D) None of these

Q.16 The differential equation for all parabolas each of which has a latus rectum '4a' and whose axes are parallel to x-axis is:

(A) of degree 2 and order 1   (B) of order 2 and degree 3

(C) $2a\dfrac{d^2 x}{dy^2} = 1$ (D) $2a\dfrac{d^2 x}{dy^2} + \left(\dfrac{dy}{dx}\right)^3 = 0$

Previous Years' Questions

Q.1 The order of the differential equation whose general solution is given by $y = (c_1 + c_2)\cos(x + c_3) - c_4 e^{x+c_5}$. where c_1, c_2, c_3, c_4, c_5 are arbitrary constants, is   *(1998)*

(A) 5   (B) 4   (C) 3   (D) 2

Q.2 A solution of the differential equation $\left(\dfrac{dy}{dx}\right)^2 - x\dfrac{dy}{dx} + y = 0$ is: *(1999)*

(A) y = 2   (B) y = 2x

(C) y = 2x – 4   (D) y = 2x² – 4

Q.3 If y(t) is a solution of $(1 + t)\dfrac{dy}{dt} - ty = 1$ and y(0) = –1, then y(1) is equal to:-   *(2003)*

(A) $-\dfrac{1}{2}$ (B) $e + \dfrac{1}{2}$ (C) $e - \dfrac{1}{2}$ (D) $\dfrac{1}{2}$

Q.4 If y = y(x) and $\dfrac{2 + \sin x}{y + 1}\left(\dfrac{dy}{dx}\right) = -\cos x$, y(0) = 1, then $y\left(\dfrac{\pi}{2}\right)$ equals   *(2004)*

(A) $\dfrac{1}{3}$ (B) $\dfrac{2}{3}$ (C) $-\dfrac{1}{3}$ (D) 1

Q.5 If $\dfrac{dy}{dx} = y(\log y - \log x + 1)$, then the solution of the equation is   *(2005)*

(A) $\log\left(\dfrac{y}{x}\right) = cx$ (B) $\log\left(\dfrac{y}{x}\right) = cy$

(C) $y\log\left(\dfrac{y}{x}\right) = cx$ (D) $x\log\left(\dfrac{y}{x}\right) = cy$

Q.6 A right circular cone with radius R and height H contains a liquid which evaporates at a rate proportional to its surface area in contact with air (proportionality constant = k > 0). Find the time after which the cone is empty.   *(2003)*

Q.7 If length of tangent at any point on the curve y = f(x) intercepted between the point and the x-axis is of length 1. Find the equation of the curve.   *(2005)*

Q.8 If a curve y = f(x) passes through the point (1, –1) and satisfies the differential equation, y(1 + xy) dx = xdy, then $f\left(-\dfrac{1}{2}\right)$ is equal   *(2016)*

(A) $-\dfrac{4}{5}$ (B) $\dfrac{2}{5}$ (C) $\dfrac{4}{5}$ (D) $-\dfrac{2}{5}$

Q.9 Let $y(x)$ be the solution of the differential equal $(x \log x) \dfrac{dy}{dx} + y = 2x\log x$, $(x \geq 1)$. Then $y(e)$ is equal to

(2015)

(A) e   (B) 0   (C) 2   (D) 2e

Q.10 If $y = \sec(\tan^{-1}x)$, then $\dfrac{dy}{dx}$ at $x = 1$ is equal to :

(2013)

(A) $\dfrac{1}{\sqrt{2}}$ (B) $\dfrac{1}{2}$ (C) 1   (D) $\sqrt{2}$

Q.11 $\dfrac{d^2x}{dy^2}$ equals

(2011)

(A) $-\left(\dfrac{d^2y}{dx^2}\right)^{-1}\left(\dfrac{dy}{dx}\right)^{-3}$

(B) $\left(\dfrac{d^2y}{dx^2}\right)\left(\dfrac{dy}{dx}\right)^{-2}$

(C) $-\left(\dfrac{d^2y}{dx^2}\right)\left(\dfrac{dy}{dx}\right)^{-3}$

(D) $\left(\dfrac{d^2y}{dx^2}\right)^{-1}$

Q.12 Solution of the differential equation $\cos x\, dy = y(\sin x - y) dx$, $0 < x < \dfrac{\pi}{2}$

(2010)

(A) $y \sec x = \tan x + c$ (B) $y \tan x = \sec x + c$

(C) $\tan x = (\sec x + c)y$ (D) $\sec x = (\tan x + c)y$

Q.13 The differential equation which represents the family of curves $y = c_1 e^{c_2 x}$, where c_1 and c_2 are arbitrary constants is

(2009)

(A) $y' = y^2$ (B) $y'' = y' y$

(C) $yy'' = y'$ (D) $yy'' = (y')^2$

Q.14 The solution of the differential equation $\dfrac{dy}{dx} = \dfrac{x+y}{x}$ satisfying the condition $y(1) = 1$ is *(2008)*

(A) $y = \log x + x$ (B) $y = x \log x + x^2$

(C) $y = xe^{(x-1)}$ (D) $y = x \log x + x$

JEE Advanced/Boards

Exercise 1

Q.1 (i) Solve $\dfrac{dy}{dx} = \dfrac{x^2 + xy}{x^2 + y^2}$

(ii) $(x^3 - 3xy^2)dx = (y^3 - 3x^2y)dy$

Q.2 Find the equation of a curve such that the projection of its ordinate upon the normal is equal to its abscissa.

Q.3 The light rays emitting from a point source situated at origin when reflected from the mirror of a search light are reflected as beam parallel to the x-axis. Show that the surface is parabolic, by first forming the differential equation and then solve it.

Q.4 The perpendicular from the origin to the tangent at any point on a curve is equal to the abscissa of the point of contact. Find the equation of the curve satisfying the above condition and which passes through (1, 1)

Q.5 Use the substitution $y^2 = a - x$ to reduce the equation $y^3 . \dfrac{dy}{dx} + x + y^2 = 0$ to homogeneous form and hence solve it.

Q.6 Solve: $\left[x \cos\dfrac{y}{x} + y \sin\dfrac{y}{x}\right]y = \left[y \sin\dfrac{y}{x} - x \cos\dfrac{y}{x}\right]x\dfrac{dy}{dx}$

Q.7 Find the curve for which any tangent intersects the y-axis at the point equidistant from the point of tangency and the origin

Q.8 Solve: $(x - y)dy = (x + y + 1)dx$

Q.9 Solve: $\dfrac{dy}{dx} = \dfrac{x + 2y - 3}{2x + y - 3}$

Q.10 Solve: $\dfrac{dy}{dx} = \dfrac{y - x + 1}{y + x + 5}$

Q.11 Solve: $\dfrac{dy}{dx} = \dfrac{x + y + 1}{2x + 2y + 3}$

Q.12 Solve: $\dfrac{dy}{dx} = \dfrac{2(y + 2)^2}{(x + y - 1)^2}$

Q.13 Show that the curve such that the distance between the origin and the tangent at an arbitrary point is equal to the distance between the origin and the normal at the same point, $\sqrt{x^2 + y^2} = ce^{\pm \tan^{-1}\frac{y}{x}}$

Q.14 If solution of differential equation $\frac{dy}{dx} - y = 1 - e^{-x}$ and $y(0) = y_0$ has a finite value. When $x \to \infty$, then find y_0.

Q.15 Let $y = y(t)$ be a solution to the differential equation $y' + 2t\,y = t^2$, then find $\lim\limits_{t \to \infty} \frac{y}{t}$

Q.16 Solve: $\dfrac{dy}{dx} + \dfrac{x}{1+x^2}\,y = \dfrac{1}{2x\left(1+x^2\right)}$

Q.17 Solve: $(1 - x^2)\dfrac{dy}{dx} + 2xy = x(1 - x^2)^{1/2}$

Q.18 (i) Find the curve such that the area of the trapezium formed by the co-ordinate axes, ordinate of an arbitrary point and the tangent at this point equals half the square of its abscissa.

(ii) A curve in the first quadrant is such that the area of the triangle formed in the first quadrant by the x-axis, a tangent to the curve at any of its point P and radius vector of the point P is 2 square units. If the curve passes through (2, 1) find the equation of the curve.

Q.19 Solve: $x(x-1)\dfrac{dy}{dx} - (x-2)y = x^3(2x-1)$

Exercise 2

Single Correct Choice Type

Q.1 A curve passes through the point $\left(1, \dfrac{\pi}{4}\right)$, and its slope at any point is given by $\dfrac{y}{x} - \cos^2\left(\dfrac{y}{x}\right)$, Then the curve has the equation, y is equal to:

(A) $y = x\tan^{-1}\left(\ell n\dfrac{e}{x}\right)$

(B) $y = x\tan^{-1}(\log + 2)$

(C) $y = \dfrac{1}{x}\tan^{-1}\left(\ell n\dfrac{e}{x}\right)$

(D) None of these

Q.2 $y = f(x)$ satisfies the differential equation $\dfrac{dy}{dx} - y = \cos x - \sin x$ with the condition that y is bounded when $x \to +\infty$. The longest interval in which f(x) is increasing in the interval

(A) $\left(\dfrac{\pi}{3}, \dfrac{\pi}{2}\right)$ (B) $\left(0, \dfrac{\pi}{2}\right)$ (C) $\left(\dfrac{\pi}{2}, \dfrac{5\pi}{6}\right)$ (D) $\left(0, \dfrac{\pi}{6}\right)$

Q.3 The real value of m for which the substitution, $y = u^m$ will transform the differential equation $2x^4y\dfrac{dy}{dx} + y^4 = 4x^6$ into a homogeneous equation is:

(A) m = 0 (B) m = 1
(C) m = 3/2 (D) m = 2/3

Q.4 The solution of the differential equation

$x^2\dfrac{dy}{dx}.\cos\dfrac{1}{x} - y\sin\dfrac{1}{x} = -1$, Where $y \to -1$ as $x \to \infty$ is

(A) $y = \sin\dfrac{1}{x} - \cos\dfrac{1}{x}$ (B) $y = \dfrac{x+1}{x\sin\dfrac{1}{x}}$

(C) $y = \cos\dfrac{1}{x} + \sin\dfrac{1}{x}$ (D) $y = \dfrac{x+1}{x\cos\dfrac{1}{x}}$

Q.5 The equation of a curve for which the product of the abscissa of point P and the intercept made by a normal at P on the x-axis equals twice the square of the radius vector of the point P, and passes through (1, 0) is:

(A) $x^2 + y^2 = x^4$ (B) $x^2 + y^2 = 2x^4$
(C) $x^2 + y^2 = 4x^4$ (D) None of these

Q.6 The order and the degree of the differential equation whose general solution is, $y = c(x - c)^2$, are respectively:

(A) 1, 1 (B) 1, 2 (C) 1, 3 (D) 2, 1

Q.7 the degree of the differential equation

$\dfrac{d^2y}{dx^2} + 3\left(\dfrac{dy}{dx}\right)^2 = x\ell n\left(\dfrac{d^2y}{dx^2}\right)$ is

(A) 1 (B) 2 (C) 3 (D) None of these

Q.8 Orthogonal trajectories of family of parabolas $y^2 = 4a(x + a)$ where 'a' is an arbitrary constant is

(A) $ax^2 = 3cy$ (B) $x^2 + y^2 = a^2$

(C) $y = ce^{-\frac{x}{2a}}$ (D) $axy = c^2$

Q.9 If the function $y = e^{4x} + 2e^{-x}$ is a solution of the differential equation $\dfrac{\dfrac{d^3y}{dx^3} - 13\dfrac{dy}{dx}}{y} = K$, then the value of K is:-

(A) 4   (B) 6   (C) 9   (D) 12

Q.10 Solution set of the equation $\dfrac{x\,dy}{dx} - y = x \cdot \dfrac{f(y/x)}{f'(y/x)}$

(A) $f\left(\dfrac{x}{y}\right) = cy$ (B) $f\left(\dfrac{y}{x}\right) = cx$

(C) $f\left(\dfrac{y}{x}\right) = cxy$ (D) None of these

Q.11 $\dfrac{dy}{dx} = \dfrac{x^2 + 2xy + y^2}{x^2 - 2xy + 2y^2}$. Let C_1 and C_2 be two of it's solutions. C_1 passes through, A(1, 2), and line through origin and A meets C_2 at B. Then slope of the tangent to the curve C_2 at B is:

(A) $\dfrac{5}{9}$ (B) $\dfrac{9}{5}$ (C) $-\dfrac{9}{5}$ (D) None of these

Q.12 The solution of the differential equation $\log\left(\dfrac{dy}{dx}\right) = 4x - 2y - 2$, $y = 1$ when $x = 1$ is:-

(A) $2e^{2y+2} = e^{4x} + e^2$

(B) $2e^{2y-2} = e^{4x} + e^4$

(C) $2e^{2y+2} = e^{4x} + e^4$

(D) $3e^{2y+2} = e^{3x} + e^4$

Multiple Correct Choice Type

Q.13 The general solution of the differential equation, $x\left(\dfrac{dy}{dx}\right) = y\ell n\left(\dfrac{y}{x}\right)$ is:

(A) $y = xe^{1-cx}$ (B) $y = xe^{1+cx}$

(C) $y = xe.xe^{cx}$ (D) $y = xe^{cx}$

where c is an arbitrary constant.

Previous Years' Questions

Q.1 Le f(x) be d fferentiable on the interval $(0, \infty)$ such that $f(1) = 1$, and $\lim\limits_{t \to x} \dfrac{t^2 f(x) - x^2 f(t)}{t - x} = 1$ for each $x > 0$. Then f(x) is: *(2007)*

(A) $\dfrac{1}{3x} + \dfrac{2x^2}{3}$ (B) $-\dfrac{1}{3x} + \dfrac{4x^2}{3}$

(C) $-\dfrac{1}{x} + \dfrac{2}{x^2}$ (D) $\dfrac{1}{x}$

Q.2 The differential equation $\dfrac{dy}{dx} = \dfrac{\sqrt{1-y^2}}{y}$ determines a family of circles with *(2007)*

(A) Variable radi and a fixed center at (0, 1)

(B) Variable radii and fixed center at (0, –1)

(C) Fixed radius of 1 and variable center along the x-axis

(D) Fixed radius of 1 and variable center a long the y-axis

Q.3 Let $y = f(x)$ be a curve passing through (1, 1) such that the triangle formed by the coordinates axes and the tangent at any point of the curve lies in the first quadrant and has area 2 unit, from the differential equation and determine all such possible curves. *(1995)*

Q.4 A and B are two separate reservoir of water. Capacity of reservoir A is double the capacity of reservoir B. Both the reservoirs are filled completely with water, their inlets are closed and then the water is released simultaneously from both the reservoirs. The rate of flow of water out of each reservoir at any instant of time is proportional to the quantity of water in the reservoir at the time. One hour after the water is released the quantity of water in reservoir A is $1\dfrac{1}{2}$ times the quantity of water in reservoir B. After how many hours do both the reservoirs have the same quantity of water? *(1997)*

Q.5 Let u(x) and v(x) satisfy the differential equation $\dfrac{du}{dx} + p(x)u = f(x)$ and $\dfrac{dv}{dx} + p(x) = g(x)$, where P(x), f(x) and g(x) are continuous functions. If $u(x_1) > v(x_1)$ for some x_1 and $f(x) > g(x)$ for all $x > x_1$, prove that any point (x, y) where $x > x_1$ does not satisfy the equations $y = u(x)$ and $y = v(x)$ *(1997)*

Q.6 A curve passing through the point (1, 1) has the property that the perpendicular distance of the origin from the normal at any point P of the curve is equal to the distance of P from the x-axis. Determine the equation of the curve. *(1999)*

Q.7 A country has food deficit of 10%. Its population grows continuously at a rate of 3% per year. Its annual food production every year is 4% more than that of the last year. Assuming that the average food requirement per person remains constant, prove that the country will become self-sufficient in food after n years, where n is the smallest integer bigger than or equal to

$$\frac{\ln 10 - \ln 9}{\ln(1.04) - (0.03)}$$ **(2000)**

Q.8 Let f: R → R be a continuous function, which satisfies $f(x) = \int_0^x f(t)dt$. Then the value of f(log 5) is .... **(2009)**

Q.9 If the function $f(x) = x^3 + e^{\frac{x}{2}}$ and $g(x) = f^{-1}(x)$, then the values of g'(1) is .... **(2009)**

Q.10 Let y'(x) + y(x)g'(x) = g(x)g'(x), y(0) = 0, x ∈ R, where f'(x) denotes $\frac{df(x)}{dx}$ and g(x) is a given non-constant differentiable function on R with g(0) =g(2) = 0. Then the value of y(2) is .... **(2011)**

Q.11 A solution curve of the differential equation $(x^2 + xy + 4x + 2y + 4)\frac{dy}{dx} - y^2 = 0$, x > 0, passes through the point (1, 3). Then the solution curve dy dx **(2016)**

(A) Intersects y = x + 2 exactly at one point.

(B) Intersects y = x + 2 exactly at two points

(C) Intersects y = (x + 2)²

(D) Does NOT intersect y = (x + 3)²

Q.12 Consider the family of all circles whose centers lie on the straight line y = x. If this family of circles is represented by the differential equation Py" + Qy' + 1 = 0, where P, Q are functions of x, y and y' $\left(\text{here } y' = \frac{dy}{dx}, y" = \frac{d^2y}{dx^2}\right)$, then which of the following statements is (are) true ? **(2015)**

(A) P = y + x

(B) P = y − x

(C) Q = 1 + y_1 + y_1^2

(D) P − Q = x + y − y' − (y')²

Q.13 The function y = f (x) is the solution of the differential equation $\frac{dy}{dx} + \frac{xy}{x^2 - 1} = \frac{x^4 + 2x}{\sqrt{1 - x^2}}$ in (-1, 1) satisfying f(0) = 0. Then $\int_{-\frac{\sqrt{3}}{2}}^{\frac{\sqrt{3}}{2}} f(x)dx$ is **(2014)**

(A) $\frac{\pi}{3} - \frac{\sqrt{3}}{2}$

(B) $\frac{\pi}{3} - \frac{\sqrt{3}}{4}$

(C) $\frac{\pi}{6} - \frac{\sqrt{3}}{4}$

(D) $\frac{\pi}{6} - \frac{\sqrt{3}}{2}$

Q.14 A curve passes through the point $\left(1, \frac{\pi}{6}\right)$, Let the slope of the curve at each point (x, y) be $\frac{y}{x} + \sec\left(\frac{y}{x}\right)$, x > 0, x > 0. Then the equation of the curve is **(2013)**

(A) $\sin\left(\frac{y}{x}\right) = \log x + \frac{1}{2}$

(B) $\text{cosec}\left(\frac{y}{x}\right) = \log x + 2$

(C) $\sec\left(\frac{2y}{x}\right) = \log x + 2$

(D) $\cos\left(\frac{2y}{x}\right) = \log x + \frac{1}{2}$

Q.15 If y(x) satisfies the differential equation y' - ytanx = 2x secx and y(0) = 0, then **(2012)**

(A) $y\left(\frac{\pi}{4}\right) = \frac{\pi^2}{8\sqrt{2}}$

(B) $y'\left(\frac{\pi}{4}\right) = \frac{\pi^2}{18}$

(C) $y\left(\frac{\pi}{3}\right) = \frac{\pi^2}{9}$

(D) $y'\left(\frac{\pi}{3}\right) = \frac{4\pi}{3} + \frac{2\pi^2}{3\sqrt{2}}$

Q.16 Let f: [1, ∞) → [2, ∞) be a differentiable function such that f(1)= 2. If $6\int_1^x f(1)dt = 3xf(x) - x^3$ for all x ≥ 1, then the value of f(2) is **(2011)**

Q.17 Let f be a real-valued differentiable function on R (the set of all real numbers) such that f(1) = 1. If the y-intercept of the tangent at any point P(x, y) on the curve y = f(x) is equal to the cube of the abscissa of P, then the value of f(−3) is equal to **(2010)**

Q.18 Interval contained in the domain of definition of non-zero solutions of the differential equation $(x - 3)^2 y' + y = 0$ **(2009)**

Q.19 Let a solution $y = y(x)$ of the differential equation

$$x\sqrt{x^2 - 1}\, dy - y\sqrt{y^2 - 1}\, dx = 0 \quad \text{satisfy } y(2) = \frac{2}{\sqrt{3}}$$

Statement-I: $y(x) = \sec\ y(x) = \sec\left(\sec^{-1} x - \dfrac{\pi}{6}\right)$ and

Statement-II: $y(x)$ is given by $\dfrac{1}{y} = \dfrac{2\sqrt{3}}{x} - \sqrt{1 - \dfrac{1}{x^2}}$

(2008)

(A) Statement-I is True, statement-II is True; statement-II is a correct explanation for statement-I

(B) Statement-I is True, statement-II is True; statement-II is NOT a correct explanation for statement-I.

(C) Statement-I is True, statement-II is False

(D) Statement-I is False, statement-II is True

Important Questions

JEE Main/Boards

Exercise 1

| Q.9 | Q.14 | Q.20 | Q.21 |
| Q.26 |

Exercise 2

| Q.3 | Q.4 | Q.8 | Q.10 |
| Q.11 | Q.16 | Q.14 |

Previous Years' Questions

| Q.3 | Q.5 | Q.8 |

JEE Advanced/Boards

Exercise 1

| Q.3 | Q.6 | Q.14 | Q.18 |
| Q.19 |

Exercise 2

| Q.1 | Q.4 | Q.5 | Q.9 |
| Q.11 | Q.13 |

Previous Years' Questions

| Q.2 | Q.4 | Q.7 |
| Q.10 |

JEE Main/Boards

Exercise 1

Q.1 Order = 1; Degree = 1

Q.2 $\sin^{-1}y + \sin^{-1}x = c$

Q.3 Order = 1, degree is not defined

Q.4 Separate the variables after factorizing

Q.5 $e^{\tan^{-1}y}$

Q.6 $r = \sin\theta + c$

Q.7 $(I.F.)y = \int (I.F.).Q$

Q.9 $y'' = -4A\cos 2x - 4B\sin 2x = -4y$

Q.12 $y\dfrac{d^2y}{dx^2} = \left(\dfrac{dy}{dx}\right)^2$

Q.13 $e^y = e^x + \dfrac{x^4}{4} + c$

Q.14 $\tan y - \cot x = c$

Q.15 $\sqrt{x^2+1} + \sqrt{y^2+1} + \dfrac{1}{2}\log\left(\dfrac{\sqrt{1+x^2}-1}{\sqrt{1+x^2}+1}\right) + c = 0$

Q.16 $\log|y| = \dfrac{x^3}{3y^3} + c$

Q.17 $y = \dfrac{(\cos x)^6}{6} - \dfrac{(\cos x)^4}{4} + e^x(x+1) + c$

Q.18 $-[\log y - \log(y-1)] = -\dfrac{1}{2}\log(1-x^2) + c$

Q.19 $y - 1 = xy$

Q.20 $x = \dfrac{y^3}{4} + \dfrac{2}{y}$

Q.21 $\dfrac{y}{y - \dfrac{1}{a}} = \left(\dfrac{x}{\left(x+\dfrac{1}{a}\right)}\right)\left(\dfrac{a^2+1}{a^2-1}\right)$

Q.22 $y\log|x| = \dfrac{-2\log|x|}{x} - \dfrac{2}{x} + c$

Q.23 $y = (x+1)\log|x+1| - x + 4$

Q.24 $\dfrac{1}{y} = 2x + \dfrac{2}{\pi}$

Q.25 $y = \sin x$

Q.26 $\dfrac{y}{x} = \log|y| + 1$

Q.27 $\tan^{-1}(\sin x) + \tan^{1}y = \dfrac{\pi}{4}$

Q.28 $y - 5\log|y + 5| = \log|x| - 6\log 5$

Q.29 $y = \dfrac{1}{2}\log(x^2 + 4x + 9) - \log 3$

Exercise 2

Single Correct Choice Type

Q.1 A	**Q.2** A	**Q.3** C	**Q.4** B	**Q.5** D	**Q.6** B
Q.7 B	**Q.8** B	**Q.9** A	**Q.10** C	**Q.11** D	**Q.12** C
Q.13 D	**Q.14** C	**Q.15** B	**Q.16** C		

Previous Years' Questions

Q.1 C	**Q.2** C	**Q.3** A	**Q.4** A	**Q.5** A	**Q.6** $T = \dfrac{dy}{dx}$

Q.7 $x^2\dfrac{d^2y}{dx^2} + x\dfrac{dy}{dx} = \pm x + c$

Q.8 C	**Q.9** C	**Q.10** A	**Q.11** C
Q.12 D	**Q.13** D	**Q.14** D	

JEE Advanced/Boards

Exercise 1

Q.1 (i) $c(x-y)^{2/3}(x^2+xy+y^2)^{1/6} = \exp\left[\dfrac{1}{\sqrt{3}}\tan^{-1}\dfrac{x+2y}{x\sqrt{3}}\right]$ where $\exp x = e^x$ (ii) $(x^2+y^2)^2 = (x^2-y^2)c$

Q.2 $\dfrac{y^2 \pm y\sqrt{y^2-x^2}}{x^2} = \log\left|\left(y\pm\sqrt{y^2-x^2}\right)\cdot\dfrac{c^2}{x^3}\right|$, where same sign has to be taken

Q.4 $x^2+y^2-2x=0$ **Q.5** $\dfrac{1}{2}\log|x^2+a^2| - \tan^{-1}\left(\dfrac{a}{x}\right) = c$, where $a = x+y^2$

Q.6 $xy\cos\dfrac{y}{x} = c$ **Q.7** $x^2+y^2 = cx$

Q.8 $e^{\,c\tan^{-1}\left(\frac{y+\frac{1}{2}}{x+\frac{1}{2}}\right)}$ **Q.9** $(x+y-2) = c(y-x)^3$

Q.10 $\tan^{-1}\dfrac{y+3}{x+2} + \log c\sqrt{(y+3)^2+(x+2)^2} = 0$ **Q.11** $x+y+\dfrac{4}{3} = ce^{3(x-2y)}$

Q. 12 $c = e^{-2\tan^{-1}\frac{y+2}{x-3}} = (y+2)$ **Q.14** $\dfrac{1}{2}$ **Q.15** $\dfrac{1}{2}$

Q.16 $y\sqrt{1+x^2} = c+\dfrac{1}{2}\log\left[\tan\dfrac{1}{2}\arctan x\right]$ another form is $y\sqrt{1+x^2} = c+\dfrac{1}{2}\log\dfrac{\sqrt{1+x^2}-1}{x}$

Q.17 $y = c(1-x^2) + \sqrt{1-x^2}$ **Q.18** (i) $y = cx^2 - x$ (ii) $xy = 2$

Q.19 $y(x-1) = x^2(x^2-x+c)$

Exercise 2

Single Correct Choice Type

Q.1 A	**Q.2** B	**Q.3** C	**Q.4** A	**Q.5** A	**Q.6** C
Q.7 D	**Q.8** C	**Q.9** D	**Q.10** B	**Q.11** B	**Q.12** C

Multiple Correct Choice Type

Q.13 B, C

Previous Years' Questions

Q.1 A	**Q.2** C	**Q.3** $x+y=2$, $xy=1$	**Q.4** $\log_{3/4}\left(\dfrac{1}{2}\right)$	**Q.6** $x^2+y^2=2x$
Q.8 0	**Q.9** 2	**Q.10** 0	**Q.11** A, C	**Q.12** B, C **Q.13** B
Q.14 A	**Q.15** A, D	**Q.16** 6	**Q.17** 9	**Q.19** C

JEE Main/Boards

Exercise 1

Sol 1: $x = \cos\left(\dfrac{dy}{dx}\right) \Rightarrow \dfrac{dy}{dx} = \cos^{-1} x$

$\therefore$ Degree = 1, order = 1

Sol 2: $\dfrac{dy}{dx} = -\sqrt{\dfrac{1-y^2}{1-x^2}}$

$\Rightarrow \displaystyle\int -\dfrac{dy}{\sqrt{1-y^2}} = \int \dfrac{dx}{\sqrt{1-x^2}}$

$\Rightarrow -\sin^{-1}y = \sin^{-1}x + c$ or $\sin^{-1}x + \sin^{-1}y = c$

Sol 3: $\dfrac{dy}{dx} + \sin\left(\dfrac{dy}{dx}\right) = 0$

Highest order derivative = 1

Degree is not defined as differential coefficient is not free from radical and fraction.

Sol 4: $\dfrac{dy}{dx} = 1 + x + y(1 + x)$ or $\dfrac{dy}{dx} = (1 + x)(1 + y)$

(Separation of variables method)

$\Rightarrow \displaystyle\int \dfrac{dy}{(1+y)} = \int (1+x)dx$

$\Rightarrow \log(1 + y) = x + \dfrac{x^2}{2} + c$

Sol 5: $(1 + y^2)dx = (\tan^{-1}y - x)dy$

$\dfrac{dx}{dy} + \dfrac{1}{(1+y^2)}x = \dfrac{\tan^{-1} y}{1+y^2}$

$\therefore$ Integrating factor $= e^{\int\left(\frac{1}{1+y^2}\right)dy} = e^{\tan^{-1} y}$

Sol 6: $\dfrac{dr}{d\theta} = \cos\theta$

$\displaystyle\int dr = \int \cos\theta\, d\theta$

$r = \sin\theta + c$

Sol 7: $\dfrac{dy}{dx} + 2y = 6e^x$

This is a linear equation

$\therefore$ Integrating factor $= e^{\int 2\, dx} = e^{2x}$

$\therefore e^{2x}\dfrac{dy}{dx} + 2e^{2x}y = 6e^{3x}$

$\Rightarrow \displaystyle\int d(e^{2x}y) = \int 6e^{3x}dx \Rightarrow e^{2x}y = \dfrac{6}{3}e^{3x} + c$

$\therefore y = 2e^x + ce^{-2x}$

Sol 8: Ellipse with their axis coincide with x-axis

$\dfrac{x^2}{a^2} + \dfrac{y^2}{b^2} = 1 \Rightarrow \dfrac{2x}{a^2} + \dfrac{2yy'}{b^2} = 0$

$\therefore y' = -\dfrac{b^2}{a^2}\dfrac{x}{y} \Rightarrow y'' = \dfrac{-b^2}{a^2}\left[\dfrac{y - xy'}{y^2}\right]$

$\Rightarrow yy'' = \dfrac{-b^2}{a^2}\left[1 - \dfrac{x}{y}y'\right] = \dfrac{-b^2}{a^2} + \left(\dfrac{b^2}{a^2}\dfrac{x}{y}\right)y'$

$\therefore yy'' + (y')^2 = \dfrac{-b^2}{a^2}$

For parabola

Equation will be $y^2 = 4ax$

$\Rightarrow 2yy' = 4a$ or $y' = \dfrac{2a}{y}$

or $2a = yy'$

$\Rightarrow (y')^2 + yy'' = 0$

Sol 9: $y = A\cos 2x + B\sin 2x$

$y' = -2A\sin 2x + 2B\cos 2x$

$y'' = -4A\cos 2x - 4B\sin 2x = -4y$

Sol 10: $y = Ax + \dfrac{B}{x}$

$y' = A - \dfrac{B}{x^2}$

$y'' = \dfrac{2B}{x^3}$

$\therefore x^2y'' + xy' - y$

$$= \frac{x^2(2B)}{x^3} + x\left(A - \frac{B}{x^2}\right) - \left(Ax + \frac{B}{x}\right)$$

$$= \frac{2B}{x} + Ax - \frac{B}{x} - Ax - \frac{B}{x} = 0$$

Hence proved.

Sol 11: $1 + 8y^2\tan x = cy^2$

$y^2(c - 8\tan x) = 1$

$\therefore 2y(c - 8\tan x)\dfrac{dy}{dx} + y^2(-8\sec^2 x) = 0$

$2y\dfrac{1}{y^2}\dfrac{dy}{dx} = 8y^2\sec^2 x$

$\therefore \dfrac{dy}{dx} = 4y^3\sec^2 x \quad$ or $\quad \cos^2 x\dfrac{dy}{dx} = 4y^3$

Sol 12: $y = Ae^{Bx}$

$\Rightarrow y' = ABe^{Bx}$

$\Rightarrow y'' = AB^2e^{Bx}$

$\Rightarrow yy'' = A^2B^2e^{2Bx} = (y')^2$

$\therefore y\dfrac{d^2y}{dx^2} = \left(\dfrac{dy}{dx}\right)^2$

Sol 13: $\dfrac{dy}{dx} = e^{x-y} + x^3e^{-y}$

$\Rightarrow \int e^y dy = \int(e^x + x^3)\ dx$

$\Rightarrow e^y = e^x + \dfrac{x^4}{4} + c$

Sol 14: $\dfrac{dy}{dx} + \dfrac{1 + \cos 2y}{1 - \cos 2x} = 0$

$\Rightarrow \int\dfrac{dy}{1 + \cos 2y} = \int\dfrac{dx}{\cos 2x - 1}$

$\Rightarrow \int\dfrac{dy}{2\cos^2 y} = -\int\dfrac{dx}{2\sin^2 x}$

$\Rightarrow \int\sec^2 y\ dy = -\int\cosec^2 x\ dx$

$\Rightarrow \tan y = \cot x + c$

$\therefore \tan y - \cot x = c$

Sol 15: $\sqrt{1 + x^2 + y^2 + x^2y^2} + xy\dfrac{dy}{dx} = 0$

$\sqrt{(1+x^2)(1+y^2)} + xy\dfrac{dy}{dx} = 0$

$\Rightarrow xy\dfrac{dy}{dx} = -\sqrt{(1+x^2)(1+y^2)}$

$\Rightarrow \int\dfrac{y}{\sqrt{1+y^2}}dy = \int\dfrac{\sqrt{(1+x^2)}}{x}dx$

$\Rightarrow$ take $1 + y^2 = t$, differentiating both sides

$2ydy = dt$

$I_1 = \dfrac{1}{2}\int\dfrac{dt}{\sqrt{t}} = \dfrac{1}{2}2t^{1/2} = \sqrt{t} = \sqrt{1+y^2}$

$\int\dfrac{\sqrt{1+x^2}}{x}dx = \int\dfrac{(1+x^2)}{x\sqrt{1+x^2}}dx$

$= \int\dfrac{1}{x\sqrt{1+x^2}}dx + \int\dfrac{x}{\sqrt{1+x^2}}dx$

$I_3 = \int\dfrac{x}{\sqrt{1+x^2}}dx = \sqrt{1+x^2}$

$I_2 = \int\dfrac{1}{x\sqrt{1+x^2}}dx$

Put $x = \tan\theta$; $dx = \sec^2\theta\ d\theta$

$= \int\dfrac{\sec^2\theta d\theta}{\tan\theta\sec\theta} = \int\cosec\theta d\theta = \log|\cosec\theta - \cot\theta|$

$= \log\left|\dfrac{1 - \cos\theta}{\sin\theta}\right| = \dfrac{1}{2}\log\left(\dfrac{1 - \cos\theta}{\sin\theta}\right)^2$

$= \dfrac{1}{2}\log\dfrac{(1 - \cos\theta)^2}{(1 - \cos^2\theta)} = \dfrac{1}{2}\log\left(\dfrac{1 - \cos\theta}{1 + \cos\theta}\right)$

or $I_2 = \dfrac{1}{2}\log\left(\dfrac{\sec\theta - 1}{\sec\theta + 1}\right) = \dfrac{1}{2}\log\left(\dfrac{\sqrt{1+x^2} - 1}{\sqrt{1+x^2} + 1}\right)$

$\therefore \sqrt{y^2 + 1} = -\left[\left(\sqrt{x^2 + 1}\right) + \dfrac{1}{2}\log\left(\dfrac{\sqrt{1+x^2} - 1}{\sqrt{1+x^2} + 1}\right) + c\right]$

or $\sqrt{x^2 + 1} + \sqrt{y^2 + 1} - \dfrac{1}{2}\log\left(\dfrac{\sqrt{1+x^2} - 1}{\sqrt{1+x^2} + 1}\right) + c = 0$

Sol 16: $\dfrac{dy}{dx} = \dfrac{x^2 y}{x^3 + y^3}$

$$\dfrac{dy}{dx} = \left(\dfrac{\left(\dfrac{x}{y}\right)^2}{\left(\dfrac{x}{y}\right)^3 + 1} \right)$$

Put $x = vy \Rightarrow 1 = v\dfrac{dy}{dx} + y\dfrac{dv}{dx}$

$$\dfrac{dy}{dx} = \dfrac{1}{v}\left(1 - \dfrac{x}{v}\dfrac{dv}{dx}\right) = \dfrac{v^2}{v^3 + 1}$$

$$\dfrac{-x}{v}\dfrac{dv}{dx} = \dfrac{v^3}{v^3 + 1} - 1 = \dfrac{-1}{v^3 + 1}$$

$$\Rightarrow \int \dfrac{(v^3 + 1)}{v} dv = \int \dfrac{dx}{x} \Rightarrow \dfrac{v^3}{3} + \log v = \log x + c$$

$$\Rightarrow \dfrac{1}{3}\left(\dfrac{x}{y}\right)^3 + \log x - \log y = \log x + c$$

$$\therefore \log y + c = \dfrac{x^3}{3y^3} \Rightarrow y = Ce^{\frac{x^3}{3y^3}}$$

Sol 17: $\dfrac{dy}{dx} = \sin^3 x \cos^3 x + xe^x$

$dy = \int (\sin^3 x \cos^3 x + xe^x) dx$

$= \int \sin x (1 - \cos^2 x) \cos^3 x\, dx + xe^x - e^x$

Put $\cos x = t \Rightarrow -\sin x\, dx = dt$

$= -\int (1 - t^2)(t^3) dt + x(e^x) - e^x$

$= \int (t^5 - t^3) dt + e^x(x - 1)$

$= \dfrac{t^6}{6} - \dfrac{t^4}{4} + e^x(x - 1) + c$

$= \dfrac{(\cos x)^6}{6} - \dfrac{(\cos x)^4}{4} + e^x(x + 1) + c$

Sol 18: $(1 - x^2)dy + xy\,dx = xy^2 dx$

$(1 - x^2)dy = (xy^2 - xy)dx = x(y^2 - y)dx$

$\int \dfrac{dy}{(y^2 - y)} = \int \dfrac{x}{1 - x^2} dx$

$-\int \left[\dfrac{1}{y} - \dfrac{1}{(y-1)}\right] dy = -\dfrac{1}{2}\int \dfrac{-2x}{1 - x^2} dx$

$\Rightarrow -[\log y - \log(y - 1)] = -\dfrac{1}{2}\log(1 - x^2) + c$

Sol 19: $x\dfrac{dy}{dx} + y = y^2$

$\Rightarrow x\dfrac{dy}{dx} = (y^2 - y)$

$\Rightarrow \int \dfrac{dy}{y^2 - y} = \int \dfrac{dx}{x} \Rightarrow -\int \left[\dfrac{1}{y} - \dfrac{1}{(y-1)}\right] dy = \log x + c$

$\Rightarrow \log(y - 1) - \log(y) = \log x + \log c$

$\Rightarrow (x - 1) = xy$

Sol 20: $(x - y^3)dy + y\,dx = 0$

$\Rightarrow x\,dy + y\,dx = y^3 dy$

$\Rightarrow \int d(xy) = \int y^3 dy$

$\Rightarrow xy = \dfrac{y^4}{4} + c \ \text{ or } x = \dfrac{y^3}{4} + \dfrac{c}{y}$

Sol 21: $y - x\dfrac{dy}{dx} = a\left(y^2 + x^2\dfrac{dy}{dx}\right)$

$\Rightarrow \dfrac{dy}{dx}(ax^2 + x) = y - ay^2$

$\therefore \int \dfrac{dy}{y - ay^2} = \int \dfrac{dx}{ax^2 + x}$

$\Rightarrow \int \dfrac{1}{y(1 - ay)} dy = \int \dfrac{1}{x(ax + 1)} dx$

$\Rightarrow \int \left(\dfrac{1}{y} + \dfrac{a}{1 - ay}\right) dy = \int \left(\dfrac{1}{x} - \dfrac{a}{ax + 1}\right) dx$

$\Rightarrow \int \left[\dfrac{1}{y} - \dfrac{1}{\left(y - \dfrac{1}{a}\right)}\right] dy = \int \left[\dfrac{1}{x} - \dfrac{1}{\left(x + \dfrac{1}{a}\right)}\right] dx$

$\log y - \log\left(y - \dfrac{1}{a}\right) = \log x - \log\left(x + \dfrac{1}{a}\right) + \log c$

$\log a - \log\left(a - \dfrac{1}{a}\right) = \log a \log.\left(a + \dfrac{1}{a}\right) + \log c$

$\log c = \log \dfrac{a^2 + 1}{a^2 - 1}$

$\Rightarrow \dfrac{y}{y - \dfrac{1}{a}} = \left(\dfrac{x}{\left(x + \dfrac{1}{a}\right)}\right)\left(\dfrac{a^2 + 1}{a^2 - 1}\right)$

Sol 22: $x\log x \dfrac{dy}{dx} + y = \dfrac{2}{x}\log x$

$\Rightarrow \dfrac{dy}{dx} + \dfrac{1}{x\log x}y = \dfrac{2}{x^2}$

I. F. $= e^{\int \frac{1}{x\log x}dx} = e^{\ell n \ell nx} = \log x$

$\therefore (\log x)y = \int \dfrac{2}{x^2}\log x\, dx$

Put $\log x = t \Rightarrow \dfrac{1}{x}dx = dt$

$\therefore (\log x)y = 2\int e^{-t}t\,dt$

$u = -t$

$dt = -du = +2\int e^u u\,du = +2[e^u(u-1)] + c$

$\therefore (\log x)y = +2[e^{-\log x}(-\log x - 1)] + c$

$\therefore (\log x)y = -\dfrac{2}{x}(\log|x|+1) + c$

Sol 23: $e^{dy/dx} = x + 1,\ y(0) = 4$

$\dfrac{dy}{dx} = \log(x+1) \Rightarrow \int dy = \int \ell n(x+1)dx$

$\int dy = \log(x+1)\int 1dx - \int\left(\dfrac{d\ell n(x+1)}{dx}\int 1.dx\right)dx$

$y = x\log(x+1) - \int \dfrac{x}{x+1}dx + c$

$= x\log(x+1) - x + \log(x+1) + c$

Or $y = (x+1)\log(x+1) - x + c$

$y(0) = 4$

$\therefore 4 = c$

$\therefore y = (x+1)\log(x+1) - x + 4$

Sol 24: $y' + 2y^2 = 0$

$\dfrac{dy}{dx} = -2y^2$

$\therefore \int -\dfrac{1}{2y^2}dy = \int dx$

$\Rightarrow -\dfrac{1}{2}\left(-\dfrac{1}{y}\right) = x + c$ or $\dfrac{1}{2y} = x + c$

$y(0) = \dfrac{\pi}{2}$

$\therefore c = \dfrac{1}{2\times\dfrac{\pi}{2}} = \dfrac{1}{\pi}$ $\therefore y = \dfrac{1}{2\left(x + \dfrac{1}{\pi}\right)}$

or $\dfrac{1}{y} = 2x + \dfrac{2}{\pi}$

Sol 25: $x\dfrac{dy}{dx} + y = x\cos x + \sin x$

Or $\dfrac{dy}{dx} + \left(\dfrac{1}{x}\right)y = \cos x + \dfrac{\sin x}{x}$

I. F. $= e^{\int \frac{1}{x}dx} = x$

$\therefore \dfrac{d(xy)}{dx} = \left(\cos x + \dfrac{\sin x}{x}\right)x$

$\therefore \int d(xy) = \int(x\cos x + \sin x)dx$

$xy = \int x\cos x\,dx + \int \sin x\,dx$

$= x\int\cos x\,dx - \int\left(\dfrac{dx}{dx}\int\cos x\,dx\right)dx + \int\sin x\,dx$

$= x\sin x - \int\sin x + \int\sin x\,dx + c$

$\therefore xy = x\sin x + c$

$y\left(\dfrac{\pi}{2}\right) = 1$

$\dfrac{\pi}{2} = \dfrac{\pi}{2} + c \quad \Rightarrow c = 0$

$\therefore y = \sin x + 0$

$\therefore y = \sin x$

Sol 26: $y^2 + x^2\dfrac{dy}{dx} = xy\dfrac{dy}{dx}$

$\therefore \dfrac{dy}{dx} = \dfrac{y^2}{xy - x^2} = \dfrac{\left(\dfrac{y}{x}\right)^2}{\left(\dfrac{y}{x}\right) - 1}$

Let $y = vx$

$\therefore \dfrac{dy}{dx} = v + x\dfrac{dv}{dx}$

$\therefore v + x\dfrac{dv}{dx} = \dfrac{v^2}{v-1}$ or $x\dfrac{dv}{dx} = \dfrac{v^2}{v-1} - v = \dfrac{v}{v-1}$

$\therefore \int\dfrac{v-1}{v}dv = \int\dfrac{dx}{x}$

$\Rightarrow v - \log v = \log x + c$ or $\dfrac{y}{x} - \log\left(\dfrac{y}{x}\right) = \log x + c$

when x = 1, y = 1

1 – log1 = log1 + c or c = 1

$\therefore \dfrac{y}{x} = \log|y| + 1$

or   y = xlog(ey)

Sol 27: $(1 + \sin^2 x)dy + (1 + y^2)\cos x\, dx = 0$

$\Rightarrow \int \dfrac{dy}{1+y^2} = -\int \dfrac{\cos x}{1+\sin^2 x}dx$

$\tan^{-1}y = -\int \dfrac{dt}{1+t^2}$ (Putting sinx = t)

$\therefore \cos x\, dx = dt$

$\therefore \tan^{-1}y = -\tan^{-1}\sin x + c$

At $x = \dfrac{\pi}{4}$, y = 0

$\therefore c = \tan^{-1}1 = \dfrac{\pi}{4}$

$\therefore \tan^{-1}\sin x + \tan^{-1}y = \dfrac{\pi}{4}$

Sol 28: xydy = (y + 5)dx

$\Rightarrow \left(\dfrac{y}{y+5}\right)dy = \dfrac{1}{x}dx$

$\Rightarrow \int\left(1 - \dfrac{5}{y+5}\right)dy = \log x + c$

$\Rightarrow y - 5\log(y + 5) = \log x + c$

since for x = 5, y = 0

$\Rightarrow 0 - 5\log 5 = \log 5 + c \Rightarrow c = -6\log 5$

$\therefore y = 5\log(y + 5) + \log x - 6\log 5$

$y = 5\log\left(\dfrac{y+5}{5}\right) + \log\left(\dfrac{x}{5}\right)$

or y – 5 log|y + 5| = log|x| – 6 log 5

Sol 29: $(x + 2)dx = (x^2 + 4x + 9)dy$

$\therefore dy = \int \dfrac{(x+2)}{(x^2+4x+9)}dx \Rightarrow y = \dfrac{1}{2}\int \dfrac{2x+4}{(x^2+4x+9)}dx$

Put $x^2 + 4x + 9 = t$

$\therefore (2x + 4)dx = dt$

$\therefore y = \dfrac{1}{2}\int \dfrac{dt}{t} = \dfrac{1}{2}\log t + c$ or $y = \dfrac{1}{2}\log(x^2 + 4x + 9) + c$

for x = 0, y = 0

$\therefore c = -\dfrac{1}{2}\log 9$

$\therefore y = \dfrac{1}{2}\log \dfrac{(x^2+4x+9)}{9}$

Or $y = \dfrac{1}{2}\log(x^2 + 4x + 9) - \log 3$

Exercise 2

Single Correct Choice Type

Sol 1: (A) $y' + y\,\phi'(x) - \phi(x)\,\phi'(x) = 0$

This is a linear equation

I. F. = $e^{\int \phi'(x)dx} = e^{\phi(x)}$

$\therefore \int d(e^{\phi(x)}.y) = \int e^{\phi(x)}\phi(x)\phi'(x)dx$

$\therefore e^{\phi(x)}y = te^t - \int e^t t\, dx$

Let $\phi(x) = t$

$\phi'(x)dx = dt$

$e^{\phi(x)}y = te^t - \int e^t dt + c = te^t - e^t + c = (\phi(x) - 1)e^{\phi(x)} + c$

$\therefore y = (\phi(x) - 1) + ce^{-\phi(x)}$

or $y = ce^{-\phi(x)} + \phi(x) - 1$

Sol 2: (A) $y\dfrac{dy}{dx} + x = c \Rightarrow \int y\, dy = \int(c - x)dx$

$\dfrac{y^2}{2} = cx - \dfrac{x^2}{2} \Rightarrow \dfrac{x^2}{2} - cx + \dfrac{y^2}{2} = 0$

$\Rightarrow x^2 - 2cx + y^2 = 0$ or $(x - c)^2 + y^2 = c^2$

$\therefore$ Circle with centre at (c, 0) and radius c.

Sol 3: (C) Parabola equation $y^2 = 4ax$

$\therefore 2y\dfrac{dy}{dx} = 4a$

or $\left(\dfrac{dy}{dx}\right)^2 + y\dfrac{d^2y}{dx^2} = 0$

Sol 4: (B) $xy\dfrac{dy}{dx} = \dfrac{(1+y^2)}{(1+x^2)}(1 + x + x^2)$

$\Rightarrow \int \dfrac{y}{1+y^2}dy = \int \dfrac{(1+x^2)+x}{(1+x^2)x}dx$

$$\Rightarrow \frac{1}{2}\log(1+y^2) = \int \frac{1}{x}dx + \int \frac{1}{1+x^2}dx$$

$$\Rightarrow \frac{1}{2}\log(1+y^2) = \log x + \tan^{-1}x + c$$

For $x = 1$, $y = 0$

$$\therefore c = -\tan^{-1}1 = -\frac{\pi}{4}$$

or $\log(1+y^2) = 2\log x + 2\tan^{-1}x - \frac{\pi}{2}$

or $\log \frac{(1+y^2)}{x^2} = 2\tan^{-1}x - \frac{\pi}{2}$

Sol 5: (D) $y = 1 + \cos x$

$y = 1 + \sin x$

using option we can see that

$y = 1 + \cos x + \sin x$ is satisfying the equation

$$\frac{dy}{dx} = \cos x - \sin x$$

$$\frac{d^2y}{dx^2} = -\sin x - \cos x \therefore \frac{d^2y}{dx^2} + y = 1$$

Sol 6: (B) $(x + 2y^3)\frac{dy}{dx} = y$

$$x\frac{dy}{dx} + 2y^3\frac{dy}{dx} = y$$

$$\Rightarrow ydx - xdy = 2y^3dy$$

$$\Rightarrow \frac{ydx - xdy}{y^2} = 2ydy$$

$$\therefore \frac{x}{y} = y^2 + c$$

Sol 7: (B) Equation of normal at P(x, y)

$$(Y - y) = -\frac{dx}{dy}(X - x)$$

OA = x-intercept = $x + y\frac{dy}{dx}$

OB = y-intercept = $y + x\frac{dx}{dy}$

$$\therefore \frac{1}{x + y\frac{dy}{dx}} + \frac{1}{y + x\frac{dx}{dy}} = 1 \text{ or } \frac{1 + \frac{dy}{dx}}{y\frac{dy}{dx} + x} = 1$$

$$\Rightarrow (y - 1)\frac{dy}{dx} = (1 - x) \text{ or } \int (y - 1)dy = \int (1 - x)dx$$

$$\Rightarrow \frac{y^2}{2} - y = x - \frac{x^2}{2} + c$$

$$\Rightarrow (y - 1)^2 + (x - 1)^2 - 2 = 2c$$

$$\therefore (y - 1)^2 + (x - 1)^2 = 2 + 2c$$

at $x = 5$, $y = 4$

$$\therefore 4^2 + 3^2 = 2 + 2c \text{ or } c = \frac{23}{2}$$

$$\therefore (y - 1)^2 + (x - 1)^2 = (5)^2$$

This is circle with centre (1, 1) and radius 5.

Sol 8: (B) X-intercept of normal = $y\frac{dy}{dx} + x = x + 1$

$$\therefore y\frac{dy}{dx} = 1; \qquad \Rightarrow \frac{y^2}{2} = x + c \Rightarrow y^2 = 2(x + c)$$

$\therefore$ This curve pass through origin

So $c = 0$

$\therefore y^2 = 2x$

$\therefore$ Latus rectum = 2

Sol 9: (A) $2x^2y\frac{dy}{dx} + 2xy^2 = \tan[(xy)^2]$

Put $xy = t$

$$\therefore x\frac{dy}{dx} + y = \frac{dt}{dx}$$

$$\therefore 2xy\left(x\frac{dy}{dx} + y\right) = \tan x^2y^2 \text{ or } 2t\frac{dt}{dx} = \tan t^2$$

$$\therefore \int \frac{2t}{\tan t^2}dt = \int dx$$

Put $t^2 = u$

$2tdt = du$

$$\therefore \int \frac{du}{\tan u} = x + c$$

or $\log \sin u = x + c$ or $\log \sin xy = x + c$

for $x = 1$, $y = \sqrt{\frac{\pi}{2}}$

$\therefore \log \sin \frac{\pi}{2} = c + 1; \quad c = -1$

$\therefore \sin(xy)^2 = e^{x-1}$

Sol 10: (C) $\frac{dm}{dt} = cm$

$\therefore \log m = c_1 t + c_2$

at $t = 0$, $m = $ maximum $= M$

$\therefore \log m = c_2$

at $t = 1$ hr, moisture content remains $\frac{M}{2}$

$\therefore \log \frac{M}{2} = c_1 + c_2$

$\therefore \log \frac{M}{2} - \log M = c_1$

$\therefore c_1 = \log \frac{1}{2}$

After $t = x$ hr moisture content remains

$100 - 9$Sol.$9\ 99 = 0.\ 1\% = \frac{0.1M}{100}$

$\therefore \log \frac{0.1M}{100} = xc_1 + c_2$

$\therefore \log \frac{0.1M}{100} - \log m = xc_1 = x\ell n \frac{1}{2}$

$\therefore x = \dfrac{\log \dfrac{0.1}{100}}{\log \dfrac{1}{2}} \approx 96 \approx 10$ hr

Sol 11: (D) Considering option taking $\phi\left(\frac{x}{y}\right) = \frac{-y^2}{x^2}$

$\frac{dy}{dx} = \frac{y}{x} - \frac{-y^2}{x^2}$

$y = vx \Rightarrow \ell v + x\frac{dv}{dx} = v - v^2$

$\Rightarrow x\frac{dv}{dx} = -v^2 \Rightarrow \int -\frac{1}{v^2} dv = \int \frac{1}{x} dx$

$\Rightarrow \frac{1}{v} = \log x + c$

or $\frac{x}{y} = \log cx$ or $y = \dfrac{x}{\log cx}$

Sol 12: (C) In 10 min. total litres run into tank $= 50 \times 10 = 500$ lt

In 100 litres there is 20 gm salt

$\therefore$ in 500 litres we have $\frac{20 \times 500}{100} = 100$ gm $= 0.1$ kg

Initially we have 10 kg of salt

$\therefore$ Total salt after 10 minutes $= 10 + 0.1 = 10.1$ kg

Sol 13: (D) (A) order 2 degree 1

(B) $y_2 = (x - xy - (\sin x)y_1)^2$

$\therefore$ order 2 degree 1

(C) $y_1 + y = (x + 1)^2$

$\therefore$ Order 1 degree 1

Sol 14: (C) $\frac{dy}{dx} = \frac{xy + y}{xy + x}$

$\therefore \frac{dy}{dx} = \frac{y}{x}\left(\frac{x+1}{y+1}\right)$

$\int \frac{(y+1)}{y} dy = \int \frac{(x+1)}{x} dx$

$y + \log y = x + \log x + c$

$\Rightarrow \log y e^y = \log x e^x A \ [\because c = \log A]$

$\therefore y e^y = A x e^x$ or $y = A x e^{x-y}$

Sol 15: (B) Degree is 2

Sol 16: (C) $y^2 = 4xa$

$\therefore 2y\frac{dy}{dx} = 4a$

$\therefore y^2 = 2y\frac{dy}{dx}x$ or $y = 2x\frac{dy}{dx}$

$\therefore$ Order 1 degree 1

or $\frac{dx}{dy} = \frac{1}{2a}y$ or $\frac{d^2x}{dy^2} = \frac{1}{2a}$

Previous Years' Questions

Sol 1: (C) Given, $y = (c_1 + c_2)\cos(x + c_1) - $

$y\frac{dy}{dx} = \left\{y\frac{d^2y}{dx^2} + \left(\frac{dy}{dx}\right)^2\right\}x$ ... (i)

$\Rightarrow y = (c_1 + c_2)\cos(x + c_3) - ce^x. \ y\frac{dy}{dx} = xy\frac{d^2y}{dx^2} + x\left(\frac{dy}{dx}\right)^2$

Now, let $c_1 + c_2 = A$, $c_3 = B$, c_4 $xy\frac{d^2y}{dx^2} + x\left(\frac{dy}{dx}\right)^2 - y\frac{dy}{dx} = 0 = c$

$\Rightarrow y = A\cos(x + B) - ce^x$ ... (ii)

On differential w.r.t. x, we get $x\frac{dy}{dx}$ ... (iii)

Again differentiating w.r.t. x, we get

$$x\frac{d^2y}{dx^2}+1.\frac{dy}{dx}+1.\frac{dy}{dx} = -A\cos(x+B) - ce^x \qquad ..(iv)$$

$$\Rightarrow x\frac{d^2y}{dx^2}+2\frac{dy}{dx} \qquad ...(v)$$

$$\Rightarrow \left[\begin{array}{l} \because xy = Ae^x + Be^{-x} + x^2 \\ \Rightarrow Ae^x + Be^{-x} = xy - x^2 \end{array} \right]$$

Again differenting w.r.t. x, we get

$$\frac{dy}{dx} = c.2(x^2+y^2)\left(2x+2y\frac{dy}{dx}\right) \qquad ...(vi)$$

$$\Rightarrow \frac{d^3y}{dx^3}+\frac{dy}{dx} = \frac{d^2y}{dx^2}+y \quad \text{[from Eq. (v)]}$$

Which is a differential equation of order 3

Sol 2: (C) (a) y = 2

$$\Rightarrow \left(x-y\frac{dy}{dx}\right)$$

On putting in equation, (i),

$0^2 - x(0) + y = 0$

$\Rightarrow y = 0$ which is not satisfied.

(b) $y = 2x \Rightarrow \left(2x+2y\frac{dy}{dx}\right)$

on putting equation (i),

$(2)^2 - x . 2 + y = 0$

$\Rightarrow 4 - 2x + y = 0$

$\Rightarrow y = 2x - 4$ which is not satisfied.

(c) $y = 2x^2 - 4$

$$\frac{dy}{dx}$$

On putting in equation (i),

$(4x)^2 - x . 4x + y = 0$

$\Rightarrow y = 0$ which is not satisfied.

Therefore, C is the answer.

Sol 3: (A) Given, $\frac{dy}{dx}$ and $y(0) = -1$

Which represents linear differential equation of first order

$$\therefore \text{ IF} = \frac{dy}{dx} = e^{-t + \log(1+t)}$$

$$= e^{-t}.(1+t)$$

Required solution is

$$ye^{-t}(1+t) = \frac{dy}{dx}(1+t)dt + c$$

$$= \frac{d^2y}{dx^2}-\frac{dy}{dx}=\frac{dy}{dx}$$

$$\Rightarrow ye^{-t}(1+t) = -e^{-t} + c$$

Since, $y(0) = -1$

$$\Rightarrow -1\,e^0(1+0) = -e^0 + c$$

$c = 0;$

Putting $t = 1$, we get $y(1) = \frac{-1}{2}$

Sol 4: (A) Given, $\dfrac{dy}{dx} = \dfrac{-\cos x(y+1)}{2+\sin x}$

$$\Rightarrow \frac{dy}{y+1} = \frac{-\cos x}{2+\sin x} - dx$$

On integrating both sides

$$\Rightarrow \log(y+1) = -\log(2+\sin x) + \log c,$$

When, $x = 0, y = 1 \Rightarrow c = 4$

$$\Rightarrow \qquad y+1 = \frac{4}{2+\sin x}$$

$$\therefore \qquad y\left(\frac{\pi}{2}\right) = \frac{4}{3}-1$$

$$\Rightarrow \qquad y\left(\frac{\pi}{2}\right) = \frac{1}{3}$$

Sol 5: (A) $x\dfrac{dy}{dx} = y(\log y - \log x + 1)$

$$\therefore \frac{dy}{dx} = \left(\frac{y}{x}\right)\left(\log\frac{y}{x}+1\right)$$

Put $\dfrac{y}{x} = t \Rightarrow y = xt \Rightarrow \dfrac{dy}{dx} = t + x\dfrac{dt}{dx}$

$$\therefore \qquad t \log dx = x\, dt$$

$$\Rightarrow \qquad \frac{dt}{t\log t} = \frac{dx}{x}$$

$$\Rightarrow \qquad \log\log t = \log x + \log c$$

$$\Rightarrow \qquad \log\left(\frac{y}{x}\right) = cx$$

Sol 6: Given, liquid evaporates at a rate proportional to its surface are.

$$\Rightarrow \frac{B}{x} \propto -S \qquad\qquad \ldots \text{(i)}$$

We know, volume of cone = $\dfrac{d^2y}{dx^2} + x\dfrac{dy}{dx}$

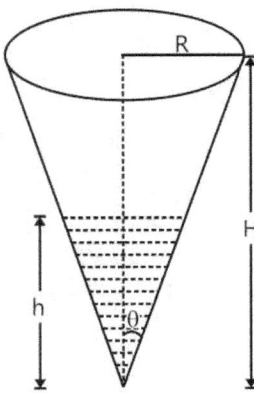

and surface area = πr^2

or $V = \dfrac{1}{3}\pi r^2 h$ and $S = \pi r^2$ ... (ii)

where $\tan\theta = \dfrac{dy}{dx}$ and $x\dfrac{d^2y}{dx^2} + 1.\dfrac{dy}{dx} + \dfrac{dy}{dx} = \tan\theta$ ... (iii)

From equation (ii) and (iii), we get

$$V = \frac{1}{3}\pi r^3 \cot\theta \text{ and } S = \pi r^2 \qquad \ldots \text{(iv)}$$

On substituting equation (iv) in equation (i), we get

$$\frac{1}{3}\cot\theta 3r^2 \frac{dr}{dt} = -k\pi r^2$$

$$\Rightarrow \cot\theta \ x^2 \frac{d^2y}{dx^2} + x\frac{dy}{dx} - y = 0 = -k\int_0^T dt$$

$$\Rightarrow \cot\theta(0 - R) = -k(T - 0)$$

$$\Rightarrow R\cos\theta = kT$$

$$\Rightarrow H = kT \text{ [from Equation (iii)]}$$

$$\Rightarrow T = y.\sqrt{x^2 + 1} = \log\left[\sqrt{x^2 + 1} - x\right]$$

$\therefore$ Required time after which the cone is empty, $T = \dfrac{dy}{dx}$

Sol 7: Since, the length of tangent = $\dfrac{dy}{dx} = 1$

$$\Rightarrow \frac{x - \sqrt{x^2 + 1}}{\sqrt{x^2 + 1 - x}} = 1 \qquad \therefore \frac{dy}{dx}$$

$$\Rightarrow \frac{dy}{dx} = \pm \frac{d^2y}{dx^2} + x\frac{dy}{dx} + y = 0$$

$$\Rightarrow \frac{dy}{dx} = \frac{a\sin(\log x)}{x} + \frac{b\cos(\log x)}{x} = \pm x + c$$

Put $y = \sin\theta$

$$\Rightarrow dy = \cos\theta d\theta \qquad \therefore x\frac{dy}{dx} = \pm x + c$$

$$\Rightarrow x\frac{d^2y}{dx^2} + 1.\frac{dy}{dx} = -\frac{a\cos(\log x)}{x} - \frac{b\sin(\log x)}{x} = \pm x + c$$

Again put $\cos\theta = t \Rightarrow -\sin\theta d\theta = dt$

$$-\int \frac{t^2}{1 + t^2}dt = \pm x + c$$

$$\therefore -x^2\frac{d^2y}{dx^2} + x\frac{dy}{dx} = \pm x + c$$

$$\Rightarrow t - \log x^2\frac{d^2y}{dx^2} + x\frac{dy}{dx} = \pm x + c$$

$$\Rightarrow x^2\frac{d^2y}{dx^2} + x\frac{dy}{dx} = \pm x + c$$

Sol 8: (C) $y(1 + xy) dx = xdy \Rightarrow ydx - xdy + xy^2dx = 0$

$$y^2 d\left(\frac{x}{y}\right) + xy^2dx = 0 \Rightarrow \frac{x}{y} + \frac{x^2}{y} = c \qquad \ldots \text{(i)}$$

Since, (1,-1) satisfies the above equation

$$-1 + \frac{1}{2} = c \Rightarrow c = -\frac{1}{2}$$

Put in (i) $x = -\dfrac{1}{2}$

$$\frac{-\frac{1}{2}}{y} + \frac{\frac{1}{4}}{2} = -\frac{1}{2} \Rightarrow \frac{-1}{2y} = \frac{-1}{2} - \frac{1}{8}$$

$$\Rightarrow \frac{1}{2y} = \frac{5}{8}; \quad \Rightarrow y = \frac{4}{5}$$

Sol 9: (C)

$$\frac{dy}{dx} + \frac{y}{x\log x} = 2 \text{ at } x = 1; y = 0$$

$$\text{I.F} = \int_e \frac{1}{x\log x}dx = e^{\log(\log x)} = \log x$$

$$\Rightarrow y(\log x) = \int 2(\log x) dx$$

$$\Rightarrow y(\log x) = 2[x\log - x] + c$$

At $x = 1$, $c = 2$ $x = e$

$$y = 2(e - e) + 2 \Rightarrow y = 2$$

Sol 10: (A)

$y = \sec(\tan^{-1} x)$

Let $\tan^{-1} x = \theta$

$x = \tan \theta$

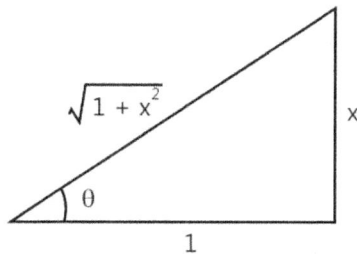

$\Rightarrow y = \sec \theta$

$\Rightarrow y = \sqrt{1 + x^2}$

$\Rightarrow \dfrac{dy}{dx} = \dfrac{1}{2\sqrt{1+x^2}} . 2x$

At $x = 1$

Therefore, $\dfrac{dy}{dx} = \dfrac{1}{\sqrt{2}}$

Sol 11: (C)

$$\dfrac{d}{dy}\left(\dfrac{dy}{dx}\right) = \dfrac{d}{dy}\left(\dfrac{1}{\left(\dfrac{dy}{dx}\right)}\right) = -\dfrac{1}{\left(\dfrac{dy}{dx}\right)^2}\dfrac{d}{dy}\left(\dfrac{dy}{dx}\right)$$

$$= -\left(\dfrac{dy}{dx}\right)^{-2}\dfrac{1}{\left(\dfrac{dy}{dx}\right)}\dfrac{d}{dx}\left(\dfrac{dy}{dx}\right) = -\left(\dfrac{d^2y}{dx^2}\right)\left(\dfrac{dy}{dx}\right)^{-3}$$

Sol 12: (D) $\cos x\, dy = y(\sin x - y)\, dx$

$\dfrac{dy}{dx} = y\tan x - y^2(\sec x)$

$\dfrac{1}{y^2}\dfrac{dy}{dx} - \dfrac{1}{y}\tan x = -\sec x$

Let $\dfrac{1}{y} = t \Rightarrow -\dfrac{1}{y^2}\dfrac{dy}{dx} = \dfrac{dt}{dx}$

$-\dfrac{dy}{dx} - t(\tan x) = -\sec x \Rightarrow \dfrac{dt}{dx} + (\tan x)t = \sec x$

I.F. $= e^{\int \tan x\, dx} = \sec x$

Solution is $t(I.F) = \int (I.F)\ \sec x\, dx$

$\dfrac{1}{y}\sec x = \tan x + c$

Sol 13: (D) $y = c_1 e^{c_2 x}$ ... (i)

$y' = c_2 c_1 e^{c_2 x}$

$y' = c_2 y$ ... (ii)

$y'' = c_2 y'$

From (ii)

$c_2 = \dfrac{y'}{y}$

So, $y'' = \dfrac{(y')^2}{y} \Rightarrow yy'' = (y')^2$

Sol 14: (D) $Y = vx$

$\dfrac{dy}{dx} = v + x\dfrac{dy}{dx}$

$v + x\dfrac{dv}{dx} = 1 + v$

$\Rightarrow dv = \dfrac{dx}{x}$

$\therefore v = \log x + c$

$\Rightarrow \dfrac{y}{x} = \log x + c$

Since, $y(i) = 1$, we have

$y = x \log x + x$

JEE Advanced/Boards

Exercise 1

Sol 1: (i) $\dfrac{dy}{dx} = \dfrac{x^2 + xy}{x^2 + y^2}$

$\dfrac{dy}{dx} = \dfrac{1 + \dfrac{y}{x}}{1 + \left(\dfrac{y}{x}\right)^2}$

Put $\dfrac{y}{x} = v$

$\therefore v + x\dfrac{dv}{dx} = \dfrac{1 + v}{1 + v^2}$

$\therefore x\dfrac{dv}{dx} = \dfrac{1 - v^3}{1 + v^2} \Rightarrow \int \dfrac{(1 + v^2)}{(1 - v^3)}dv = \int \dfrac{dx}{x}$

$\Rightarrow \int \dfrac{1}{(1 - v^3)}dv + \left(-\dfrac{1}{3}\right)\int \dfrac{-3v^2}{(1 - v^3)}dv = \log x$

⇒ $\dfrac{1}{3}\int\left(\dfrac{1}{(1-v)} + \dfrac{v+2}{(v^2+v+1)}\right)dv$

$-\dfrac{1}{3}\log(1-v^3) = \log x + c$

$= -\dfrac{1}{3}\log(1-v) + \dfrac{1}{6}\int\left(\dfrac{2v+1+3}{v^2+v+1}\right)dv$

$= \dfrac{1}{3}\log(x^3-y^3) - \log x + \log x$

$-\dfrac{1}{3}\log(1-v)\dfrac{3}{6}\int\dfrac{dv}{\left(v+\dfrac{1}{2}\right)^2 + \left(\dfrac{\sqrt{3}}{2}\right)^2}$

$= \dfrac{1}{3}\log(x^3-y^3) + \dfrac{1}{3}\log(x-y) - \dfrac{1}{3}\log x + c$

$\Rightarrow \dfrac{1}{6}\log(y^2+xy+x^2) - \dfrac{1}{3}\log x + \dfrac{1}{2}\times\dfrac{2}{\sqrt{3}}\;\tan^{-1}\left(\dfrac{v+\dfrac{1}{2}}{\dfrac{\sqrt{3}}{2}}\right)$

$= \dfrac{1}{3}\log(x^3-y^3) + \dfrac{1}{3}\log(x-y) - \dfrac{1}{3}\log x + c$

$\Rightarrow \dfrac{1}{\sqrt{3}}\tan^{-1}\dfrac{(2y+x)}{\sqrt{3}x}$

$= \log(x^3-y^3)^{1/3}(x-y)^{1/3}(y^2+xy+x^2)^{-1/6} + c$

$\therefore (x-y)^{2/3}(y^2+xy+x^2)^{1/6} = e^{\frac{1}{\sqrt{3}}\tan^{-1}\left(\frac{2y+x}{\sqrt{3}x}\right)}$

(ii) $\dfrac{dy}{dx} = \dfrac{1-3\left(\dfrac{y}{x}\right)^2}{\left(\dfrac{y}{x}\right)^3 - 3\left(\dfrac{y}{x}\right)}$

Put $\dfrac{y}{x} = v$

$v + x\dfrac{dv}{dx} = \dfrac{1-3v^2}{v^3-3v}$

or $x\dfrac{dv}{dx} = \dfrac{1-v^4}{v^3-3v}$

$\therefore \int\dfrac{v^3-3v}{1-v^4}dv = \int\dfrac{dx}{x}$

$\Rightarrow \left(-\dfrac{1}{4}\right)\int\dfrac{-4v^3}{1-v^4}dv - 3\int\dfrac{v}{1-v^4}dv = \log x + c$

$\Rightarrow -\dfrac{1}{4}\log(1-v^4) - \dfrac{3}{2}\int\dfrac{dt}{(1-t^2)} - \log x + c$

$\Rightarrow -\dfrac{1}{4}\log(1-v^4) - \dfrac{3}{2}\times\dfrac{1}{2}\log\left(\dfrac{1+t}{1-t}\right) = \log x + c$

$-\dfrac{1}{4}\log(x^4-y^4) + \log x - \dfrac{3}{4}\ell n\left(\dfrac{x^2+y^2}{x^2-y^2}\right) = \log x + c$

$\therefore \log\left(\dfrac{x^2+y^2}{x^2-y^2}\right)^{3/4}\log(x^4-y^4)^{1/4} = \log c$

Or $(x^2+y^2)(x^2-y^2)^{-1/2} = c$

Or $(x^2+y^2)^2 = (x^2-y^2)c$

Sol 2: Projection of ordinate on normal

$y\cos\theta$

$y\cos\theta = x$

$\cos\theta = \dfrac{x}{y}$

$1 - \sin^2\theta = \dfrac{x^2}{y^2} \Rightarrow \sin\theta = \sqrt{\dfrac{y^2-x^2}{y^2}}$

$\therefore \tan\theta = \dfrac{dy}{dx} = \dfrac{\sqrt{y^2-x^2}}{x} = \sqrt{\left(\dfrac{y}{x}\right)^2 - 1}$

$y = vx$

$v + x\dfrac{dv}{dx} = \sqrt{v^2-1}$

$\therefore \int\dfrac{1}{\sqrt{v^2-1}-v}dv = \log x + c$

$\Rightarrow -\int\left(\sqrt{v^2-1}+v\right)dv = \log x + c$

$\Rightarrow -\dfrac{v^2}{2} - \int\sqrt{v^2-1}\,dv = \log x + c$

$\Rightarrow -\dfrac{v^2}{2} - \left[\dfrac{v}{2}\sqrt{v^2-1} - \dfrac{1}{2}\log[v+\sqrt{v^2-1}]\right]$

$= \log x + c$

$\Rightarrow \dfrac{y^2 \pm y\sqrt{y^2-x^2}}{x^2} = \log\left[\left[y\pm\sqrt{y^2-x^2}\right]\cdot\dfrac{c^2}{x^3}\right] + c$

Sol 3: $\therefore 2\tan\theta = -\dfrac{0 + \dfrac{y}{x}}{1 + 0 \times \dfrac{y}{x}}$

or $\tan\theta = \dfrac{dy}{dx} = +\dfrac{y}{2x}$

$\therefore 2\log y = \log x + c$

$\therefore y^2 = cx$

$\therefore$ This is a parabola

Sol 4: Equation of tangent

$(Y - y)$

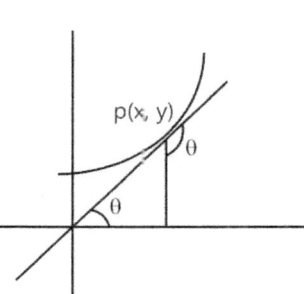

$= \dfrac{dy}{dx}(X - x)$

$\Rightarrow X\dfrac{dy}{dx} - Y + y - x\dfrac{dy}{dx} = 0$

Distance from origin

$\Rightarrow \dfrac{y - x\dfrac{dy}{dx}}{\sqrt{1 + \left(\dfrac{dy}{dx}\right)^2}} = x$ or $y^2 - 2xy\dfrac{dy}{dx}$

$= \left[1 + \left(\dfrac{dy}{dx}\right)^2\right]x^2 - x^2\left(\dfrac{dy}{dx}\right)^2$

or $\dfrac{dy}{dx} = \dfrac{y^2 - x^2}{2xy} = \dfrac{1}{2}\left(\dfrac{y}{x} - \dfrac{x}{y}\right)$

$\dfrac{y}{x} = v$ or $v + x\dfrac{dv}{dx} = \dfrac{1}{2}\left(v - \dfrac{1}{v}\right)$

$x\dfrac{dv}{dx} = -\dfrac{v}{2} - \dfrac{1}{2v}$ or $\int \dfrac{1}{-\dfrac{1}{2}\left(\dfrac{v^2+1}{v}\right)}dv = \int \dfrac{dx}{x}$

$= -\int\left(\dfrac{2v}{v^2 + 1}\right)dv = \log x + c$

Or $-\log(v^2 + 1) = \log x + c$ or $\log(x^2 + y^2) + 2\log x$

$= \log x + c$ or $\log x - \log(x^2 + y^2) = c$

for $(1, 1)$

$c = -\log 2$

$\therefore \log 2x = \log(x^2 + y^2)$

or $x^2 + y^2 - 2x = 0$

Sol 5: $y^3\dfrac{dy}{dx} + x + y^2 = 0$

$y^2 + x = a \ \therefore \ \dfrac{da}{dx} = 2y\dfrac{dy}{dx} + 1$

$\Rightarrow 2y^3\dfrac{dy}{dx} + y^2 = y^2\dfrac{da}{dx} = (a - x)\dfrac{da}{dx}$

$\therefore x + (a - x)\dfrac{da}{dx} = 0$

$\dfrac{da}{dx} = \dfrac{x}{x - a} = \dfrac{1}{1 - \dfrac{a}{x}}$ $\qquad \because a = vx$

$v + x\dfrac{dv}{dx} = \dfrac{1}{1 - v} = \int \dfrac{(1 - v)}{(v^2 - v + 1)}dv = \log x + c$

$= \int \dfrac{1}{\left(v - \dfrac{1}{2}\right)^2 + \left(\dfrac{\sqrt{3}}{2}\right)^2}dv + \int \dfrac{(-v)}{(v^2 - v + 1)}dv = \log x + c$

$= \dfrac{2}{\sqrt{3}}\tan^{-1}\dfrac{\left(v - \dfrac{1}{2}\right)}{\left(\dfrac{\sqrt{3}}{2}\right)} - \dfrac{1}{2}\log(v^2 - v + 1)$

$- \dfrac{1}{\sqrt{3}}\tan^{-1}\dfrac{v - \dfrac{1}{2}}{\dfrac{\sqrt{3}}{2}} = \log x + c$

$\Rightarrow \dfrac{1}{2}\log|x^2 + a^2| - \tan^{-1}\left(\dfrac{a}{x}\right) = c$

Where $a = x + y^2$

Sol 6: $\dfrac{dy}{dx} = \dfrac{y}{x}\left[\dfrac{\cos\dfrac{y}{x} + \dfrac{y}{x}\sin\dfrac{y}{x}}{\dfrac{y}{x}\sin\dfrac{y}{x} - \cos\dfrac{y}{x}}\right]$

Put $\dfrac{y}{x} = v$

$v + x\dfrac{dv}{dx} = \dfrac{v\cos v + v^3\sin v}{v\sin v - \cos v}$

or $x\dfrac{dv}{dx} = \dfrac{2v\cos v}{v\sin v - \cos v}$

or $\int \dfrac{v\sin v - \cos v}{2v\cos v}dv = \int \dfrac{dx}{x}$

$\dfrac{1}{2}\left[\int \tan v\, dv - \int \dfrac{1}{v}dv\right] = \log x + c$

$\frac{1}{2}\log|\sec v| - \frac{1}{2}\log v = \log x + c$

$\log\left|\frac{\sec v}{v}\right| = \log x^2 + 2\log c$

or $\log\left|\frac{\sec \frac{y}{x}}{\frac{y}{x} \times x^2}\right| = \log c^2$ or $xy\cos\frac{y}{x} = c$

Sol 7: Equation of tangent

$y - \frac{dy}{dx}x + x\frac{dy}{dx} - y = 0$

Intercept at y-axis $\Rightarrow Y = y - x\frac{dy}{dx}$

$\therefore y - x\frac{dy}{dx} = \sqrt{x^2 + x^2\left(\frac{dy}{dx}\right)^2}$

or $y^2 - 2xy\frac{dy}{dx} = x^2$ or $\frac{dy}{dx} = \frac{y^2 - x^2}{2xy}$

$\Rightarrow x^2 + y^2 = cx$

Sol 8: $(x - y)dy = (x + y + 1)dx$

$\frac{dy}{dx} = \frac{x + y + 1}{x - y}$

Put $x = X + h$

$y = Y + k$

$X + Y + h + k + 1$

$X - Y + h - k$

$\therefore h + k + 1 = 0$

$h - k = 0$

$\Rightarrow h = k = -\frac{1}{2}$ $\qquad \therefore \frac{dy}{dx} = \frac{X + Y}{X - Y}$

Put $\frac{Y}{X} = v$

$\therefore v + X\frac{dv}{dx} = \frac{1 + v}{1 - v}$

or $X\frac{dv}{dX} = \frac{1 + v^2}{1 - v}$ or $\int\left(\frac{1 - v}{1 + v^2}\right)dv = \log X + c$

$\Rightarrow \tan^{-1}v - \frac{1}{2}\log(1 + v^2) = \log X + c$

$\Rightarrow \tan^{-1}\frac{Y}{X} - \frac{1}{2}\log\left(1 + \frac{Y^2}{X^2}\right) = \log X + c$

or $\tan^{-1}\frac{Y}{X} - \log\sqrt{X^2 + Y^2} = c$

or $\sqrt{x^2 + y^2} = e^{c\tan^{-1}\frac{y}{x}}$

or $\sqrt{\left(x + \frac{1}{2}\right)^2 + \left(y + \frac{1}{2}\right)^2} = e^{c\tan^{-1}\left(\frac{y + \frac{1}{2}}{x + \frac{1}{2}}\right)}$

Sol 9: $\frac{dy}{dx} = \frac{x + 2y - 3}{2x + y - 3}$

$x = X + h$

$y = Y + k$

$\therefore h + 2k - 3 = 0$

$2h + k - 3 = 0$

$\therefore h = 1, k = 1$

$\therefore x = X + 1, y = Y + 1$

$\frac{dY}{dX} = \frac{X + 2Y}{2X + Y}$

$Y = vX$

$V + X\frac{dv}{dx} = \frac{1 + 2v}{2 + v}$

or $X\frac{dv}{dx} = \frac{1 - v^2}{2 + v}$ or $\int\left(\frac{2 + v}{1 - v^2}\right)dv = \log X + c$

$\log\frac{1 + v}{1 - v} + \left(-\frac{1}{2}\right)\log(1 - v^2) = \log cX$

or $\log\frac{X + Y}{X - Y} - \frac{1}{2}\log(X^2 - Y^2) + \log X = \log X + c$

or $\log\left(\frac{X + Y}{X - Y} \times \frac{1}{\sqrt{X^2 - Y^2}}\right) = c$

$\therefore \sqrt{X + Y} = (X - Y)^{3/2}C$

or $X + Y = (X - Y)^3 C$

$(X + Y - 2) = c(X - Y)^3$

or $(X + Y - 2) = c(Y - X)^3$

Sol 10: $\frac{dy}{dx} = \frac{y - x + 1}{y + x + 5}$

$x = X + h, y = Y + k$

$h + k + 5 = 0$

$k - h + 1 = 0$

$\therefore$ k = -3

h = -2

$\therefore \dfrac{dy}{dx} = \dfrac{Y-X}{Y+X}$

Put y = vX

$\therefore$ v + X $\dfrac{dv}{dX} = \dfrac{v-1}{v+1}$

$X\dfrac{dv}{dx} = \dfrac{-1-v^2}{1+v} = -\dfrac{(1+v^2)}{1+v}$

or $-\displaystyle\int \dfrac{(1+v)}{1+v^2}dv = \int \dfrac{dX}{X}$

$\Rightarrow -\tan^{-1}v - \dfrac{1}{2}\log(1+v^2) = \log X + c$

$\Rightarrow -\tan^{-1}\dfrac{Y}{X} - \log\sqrt{(X^2+Y^2)} = c$

$\therefore \tan^{-1}\dfrac{y+3}{x+2} + \log c\sqrt{(x+2)^2+(y+3)^2} = 0$

Sol 11: $\dfrac{dy}{dx} = \dfrac{x+y+1}{2(x+y)+3}$

x + y = v

$1 + \dfrac{dy}{dx} = \dfrac{dv}{dx}$

$\therefore \dfrac{dv}{dx} - 1 = \dfrac{v+1}{2v+3}$

$\dfrac{dv}{dx} = \dfrac{3v+y}{2v+3} \Rightarrow \displaystyle\int \dfrac{2v+3}{3v+4}dv = \int dx$

or $\displaystyle\int \left(\dfrac{\frac{2}{3}(3v+4)+\frac{1}{3}}{3v+4}\right)dv = x + c$

$\dfrac{2}{3}v + \dfrac{1}{9}\log\left(v+\dfrac{4}{3}\right) = x + c$

or $\dfrac{2}{3}(x+y) + \dfrac{1}{9}\log(3(x+y)+4) = x + c$

or $\dfrac{1}{9}\log\left(x+y+\dfrac{4}{3}\right) = \dfrac{x}{3} - \dfrac{2}{3}y + c$

$\log\left(x+y+\dfrac{4}{3}\right) = 3(x-2y) + \log c$

or $x + y + \dfrac{4}{3} = ce^{3(x-2y)}$

Sol 12: $\dfrac{dy}{dx} = \dfrac{2(y+2)^2}{(x+y-1)^2}$

x = X + h   y = Y + k

k + 2 = 0 $\therefore$ k = -2

h + k - 1 = 0 ard h = 3

$\therefore \dfrac{dY}{dX} = \dfrac{2Y^2}{(X+Y)^2} = \dfrac{2\left(\dfrac{Y}{X}\right)^2}{\left(1+\dfrac{Y}{X}\right)^2}$

Putting $\dfrac{Y}{X} = v$

$v + X\dfrac{dv}{dx} = \dfrac{2v^2}{(1+v)^2}$

$\therefore X\dfrac{dv}{dx} = \dfrac{2v^2-v(1+v^2+2v)}{(1+v)^2} = -\dfrac{v(1+v^2)}{(1+v)^2}$

or $\displaystyle\int -\dfrac{(1+v)^2}{v(1+v^2)}dv = \log X + c$

$-\displaystyle\int \dfrac{1+v^2+2v}{v(1+v^2)}dv = \log X + c$

$\Rightarrow -\displaystyle\int \left(\dfrac{1}{v} + \dfrac{2}{(1+v^2)}\right)dv = \log X + c$

$-\displaystyle\int \left(\dfrac{1}{v} + \dfrac{2}{(1+v^2)}\right)dv = \log X + c$

$\Rightarrow -\log v - 2\tan^{-1}v = \log X + c$

$\therefore \log Y + 2\tan^{-1}\dfrac{Y}{X} = C$

or $Y = ce^{-2\tan^{-1}\frac{Y}{X}}$ or $(y+2) = ce^{-2\tan^{-1}\frac{y+2}{x-3}}$

Sol 13: Equat on of tangent

$Y - y = X\dfrac{dy}{dx} - x\dfrac{dy}{dx}$

Equation of normal

$Y - y = -\dfrac{dx}{dy}(X-x)$

Distance of tangent from origin = $\dfrac{y-x\dfrac{dy}{cx}}{\sqrt{1+\left(\dfrac{dx}{dy}\right)^2}}$

4.55

Distance of normal from origin = $\dfrac{x\dfrac{dx}{dy}+y}{\sqrt{1+\left(\dfrac{dy}{dx}\right)^2}}$

$\therefore\ y-x\dfrac{dy}{dx}=\pm\left(x\dfrac{dx}{dy}+y\right)\dfrac{dy}{dx}$

$\therefore\ y-x\dfrac{dy}{dx}=\pm\left(x+y\dfrac{dy}{dx}\right)$

or $\dfrac{dy}{dx}=\dfrac{y-x}{y+x}$ or $\dfrac{dy}{dx}=-\left(\dfrac{y+x}{y-x}\right)=\dfrac{y+x}{x-y}$

Put $y=vx$

$v+x\dfrac{dv}{dx}=\dfrac{v-1}{v+1}$ or $v+x\dfrac{dv}{dx}=-\left(\dfrac{v+1}{v-1}\right)$

$x\dfrac{dv}{dx}=-\dfrac{1-v^2}{v+1}$ $x\dfrac{dv}{dx}=-\dfrac{(v^2+1)}{v-1}$

or $\displaystyle\int-\dfrac{(v+1)}{(1+v^2)}dv=\log x+c$

or $\displaystyle\int-\dfrac{(v-1)}{(v^2+1)}dv=\log x+c$

$\Rightarrow -\dfrac{1}{2}\log(1+v^2)\pm\tan^{-1}v=\log x+c$

$-\log\sqrt{(x^2+y^2)}\pm\tan^{-1}\dfrac{y}{x}=\log c$

or $\sqrt{x^2+y^2}=ce^{\pm\tan^{-1}\frac{y}{x}}$

Sol 14: $\dfrac{dy}{dx}-y=1-e^{-x}$

If $=e^{\int-1dx}=e^{-x}$

$(e^{-x}y)=\displaystyle\int\left(e^x-e^{-2x}\right)dx$

$e^{-x}y=-e^{-x}+\dfrac{1}{2}e^{-2x}+c$

for $x=0$

$y_0=-1+\dfrac{1}{2}+c$

$\therefore\ c=y_0+\dfrac{1}{2}$

$\therefore\ y=-1+\dfrac{1}{2}e^{-x}+ce^x$

For $x\to\infty$ and y to be finite

$c=0$

$\therefore\ y_0+\dfrac{1}{2}=0$

$y_0=-\dfrac{1}{2}$

Sol 15: $y'+2ty=t^2$

I. F. $=e^{\int 2tdt}=e^{t^2}$

$\therefore\ e^{t^2}y=\displaystyle\int t^2e^{t^2}dt$

$\therefore\ y=\dfrac{1}{e^{t^2}}\displaystyle\int t^2e^{t^2}dt$

$\displaystyle\lim_{t\to\infty}\dfrac{y}{t}=\lim_{t\to\infty}\dfrac{1}{te^{t^2}}\int t^2e^{t^2}dt=\lim_{t\to\infty}\dfrac{t^2e^{t^2}}{e^{t^2}+2t^2e^{t^2}}$

$=\displaystyle\lim_{t\to\infty}\dfrac{t^2}{1+2t^2}=\lim_{t\to\infty}\dfrac{1}{\dfrac{1}{t^2}+2}=\dfrac{1}{2}$

Sol 16: $\dfrac{dy}{dx}+\dfrac{x}{1+x^2}y=\dfrac{1}{2x(1+x^2)}$

I. F. $=e^{\int\frac{x}{1+x^2}dx}=e^{\frac{1}{2}\ell n(1+x^2)}=\sqrt{1+x^2}$

$\therefore\ \left(\sqrt{1+x^2}\right)y=\displaystyle\int\dfrac{1}{2x\sqrt{1+x^2}}dx$

$\Rightarrow$ Put $x=\tan\theta$

$\Rightarrow dx=\sec^2\theta d\theta=\displaystyle\int\dfrac{\sec^2\theta d\theta}{2\tan\theta\sec\theta}$

$=\dfrac{1}{2}\displaystyle\int\text{cosec}\,\theta d\theta=\dfrac{1}{2}\log|\text{cosec}\,\theta-\cot\theta|=\dfrac{1}{2}\log\left|\dfrac{1-\cos\theta}{\sin\theta}\right|$

$=\dfrac{1}{2}\log\tan\left(\dfrac{\theta}{2}\right)+c$

$\therefore\ \left(\sqrt{1+x^2}\right)y=\dfrac{1}{2}\left\{\log\tan\left(\dfrac{1}{2}\tan^{-1}x\right)+c\right\}$

Sol 17: $\dfrac{dy}{dx}+\dfrac{2x}{(1-x^2)}y=\dfrac{x}{(1-x^2)^{1/2}}$

I. F. $=e^{\int\frac{2x}{1-x^2}dx}=e^{-\log(1-x^2)}=\dfrac{1}{(1-x^2)}$

$\therefore\ \left(\dfrac{1}{1-x^2}\right)y=\displaystyle\int\dfrac{x}{(1-x^2)^{3/2}}dx$

$=\left(-\dfrac{1}{2}\right)\displaystyle\int\dfrac{-2x}{(1-x^2)^{3/2}}dx=\left(-\dfrac{1}{2}\right)\int\dfrac{dt}{t^{3/2}}$

4.56

Put $1 - x^2 = t$

$-2x\,dx = dt$

$= \left(-\dfrac{1}{2}\right)\dfrac{1}{\left(-\dfrac{1}{2}\right)} \times \dfrac{1}{t^{1/2}} = \dfrac{1}{\sqrt{1-x^2}} + c$

$y = \sqrt{1-x^2} + c(1 - x^2)$

Sol 18: (i) Equation of tangent

$(Y - y) = \dfrac{dy}{dx}(X - x)$

y-intercept

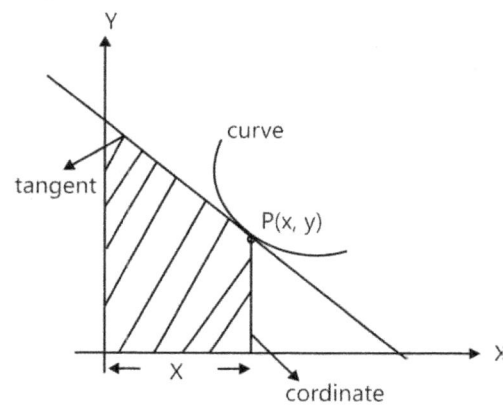

curve

tangent

P(x, y)

X

cordinate

$\Rightarrow Y = y - x\dfrac{dy}{dx}$

$\therefore \quad A = \dfrac{1}{2}x\left[y + y - x\dfrac{dy}{dx}\right] = \dfrac{1}{2}x^2$

or $y + y - x\dfrac{dy}{dx} = x$

$x\dfrac{dy}{dx} - 2y + x = 0$

or $\dfrac{dy}{dx} - \dfrac{2}{x}y = -1$

I. F. $= e^{\int -\frac{2}{x}dx} = e^{-2\log x} = \dfrac{1}{x^2}$

$\dfrac{1}{x^2}y = \int -\dfrac{1}{x^2}dx = \dfrac{1}{x} + c$

$\therefore y = x + cx^2$

(ii) x-interupt because of tangent

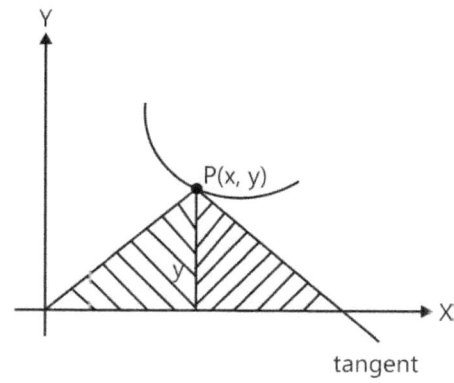

tangent

$= x - y\dfrac{dx}{dy}$

$\therefore A = \dfrac{1}{2} \times y \times \left(x - y\dfrac{dx}{dy}\right) = 2$

$xy - y^2\dfrac{dx}{dy} = 4$

or $\dfrac{dx}{dy} - \dfrac{x}{y} + \dfrac{4}{y^2} = 0$

or $\dfrac{dx}{dy} - \dfrac{x}{y} = \dfrac{-4}{y^2}$

I. F. $= e^{\int -\frac{1}{y}dy} = \dfrac{1}{y}$

$\left(\dfrac{1}{y}.x\right) = \int -\dfrac{4}{y^3}dy = \dfrac{-4y^{-2}}{-2} + c$

$\dfrac{x}{y} = \dfrac{2}{y^2} + c$

$x = \dfrac{2}{y} + cy$

For $x = 2$, $y = 1$

$\therefore 2 = 2 + c \times 1 \Rightarrow c = 0$

$\therefore xy = 2$

Sol 19: $x(x - 1)\dfrac{dy}{dx} - (x - 2)y = x^3(2x - 1)$

$\dfrac{dy}{dx} - \dfrac{(x-2)}{(x-1)x}y = \dfrac{x^2(2x-1)}{(x-1)}$

I. F. $= e^{-\int \frac{(x-2)}{(x-1)x}dx} = e^{-\int\left(\frac{1}{x} - \frac{1}{x(x-1)}\right)dx} = e^{-\int\left(\frac{1}{x} - \frac{1}{x-1} + \frac{1}{x}\right)dx}$

$= e^{-\int\left(\frac{2}{x} - \frac{1}{x-1}\right)dx} = e^{-[2\log x - \log(x-1)]} = \dfrac{x-1}{x^2}$

4.57

$\therefore \dfrac{(x-1)}{x^2}y = \int(2x-1)dx$

$\therefore \dfrac{(x-1)y}{x^2} = x^2 - x + c$

$y(x-1) = x^2(x^2 - x + c)$

Exercise 2

Single Correct Choice Type

Sol 1: (A) $\dfrac{dy}{dx} = \dfrac{y}{x} - \cos^2\left(\dfrac{y}{x}\right)$

Let $y = vx$

$\therefore \dfrac{dy}{dx} = v + x\dfrac{dv}{dx}$

$\therefore v + x\dfrac{dv}{dx} = v - \cos^2 v$

or $\int \sec^2 v\, dv = \int -\dfrac{1}{x}dx \Rightarrow \tan v = -\log x + c$

or $\tan\dfrac{y}{x} = -\log x + c$

$\therefore$ This curve is passing through $\left(1, \dfrac{\pi}{4}\right)$

$\therefore \tan\dfrac{\pi}{4} = c \Rightarrow c = 1$

$\therefore y = x\tan^{-1}(1 - \log x)$

or $y = x\tan^{-1}\left(\log\dfrac{e}{x}\right)$

Sol 2: (B) $\dfrac{dy}{dx} - y = \cos x - \sin x$

I. F. $= e^{\int -1 dx} = e^{-x}$

$\therefore \int d(e^{-x}y) = \int(\cos x - \sin x)e^{-x}dx$

$e^{-x}y = -\cos xe^{-x} - \int(-\sin x)(-e^{-x})dx - \int \sin xe^{-x}dx$

$= -\cos xe^{-x} - 2\int \sin xe^{-x}dx = -\cos xe^{-x} - 2$

$\left[-\sin xe^{-x} + \int \cos xe^{-x}dx\right]$

$= -\cos xe^{-x} + 2\sin xe^{-x} - 2\int \cos xe^{-x}dx$

Also

$e^{-x}y = \int \cos xe^{-x}dx - \int \sin xe^{-x}dx$

$= \int \cos xe^{-x}dx - \left[-\sin xe^{-x} + \int \cos xe^{-x}dx\right]$

$e^{-x}y = \sin xe^{-x} + c$

$\therefore y = \sin x + ce^{+x}$

As $x \to \infty$

$\therefore y \to \sin x$

$\therefore y = \sin x$

$\therefore y = f(x)$ is increasing in $\left[0, \dfrac{\pi}{2}\right]$

Sol 3: (C) $2x^4y\dfrac{dy}{dx} + y^4 = 4x^6$

$\dfrac{dy}{dx} = \dfrac{4x^6 - y^4}{2x^4y}$

$y = u^m$

$\dfrac{dy}{dx} = mu^{m-1}\dfrac{du}{dx}$

$\therefore mu^{m-1}\dfrac{du}{dx} = \dfrac{4x^6 - u^{4m}}{2x^4u^m}$

$u = x$

$\therefore mx^{m-1}\dfrac{dx}{dx} = \dfrac{4x^6 - x^{4m}}{2x^4x^m} = \dfrac{x^6(4 - x^{4m-6})}{2x^{4+m}}$

$\therefore 6 = 4 + 2m - 1$

$\therefore m = \dfrac{3}{2}$

Sol 4: (A) $x^2\dfrac{dy}{dx}\cos\dfrac{1}{x} - y\sin\dfrac{1}{x} = -1$

$\therefore \dfrac{dy}{dx} - \dfrac{\tan\left(\dfrac{1}{x}\right)}{(x^2)}y = \dfrac{-\sec\dfrac{1}{x}}{(x^2)}$

$\therefore$ I. F. $= e^{\int -\frac{\tan\left(\frac{1}{x}\right)}{x^2}dx}$

Put $\dfrac{1}{x} = t$

$\Rightarrow -\dfrac{1}{x^2}dx = dt$

$\therefore$ I. F. $= e^{\int \tan t\, dt} = e^{\ell n\sec t} = \sec t = \sec\dfrac{1}{x}$

$\therefore \int d\left(\sec\dfrac{1}{x}y\right) = -\int \dfrac{\sec^2\dfrac{1}{x}}{x^2}dx$

4.58

$\Rightarrow \sec\dfrac{1}{x}y = +\displaystyle\int \sec^2 t\,dt$ Put $\dfrac{1}{x} = t$

$-\dfrac{1}{x^2} = dx = dt$

$\sec\dfrac{1}{x}y = \tan\dfrac{1}{x} + c\cos\dfrac{1}{x}$

at $x \to \infty,\ y \to -1$

$-1 = 0 + c \therefore = -1$

$\therefore y = \sin\dfrac{1}{x} - \cos\dfrac{1}{x}$

Sol 5: (A) $P(x, y)$

Equation of nor mal $(Y - y) = -\dfrac{dx}{dy}(X - x)$

$\therefore$ x-axis intercept $= x + y\dfrac{dy}{dx}$

$|r| = \sqrt{x^2 + y^2}$, $r = xi + yj$

$\therefore x\left(x + y\dfrac{dy}{dx}\right) = 2(x^2 + y^2)$ [given]

$\therefore xy\dfrac{dy}{dx} = x^2 + 2y^2$

or $\dfrac{dy}{dx} = \dfrac{x}{y} + \dfrac{2y}{x}$

Put $y = vx$

$\therefore \dfrac{dy}{dx} = v + x\dfrac{dv}{dx}$

$\therefore v + x\dfrac{dv}{dx} = \dfrac{1}{v} + 2v$

or $x\dfrac{dv}{dx} = \left(v + \dfrac{1}{v}\right)$

$\therefore \displaystyle\int \dfrac{v}{v^2 + 1}dv = \int \dfrac{dx}{x}$ or $\dfrac{1}{2}\log(v^2 + 1)$

$= \ell nx + c$

or $\dfrac{1}{2}\log(x^2 + y^2) = 2\log x + c$

for $x = 1,\ y = 0$

$\therefore c = 0$

$\therefore x^2 + y^2 = (x^2)^2 = x^4$

Sol 6: (C) $y = c(x - c)^2$

$\dfrac{dy}{dx} = 2c(x - c)$

$\left(\dfrac{dy}{dx}\right)^2 = 4c^2(x - c)^2 = 4cy$

$\therefore c = \dfrac{(y')^2}{4y}$

$\therefore y = \dfrac{(y')^2}{4y}\left(x - \dfrac{(y')^2}{4y}\right)^2$

$4y^2 = \left(x(y') - \dfrac{(y')^3}{4y}\right)^2$

$\therefore$ Degree $= 3$
Order $= 1$

Sol 7: (D) $\dfrac{d^2y}{dx^2} + 3\left(\dfrac{dy}{dx}\right)^2 = x\log\left(\dfrac{d^2y}{dx^2}\right)$

This equation is not a polynomial equation in y′, y″ so degree of such a differential equation cannot be determined.

Sol 8: (C) $y^2 = 4a(x + a)$

$2y\dfrac{dy}{dx} = 4a$

Change $\dfrac{dy}{dx} \to -\dfrac{dx}{dy}$

$\therefore -2y\dfrac{dx}{dy} = 4a$

$\therefore \displaystyle\int \dfrac{dy}{y} = \int -\dfrac{1}{2a}dx$

$\Rightarrow \log y = -\dfrac{x}{2a} + c$

or $\log cy = -\dfrac{x}{2a}$ or $y = ce^{-\frac{x}{2a}}$

Sol 9: (D) $\dfrac{dy}{dx} = 4e^{4x} - 2e^{-x}$

$\dfrac{d^2y}{dx^2} = 16e^{4x} + 2e^{-x}$ and $\dfrac{d^3y}{dx^3} = 64e^{4x} - 2e^{-x}$

$\therefore \dfrac{\dfrac{d^3y}{dx^3} - 13\dfrac{dy}{dx}}{y} = \dfrac{64e^{4x} - 2e^{-x} - 13(4e^{4x} - 2e^{-x})}{e^{4x} + 2e^{-x}}$

$= \dfrac{12e^{4x} + 24e^{-x}}{4x + 2e^{-x}} = 12$

Sol 10: (B) $\dfrac{dy}{dx} - \left(\dfrac{1}{x}\right)y = \dfrac{f\left(\dfrac{y}{x}\right)}{f'\left(\dfrac{y}{x}\right)}$

I. F. $e^{\int -\frac{1}{x}dx} = \dfrac{1}{x}$ ∴ $d\left(\dfrac{1}{x}y\right) = \dfrac{1}{x}\dfrac{f\left(\dfrac{y}{x}\right)}{f'\left(\dfrac{y}{x}\right)}$

or $\displaystyle\int \dfrac{f'\left(\dfrac{y}{x}\right)d\left(\dfrac{y}{x}\right)}{f\left(\dfrac{y}{x}\right)} = \int \dfrac{1}{x}dx$

$\Rightarrow \log f\left(\dfrac{y}{x}\right) = \log x + c$

∴ $f\left(\dfrac{y}{x}\right) = cx$

Sol 11: (B) $\dfrac{dy}{dx} = \dfrac{x^2 + 2xy + y^2}{x^2 - 2xy + 2y^2}$

$\dfrac{dy}{dx} = \dfrac{1 + 2\dfrac{y}{x} + \left(\dfrac{y}{x}\right)^2}{1 - \dfrac{2y}{x} + 2\left(\dfrac{y}{x}\right)^2}$

Put y = vx

∴ $v + x\dfrac{dv}{dx} = \dfrac{1 + 2v + v^2}{1 - 2v + 2v^2}$

$x\dfrac{dv}{dx} = \dfrac{1 + 2v + v^2 - v + 2v^2 - 2v^3}{1 - 2v + 2v^2}$

$\Rightarrow x\dfrac{dv}{dx} = \dfrac{1 + v + 3v^2 - 2v^3}{1 - 2v + 2v^2}$

∴ $\displaystyle\int \dfrac{(1 - 2v + 2v^2)}{1 + v + 3v^2 - 2v^3}dv = \int \dfrac{dx}{x}$

$\dfrac{1}{3}\displaystyle\int \left(\dfrac{6v^2 - 6v - 1 + 4}{1 + v + 3v^2 - 2v^3}\right)dv = \log x$

$\dfrac{1}{3}\log(1 + v + 3v^2 - 2v^3) + \dfrac{4}{3}\displaystyle\int \dfrac{dv}{(1 + v + 3v^2 - 2v^3)} = \log x$

Rather than solving this integration we can solve this problem in another method

Line joining origin and A(1, 2)

$\Rightarrow (y - 0) = 2(x - 0)$

y = 2x

Let $\dfrac{dy}{dx}$ for y = 2x

$\dfrac{dy}{dx} = \dfrac{x^2 + 2x.2x + (2x)^2}{x^2 - 2x.2x + (2x)^2} = \dfrac{9}{5}$

∴ If y = 2x cuts c_1 or c_2 $\dfrac{dy}{dx} = \dfrac{9}{5}$

∴ If y = 2x cuts c_2 at b then also slope of tangent at B will be equal to $\dfrac{9}{5}$.

Sol 12: (C) $\ell n\left(\dfrac{dy}{dx}\right) = 4x - 2y - 2$

$\dfrac{dy}{dx} = e^{4x - 2y - 2} = \dfrac{e^{4x}}{e^{2y}e^2}$

∴ $\displaystyle\int e^{2y}dy = \int \dfrac{e^{4x}}{e^2}dx \Rightarrow \dfrac{1}{2}e^{2y} = \dfrac{1}{4}\dfrac{e^{4x}}{e^2}dx$

For x = 1, y = 1

$\dfrac{e^2}{2} = \dfrac{1}{4}\dfrac{e^4}{e^2} + c \Rightarrow c = \dfrac{1}{4}e^2$

∴ $\dfrac{1}{2}e^{2y} = \dfrac{1}{4}e^{4x - 2} + \dfrac{1}{4}e^2$

or $2e^{2y} = \dfrac{e^{4x}}{e^2} + e^2$ or $2e^{2y+2} = e^{4x} + e^4$

Multiple Correct Choice Type

Sol 13: (B, C) $x\left(\dfrac{dy}{dx}\right) = y\log\left(\dfrac{y}{x}\right)$

$\dfrac{dy}{dx} = \dfrac{y}{x}\log\left(\dfrac{y}{x}\right)$

Put y = vx

∴ $v + x\dfrac{dv}{dx} = v\log v$

∴ $x\dfrac{dv}{dx} = v(\log v - 1)$

or $\displaystyle\int \dfrac{1dv}{V(\log v - 1)} = \int \dfrac{1}{x}dx$

$\log(\log v - 1) = \log x + c$ y = xe. xe^{cx}

C = log c

or $\log v - 1 = e^{\ell n(xc)}$

4.60

Previous Years' Questions

Sol 1: (A) Given that, $\lim\limits_{t \to x} \dfrac{t^2 f(x) - x^2 f(t)}{t-x} = 1$

$\Rightarrow x2f'(x) - 2xf(x) + 1 = 0$

$\Rightarrow x2f'(x) - cx^2 + \dfrac{1}{3x}$

Since, $f(1) = 1, 1 = c + \dfrac{1}{3}$

$\Rightarrow c = \dfrac{2}{3}$

Hence, $f(x) = \dfrac{2}{3}x^2 + \dfrac{1}{3x}$

Sol 2: (C) Given that, $\dfrac{dy}{dx} = \dfrac{\sqrt{1-y^2}}{y}$

$\Rightarrow \int \dfrac{y}{\sqrt{1-y^2}}\, dy = \int dx$

$\Rightarrow -\sqrt{1-y^2} = x + c$

$\Rightarrow (x+c)^2 + y^2 = 1$

Here, centre is $(-c, 0)$; radius $= \sqrt{c^2 - c^2 + 1} = 1$

Sol 3: Equation of tangent to the curve $y = f(x)$ at point (x, y) is

$Y - y = f'(x)(X - x)$ ... (i)

The line (i) meets the x-axis at $P\left(x - \dfrac{y}{f'(x)}, 0\right)$

And the y-axis at $Q(0, y - xf'(x))$.

Area of ΔOPQ is

$\dfrac{1}{2}(OP)(OQ) = \dfrac{1}{2}\left(x - \dfrac{y}{f'(x)}\right)(y - xf'(x)) = -\dfrac{(y - xf'(x))^2}{2f'(x)}$

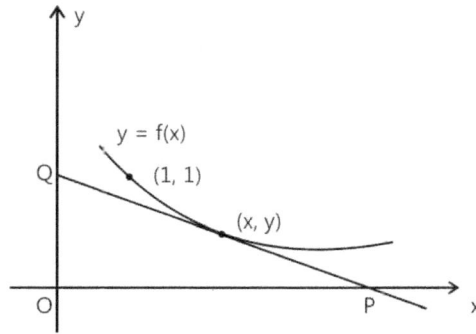

We are given that area of $\Delta OPQ = 2$, therefore,

$-\dfrac{(y - xf'(x))^2}{2f'(x)} = 2$

$\Rightarrow (y - xf'(x))^2 + 4f'(x) = 0$

$\Rightarrow (y - px)^2 + 4p = 0$ ... (ii)

Where $p = f'(x) = dy/dx$.

Since, $OQ > 0$, $y - xf'(x) > 0$. Also, note that $p = f'(x) < 0$.

We can write (ii) as $y - px = 2\sqrt{-p}$

$\Rightarrow y = px + 2\sqrt{-p}$ ... (iii)

Differentiating (iii) with respect to x, we get

$p = \dfrac{dy}{dx} = p + \dfrac{dp}{dx} x + 2\left(\dfrac{1}{2}\right)(-p)^{\frac{-1}{2}}(-1)\dfrac{dp}{dx}$

$\Rightarrow \dfrac{dp}{dx} x - (-p)^{\frac{-1}{2}}\dfrac{dp}{dx} = 0$

$\Rightarrow \dfrac{dp}{dx}[x - (-p)^{\frac{-1}{2}}] = 0$

$\Rightarrow \dfrac{dp}{dx} = 0$ or $x = (-p)^{\frac{-1}{2}}$

If $\dfrac{dp}{dx} = 0$, then $p = c$ where $c < 0$ $[\because p < 0]$

Putting this value in (iii) we get

$y = cx + 2\sqrt{-c}$ ... (iv)

This curve will pass through $(1, 1)$ if

$1 = c + 2\sqrt{-c}$

$\Rightarrow -c - 2\sqrt{-c} + 1 = 0$

$\Rightarrow (\sqrt{-c} - 1)^2 = 0$ or $\sqrt{-c} = 1$

$\Rightarrow -c = 1$ or $c = -1$

Putting the value of c in (iv) we get

$y = -x + 2$ or $x + y = 2$

Next, putting $x = (-p)^{\frac{-1}{2}}$ x or $-p = x-2$ in (iii) we get

$y = \dfrac{-x}{x^2} + 2\left(\dfrac{1}{x}\right) = \dfrac{1}{x}$

$\Rightarrow xy = 1$ $\qquad (x > 0, y > 0)$

Thus, the two required curves are $x + y = 2$ and $xy = 1$, $(x > 0, y > 0)$

Sol 4: $\dfrac{1}{e^y} + \dfrac{x^3}{3} \propto V$ for each reservoir

$\dfrac{dy}{dx} = \dfrac{1}{b}\left(\dfrac{dv}{dx} - a\right) \propto -V_A \Rightarrow \dfrac{dv}{a+bf(v)} = -K_1 V_A$

(K₁ is the proportional constant)

$(K_1$ is the proportional constant)

$$\Rightarrow \int_{V_A}^{V_A'} \frac{dV_A}{V_A} = -K_1 \int_0^1 dt$$

$$\Rightarrow \log \frac{dy}{dx} = a^2 = -K_1 t$$

$$\Rightarrow \left(\frac{dt}{dx} - 1\right) \qquad \ldots (i)$$

Similarly for B,

$$\frac{dt}{dx} = \frac{a^2}{t^2} + 1 \qquad \ldots (ii)$$

On dividing equation (i) by (ii), we get

$$\frac{a^2 + t^2}{t^2}$$

It is given that at $t = 0$, $V_A = 2V_B$ and at $t = \int \frac{t^2 dt}{t^2 + a^2}$,

$$V_A' = \frac{3}{2} V_B'$$

Thus, $\dfrac{dy}{dx}$

$$\Rightarrow \frac{x + y - 1}{\sqrt{x + y + 1}} \qquad \ldots (iii)$$

Now, let at $t = t_0$ both the reservoirs have some quantity of water. Then, $V_A' = V_B'$

From equation (iii),

$$\left(2t \frac{dt}{dx} - 1\right)$$

$$\Rightarrow \frac{t^2 - 2}{t}$$

$t_0 = \log_{3/4}(1/2)$

Sol 5: Let $w(x) = u(x) - v(x)$ ... (i)

and $h(x) = f(x) - g(x)$

On differentiation equation (i) w.r.t. x

$$\frac{2tdt}{dx} = \frac{t^2 + t - 2}{t}$$

$$= \{f(x) - p(x).u(x)\} - \{g(x)$$

$$- p(x)v(x)\} \text{ (given)}$$

$$= \{f(x) - g(x)\} - p(x)[u(x) - v(x)]$$

$$\Rightarrow \int \frac{2t^2 dt}{(t+2)(2t-2)} dt = h(x) - p(x).w(x) \qquad \ldots (ii)$$

$\dfrac{dy}{dx} + p(x)w(x) = h(x)$ which is linear differential equation.

The integrating factor is given by

$$IF = \frac{dt}{dx} = r(x) \text{(let)}$$

On multiplying both sides of equation (ii) of r(x), we get

$$r(x). \int \frac{dt}{8\cos t + 10} = \int dx + p(x)(r(x))w(x) = r(x).h(x)$$

$$\Rightarrow \frac{y}{x} = r(r).h(x)$$

$$\frac{xdy - ydx}{x^2}$$

Now, $r(x) = \dfrac{y}{x} > 0, \forall x$

And $h(x) = f(x) - g(x) > 0$ for $x > x_1$

Thus,

$$\int \frac{dr}{r^2} = \int \cos\theta \, d\theta, \forall x > x_1$$

$r(x)w(x)$ increases on the interval $[x, \Rightarrow]$

Therefore, for all $x > x_1$

$$r(x)w(x) > r(x_1) \, w(x_1) > 0$$

$[\because r(x_1) > 0$ and $u(x_1) > v(x_1)]$

$$\Rightarrow w(x) > 0 \,\, \forall \, x > x_1$$

$$\Rightarrow u(x) > \forall \,\, (x)A \,\, x > x_1$$

$[\because r(x) > 0]$

Hence, there cannot exist a point (x, y) such the $x > x_1$ and $y = u(x)$ and $y = v(x)$

Sol 6: Equation of normal at point (x, y) is

$$Y - y = -\frac{-1}{r} = \sin\theta + c \,\, (X - x)$$

Distance of perpendicular from the origin to Eq. (i)

$$= \frac{x + y\frac{dy}{dx}}{x\frac{dy}{dx} - y} = \sqrt{\frac{1 - x^2 - y^2}{x^2 + y^2}}$$

Also, distance between P and x-axis is $|y|$

$$\therefore \frac{xdx + ydy}{xdy - ydx} = \sqrt{\frac{1 - (x^2 + y^2)}{x^2 + y^2}}$$

4.62

$$\Rightarrow y^2 + \frac{dx}{dy} \cdot x^2 + 2xy \int \frac{dt}{\sqrt{1-r^2}}$$

$$= y^2 \sqrt{x^2+y^2}$$

$$\Rightarrow \frac{y}{x}(x^2-y^2) + 2xy \frac{xdx+ydy}{\sqrt{x^2+y^2}} = 0$$

$$\Rightarrow \frac{ydr-xdy}{x^2}$$

$$\Rightarrow \frac{rdr}{r^2 d\theta} \text{ or } \frac{r^2}{r\cos\theta}$$

But $\int \sec\theta \, d\theta = \int \frac{dr}{r} \Rightarrow x = c,$

where c is a constant.

Since, curve passes through (1, 1) we get the equation of the curve as x = 1

The equation $\frac{dy}{dx} = \frac{f(x,y)}{g(x,y)}$ is a homogeneous equation

Put $y = vx, \Rightarrow \frac{dy}{dx} = \frac{f(x,y)}{g(x,y)}$

$$v + x \frac{dy}{dx}$$

$$\Rightarrow \frac{dv}{dx}$$

$$= \frac{dv}{dx} = -\frac{dx}{x} = \frac{f(x,y)}{g(x,y)}$$

$$\Rightarrow \int \frac{dx}{x} = \int \frac{dv}{f(v)-v}$$

$$\Rightarrow c_1 - \log(v^2 + 1) = \log|x|$$

$$\Rightarrow \log|x| (v^2 + 1) = c_1$$

$$\Rightarrow \int_c^r y\,dx = \frac{y^3}{x}$$

$$\Rightarrow x^2 + y^2 = \pm \frac{x.3y^2 y' - y^3.1}{x^2} x$$

or $x^2 + y^2 = \pm e^c x$

is passing through (1, 1)

$\therefore 1 + 1 = \pm e^c.1 \Rightarrow \pm e^c = 2$

Hence, required curve is $x^2 + y^2 = 2x$

Sol 7: Let X_0 be initial population of the country and Y_0 be its initial food production.

Let the average consumption be a unit. Therefore, food required initially aX_0. It is given

$$y_p = aX_0\left(\frac{90}{100}\right) = 0.9aX_0 \qquad \text{... (i)}$$

Let X be the population of the country in year t

They, $\frac{dX}{dt}$ = rate of change of population

$$= \frac{3}{100} X = 0.03 X$$

$$\Rightarrow \frac{dX}{X} dt$$

$$\Rightarrow \int \frac{dX}{X} = \int 0.03 dt$$

$$\Rightarrow \log X = 0.03t + c$$

$$\Rightarrow X = A.e^{0.03t} \text{ when } A = e^c$$

At t = 0, X = X_0 thus X_0 = A

$$\therefore X = X_0 e^{0.03t}$$

Let Y be the food production in year t.

Then $Y = Y_0\left(1 + \frac{4}{100}\right)^t$

$$= 0.9aX_0(1.04)^t$$

$\because Y_0 = 0.9aX_0$ [from Eq. (i)]

Food consumption in the year t is $aX_0 e^{0.03t}$

Again, $Y - X \geq 0$ (given)

$$\Rightarrow 0.9X_0 a(1.04)^t > aX_0 e^{0.03t}$$

$$\Rightarrow \frac{(104)^t}{e^{0.03t}} > \frac{1}{0.9} = \frac{10}{9}$$

Taking log on both sides, we get

$$t[\log(1.04) - 0.03]\log 10 - \log 9$$

$$\Rightarrow t \geq \frac{\log 10 - \log 9}{\log(1.04) - 0.03}$$

Thus, the least integral values of the year n, when the country becomes self-sufficient, is the smallest integer greater than or equal to

$$\frac{\log 10 - \log 9}{\log(1.04) - 0.03}$$

Sol 8: from given integral equation $f(0) = 0$

Also, differentiation the given integral equation w.r.t. x

$f'(x) = f(x)$

If $f(x) \neq 0$ $\dfrac{f'(x)}{f(x)} = 1$

$\Rightarrow \log f(x) = x + C$

$\Rightarrow f(x) = e^c e^x$

$\because f(0) = 0 \Rightarrow e^c = 0$,

a contradiction

$\therefore f(x) = 0, \forall x \in R$

$\Rightarrow f(\log 5) = 0$

Alternate Solution

Given $f(x) = \int_0^x f(t)\,dt$

$\Rightarrow f(0) = 0$

And $f'(x) = f(x)$

It $f(x) \neq 0$

$\Rightarrow \dfrac{f'(x)}{f(x)} = 1$

$\Rightarrow \log f'(x) = x + C$

$\Rightarrow f(x) = e^c . e^x$

$\because f(0) = 0$

$\Rightarrow e^c = 0$, a contradiction

$\therefore f(x) = 0 \,\forall\, x \in R$

$\Rightarrow f(\log 5) = 0$

Sol 9: given, $g\{f(x)\} = x$

$\Rightarrow g'\{f(x)\}f'(x) = 1$ ... (i)

If $f(x) = 1 \Rightarrow x = 0$, $f(0) = 1$

Substitute $x = 0$ in eq. (1), we get

$g'(1) = \dfrac{1}{f'(0)}$

$\Rightarrow g'(1) = 2$

$\left\{ \begin{array}{l} \because f'(x) = 3x^2 + \dfrac{1}{2}e^{x/2} \\ \qquad \Rightarrow f'(0) = \dfrac{1}{2} \end{array} \right\}$

Alternate solution

Given, $f(x) = x^3 + e^{x/2}$

$\Rightarrow f'(x) = 3x^2 + \dfrac{1}{2}e^{x/2}$

For $x = 0$, $f(0) = 1$, $f'(0) = \dfrac{1}{2}$ and $g(x) = f^{-1}(x)$

Replacing x by f(x), we have

$g(f(x)) = x$

$\Rightarrow g'(f(x)).f'(x) = 1$

Put $x = 0$, we get

$g'(1) = \dfrac{1}{f'(0)} = 2$

Sol 10: $\dfrac{dy}{dx} + y.g'(x) = g(x)g'(x)$

I.F. $= e^{\int g'(x)dx} = e^{g(x)}$

$\therefore$ Solution is $y(e^{g(x)})$

$= \int g(x).g'(x).e^{g(x)}dx + C$

Put $g(x) = t$, $g'(x) = dx = dt$

$y(e^{g(x)}) = \int t.e^t dt + C$

$= t.e^t - \int 1.e^t + C$

$= t.e^t - e^t + C$

$ye^{g(x)} = (g(x) - 1)e^{g(x)} + C$ ... (i)

Given, $y(0) = 0$, $g(0) = g(2) = 0$

$\therefore$ Equation (i) becomes

$y(0).e^{g(0)} = (g(0) - 1).e^{g(0)} + C$

$\Rightarrow 0 = (-1).1 + C \Rightarrow C = 1$

$\therefore y(x).e^{g(x)} = (g(x)-1)e^{g(x)} + 1$

$\Rightarrow y(2).e^{g(2)} = (g(2)-1)e^{g(2)} + 1$,

$\Rightarrow y(2).1 = (-1).1 + 1$

$y(2) = 0$

Sol 11: (A, C) $(x+2)^2 + y(x+2) = y^2 . \dfrac{dx}{dy}$

$\Rightarrow \dfrac{dx}{dy} = \dfrac{(x+2)^2}{y^2} + \dfrac{x+2}{y}$

$\Rightarrow \dfrac{1}{(x+2)^2}\dfrac{dx}{dy} = \dfrac{1}{y^2} + \dfrac{1}{y(x+2)}$

$$\therefore \frac{1}{(x+2)^2}\frac{dx}{dy} - \frac{1}{(x+2)y} = \frac{1}{y^2}$$

$$-\frac{dt}{dy} - \frac{t}{y} = \frac{1}{y^2}$$

$$\therefore \text{Put } \frac{1}{x+2} = t, \quad -\frac{1}{(x+2)^2}\frac{dx}{dy} = \frac{dt}{dy}$$

$$\Rightarrow \frac{dt}{dy} + \frac{t}{y} = -\frac{1}{y^2} \qquad \text{I.F} = e^{\int \frac{1}{y}dy} = y$$

$$t.y = C + \int y\left(-\frac{1}{y^2}\right)dy$$

$$t.y = C - \log y$$

$$\therefore \frac{1}{x+2}.y = C - \log y$$

It passes $(1, 3) \Rightarrow 1 = C + \log 3 \Rightarrow C = 1 + \log(3)$

$$\frac{y}{x+2} = 1 + \log 3 - \log y$$

[A] option is correct.

For Option (C)

$$\frac{(x+2)^2}{(x+2)} = 1 + \log\left(\frac{y}{3}\right)$$

$$x + 1 - \log\left(\frac{3}{y}\right)$$

$$\therefore y = 3e^{-x-1}$$

$$\Rightarrow \text{Intersect}$$

For Option (D)

$$\frac{(x+3)^2}{4+2} - 1 = -\log\left(\frac{(x+3)^2}{3}\right)$$

$$\therefore \frac{(x+3)^2 - 1}{x+2} = -\log\left\{\frac{(x+3)^2}{3}\right\}$$

$$3e^{\left(\frac{(x+3)^2 - 1}{-x-2}\right)} = (x+3)^2$$

$\Rightarrow$ Will intersect.

$\Rightarrow$ (D) is not correct.

Sol 12: (B, C) Let centre of the circle is (a, a) and radius 'r'

Now equation of circle is $(x - a)^2 + (y - a)^2 = r^2$

$$\Rightarrow x^2 + y^2 - 2ax - 2ay + 2a^2 - r^2 = 0 \qquad \text{... (i)}$$

Differentiation w.r.t. x we get

$$X + yy_1 - a - ay_1 = 0 \qquad \text{... (ii)}$$

$$\Rightarrow a \frac{x + yy_1}{1 + y_1} \qquad \text{... (iii)}$$

Differentiation once again equation (ii) w.r.t. x we get

$$1 + yy_2 + y_1^2 - ay_2 = 0 \qquad \text{... (iv)}$$

Using (iii) is (iv) we have

$$\left(1 + yy_2 + y_1^2\right) - \left(\frac{x + yy_1}{1 + y_1}\right)y_2 = 0$$

$$\Rightarrow 1 + (1 + y_1 + y_1^2)y_1 + (y - x)y_2 = 0$$

Hence, $p = y - x$ and $Q = 1 + y_1 + y_1^2$

Sol 13: (B) $\dfrac{dy}{dx} + \dfrac{x}{x^2-1}y = \dfrac{x^4 + 2x}{\sqrt{1-x^2}}$

This is a linear differential equation

$$\text{I.F.} = e^{\int \frac{x}{x^2-1}dx} = e^{\frac{1}{2}\ln|x^2-1|} = \sqrt{1-x^2}$$

$\Rightarrow$ solution is

$$y\sqrt{1-x^2} = \int \frac{x(x^3+2)}{\sqrt{1-x^2}}.\sqrt{1-x^2}dx$$

$$\text{or } y\sqrt{1-x^2} = \int\left(x^4 + 2x\right)dx = \frac{x^5}{5} + x^2 + c$$

$f(0) = 0 \Rightarrow C = 0$

$$\Rightarrow f(x)\sqrt{1-x^2} = \frac{x^5}{5} + x^2$$

Now, $\displaystyle\int_{-\sqrt{3}/2}^{\sqrt{3}/2} f(x)dx = \int_{-\sqrt{3}/2}^{\sqrt{3}/2} \frac{x^2}{\sqrt{1-x^2}}dx$ (Using property)

$$= 2\int_0^{\sqrt{3}/2} \frac{x^2}{\sqrt{1-x^2}}dx = 2\int_0^{\pi/3} \frac{\sin^2\theta}{\cos\theta}\cos\theta d\theta \quad \text{(Taking } x = \sin\theta)$$

$$= 2\int_0^{\pi/3}\sin^2\theta d\theta = 2\left[\frac{\theta}{2} - \frac{\sin 2\theta}{4}\right]_0^{\pi/3}$$

$$= 2\left(\frac{\pi}{2}\right) - 2\left(\frac{\sqrt{3}}{8}\right) = \frac{\pi}{3} - \frac{\sqrt{3}}{4}$$

Sol 14: (A) $\dfrac{dy}{dx} = \dfrac{y}{x} + \sec\dfrac{y}{x}$ Let $y = vx$

$\Rightarrow \dfrac{dv}{\sec v} = \dfrac{dx}{x}$

$\displaystyle\int \cos v\,dv = \int \dfrac{dx}{x}$

$\Rightarrow \sin v = \ln x + c$

$\Rightarrow \sin\left(\dfrac{y}{x}\right) = \ln x + c$

The curve passes through $\left(1, \dfrac{\pi}{6}\right)$

$\Rightarrow \sin\left(\dfrac{y}{x}\right) = \ln x + \dfrac{1}{2}$

Sol 15: (A, D) $\dfrac{dy}{dx} - y\tan x = 2x\sec x$

$\Rightarrow \cos x\dfrac{dy}{dx} + (-\sin x)y = 2x$

$\Rightarrow \dfrac{d}{dx}(y\cos x) = 2x$

$\Rightarrow y(x)\cos x = x^2 + c$, where $c = 0$ since $y(0) = 0$

when $x = \dfrac{\pi}{4}$, $y\left(\dfrac{\pi}{4}\right) = \dfrac{\pi^2}{8\sqrt{2}}$, when $x = \dfrac{\pi}{3}$, $y\left(\dfrac{\pi}{3}\right) = \dfrac{2\pi^2}{9}$

when $x = \dfrac{\pi}{4}$, $y'\left(\dfrac{\pi}{4}\right) = \dfrac{\pi^2}{8\sqrt{2}} + \dfrac{\pi}{\sqrt{2}}$

when $x = \dfrac{\pi}{3}$, $y'\left(\dfrac{\pi}{3}\right) = \dfrac{2\pi^2}{3\sqrt{2}} + \dfrac{4\pi}{3}$

Sol 16:

$6\displaystyle\int_1^x f(t)dt = 3xf(x) - x^3 \Rightarrow 6f(x) = 3f(x) + 3xf'(x) - 3x^2$

$\Rightarrow 3f(x) = 3xf'(x) - 3x^2 \Rightarrow xf'(x) - f(x) = x^2$

$\Rightarrow x\dfrac{dy}{dx} - y = x^2 \Rightarrow \dfrac{dy}{dx} - \dfrac{1}{x}y = x$ ... (i)

I.F $= e\displaystyle\int^{-\frac{1}{x}x} = e - \log_e x$

Multiplying (i) both sides by $\dfrac{1}{x}$

$\dfrac{1}{x}\dfrac{dy}{dx} - \dfrac{1}{x^2}y = 1 \Rightarrow \dfrac{d}{dx}\left(y.\dfrac{1}{x}\right) = 1$

Integrating

$\dfrac{y}{x} = x + c$

Put $x = 1$, $y = 2$

$\Rightarrow 2 = 1 + c \Rightarrow c = 1 \Rightarrow y = x^2 + x$

$\Rightarrow f(x) = x^2 + x \Rightarrow f(2) = 6$

Note: If we put $x = 1$ in the given equation we get $f(1) = 1/3$.

Sol 17: $Y - y = m(X - x)$

y-intercept ($x = 0$)

$y = y - mXS$

Given that $y - mx = x^3$ $\Rightarrow$ $x\dfrac{dy}{dx} - y = -x^3$

$\Rightarrow$ $\dfrac{dy}{dx} - \dfrac{y}{x} = -x^2$

Integrating factor $e^{-\int \frac{1}{x}dx} = \dfrac{1}{x}$

$\therefore$ Solution is $y.\dfrac{1}{x} = \displaystyle\int \dfrac{1}{x}.(-x^2)dx$

$\Rightarrow$ $f(x) = y = -\dfrac{x^3}{2} + cx$

Given $f(1) = 1$ $\Rightarrow$ $c = \dfrac{3}{2}$

$\therefore$ $f(x) = -\dfrac{x^3}{2} + \dfrac{3x}{2}$ $\Rightarrow f(-3) = 9$

Sol 18: $(x-3)^2 \dfrac{dy}{dx} + y = 0$

$\displaystyle\int \dfrac{dx}{(x-3)^2} = -\int \dfrac{dy}{y}$

$\Rightarrow \dfrac{1}{x-3} = \ln|y| + c$

so domain is $R - \{3\}$.

Sol 19: (C) $\displaystyle\int \dfrac{dx}{x\sqrt{x^2-1}} = \int \dfrac{dy}{y\sqrt{y^2-1}}$

$\sec^{-1}2 = \sec^{-1}\left(\dfrac{2}{\sqrt{3}}\right) + c$

$\Rightarrow c = \dfrac{\pi}{3} - \dfrac{\pi}{6} = \dfrac{\pi}{6}$

$\Rightarrow \sec^{-1} x = \sec^{-1} y + \dfrac{\pi}{6}$

$\Rightarrow y = \sec\left(\sec^{-1} x - \dfrac{\pi}{6}\right)$

$\Rightarrow \cos^{-1}\dfrac{1}{x} = \cos^{-1}\dfrac{1}{y} + \dfrac{\pi}{6}$

$\Rightarrow \cos^{-1}\dfrac{1}{y} = \cos^{-1}\dfrac{1}{x} - \cos^{-1}\left(\dfrac{\sqrt{3}}{2}\right)$

$\Rightarrow \dfrac{1}{y} = \dfrac{\sqrt{3}}{2x} - \sqrt{1 - \dfrac{1}{x^2}}\left(\dfrac{1}{2}\right)$

$\Rightarrow \dfrac{2}{y} = \dfrac{\sqrt{3}}{x} - \sqrt{1 - \dfrac{1}{x^2}}$